CW00539500

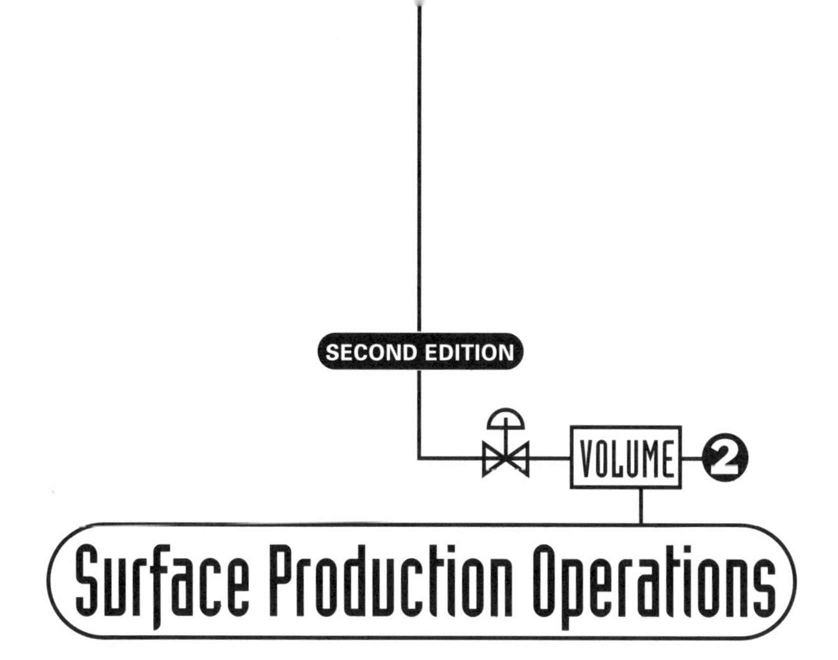
SECOND EDITION
VOLUME
2
Surface Production Operations

DISCLAIMER

This text contains descriptions, statements, equations, procedures, methodology, interpretations, and other written matter and information, hereinafter collectively called "contents," that have been carefully considered and prepared as a matter of general information. The contents are believed to reliably represent situations and conditions that have occurred or could occur, but are not represented or guaranteed as to the accuracy or application to other conditions or situations. There are many variable conditions in production facility design and related situations, and the authors have no knowledge or control of their interpretation. Therefore, the contents and all interpretations and recommendations made in connection herewith are presented solely as a guide for the user's consideration, investigation, and verification. No warranties of any kind, whether expressed or implied, are made in connection therewith. The user is specifically cautioned, reminded, and advised that any use or interpretation of the contents and resulting use or application thereof are made at the sole risk of the user. In production facility design there are many proprietary designs and techniques. We have tried to show designs and techniques in a generic nature where possible. The user must assure himself that in actual situations it is appropriate to use this generic approach. If the actual situation differs from the generic situation in design or lies outside the bounds of assumptions used in the various equations, the user must modify the information contained herein accordingly.

In consideration of these premises, any user of the contents agrees to indemnify and hold harmless the authors and publisher from all claims and actions for loss, damages, death, or injury to persons or property.

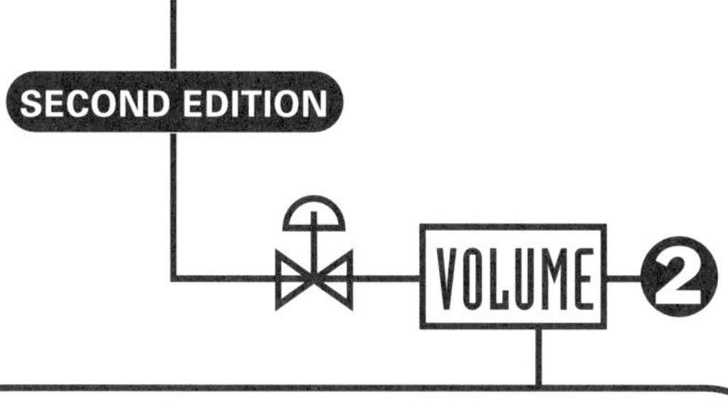

Surface Production Operations

Design of Gas-Handling Systems and Facilities

Ken Arnold
Maurice Stewart

SECOND EDITION

Surface Production Operations

VOLUME 2

Design of Gas-Handling Systems and Facilities

Originally published by Gulf Publishing Company, Houston, TX.

For information, please contact:
Manager of Special Sales
Butterworth–Heinemann
225 Wildwood Avenue
Woburn, MA 01801–2041
Tel: 781-904-2500
Fax: 781-904-2620
For information on all Butterworth–Heinemann publications available, contact our World Wide Web home page at:
http://www.bh.com

10 9 8 7 6 5 4 3

Library of Congress Cataloging-in-Publication Data

Arnold, Ken, 1942–
Design of gas-handling systems and facilities / Ken Arnold, Maurice Stewart. — 2nd ed.
p. cm. — (Surface production operations ; v. 2)
Includes index.
ISBN 0-88415-822-5 (alk. paper)
1. Natural gas—Equipment and supplies. 2. Gas wells—Equipment and supplies. I. Stewart, Maurice. II. Title. III. Series.
TN880.A69 1999
665.7—dc21 99-20405
CIP

Printed in the United States of America.

Printed on acid-free paper (∞).

Contents

CHAPTER 17

Acknowledgments

We would like to thank the following individuals who have contributed to the preparation of this edition. Without their help, this edition would not have been possible. Both of us are indebted to the many people at Paragon, Shell, and other companies who have aided, instructed, critiqued, and provided us with hours of argument about the various topics covered in this volume. In particular we would like to thank Folake A. Ayoola, K. S. Chiou, Lei Tan, Dennis A. Crupper, Kevin R. Mara, Conrad F. Anderson, Lindsey S. Stinson, Douglas L. Erwin, John H. Galey, Lonnie W. Shelton, Mary E. Thro, Benjamin T. Banken, Jorge Zafra, Santiago Pacheco, and Dinesh P. Patel.

We also wish to acknowledge Lukman Mahfoedz, Fiaz Shahab, Holland Simanjuntak, Richard Simanjuntak, Richard Sugeng, Abdul Wahab, Adolf Pangaribuan of VICO, and Allen Logue and Rocky Buras of Glytech for providing source material, suggestions, and criticism of the chapters on heat exchangers, dehydration, condensate stabilization, and surface safety systems. A final thank you to Denise Christesen for her coordinating efforts and abilities in pulling this all together for us.

Preface

As teachers of production facility design courses in petroleum engineering programs at University of Houston and Louisiana State University, we both realized there was no single source that could be used as a text in this field. We found ourselves reproducing pages from catalogs, reports, projects we had done, etc., to provide our students with the basic information they needed to understand the lectures and carry out their assignments. Of more importance, the material that did exist usually contained nomographs, charts, and rules of thumb that had no reference to the basic theories and underlying assumptions upon which they were based. Although this text often relies and builds upon information that was presented in *Surface Production Operations, Volume 1: Design of Oil-Handling Systems and Facilities,* it does present the basic concepts and techniques necessary to select, specify, and size gas-handling, -conditioning, and -processing equipment.

This volume, which covers about one semester's work or a two-week short course, focuses on areas that primarily concern gas-handling, -conditioning, and -processing facilities. Specific areas included are process selection, hydrate prevention, condensate stabilization, compression, dehydration, acid gas treating, and gas processing. As was the case with Volume 1, this text covers topics that are common to both oil- and gas-handling production facilities, such as pressure relief systems; surface safety systems; valves, fittings, and piping details; prime movers; and electrical considerations.

Throughout the text, we have attempted to concentrate on what we perceive to be modern and common practices. We have either personally been involved in the design and troubleshooting of facilities throughout the world or have people in our organizations who have done so, and undoubtedly we are influenced by our own experience and prejudices. We apologize if we left something out or have expressed opinions about equipment types that differ from your experiences. We have learned much from our students' comments about such matters and would appreciate receiving yours for future revisions/editions.

Ken E. Arnold, P.E.
Houston, Texas

Maurice I. Stewart, Ph.D., P.E.
Metairie, Louisiana

CHAPTER 1

*Overview of Gas-Handling Facilities**

The objective of a gas-handling facility is to separate natural gas, condensate, or oil and water from a gas-producing well and condition these fluids for sales or disposal. This volume focuses primarily on conditioning natural gas for sales. Gas sweetening, the removal of corrosive sulfur compounds from natural gas, is discussed in Chapter 7; methods of gas dehydration are the subject of Chapter 8, and gas processing to extract natural gas components is discussed in Chapter 9. Condensate stabilization, the process of flashing the lighter hydrocarbons to gas in order to stabilize the heavier components in the liquid phase, is the topic of Chapter 6. Treating the condensate or oil and water after the initial separation from the natural gas is covered in Volume 1.

Figure 1-1 is a block diagram of a production facility that is primarily designed to handle gas wells. The well flow stream may require heating prior to initial separation. Since most gas wells flow at high pressure, a

---

*Reviewed for the 1999 edition by Folake A. Ayoola of Paragon Engineering Services, Inc.

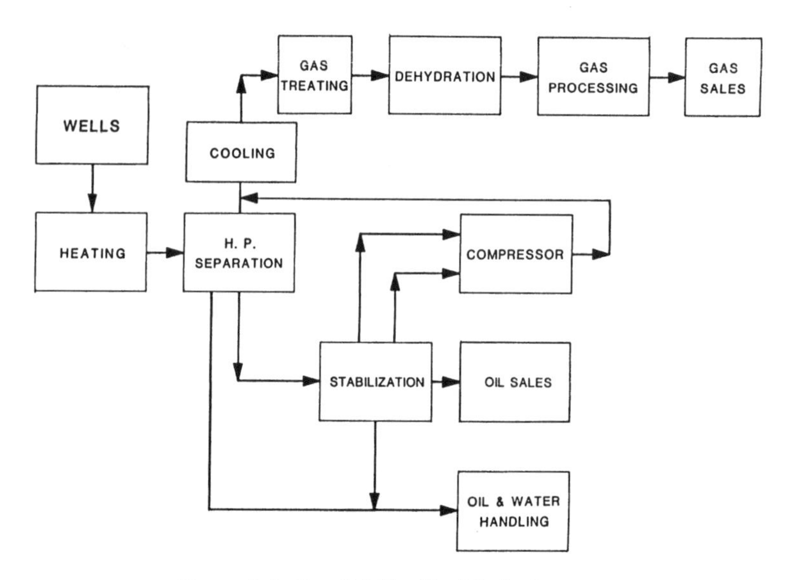

Figure 1-1. Gas field facility block diagram.

choke is installed to control the flow. When the flow stream is choked, the gas expands and its temperature decreases. If the temperature gets low enough, hydrates (a solid crystalline-like "ice" matter) will form. This could lead to plugging, so the gas may have to be heated before it can be choked to separator pressure. Low-temperature exchange (LTX) units and indirect fired heaters are commonly used to keep the well stream from plugging with hydrates.

It is also possible that cooling may be necessary. Some gas reservoirs may be very deep and very hot. If a substantial amount of gas and liquid is being produced from the well, the flowing temperature of the well could be very hot even after the choke. In this case, the gas may have to be cooled prior to compression, treating, or dehydration. Separation and further liquid handling might be possible at high temperatures, so the liquids are normally separated from the gas prior to cooling to reduce the load on the cooling equipment. Heat exchangers are used to cool the gas and also to cool or heat fluids for treating water from oil, regenerating glycol and other gas treating fluids, etc.

In some fields, it may be necessary to provide heat during the early life of the wells when flowing-tubing pressures are high and there is a high temperature drop across the choke. Later on, if the wells produce more liquid and the flowing-tubing pressure decreases, it may be necessary to cool the gas. Liquids retain the reservoir heat better and have less of a temperature drop associated with a given pressure drop than gas.

Typically, in a gas facility, there is an initial separation at a high pressure, enabling reservoir energy to move the gas through the process to sales. It is very rare that the flowing-tubing pressure of a gas well, at least initially, is less than the gas sales pressure. With time, the flowing-tubing pressure may decline and compression may be needed prior to further handling of the gas. The initial separation is normally three-phase, as the separator size is dictated by gas capacity. That is, the separator will normally be large enough to provide sufficient liquid retention time for three-phase separation if it's to be large enough to provide sufficient gas capacity. Selection and sizing of separators are described in Volume 1.

Liquid from the initial separator is stabilized either by multistage flash separation or by using a "condensate stabilization" process. Stabilization of the hydrocarbon liquid refers to the process of maximizing the recovery of intermediate hydrocarbon components (C_3 to C_6) from the liquid. Multistage flash stabilization is discussed in Volume 1. "Condensate stabilization," which refers to a distillation process, is discussed in this volume.

Condensate and water can be separated and treated using processes and equipment described in Volume 1.

Depending on the number of stages, the gas that flashes in the lower pressure separators can be compressed and then recombined with the gas from the high-pressure separator. Both reciprocating and centrifugal compressors are commonly used. In low-horsepower installations, especially for compressing gas from stock tanks (vapor recovery), rotary and vane type compressors are common.

Gas transmission companies require that impurities be removed from gas they purchase. They recognize the need for removal for the efficient operation of their pipelines and their customers' gas-burning equipment. Consequently, contracts for the sale of gas to transmission companies always contain provisions regarding the quality of the gas that is delivered to them, and periodic tests are made to ascertain that requirements are being fulfilled by the seller.

Acid gases, usually hydrogen sulfide (H_2S) and carbon dioxide (CO_2), are impurities that are frequently found in natural gas and may have to be removed. Both can be very corrosive, with CO_2 forming carbonic acid in the presence of water and H_2S potentially causing hydrogen embrittlement of steel. In addition, H_2S is extremely toxic at very low concentrations. When the gas is sold, the purchaser specifies the maximum allowable concentration of CO_2 and H_2S. A normal limit for CO_2 is between 2 and 4 volume percent, while H_2S is normally limited to ¼ grain per 100 standard cubic feet (scf) or 4 ppm by volume.

Another common impurity of natural gas is nitrogen. Since nitrogen has essentially no calorific value, it lowers the heating value of gas. Gas purchasers may set a minimum limit of heating value (normally approximately 950 Btu/scf). In some cases it may be necessary to remove the nitrogen to satisfy this requirement. This is done in very low temperature plants or with permeable membranes. These processes are not discussed in this volume.

Natural gas produced from a well is usually saturated with water vapor. Most gas treating processes also leave the gas saturated with water vapor. The water vapor itself is not objectionable, but the liquid or solid phase of water that may occur when the gas is compressed or cooled is very troublesome. Liquid water accelerates corrosion of pipelines and other equipment; solid hydrates that can form when liquid water is present plug valves, fittings, and sometimes the pipeline itself; liquid water accumulates in low points of pipeline, reducing the capacity of the lines. Removal of the water vapor by dehydration eliminates these possible difficulties and is normally required by gas sales agreements. When gas is dehydrated its dewpoint (the temperature at which water will condense from the gas) is lowered.

A typical dehydration specification in the U.S. Gulf Coast is 7 lb of water vapor per MMscf of gas (7 lb/MMscf). This gives a dew point of around 32°F for 1,000 psi gas. In the northern areas of the U.S. and Canada the gas contracts require lower dew points or lower water vapor concentrations in the gas. Water vapor concentrations of 2–4 lb/MMscf are common. If the gas is to be processed at very low temperatures, as in a cryogenic gas plant, water vapor removal down to 1 ppm may be required.

Often the value received for gas depends on its heating value. However, if there is a market for ethane, propane, butane, etc., it may be economical to process these components from the gas even though this will lower the heating value of the gas. In some cases, where the gas sales pipeline supplies a residential or commercial area with fuel, and there isno plant to extract the high Btu components from the gas, the sales contract may limit the Btu content of the gas. The gas may then have to be processed to minimize its Btu content even if the extraction process by itself is not economically justified.

Table 1-1
Example Field

Q_g	— Gas flow rate (Total 10 wells)	100 MMscfd
SIBHP	— Shut-in bottom-hole pressure	8,000 psig
SITP	— Shut-in tubing pressure	5,000 psig
Initial FTP	— Initial flowing-tubing pressure	4,000 psig
Final FTP	— Final flowing-tubing pressure	1,000 psig
Initial FTT	— Initial flowing-tubing temperature	120°F
Final FTT	— Final flowing-tubing temperature	175°F
BHT	— Bottom-hole temperature	224°F

Separator Gas Composition (1,000 psia)

Component	Mole %
CO_2	4.03
N_2	1.44
C_1	85.55
C_2	5.74
C_3	1.79
i C_4	0.41
n C_4	0.41
i C_5	0.20
n C_5	0.13
C_6	0.15
C_7^+	0.15
H_2S	19 ppm

For C_7^+; mol. wt. = 147, P_c = 304 psia, T_c = 1,112°R
Condensate — 60 bbl/MMscf, 52.3 °API
Initial free-water production — 0 bbl/MMscf
Final free-water production — 15 bbl/MMscf (at surface conditions)
Gas sales requirements — 1,000 psi, 7 lb/MMscf, ¼ grain H_2S, 2% CO_2

Chapter 9 discusses the refrigeration and cryogenic processes used to remove specific components from a gas stream, thereby reducing its Btu content.

Throughout the process in both oil and gas fields, care must be exercised to assure that the equipment is capable of withstanding the maximum pressures to which it could be subjected. Volume 1 discusses procedures for determining the wall thickness of pipe and specifying classes of fittings. This volume discusses procedures for choosing the wall thickness of pressure vessels. In either case, the final limit on the design pressure (maximum allowable working pressure) of any pipe/equipment system is set by a relief valve. For this reason, a section on pressure relief has been included.

Since safety considerations are so important in any facility design, Chapter 14 has been devoted to safety analysis and safety system design. (Volume 1, Chapter 13 discusses the need to communicate about a facility design by means of flowsheets and presents general comments and several examples of project management.)

Table 1-1 describes a gas field. The example problems that are worked in many of the sections of this text are for sizing the individual pieces of equipment needed for this field.

CHAPTER

2

*Heat Transfer Theory**

Many of the processes used in a gas-handling production facility require the transfer of heat. This will be necessary for heating and cooling the gas, as well as for regenerating the various substances used in gas treating and processing. This chapter discusses the procedures used to calculate the rate at which heat transfer occurs and to calculate the heat duty required to heat or cool gas or any other substance from one temperature to another. Subsequent chapters will discuss the detail design of shell and tube heat exchangers and water bath heaters.

Chapter 3 of Volume 1 discusses many of the basic properties of gas and methods presented for calculating them. Chapter 6 of Volume 1 contains a brief discussion of heat transfer and an equation to estimate the heat required to change the temperature of a liquid. This chapter discusses heat transfer theory in more detail. The concepts discussed in this chapter can be used to predict more accurately the required heat duty for oil treating, as well as to size heat exchangers for oil and water.

---

*Reviewed for the 1999 edition by K. S. Chiou of Paragon Engineering Services, Inc.

MECHANISMS OF HEAT TRANSFER

There are three distinct ways in which heat may pass from a source to a receiver, although most engineering applications are combinations of two or three. These are conduction, convection, and radiation.

Conduction

The transfer of heat from one molecule to an adjacent molecule while the particles remain in fixed positions relative to each other is conduction. For example, if a piece of pipe has a hot fluid on the inside and a cold fluid on the outside, heat is transferred through the wall of the pipe by conduction. This is illustrated in Figure 2-1. The molecules stay intact, relative to each other, but the heat is transferred from molecule to molecule by the process of conduction. This type of heat transfer occurs in solids or, to a much lesser extent, within fluids that are relatively stagnant.

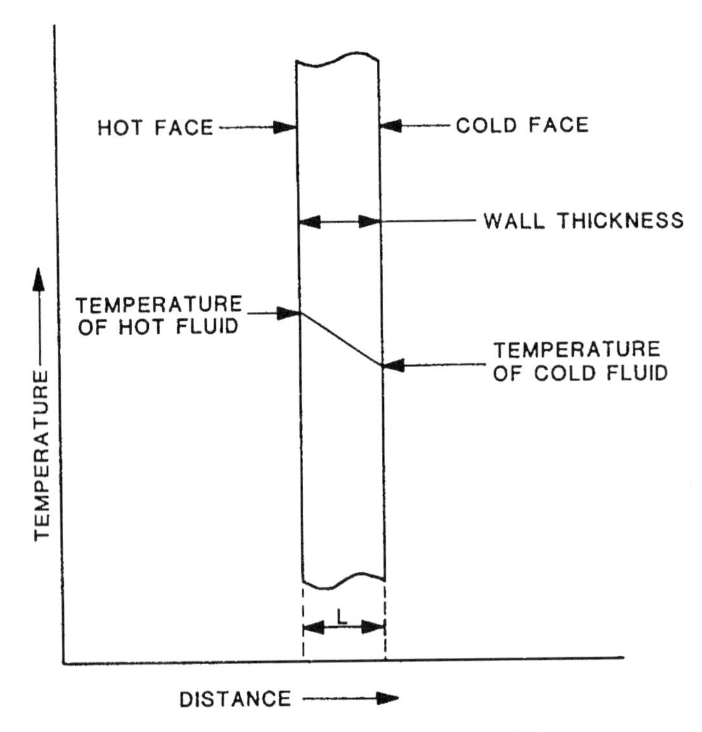

Figure 2-1. Heat flow through a solid.

The rate of flow of heat is proportional to the difference in temperature through the solid and the heat transfer area of the solid, and inversely proportional to the thickness of the solid. The proportionality constant, k, is known as the thermal conductivity of the solid. Thus, the quantity of heat flow may be expressed by the following equation:

$$q = kA(\Delta T)/L \tag{2-1}$$

where q = heat transfer rate, Btu/hr
A = heat transfer area, ft^2
ΔT = temperature difference, °F
k = thermal conductivity, Btu/hr-ft-°F
L = distance heat energy is conducted, ft

The thermal conductivity of solids has a wide range of numerical values, depending upon whether the solid is a relatively good conductor of heat, such as metal, or a poor conductor, such as glass-fiber or calcium silicate. The latter serves as insulation.

Convection

The transfer of heat within a fluid as the result of mixing of the warmer and cooler portions of the fluid is convection. For example, air in contact with the hot plates of a radiator in a room rises and cold air is drawn off the floor of the room. The room is heated by convection. It is the mixing of the warmer and cooler portions of the fluid that conducts the heat from the radiator on one side of a room to the other side. Another example is a bucket of water placed over a flame. The water at the bottom of the bucket becomes heated and less dense than before due to thermal expansion. It rises through the colder upper portion of the bucket transferring its heat by mixing as it rises.

A good example of convection in a process application is the transfer of heat from a fire tube to a liquid, as in an oil treater. A current is set up between the cold and the warm parts of the water transferring the heat from the surface of the fire tube to the bulk liquid.

This type of heat transfer may be described by an equation that is similar to the conduction equation. The rate of flow of heat is proportional to the temperature difference between the hot and cold liquid, and the heat transfer area. It is expressed:

$$q = hA(\Delta T) \tag{2-2}$$

where q = heat transfer rate, Btu/hr
A = heat transfer area, ft^2
ΔT = temperature difference, °F
h = film coefficient, Btu/hr-ft^2-°F

The proportionality constant, h, is influenced by the nature of the fluid and the nature of the agitation and is determined experimentally. If agitation does not exist, h is only influenced by the nature of the fluid and is called the film coefficient.

Radiation

The transfer of heat from a source to a receiver by radiant energy is radiation. The sun transfers its energy to the earth by radiation. A fire in a fireplace is another example of radiation. The fire in the fireplace heats the air in the room and by convection heats up the room. At the same time, when you stand within line of sight of the fireplace, the radiant energy coming from the flame of the fire itself makes you feel warmer than when you are shielded from the line of sight of the flame. Heat is being transferred both by convection and by radiation from the fireplace.

Most heat transfer processes in field gas processing use a conduction or convection transfer process or some combination of the two. Radiant energy from a direct flame is very rarely used. However, radiant energy is important in calculating the heat given off by a flare. A production facility must be designed to relieve pressure should an abnormal pressure situation develop. Many times this is done by burning the gas in an atmospheric flare. One of the criteria for determining the height and location of a flare is to make sure that radiant energy from the flare is within allowable ranges. Determining the radiation levels from a burning flare is not covered in this text. *API Recommended Practice 521, Guide for Pressure Relief and Depressuring Systems* provides a detailed description for flare system sizing and radiation calculation.

Some gas processes use direct fired furnaces. Process fluid flows inside tubes that are exposed to a direct fire. In this case radiant energy is important. Furnaces are not as common as other devices used in production facilities because of the potential fire hazard they represent. Therefore, they are not discussed in this volume.

Multiple Transfer Mechanisms

Most heat transfer processes used in production facilities involve combinations of conduction and convection transfer processes. For example, in heat exchangers the transfer of heat energy from the hot fluid to the cold fluid involves three steps. First, the heat energy is transferred from the hot fluid to the exchanger tube, then through the exchanger tube wall, and finally from the tube wall to the cold fluid. The first and third steps are convection transfer processes, while the second step is conduction process.

To calculate the rate of heat transfer in each of the steps, the individual temperature difference would have to be known. It is difficult to measure accurately the temperatures at each boundary, such as at the surface of the heat exchanger tube. Therefore, in practice, the heat transfer calculations are based on the overall temperature difference, such as the difference between the hot and cold fluid temperatures. The heat transfer rate is expressed by the following equation, similar to the conductive/convective transfer process:

$$q = UA(\Delta T) \tag{2-3}$$

where q = overall heat transfer rate, Btu/hr
U = overall heat transfer coefficient, Btu/hr-ft^2-°F
A = heat transfer area, ft^2
ΔT = overall temperature difference, °F

Examples of overall heat transfer coefficient and overall temperature difference calculations are discussed in the following sections.

Overall Temperature Difference

The temperature difference may not remain constant throughout the flow path. Plots of temperature vs. pipe length for a system of two concentric pipes in which the annular fluid is cooled and the pipe fluid heated are shown in Figures 2-2 and 2-3. When the two fluids travel in opposite directions, as in Figure 2-2, they are in countercurrent flow. When the fluids travel in the same direction, as in Figure 2-3, they are in co-current flow.

The temperature of the inner pipe fluid in either case varies according to one curve as it proceeds along the length of the pipe, and the temperature of the annular fluid varies according to another. The temperature difference at any point is the vertical distance between the two curves.

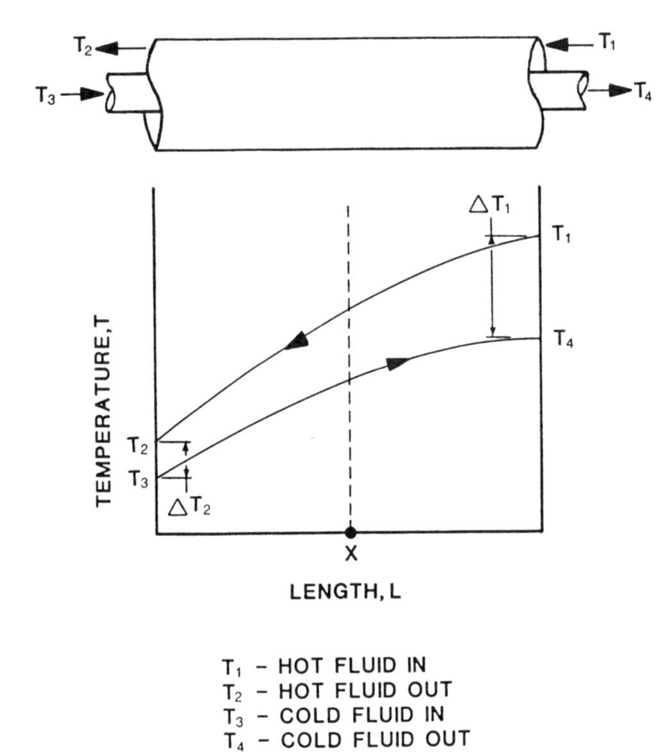

Figure 2-2. Change in ΔT over distance, counter-current flow of fluids.

Since the temperature of both fluids changes as they flow through the exchanger, an "average" temperature difference must be used in Equation 2-3. Normally a log mean temperature difference is used and can be found as follows:

$$\text{LMTD} = \frac{\Delta T_1 - \Delta T_2}{\log_e (\Delta T_1 / \Delta T_2)} \qquad (2\text{-}4)$$

where LMTD = log mean temperature difference, °F
ΔT_1 = larger terminal temperature difference, °F
ΔT_2 = smaller terminal temperature difference, °F

Although two fluids may transfer heat in either counter-current or co-current flow, the relative direction of the two fluids influences the value of the LMTD, and thus, the area required to transfer a given amount of

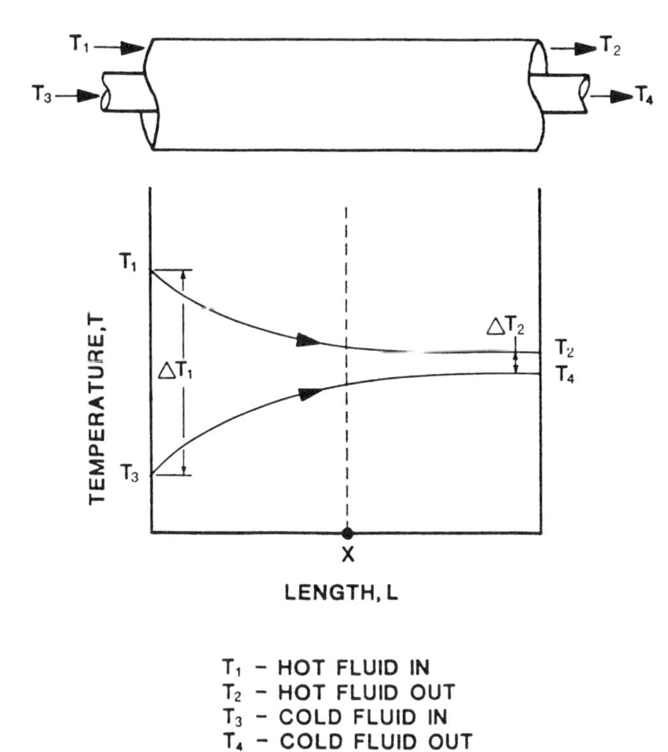

Figure 2-3. Change in ΔT over distance, co-current flow of fluids.

heat. The following example demonstrates the thermal advantage of using counter-current flow.

Given: A hot fluid enters a concentric pipe at a temperature of 300°F and is to be cooled to 200°F by a cold fluid entering at 100°F and heated to 150°F.

Co-current Flow:

Side	Hot Fluid °F	Cold Fluid °F	ΔT °F
Hot Fluid Inlet	300	100	200
Hot Fluid Outlet	200	150	50

$$\text{LMTD} = \frac{200 - 50}{\log_e (200/50)} = 108.2°\text{F}$$

Counter-current Flow:

Side	Hot Fluid °F	Cold Fluid °F	ΔT °F
Hot Fluid Inlet	300	150	150
Hot Fluid Outlet	200	100	100

$$LMTD = \frac{150 - 100}{\log_e (150/100)} = 123.3°F$$

Equation 2-4 assumes that two fluids are exchanging heat energy while flowing either co-current or counter-current to each other. In many process applications the fluids may flow part of the way in a co-current and the remainder of the way in a counter-current direction. The equations must be modified to model the actual flow arrangement. For preliminary sizing of heat transfer areas required, this correction factor can often be ignored. Correction factors for shell and tube heat exchangers are discussed in Chapter 3.

Overall Heat Transfer Coefficient

The overall heat transfer coefficient is a combination of the internal film coefficient, the tube wall thermal conductivity and thickness, the external film coefficient, and fouling factors. That is, in order for the energy to be transferred through the wall of the tube it has to pass through a film sitting on the inside wall of the tube. That film produces a resistance to the heat transfer, which is represented by the inside film coefficient for this convective heat transfer. It then must pass through the wall of the tube by a conduction process which is controlled by the tube-wall's thermal conductivity and tube-wall thickness. The transfer of heat from the outside wall of the tube to the bulk of the fluid outside is again a convective process. It is controlled by the outide film coefficient. All of these resistances are added in series, similar to a series of electrical resistance, to produce an overall resistance. The heat transfer coefficient is similar to the electrical conductance, and its reciprocal is the resistance.

Therefore, the following equation is used to determine the overall heat transfer coefficient for use in Equation 2-3.

$$\frac{1}{U} = \frac{1}{h_i A_i / A_o} + \frac{1}{k / L} + \frac{1}{h_o} + R_i + R_o \qquad (2\text{-}5)$$

where h_i = inside film coefficient, Btu/hr-ft^2-°F
h_o = outside film coefficient, Btu/hr-ft^2-°F
k = pipe wall thermal conductivity, Btu/hr-ft-°F
L = pipe wall thickness, ft
R_i = inside fouling resistance, hr-ft^2-°F/Btu
R_o = outside fouling resistance, hr-ft^2-°F/Btu
A_i = pipe inside surface area, ft^2/ft
A_o = pipe outside surface area, ft^2/ft

R_i and R_o are fouling factors. Fouling factors are normally included to allow for the added resistance to heat flow resulting from dirt, scale, or corrosion on the tube walls. The sum of these fouling factors is normally taken to be 0.003 hr-ft^2-°F/Btu, although this value can vary widely with the specific service.

Equation 2-5 gives a value for "U" based on the outside surface area of the tube, and therefore the area used in Equation 2-3 must also be the tube outside surface area. Note that Equation 2-5 is based on two fluids exchanging heat energy through a solid divider. If additional heat exchange steps are involved, such as for finned tubes or insulation, then additional terms must be added to the right side of Equation 2-5. Tables 2-1 and 2-2 have basic tube and coil properties for use in Equation 2-5 and Table 2-3 lists the conductivity of different metals.

Inside Film Coefficient

The inside film coefficient represents the resistance to heat flow caused by the change in flow regime from turbulent flow in the center of the tube to laminar flow at the tube surface. The inside film coefficient can be calculated from:

$$\frac{h_i D_i}{k} = 0.022 \left(\frac{D_i G}{\mu_e}\right)^{0.8} \left(\frac{C\mu_e}{k}\right)^{0.4} \left(\frac{\mu_e}{\mu_{ew}}\right)^{0.16} \qquad (2\text{-}6)$$

Table 2-1
Characteristics of Tubing

Tube OD In.	B.W.G. Gauge	Thickness In.	Internal Area In.^{2}	Ft^2 External Surface Per Ft Length	Ft^2 Internal Surface Per Ft Length
¼	22	.028	.0295	.0655	.0508
¼	24	.022	.0333	.0655	.0539
¼	26	.018	.0360	.0655	.0560
¼	27	.016	.0373	.0655	.0570
⅜	18	.049	.0603	.0982	.0725
⅜	20	.035	.0731	.0982	.0798
⅜	22	.028	.0799	.0982	.0835
⅜	24	.022	.0860	.0982	.0867
½	16	.065	.1075	.1309	.0969
½	18	.049	.1269	.1309	.1052
½	20	.035	.1452	.1309	.1126
½	22	.028	.1548	.1309	.1162
⅝	12	.109	.1301	.1636	.1066
⅝	13	.095	.1486	.1636	.1139
⅝	14	.083	.1655	.1636	.1202
⅝	15	.072	.1817	.1636	.1259
⅝	16	.065	.1924	.1636	.1296
⅝	17	.058	.2035	.1636	.1333
⅝	18	.049	.2181	.1636	.1380
⅝	19	.042	.2298	.1636	.1416
⅝	20	.035	.2419	.1636	.1453
¾	10	.134	.1825	.1963	.1262
¾	11	.120	.2043	.1963	.1335
¾	12	.109	.2223	.1963	.1393
¾	13	.095	.2463	.1963	.1466
¾	14	.083	.2679	.1963	.1529
¾	15	.072	.2884	.1963	.1587
¾	16	.065	.3019	.1963	.1623
¾	17	.058	.3157	.1963	.1660
¾	18	.049	.3339	.1963	.1707
¾	20	.035	.3632	.1963	.1780
1	8	.165	.3526	.2618	.1754
1	10	.134	.4208	.2618	.1916
1	11	.120	.4536	.2618	.1990
1	12	.109	.4803	.2618	.2047
1	13	.095	.5153	.2618	.2121
1	14	.083	.5463	.2618	.2183
1	15	.072	.5755	.2618	.2241
1	16	.065	.5945	.2618	.2278

Table 2-1 (Continued)
Characteristics of Tubing

Tube OD In.	B.W.G. Gauge	Thickness In.	Internal Area In.^{2}	Ft^2 External Surface Per Ft Length	Ft^2 Internal Surface Per Ft Length
1	18	.049	.6390	.2618	.2361
1	20	.035	.6793	.2618	.2435
1¼	7	.180	.6221	.3272	.2330
1¼	8	.165	.6648	.3272	.2409
1¼	10	.134	.7574	.3272	.2571
1¼	11	.120	.8012	.3272	.2644
1¼	12	.109	.8365	.3272	.2702
1¼	13	.095	.8825	.3272	.2775
1¼	14	.083	.9229	.3272	.2838
1¼	16	.065	.9852	.3272	.2932
1¼	18	.049	1.042	.3272	.3016
1¼	20	.035	1.094	.3272	.3089
1½	10	.134	1.192	.3927	.3225
1½	12	.109	1.291	.3927	.3356
1½	14	.083	1.398	.3927	.3492
1½	16	.065	1.474	.3927	.3587
2	11	.120	2.433	.5236	.4608
2	12	.109	2.494	.5236	.4665
2	13	.095	2.573	.5236	.4739
2	14	.083	2.642	.5236	.4801

where h_i = inside film heat transfer coefficient, Btu/hr-ft^2-°F
D_i = tube inside diameter, ft
k = fluid thermal conductivity, Btu/hr-ft-°F
G = mass velocity of fluid, lb/hr-ft^2
C = fluid specific heat, Btu/lb-°F
μ_e = fluid viscosity, lb/hr-ft
μ_{ew} = fluid viscosity at tube wall, lb/hr-ft

(The viscosity of a fluid in lb/hr-ft is its viscosity in centipoise times 2.41.)

The bulk fluid temperature at which the fluid properties are obtained should be the average temperature between the fluid inlet and outlet temperatures. The viscosity at the tube wall should be the fluid viscosity at the arithmetic average temperature between the inside fluid bulk temper-

(text continued on page 20)

Table 2-2
Pipe Coil Data

Nom. Size in.	Sch. No.	OD in.	ID in.	Internal Surface Area (ft²/ft)	External Surface Area (ft²/ft)
1	S40	1.315	1.049	0.275	0.344
	X80		0.957	0.251	
	160		0.815	0.213	
	XX		0.599	0.157	
2	S40	2.375	2.067	0.541	0.622
	X80		1.939	0.508	
	160		1.687	0.442	
	XX		1.503	0.394	
2½	XXX	2.875	1.375	0.360	0.753
3	S40	3.50	3.068	0.803	0.916
	X80		2.900	0.759	
	160		2.624	0.687	
	XX		2.300	0.602	
4	S40	4.50	4.026	1.054	1.19
	X80		3.826	1.002	
	160		3.438	0.900	
	XX		3.152	0.825	

Table 2-3
Thermal Conductivity of Metals at 200°F

Material	Conductivity Btu/hr-ft-°F
Aluminum (annealed)	
Type 1100-0	126
Type 3003-0	111
Type 3004-0	97
Type 6061-0	102
Aluminum (tempered)	
Type 1100 (all tempers)	123
Type 3003 (all tempers)	96
Type 3004 (all tempers)	97
Type 6061-T4 & T6	95
Type 6063-T5 & T6	116
Type 6063-T42	111
Cast iron	31
Carbon steel	30

Table 2-3 (Continued)
Thermal Conductivity of Metals at 200°F

Material	Conductivity Btu/hr-ft-°F
Carbon moly (½%) steel	29
Chrom moly steels	
1% Cr, ½% Mo	27
2¼% Cr, 1% Mo	25
5% Cr, ½% Mo	21
12% Cr	14
Austenitic stainless steels	
18% Cr, 8% Ni	9.3
25% Cr, 20% Ni	7.8
Admiralty	70
Naval brass	71
Copper	225
Copper & nickel alloys	
90% Cu, 10% Ni	30
80% Cu, 20% Ni	22
70% Cu, 30% Ni	18
30% Cu, 70% Ni alloy 400	15
Muntz	71
Aluminum bronze alloy D	46
alloy E	22
Copper silicon alloy B	33
Alloy A, C, D	21
Nickel	38
Nickel-chrome-iron	
alloy 600	9.4
Nickel-iron-chrome	
alloy 800	
Ni-Fe-Cr-Mo-Cu	
alloy 825	7.1
Ni-Mo alloy B	7.0
Ni-Mo-Cr alloy C-276	6.4
Cr-Mo alloy XM-27	11.3
Zirconium	12.0
Titanium grade 3	11.3
Cr-Ni-Fe-Mo-Cu-Cb alloy 20 CB	7.6

(text continued from page 17)

ature and the tube wall temperature. The tube wall temperature may be approximated by taking the arithmetic average between the inside fluid bulk temperature and the outside fluid bulk temperature.

The thermal conductivity of natural and hydrocarbon gases is given in Figure 2-5. The value from Figure 2-4 is multiplied by the ratio of k/k_A from Figure 2-5.

The thermal conductivity of hydrocarbon liquids is given in Figure 2-6. The viscosity of natural gases and hydrocarbon liquids is discussed in Volume 1.

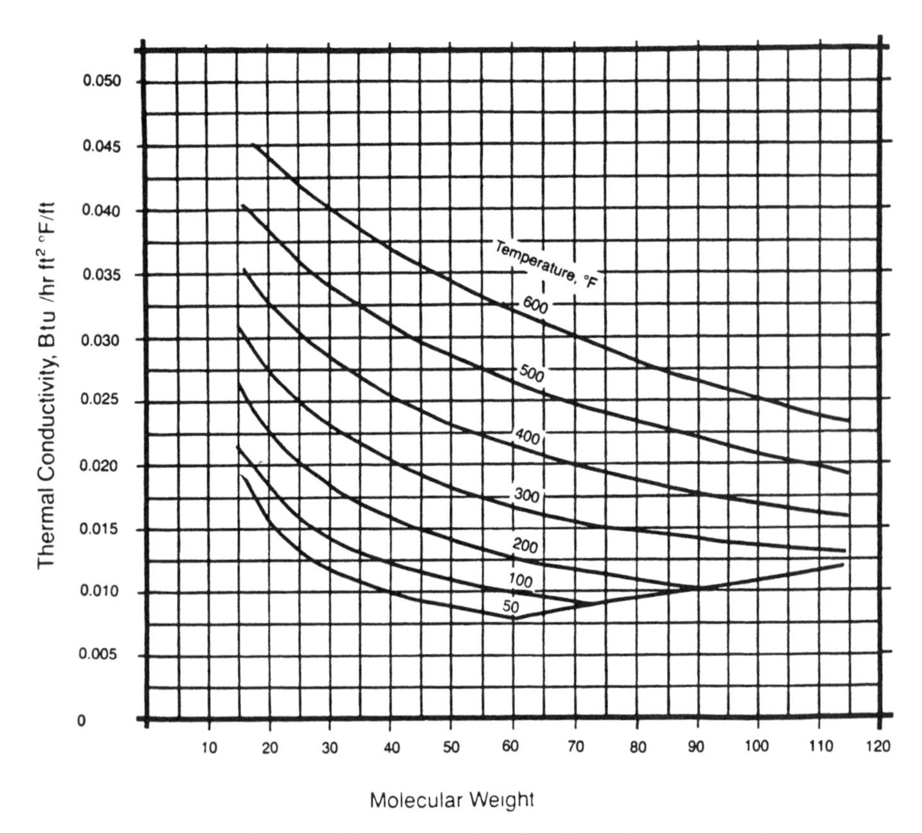

Figure 2-4. Thermal conductivity of natural and hydrocarbon gases at 1 atmosphere, 14.696 psia. (From Gas Processors Suppliers Association, *Engineering Data Book*, 10th Edition.)

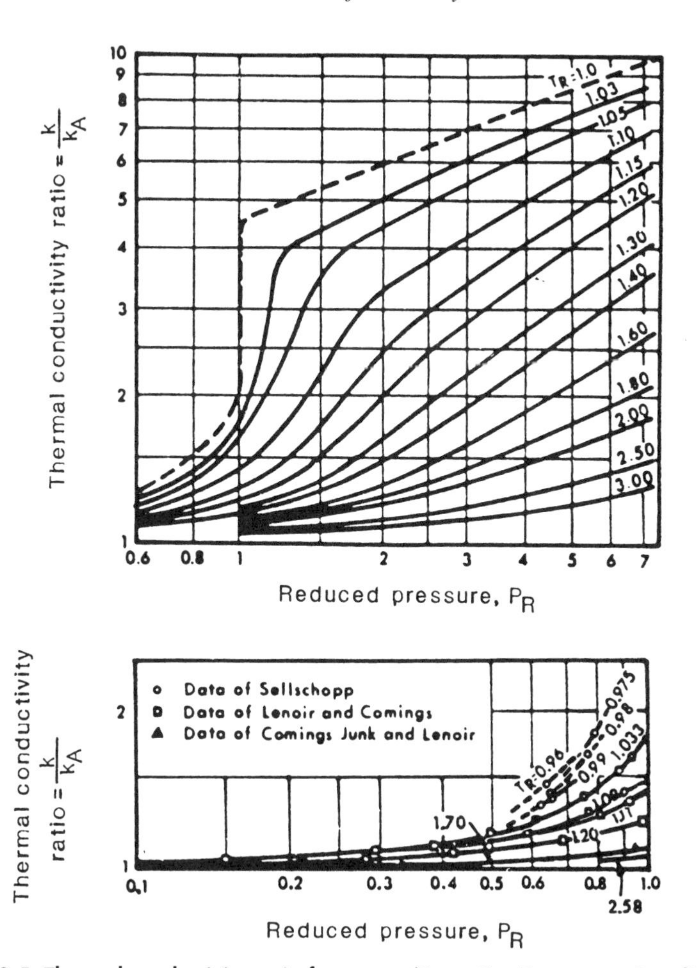

Figure 2-5. Thermal conductivity ratio for gases. (From Gas Processors Suppliers Association, *Engineering Data Book,* 10th Edition.)

The mass velocity of a fluid in pounds per hour per square foot can be calculated from

$$G = 18.6 \frac{Q_1 (SG)}{D^2} \tag{2-7}$$

$$G = 4{,}053 \frac{Q_g S}{D^2} \tag{2-8}$$

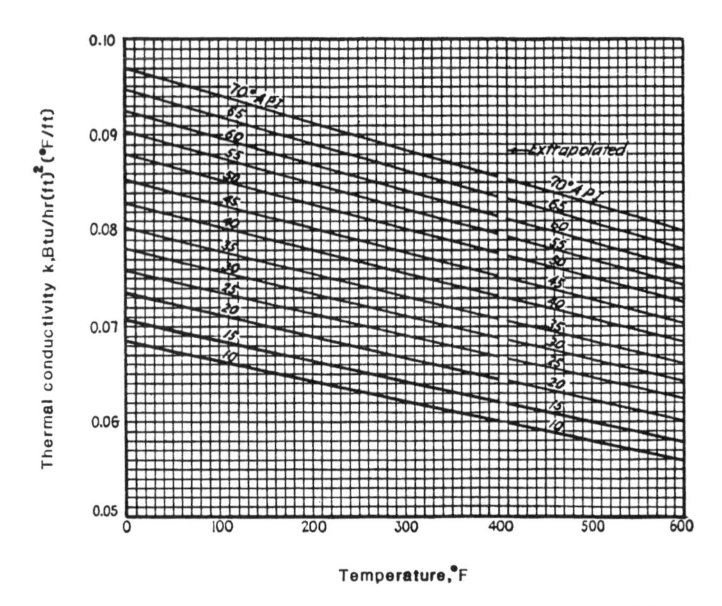

Figure 2-6. Thermal conductivities of hydrocarbon liquids. (Adapted from Natl. Bur. Stds. Misc. Pub. 97, reprinted from *Process Heat Transfer* by Kern, McGraw-Hill, Co. © 1950.)

where Q_l = liquid flow rate per tube, Bpd
Q_g = gas flow rate per tube, MMscfd
SG = liquid specific gravity relative to water
S = gas specific gravity relative to air
D = tube inside diameter, ft

The specific heat of natural gas and hydrocarbon liquids can be calculated using procedures described later in this text (see pp. 41 and 42).

The physical properties and the optimum temperature range for various heat transfer fluids are given in Table 2-4. Graphs showing more detailed physical properties and heat transfer coefficient at various conditions, such as those shown in Figures 2-7 through 2-9, can be obtained directly from manufacturers. A personal computer program for obtaining detailed physical properties of Therminol and for computing heat transfer coefficients and pressure drops in a wide variety of tube sizes and flow conditions using Therminol as the heat transfer fluid is available from Monsanto Company.

(*text continued on page 28*)

Table 2-4
Properties of Some Heat Transfer Fluids

Product Name	Therminol 55	Therminol 59	Therminol 66	Therminol VP-1	Syltherm 800	UCON HTF 500	Dowtherm 4000	
Application	Medium Temp. Range	Wide Temp. Range	High Temp. Low Press.	Ultra High Temp.	High Temp. Long Life	High Temp. Water Soluble	Water-based Heating/Cooling	
Composition	Alkylated Aromatics	Alkylated Aromatics	Modified Terphenyl	Byphenyl and Dipheynl Oxide	Polydimethyl Siloxane	Polyalkylene Glycol Polymer	Inhibited Glycol-based	
Max. Temperature								
Film	635°F	650°F	705°F	800°F	800°F			
Bulk	550°F	600°F	650°F	750°F	750°F			
Optimum Use Temp. Range, °F	–15°F to 550°F	–50°F to 600°F	30°F to 650°F	54°F to 750°F	–40°F to 750°F	0°F to 500°F	–60°F to 350°F	
Flash Point, COC	350°F	280°F	363°F	255°F	320°F	575°F		
Fire Point, COC	425°F	325°F	414°F	260°F		600°F		
Autoignition Temp. (ASTM)	690°F	760°F	750°F	1,150°F	725°F	750°F		
Density, lb/gal								
75°F	7.25	8.11	8.40	8.86	7.79	8.60	**40°F**	9.26
240°F	6.74	7.55	7.88	8.23	7.11	7.98	**180°F**	8.84
400°F	6.22	6.98	7.36	7.59	6.43	7.39	**250°F**	8.47
600°F	5.69†	6.18	6.64	6.68	5.45	7.00††	**350°F**	8.12
Heat Capacity, Btu/lb-°F								
75°F	0.459	0.405	0.377	0.372	0.386	0.47	**40°F**	0.762
240°F	0.537	0.476	0.452	0.435	0.423	0.53	**180°F**	0.835
400°F	0.612	0.547	0.528	0.492	0.460	0.56	**250°F**	0.872
600°F	0.682†	0.640	0.628	0.563	0.505	0.57††	**350°F**	0.925

(table continued on next page)

Table 2-4 (Continued)
Properties of Some Heat Transfer Fluids

Product Name	Therminol 55	Therminol 59	Therminol 66	Therminol VP-1	Syltherm 800	UCON HTF 500	Dowtherm 4000	
Thermal Conductivity, Btu/hr-ft-°F								
75°F	0.0740	0.0700	0.0679	0.0786	0.0776	0.097	**40°F**	0.212
240°F	0.0678	0.0656	0.0650	0.0727	0.0676	0.088	**180°F**	0.238
400°F	0.0618	0.0600	0.0608	0.0654	0.0580	0.080	**250°F**	0.241
600°F	0.0561†	0.0513	0.0535	0.0540	0.0459	0.075††	**350°F**	0.233
Viscosity, cps								
75°F	34.5	6.17	100.4	3.900	9.4	113.0	**40°F**	6.80
240°F	2.16	1.08	2.630	0.823	2.48	8.6	**180°F**	0.94
400°F	0.718	0.461	0.825	0.383	1.01	3.0	**250°F**	0.52
600°F	0.366†	0.231	0.379	0.206	0.420	2.0††	**350°F**	0.27
Pour Point	−65°F	−90°F	−25°F	54°F*	−40°F**	−55°F	−34°F**	
Vapor Pressure, psia								
240°F	0.010	0.102	0.014	0.151	1.0		Boiling Point:	
400°F	0.360	2.14	0.370	3.94	15.0		225°F	
600°F	3.74†	23.6	6.24	45.7	87.0	<0.1 mmHg††		
Equivalent Product	Dowtherm T	Dowtherm J	Dowtherm HT	Dowtherm A				

† @ 550°F
**Crystallizing Point*
†† @ 500°F
***Freezing Point*
Therminol is a registered trademark of Monsanto Company. Dowtherm and Syltherm are registered trademarks of Dow Chemical Company. UCON is a registered trademark of Union Carbide.

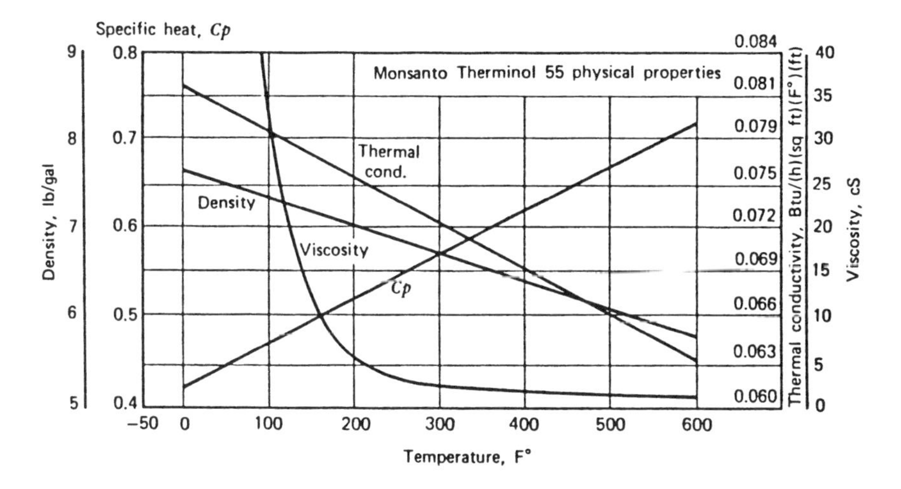

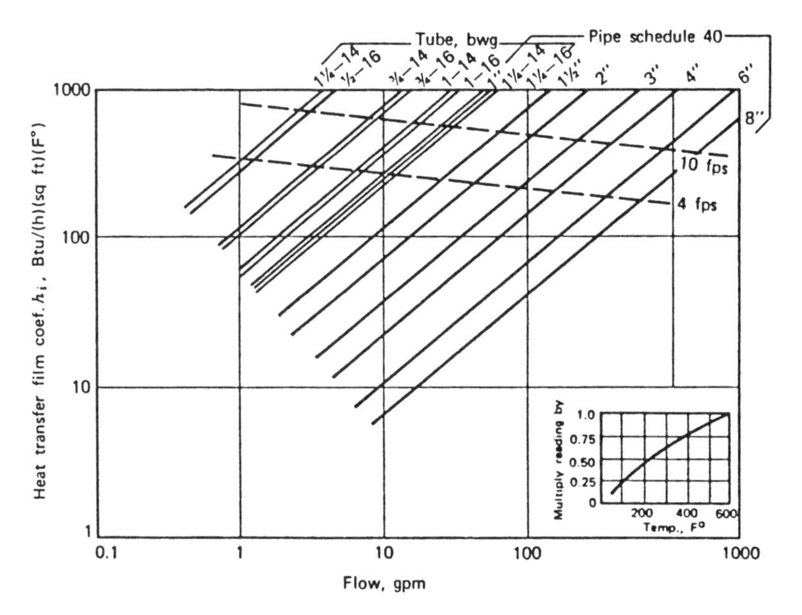

Figure 2-7. Heat transfer coefficient, Monsanto Therminol 55. Use range 0°F to 600°F; maximum film temperature 635°F. (From *Practical Heat Recovery,* John L. Boyen, John Wiley & Sons, Inc., © 1975.)

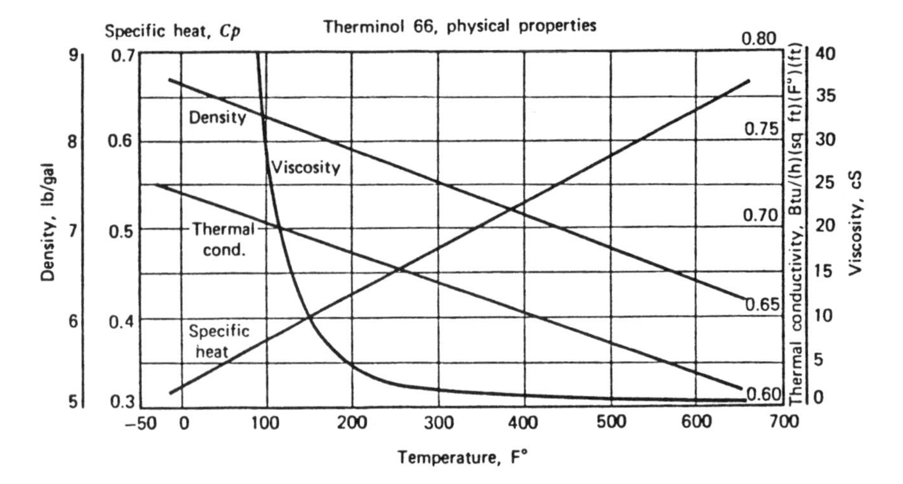

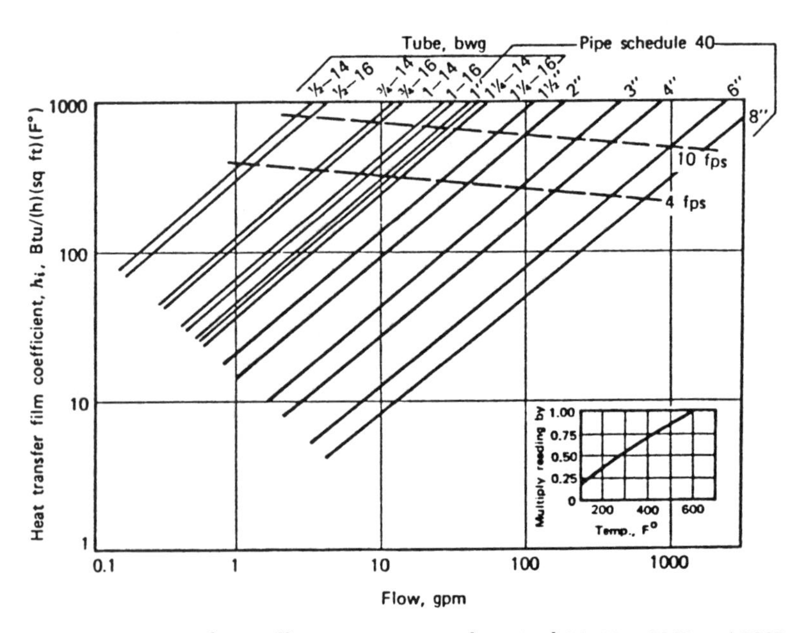

Figure 2-8. Heat transfer coefficient, Monsanto Therminol 66. Use 20°F to 650°F; maximum film temperature 705°F. (From *Practical Heat Recovery*, John L. Boyen, John Wiley & Sons, Inc., © 1975.)

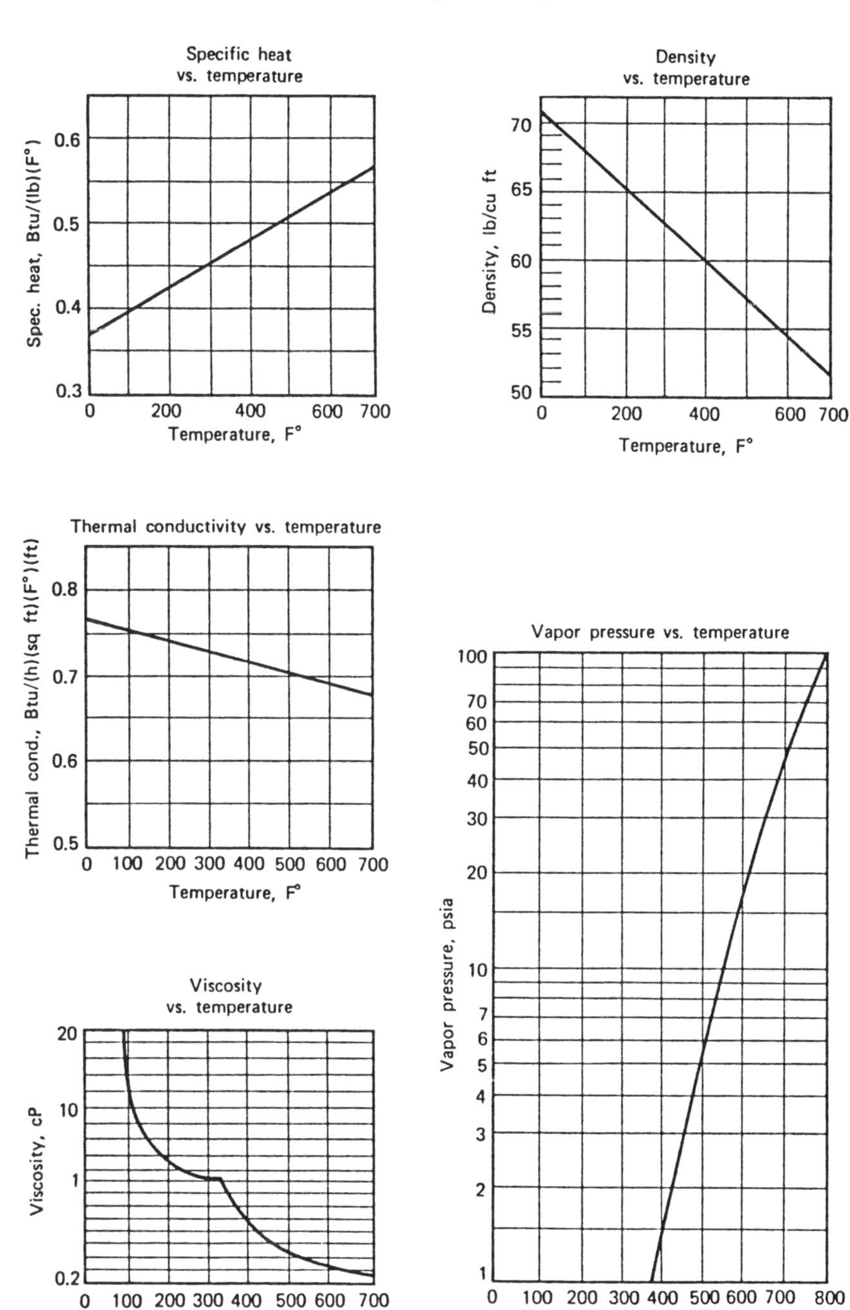

Figure 2-9. Physical properties Dowtherm "G" versus temperature. (From *Practical Heat Recovery,* John L. Boyen, John Wiley & Sons, Inc., © 1975.)

(text continued from page 22)

Figure 2-10 gives the thermal conductivity, viscosity, and specific heat of water.

Outside Film Coefficient (in a Liquid Bath)

The outside film coefficient for a process coil in a liquid bath heater is the result of natural or free convection. Temperature variations in the fluid cause density variations. These density variations in turn cause the fluid to circulate, which produces the free convective heat transfer. For horizontal pipes and tubes spaced more than one diameter apart, the following equation, based on empirical studies, may be used to determine the film coefficient.

$$h_o = 116\left[\frac{k^3\, C\, \rho^2\, \beta\, \Delta T}{\mu d_o}\right]^{0.25} \tag{2-9}$$

where h_o = outside film coefficient, Btu/hr-ft^2-°F
k = bath fluid thermal conductivity, Btu/hr-ft-°F
C = bath fluid specific heat, Btu/lb-°F
ρ = bath fluid density, lb/ft^3
β = bath fluid coefficient of thermal expansion, 1/°F
μ = bath fluid viscosity, cp
ΔT = average temperature difference between the fluid in the coil and the liquid bath, °F
d_o = pipe outside diameter, in.

The fluid properties can be determined as before, except for the bath fluid coefficient of thermal expansion, which is given in Table 2-5. The density of water is one divided by the specific volume given in steam tables (Table 2-6).

(text continued on page 33)

Table 2-5
Coefficients of Thermal Expansion

Fluid	Coefficient (1/°F)
Water	0.0024
Dowtherms	0.00043
Therminols	0.00039
Mobiltherms	0.00035

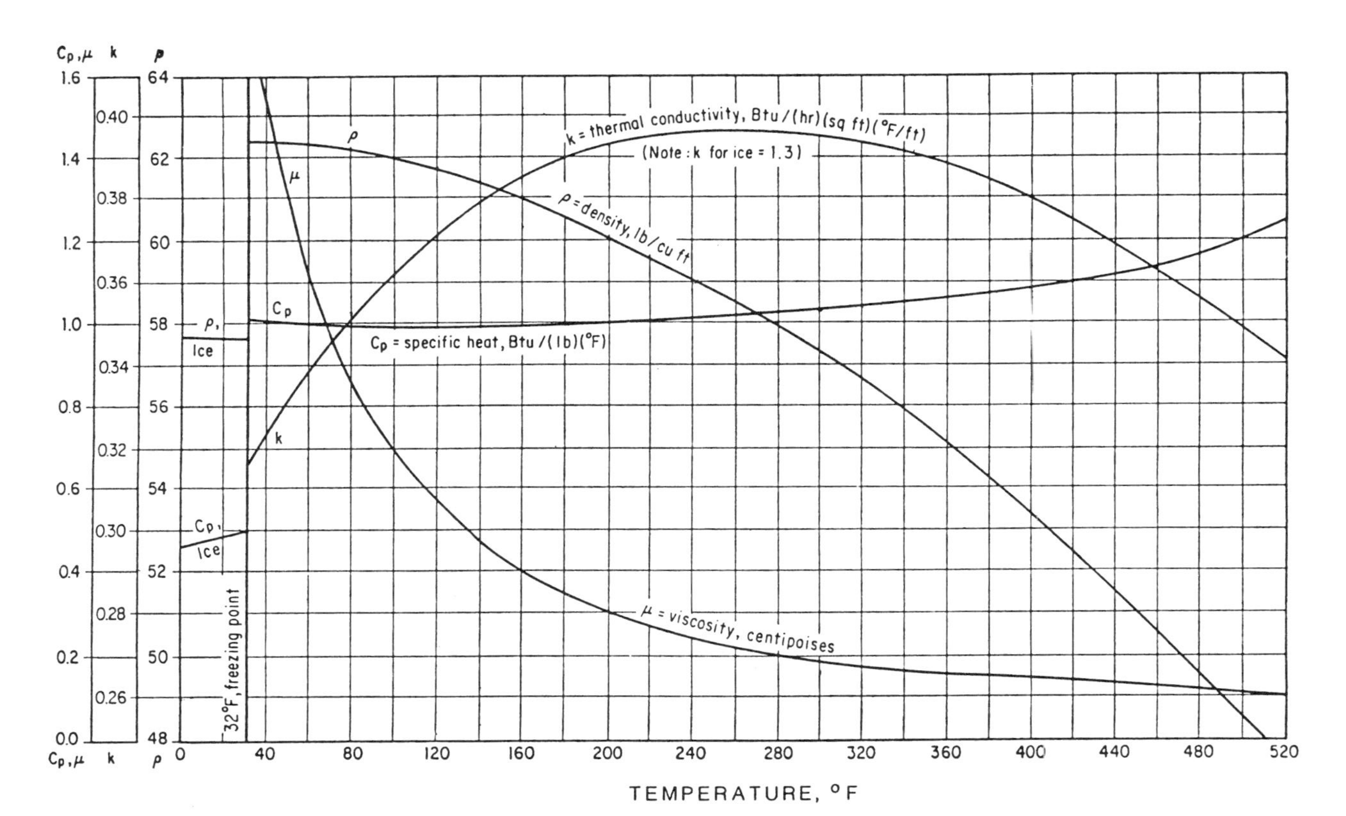

Figure 2-10. Physical properties of water.

Table 2-6
Properties of Dry Saturated Steam

Temp. F t	Abs. Press. psia P	Specific Volume Sat. Liquid v_f	Evap v_{fg}	Sat. Vapor v_g	Enthalpy Sat. Liquid h_f	Evap. h_{fg}	Sat. Vapor h_g	Entropy Sat. Liquid s_f	Evap. s_{fg}	Sat. Vapor s_g	Temp. F t
32	0.08854	0.01602	3306	3306	0.00	1075.8	1075.8	0.0000	2.1877	2.1877	32
35	0.09995	0.01602	2947	2947	3.02	1074.1	1077.1	0.0061	2.1709	2.1770	35
40	0.12170	0.01602	2444	2444	8.05	1071.3	1079.3	0.0162	2.1435	2.1597	40
45	0.14752	0.01602	2036.4	2036.4	13.06	1068.4	1081.5	0.0262	2.1167	2.1429	45
50	0.17811	0.01603	1703.2	1703.2	18.07	1065.6	1083.7	0.0361	2.0903	2.1264	50
60	0.2563	0.01604	1206.6	1206.7	28.06	1059.9	1088.0	0.0555	2.0393	2.0948	60
70	0.3631	0.01606	867.8	867.9	38.04	1054.3	1092.3	0.0745	1.9902	2.0647	70
80	0.5069	0.01608	633.1	633.1	48.02	1048.6	1096.6	0.0932	1.9428	2.0360	80
90	0.6982	0.01610	468.0	468.0	57.99	1042.9	1100.9	0.1115	1.8972	2.0087	90
100	0.9492	0.01613	350.3	350.4	67.97	1037.2	1105.2	0.1295	1.8531	1.9826	100
110	1.2748	0.01617	265.3	265.4	77.94	1031.6	1109.5	0.1471	1.8106	1.9577	110
120	1.6924	0.01620	203.25	203.27	87.92	1025.8	1113.7	0.1645	1.7694	1.9339	120
130	2.2225	0.01625	157.32	157.34	97.90	1020.0	1117.9	0.1816	1.7296	1.9112	130
140	2.8886	0.01629	122.99	123.01	107.89	1014.1	1122.0	0.1984	1.6910	1.8894	140
150	3.718	0.01634	97.06	97.07	117.89	1008.2	1126.1	0.2149	1.6537	1.8685	150
160	4.741	0.01639	77.27	77.29	127.89	1002.3	1130.2	0.2311	1.6174	1.8485	160
170	5.992	0.01645	62.04	62.06	137.90	996.3	1134.2	0.2472	1.5822	1.8293	170
180	7.510	0.01651	50.21	50.23	147.92	990.2	1138.1	0.2630	1.5480	1.8109	180
190	9.339	0.01657	40.94	40.96	157.95	984.1	1142.0	0.2785	1.5147	1.7932	190
200	11.526	0.01663	33.62	33.64	167.99	977.9	1145.9	0.2938	1.4824	1.7762	200
210	14.123	0.01670	27.80	27.82	178.05	971.6	1149.7	0.3090	1.4508	1.7598	210
212	14.696	0.01672	26.78	26.80	180.07	970.3	1150.4	0.3120	1.4446	1.7566	212

Table 2-6 (Continued)
Properties of Dry Saturated Steam

Temp. F t	Abs. Press. psia P	Specific Volume: Sat. Liquid v_f	Specific Volume: Evap v_{fg}	Specific Volume: Sat. Vapor v_g	Enthalpy: Sat. Liquid h_f	Enthalpy: Evap. h_{fg}	Enthalpy: Sat. Vapor h_g	Entropy: Sat. Liquid s_f	Entropy: Evap. s_{fg}	Entropy: Sat. Vapor s_g	Temp. F t
20	17.186	0.01677	23.13	23.15	188.13	965.2	1153.4	0.3239	1.4201	1.7440	220
230	20.780	0.01684	19.365	19.382	198.23	958.8	1157.0	0.3387	1.3901	1.7288	230
240	24.969	0.01692	16.306	16.323	208.34	952.2	1160.5	0.3531	1.3609	1.7140	240
250	29.825	0.01700	13.804	13.821	216.48	945.5	1164.0	0.3675	1.3323	1.6998	250
260	35.429	0.01709	11.746	11.763	228.64	938.7	1167.3	0.3817	1.3043	1.6860	260
270	41.858	0.01717	10.044	10.061	238.84	931.8	1170.6	0.3958	1.2769	1.6727	270
280	49.203	0.01726	8.628	8.645	249.06	924.7	1173.8	0.4096	1.2501	1.6597	280
290	57.556	0.01735	7.444	7.461	259.31	917.5	1176.8	0.4234	1.2238	1.6472	290
300	67.013	0.01745	6.449	6.466	269.59	910.1	1179.7	0.4369	1.1980	1.6350	300
310	77.68	0.01755	5.609	5.626	279.92	902.6	1182.5	0.4504	1.1727	1.6231	310
320	89.66	0.01765	4.896	4.914	290.28	894.9	1185.2	0.4637	1.1478	1.6115	320
330	103.06	0.01776	4.289	4.307	300.68	887.0	1187.7	0.4769	1.1233	1.6002	330
340	118.01	0.01787	3.770	3.788	311.13	879.0	1190.1	0.4900	1.0992	1.5891	340
350	134.63	0.01799	3.324	3.342	321.63	870.7	1192.3	0.5029	1.0754	1.5783	350
360	153.04	0.01811	2.939	2.957	332.18	862.2	1194.4	0.5158	1.0519	1.5677	350
370	173.37	0.01823	2.606	2.625	342.79	853.5	1196.3	0.5286	1.0287	1.5573	370
380	195.77	0.01836	2.317	2.335	353.45	844.6	1198.1	0.5413	1.0059	1.5471	380
390	220.37	0.01850	2.0651	2.0836	364.17	835.4	1199.6	0.5539	0.9832	1.5371	390

(table continued on next page)

Table 2-6 (Continued)
Properties of Dry Saturated Steam

Temp. F t	Abs. Press. psia P	Specific Volume			Enthalpy			Entropy			Temp. F t
		Sat. Liquid v_f	Evap v_{fg}	Sat. Vapor v_g	Sat. Liquid h_f	Evap. h_{fg}	Sat. Vapor h_g	Sat. Liquid s_f	Evap. s_{fg}	Sat. Vapor s_g	
400	247.31	0.01864	1.8447	1.8633	374.97	826.0	1201.0	0.5664	0.9608	1.5272	400
410	276.75	0.01878	1.6512	1.6700	385.83	816.3	1202.1	0.5788	0.9386	1.5174	410
420	308.83	0.01894	1.4811	1.5000	396.77	806.3	1203.1	0.5912	0.9166	1.5078	420
430	343.72	0.01910	1.3308	1.3499	407.79	796.0	1203.8	0.6035	0.8947	1.4982	430
440	381.59	0.01926	1.1979	1.2171	418.90	785.4	1204.3	0.6158	0.8730	1.4887	440
450	422.6	0.0194	1.0799	1.0993	430.1	774.5	1204.6	0.6280	0.8513	1.4793	450
460	466.9	0.0196	0.9748	0.9944	441.4	763.2	1204.6	0.6402	0.8298	1.4700	460
470	514.7	0.0198	0.8811	0.9009	452.8	751.5	1204.3	0.6523	0.8083	1.4606	470
480	566.1	0.0200	0.7972	0.8172	464.4	739.4	1203.7	0.6645	0.7868	1.4513	480
490	621.4	0.0202	0.7221	0.7423	476.0	726.8	1202.8	0.6766	0.7653	1.4419	490
500	680.8	0.0204	0.6545	0.6749	487.8	713.9	1201.7	0.6887	0.7438	1.4325	500
520	812.4	0.0209	0.5385	0.5594	511.9	686.4	1198.2	0.7130	0.7006	1.4136	520
540	962.5	0.0215	0.4434	0.4649	536.6	656.6	1193.2	0.7374	0.6568	1.3942	540
560	1133.1	0.0221	0.3647	0.3868	562.2	624.2	1186.4	0.7621	0.6121	1.3742	560
580	1325.8	0.0228	0.2989	0.3217	588.9	588.4	1177.3	0.7872	0.5659	1.3532	580
600	1542.9	0.0236	0.2432	0.2668	617.0	548.5	1165.3	0.8131	0.5176	1.3307	600
620	1786.6	0.0247	0.1955	0.2201	646.7	503.6	1150.3	0.8398	0.4684	1.3062	620
640	2059.7	0.0260	0.1538	0.798	678.6	452.0	1130.5	0.8679	0.4110	1.2789	640
660	2365.4	0.0278	0.1165	0.1442	714.2	390.2	1104.4	0.8987	0.3485	1.2472	660
680	2708.1	0.0305	0.0810	0.1115	757.3	309.9	1067.2	0.9351	0.2719	1.2071	680
700	3093.7	0.0369	0.0392	0.0761	823.3	172.1	995.4	0.9905	0.1484	1.1389	700
705.4	3206.2	0.0503	0	0.0503	902.7	0	902.7	1.0580	0	1.0580	705.4

Data in the steam tables abstracted by permission from Thermodynamic Properties of Steam *by J. H. Keenan and F. G. Keyes, published by John Wiley & Sons, Inc., 1936.*

(*text continued from page 28*)

Outside Film Coefficient (Shell-and-Tube Exchangers)

For shell-and-tube heat exchangers with shell-side baffles, the shell-side fluid flow is perpendicular to the tubes. In this arrangement, the outside film coefficient can be calculated from the following equation:

$$\frac{h_o D}{k} = 0.6\,K\left(\frac{C\mu_e}{k}\right)^{0.33}\left(\frac{DG_{max}}{\mu_e}\right)^{0.6} \qquad (2\text{-}10)$$

where h_o = outside film coefficient, Btu/hr-ft^2- °F
D = tube outside diameter, ft
k = fluid thermal conductivity, Btu/hr-ft-°F
G_{max} = maximum mass velocity of fluid, lb/hr-ft^2
C = fluid specific heat, Btu/lb °F
μ_e = fluid viscosity, lb/hr-ft
K = coefficient from Table 2-7

Table 2-7
Value of K for Fluid Flow Perpendicular to a Bank of Staggered Tube "N" Rows Deep

N	1	2	3	4	5	6	10
K	0.24	0.27	0.29	0.30	0.31	0.32	0.33

Approximate Overall Heat Transfer Coefficient

The calculation of overall heat transfer coefficient U using the equations previously presented can be rather tedious. Heat transfer specialists have computer programs to calculate this value. There are some quick approximation techniques. Table 2-8 comes from the Gas Processors Suppliers Association's *Engineering Data Book* and gives an approximate value of U for shell and tube heat exchangers.

It can be seen from the table that exchanging water with 100-psi gas gives a low U value. Thus, a high surface area is required. Exchanging water with 1,000 psi gas gives a much higher U value and less surface area can be used in the exchanger. If water is being exchanged with

Table 2-8
Typical Bare-Tube Overall Heat-Transfer Coefficients, U for Shell and Tube Heat Exchangers Btu/hr-ft²-°F

Service	U
Water with 100-psi gas	35–40
Water with 300-psi gas	40–50
Water with 700-psi gas	60–70
Water with 1,000-psi gas	80–100
Water with kerosine	80–90
Water with MEA	130–150
Water with air	20–25
Water with water	180–200
Oil with oil	80–100
C_3 with C_3 liquid	110–130
Water condensers with C_3, C_4	125–135
Water condensers with naphtha	70–80
Water condensers with still ovhd.	70–80
Water condensers with amine	100–110
Reboilers with steam	140–160
Reboilers with hot oil	90–120
100-psi gas with 500-psi gas	50–70
1,000-psi gas with 1,000-psi gas	60–80
1,000-psi gas chiller (gas-C_3)	60–80
MEA exchanger	120–130

Maximum boiling film transfer coefficient
Hydrocarbons—300 to 500 Btu/hr-ft²-°F
Water—2,000 Btu/hr-ft²-°F

water, a very high U value is achieved. Water with air gives a very low U value. Oil with oil is much less than water with water, because of the viscous nature of the oil.

The approximate U values in the table do not differentiate between tube-side and shell-side fluids. Which fluid is on the inside of the tubes and which is on the outside does make a difference to the U value. This is beyond the accuracy of the table.

Figure 2-11 gives values of U for exchange from a water bath to a natural gas stream in a coil. Figure 2-12 is a nomograph for crude oil streams heated with a water bath.

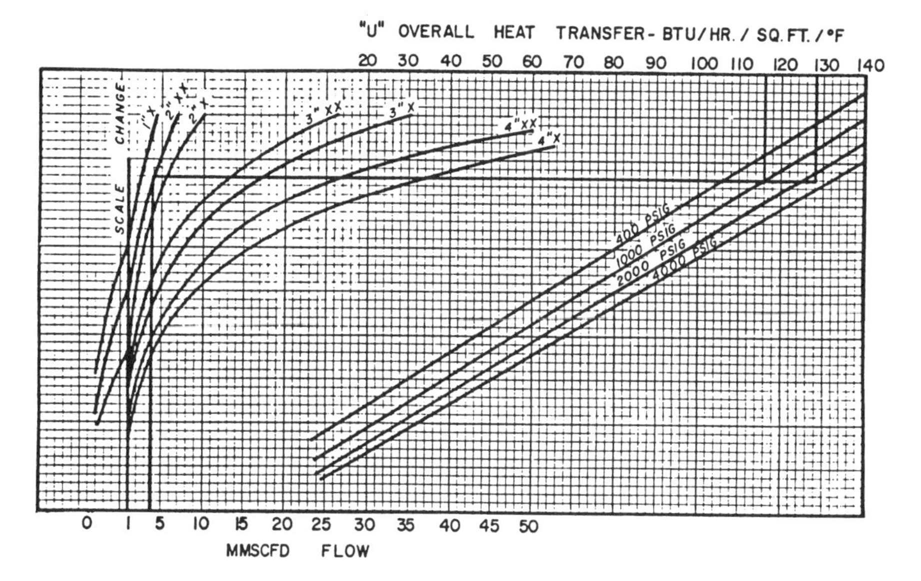

Figure 2-11. Values of U for exchange from water bath to natural gas stream in a coil. (*Courtesy of Smith Industries, Inc.*)

PROCESS HEAT DUTY

The process heat duty is the heat required to be added or removed from the process fluids to create the required change in temperature. This can be in the form of sensible heat, latent heat, or both.

Sensible Heat

The amount of heat absorbed or lost by a substance that causes a change in the temperature of the substance is sensible heat. It is called sensible heat because it can be measured by the change in temperature it causes. For example, as heat is added to a piece of steel the temperature of that steel increases and can be measured. The general equation for calculating sensible heat is:

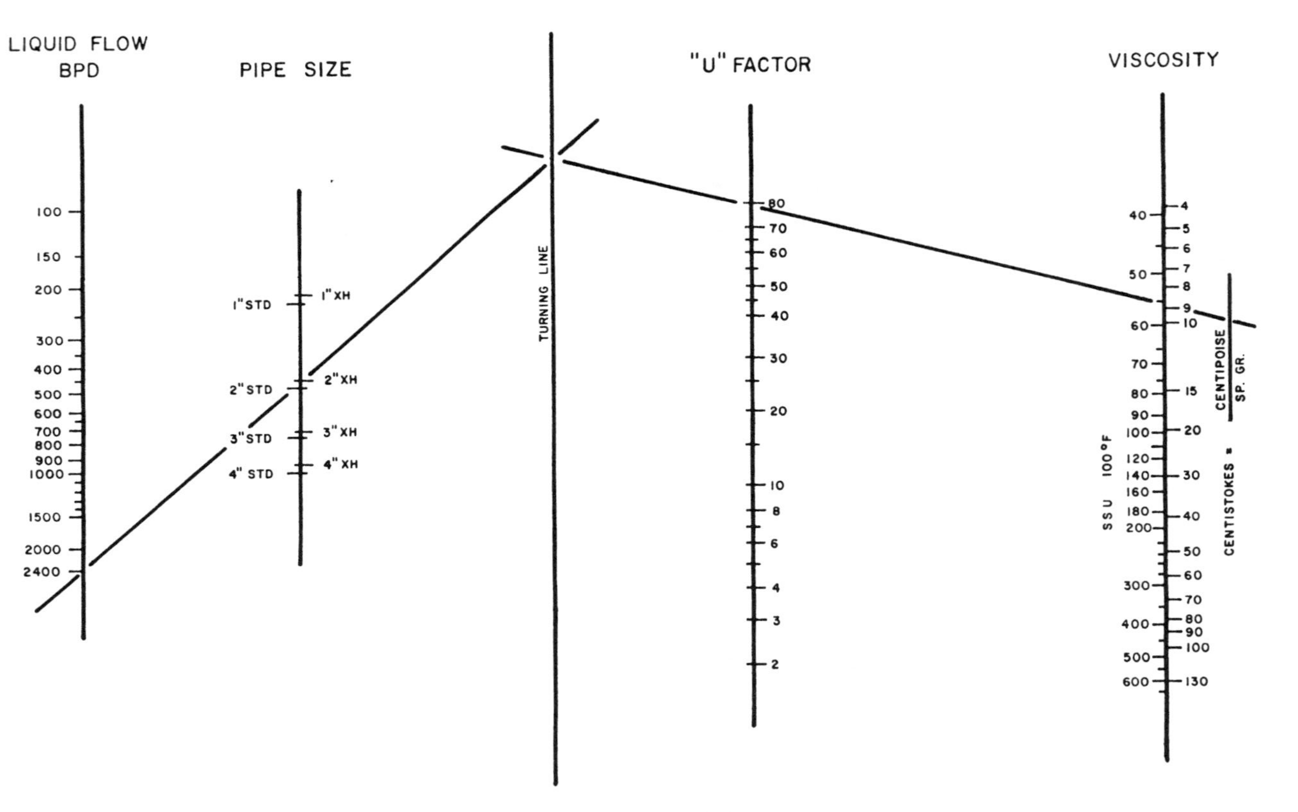

Figure 2-12. Crude oil streams heated with water bath heaters. *(Courtesy of Smith Industries, Inc.)*

$$q_{sh} = W\,(C)\,(T_2 - T_1) \tag{2-11}$$

where q_{sh} = sensible heat duty, Btu/hr
W = mass flow rate, lb/hr
C = specific heat of the fluid, Btu/lb-°F
T_1 = initial temperature, °F
T_2 = final temperature, °F

The specific heat of hydrocarbon vapors and liquids is given by Figures 2-13 and 2-14.

In Chapter 6 of Volume 1, it was assumed that C = 0.5 Btu/lb-°F for crude oil. It can be seen from Figure 2-13 that this is true for the range of treating temperatures and crude gravities normally encountered in oil treating.

Latent Heat

The amount of heat energy absorbed or lost by a substance when changing phases is called "latent heat." When steam is condensed to

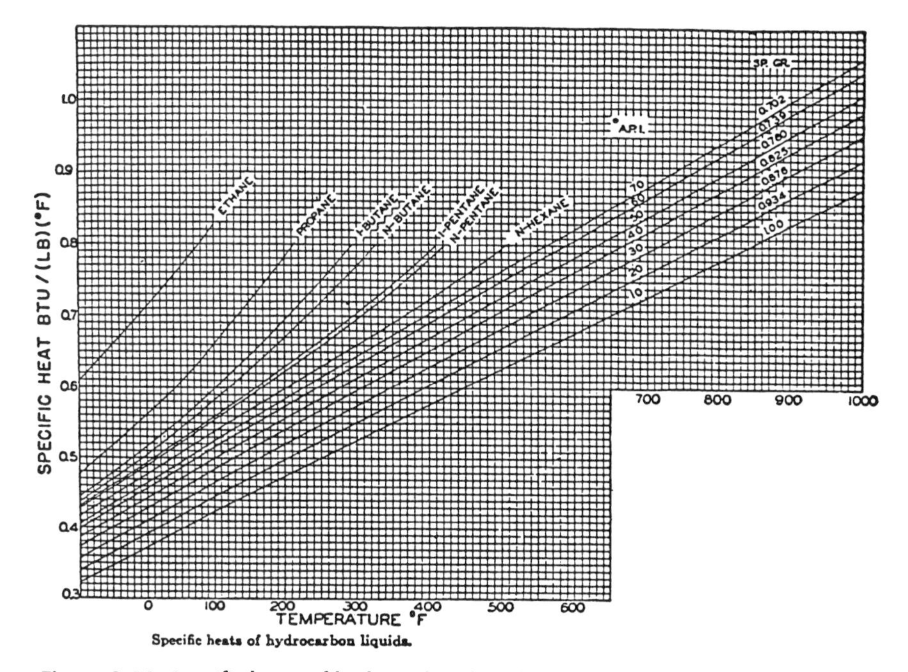

Figure 2-13. Specific heats of hydrocarbon liquids. (From Holcomb and Brown, *Ind. Eng. Chem.*, 34, 595, 1942; reprinted from *Process Heat Transfer*, Kern, McGraw-Hill Co., © 1950.)

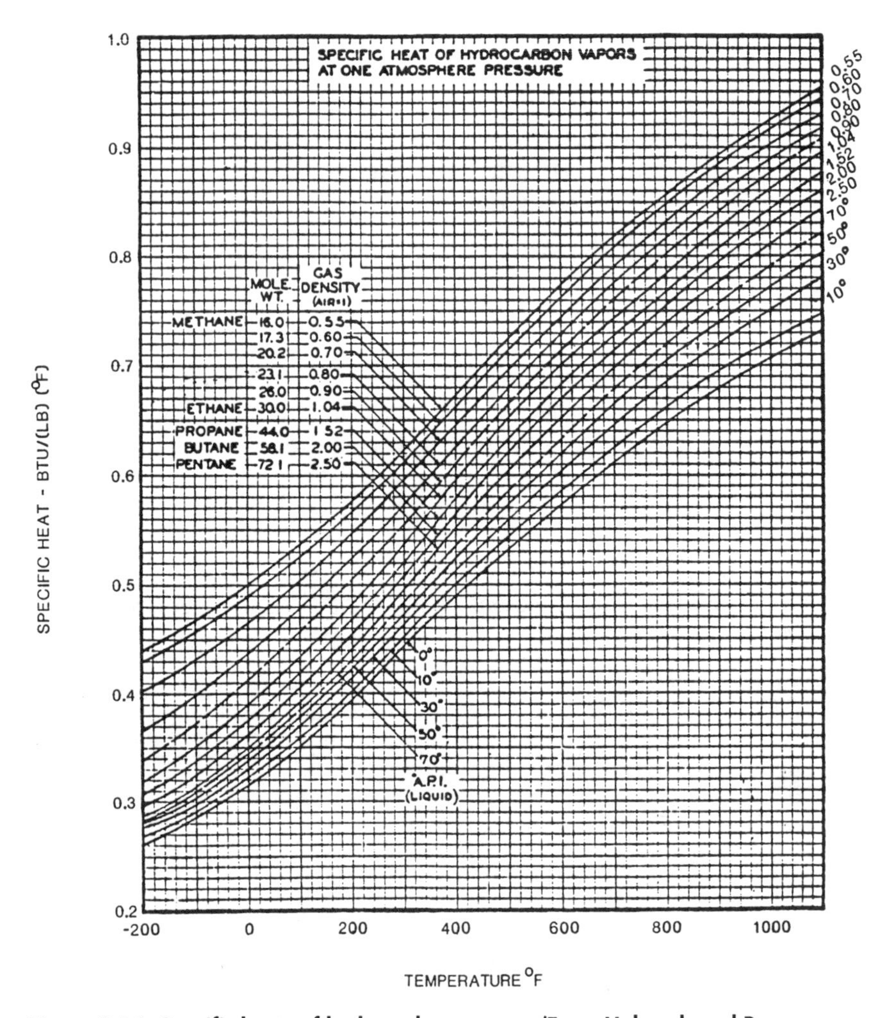

Figure 2-14. Specific heats of hydrocarbon vapors. (From Holcomb and Brown, *Ind. Eng. Chem.*, 34, 595, 1942; reprinted from *Process Heat Transfer*, Kern, McGraw-Hill Co., © 1950.)

water, the temperature doesn't change, but heat must be extracted from the steam as it goes through a phase change to water. To change water to steam, heat must be added. When a substance changes from a solid to a liquid or from a liquid to a vapor, the heat absorbed is in the form of latent heat. This heat energy is referred to as latent heat because it cannot be sensed by measuring the temperature.

$$q_{lh} = W(\lambda) \tag{2-12}$$

where q_{lh} = latent heat duty, Btu/hr
W = mass flow rate, lb/hr
λ = latent heat, Btu/lb

The latent heat of vaporization for hydrocarbon compounds is given in Table 2-9. The latent heat of vaporization of water is given by h_{fg} in the steam table (Table 2-6).

Heat Duty for Multiphase Streams

When a process stream consists of more than one phase, the process heat duty can be calculated using the following equation:

$$q_p = q_g + q_o + q_w \tag{2-13}$$

where q_p = overall process heat duty, Btu/hr
q_g = gas heat duty, Btu/hr
q_o = oil heat duty, Btu/hr
q_w = water heat duty, Btu/hr

Table 2-9
Latent Heat of Vaporization

Compound	Heat of Vaporization, 14.696 psia at Boiling Point, Btu/lb
Methane	219.22
Ethane	210.41
Propane	183.05
n-Butane	165.65
Isobutane	157.53
n-Pentane	153.59
Isopentane	147.13
Hexane	143.95
Heptane	136.01
Octane	129.53
Nonane	123.76
Decane	118.68

Natural Gas Sensible Heat Duty at Constant Pressure

The sensible heat duty for natural gas at constant pressure is:

$$q_g = 41.7\ Q_g C_g\ (T_2 - T_1) \quad (2\text{-}14)$$

where Q_g = gas flow rate, MMscfd
C_g = gas heat capacity, Btu/Mscf °F
T_1 = inlet temperature, °F
T_2 = outlet temperature, °F

Heat capacity is determined at atmospheric conditions and then corrected for temperature and pressure based on reduced pressure and temperature.

$$C_g = 2.64\ [29\ S\ (C) + \Delta C_p] \quad (2\text{-}15)$$

where C = gas specific heat at one atmosphere pressure, Btu/lb-°F (Figure 2-14)
ΔCp = correction factor
S = gas specific gravity

The correction factor ΔC_p is obtained from Figure 2-15 where:

$$P_r = P/P_c \quad (2\text{-}16)$$

$$T_r = T_a/T_c \quad (2\text{-}17)$$

where P_r = gas reduced pressure
P = gas pressure, psia
P_c = gas pseudo critical pressure, psia
T_r = gas reduced temperature
T_a = gas average temperature, °R = $1/2\ (T_1 + T_2)$
T_c = gas pseudo critical temperature, °R

The gas pseudo critical pressures and temperatures can be approximated from Figure 2-16 or they can be calculated as weighted averages of the critical temperatures and pressures of the various components on a

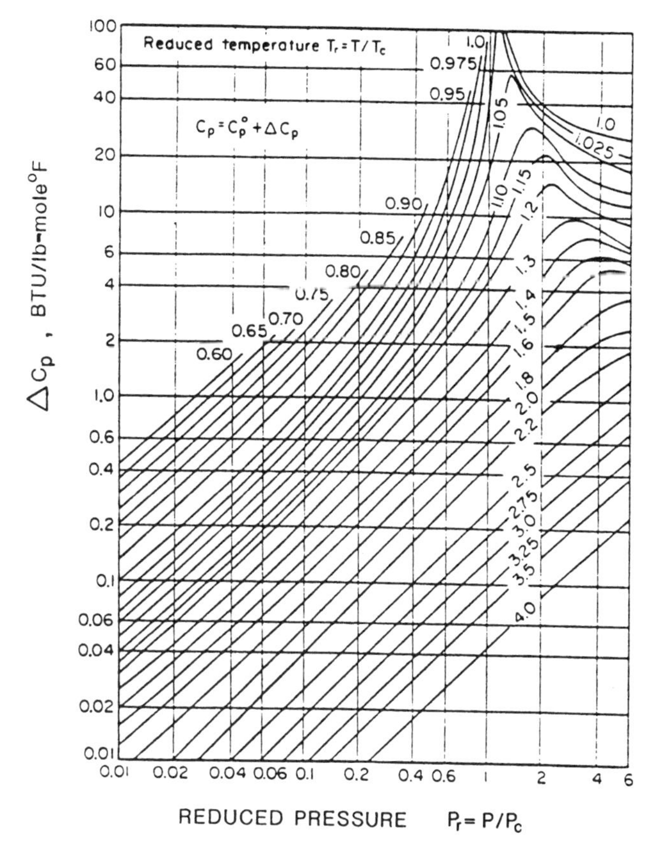

Figure 2-15. Heat capacity correction factor. (From *Chemical Engineer's Handbook*, 5th Edition, R. Perry and C. Chilton, McGraw-Hill Co., © 1973.)

mole fraction basis. Table 2-10 shows a calculation for the gas stream in our example field. For greater precision, a correction for H_2S and CO_2 content may be required. Refer to the Gas Processors Suppliers Association's *Engineering Data Book* or other text for a correction procedure.

Oil Sensible Heat Duty

The sensible heat duty for the oil phase is:

$$q_o = 14.6\ SG\ Q_o\ C_o\ (T_2 - T_1) \tag{2-18}$$

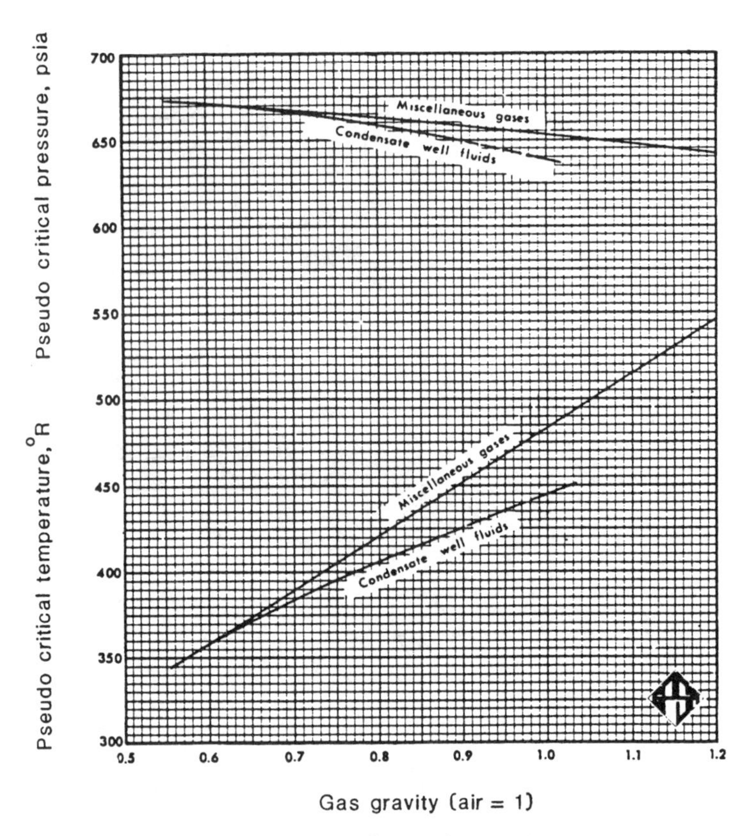

Figure 2-16. Pseudo critical properties of natural gases. (From Gas Processors Suppliers Association, *Engineering Data Handbook*, 9th Edition.)

where Q_o = oil flow rate, bpd
SG = oil specific gravity
C_o = oil specific heat, Btu/lb-°F (Figure 2-13)
T_1 = initial temperature, °F
T_2 = final temperature, °F

Water Sensible Heat Duty

The duty for heating free water may be determined from the following equation by assuming a water specific heat of 1.0 Btu/lb-°F.

$$q_w = 14.6\ Q_w\ (T_2 - T_1) \tag{2-19}$$

where Q_w = water flow rate, bpd

Table 2-10
Estimate of Specific Gravity, Pseudo Critical Temperature and Pseudo Critical Pressure for the Example Field

	A Mole % Gas Composition	B Molecular Weight	C Critical Temp. °R	D Critical psia
CO_2	4.03	44.010	547.87	1071.0
N_2	1.44	28.013	227.3	493.0
H_2S	0.0019	34.076	672.6	1036.0
C_1	85.55	16.043	343.37	667.8
C_2	5.74	30.070	550.09	707.0
C_3	1.79	44.097	666.01	616.3
iC_4	0.41	58.124	734.98	529.1
nC_4	0.41	58.124	765.65	550.7
iC_5	0.20	72.151	829.10	490.4
nC_5	0.13	72.151	845.70	488.6
C_6	0.15	86.178	913.70	436.9
C_7^+	0.15	147	1112.0	304
Computed Value	100.00	19.48	374.6	680.5
Computation	Sum (A_i)	$\frac{\text{Sum }(A_i \times B_i)}{\text{Sum }(A_i)}$	$\frac{\text{Sum }(A_i \times C_i)}{\text{Sum }(A_i)}$	$\frac{\text{Sum }(A_i \times D_j)}{\text{Sum }(A_i)}$

$$\text{Specific Gravity} = \frac{19.48}{29} = 0.67$$

Heat Duty and Phase Changes

If a phase change occurs in the process stream for which heat duties are being calculated, it is best to perform a flash calculation and determine the heat loss or gain by the change in enthalpy. For a quick hand approximation it is possible to calculate sensible heat for both the gas and liquid phases of each component. The sum of all the latent and sensible heats is the approximate total heat duty.

Heat Lost to Atmosphere

The total heat duty required to raise a substance from one temperature to another temperature must include an allowance for heat lost to the atmosphere during the process. For example, if the process fluid flows through a coil in a water bath, not only is the water bath exchanging heat with the process fluid, but it is also exchanging heat with the surrounding atmosphere.

The heat lost to the atmosphere can be calculated in the same manner as any other heat exchange problem using Equation 2-3. The overall heat transfer coefficient may be calculated from a modification of Equation 2-5. By assuming that the inside film coefficient is very large compared to the outside film coefficient, by adding a factor for conduction losses through insulation, and by eliminating fouling factors to be conservative, Equation 2-5 becomes:

$$\frac{1}{U}=\frac{1}{h_o}+\frac{\Delta X_1}{K_1}+\frac{\Delta X_2}{K_2} \qquad (2\text{-}20)$$

where h_o = outside film coefficient Btu/hr-ft^2-°F
= $1 + 0.22\ V_w (V_w < 16$ ft/sec)
= $0.53\ V_w^{0.8} (V_w > 16$ ft/sec)
V_w = wind velocity (ft/sec)
= $1.47 \times$ (mph)
ΔX_1 = shell thickness, ft
K_1 = shell thermal conductivity Btu/hr-ft-°F
≅ 30 for carbon steel (Table 2-3)
ΔX_2 = insulation thickness, ft
K_2 = insulation thermal conductivity, Btu/hr-ft-°F
≅ 0.03 for mineral wool

For preliminary calculations it is sometimes assumed that the heat lost to atmosphere is approximately 5–10 % of the process heat duty for uninsulated equipment and 1–2% for insulated equipment.

Heat Transfer from a Fire Tube

A fire tube contains a flame burning inside a piece of pipe which is in turn surrounded by the process fluid. In this situation, there is radiant and convective heat transfer from the flame to the inside surface of the fire tube, conductive heat transfer through the wall thickness of the tube, and convective heat transfer from the outside surface of that tube to the oil being treated. It would be difficult in such a situation to solve for the heat transfer in terms of an overall heat transfer coefficient. Rather, what is most often done is to size the fire tube by using a heat flux rate. The heat flux rate represents the amount of heat that can be transferred from the fire tube to the process per unit area of outside surface of the fire tube. Common heat flux rates are given in Table 2-11.

Table 2-11
Common Heat Flux Rates

Medium Being Heated	Design Flux Rate Btu/hr-ft^2
Water	10,000
Boiling water	10,000
Crude oil	8,000
Heat medium oils	8,000
Glycol	7,500
Amine	7,500

The required fire tube area is thus given by:

$$\text{Surface area of fire tube} = \frac{\text{Heat duty including losses (Btu/hr)}}{\text{Design flux rate (Btu/hr-ft}^2\text{)}} \quad (2\text{-}21)$$

For example, if total heat duty (sensible heat, latent heat duty, heat losses to the atmosphere) was 1 MMBtu/hr and water was being heated, a heat flux of 10,000 Btu/hr-ft^2 would be used and 100 ft^2 of fire tube area would be required.

Table 2-12
Standard Burner Sizes and Minimum Diameter

Btu/hr	Minimum Diameter-in.
100,000	2.5
250,000	3.9
500,000	5.5
750,000	6.7
1,000,000	7.8
1,500,000	9.5
2,000,000	11.0
2,500,000	12.3
3,000,000	13.5
3,500,000	14.6
4,000,000	15.6
5,000,000	17.4

For natural draft fire tubes, the minimum cross-sectional area of the fire tube is set by limiting the heat release density to 21,000 Btu/hr-in^2. At heat release densities above this value, the flame may become unstable because of insufficient air. Using this limit, a minimum fire tube diameter is established by:

$$d^2 = \frac{\text{Burner heat release density (Btu/hr)}}{16{,}500} \qquad (2\text{-}22)$$

where d = minimum fire tube diameter, in.

In applying Equation 2-22 note that the burner heat release density will be somewhat higher than the heat duty, including losses used in Equation 2-21, as a standard burner size will be chosen slightly larger than that required. Standard burner sizes and minimum fire tube diameters are included in Table 2-12.

CHAPTER

3

Heat Exchangers*

HEAT EXCHANGERS

Heat exchangers used in gas production facilities are shell-and-tube, double-pipe, plate-and-frame, bath-type, forced-air, or direct-fired. In this chapter we will discuss the basic concepts for sizing and selecting heat exchangers. This is just a brief overview of this complex subject and is meant to provide the reader with a basis upon which to discuss specific sizing and selection details with heat exchange experts in engineering companies and with vendors.

Bath-type heat exchangers can be either direct or indirect. In a direct bath exchanger, the heating medium exchanges heat directly with the fluid to be heated. The heat source for bath heaters can be a coil of a hot heat medium or steam, waste heat exhaust from an engine or turbine, or heat from electric immersion heaters. An example of a bath heater is an emulsion heater-treater of the type discussed in Volume 1. In this case, a fire tube immersed in the oil transfers heat directly to the oil bath. The calculation of heat duties and sizing of fire tubes for this type of heat exchanger can be calculated fom Chapter 2.

---

*Reviewed for the 1999 edition by Lei Tan of Paragon Engineering Services, Inc.

In an indirect bath heat exchanger, the heating medium provides heat to an intermediary fluid, which then transfers the heat to the fluid being heated. An example of this is the common line heater used on many gas well streams to keep the temperature above the hydrate formation temperature. A fire tube heats a water bath, which provides heat to the well stream flowing through a coil immersed in the bath. Details pertaining to design of indirect bath heaters are presented in Chapter 5.

SHELL-AND-TUBE EXCHANGERS

Shell-and-tube heat exchangers are cylindrical in shape, consisting of a bundle of parallel tubes surrounded by an outer casing (shell). Both the tube bundle and the shell are designed as pressure containing elements in accordance with the pressure and temperature requirements of the fluids that flow through each of them. The tube-side fluid is isolated from the shell-side fluid either by gasketed joints or by permanent partitions. The Standards of Tubular Exchanger Manufacturers Association (TEMA) define the various types of shell and tube exchangers, as well as design and construction practices.

The shell-and-tube exchanger is by far the most common type of heat exchanger used in production operations. It can be applied to liquid/liquid, liquid/vapor, or vapor/vapor heat transfer services. The TEMA standards define the design requirements for virtually all ranges of temperature and pressure that would be encountered in an oil or gas production facility.

The simplest type of shell-and-tube heat exchanger is shown in Figure 3-1. The essential parts are a shell (1), equipped with two nozzles and having tube sheets (2) at both ends, which also serve as flanges for the attachment of the two channels or heads (3) and their respective channel covers (4). The tubes are expanded into both tube sheets and are equipped with transverse baffles (5) on the shell side for support. The calculation of the effective heat transfer surface is based on the distance between the inside faces of the tube sheets instead of the overall tube length.

The shell-and-tube exchanger shown in Figure 3-1 is considered to operate in counter-current flow, since the shell fluid flows across the outside of the tubes. Often, in order to maintain a high enough tube velocity to avoid laminar flow and to increase heat transfer, the design is modified so that the tube fluid passes through a fraction of the tubes in two or more successive "passes" from head to head. An example of a two-pass fixed-tube exchanger is shown in Figure 3-2.

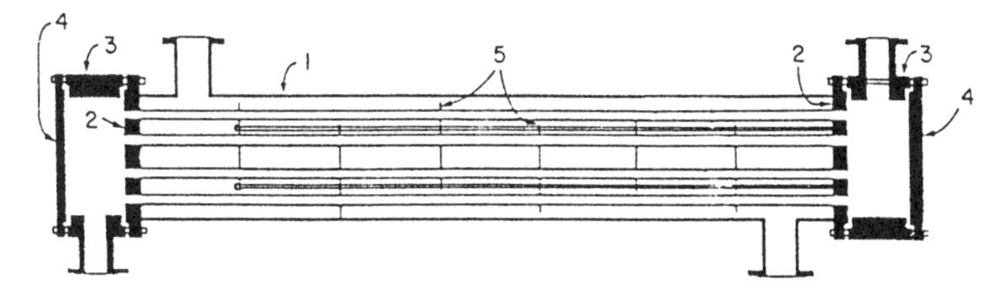

Figure 3-1. Fixed head shell-and-tube heat exchanger.

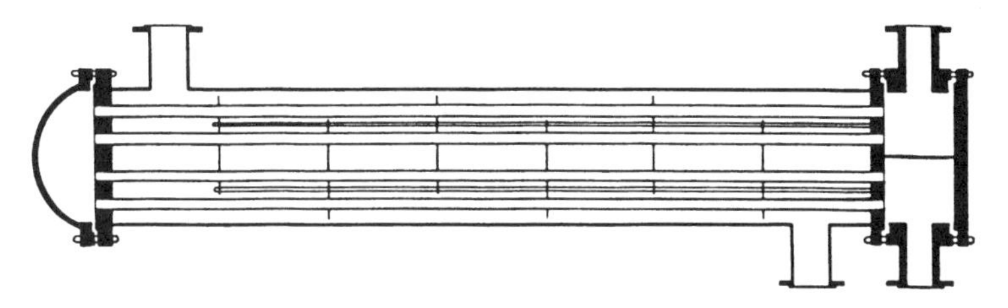

Figure 3-2. Fixed head 1-2 shell-and-tube heat exchanger.

An exchanger in which the shell-side fluid flows in one shell pass and the tube fluid in two or more passes is called a 1-2 exchanger. A single channel is employed with a partition to permit the entry and exit of the tube fluid from the same channel. At the opposite end of the exchanger a bonnet is provided to permit the tube fluid to cross from the first to the second pass. As with all fixed-tubesheet exchangers, the outsides of the tubes are inaccessible for inspection or mechanical cleaning. The insides of the tubes can be cleaned in place by removing only the channel cover and using a rotary cleaner or wire brush.

Baffles

Shell-and-tube exchangers contain several types of baffles to help direct the flow of both tube-side and shell-side fluids. Pass partition baffles force the fluid to flow through several groups of parallel tubes. Each of these groups of tubes is called a "pass," since it passes the fluid from one head to another. By adding pass partition baffles on each end, the tube-side fluid can be forced to take as many passes through the exchanger as desired.

Transverse baffles support the tubes that pass through holes in the baffle. The transverse baffle cannot go all the way across the cross-section of the shell, because the fluid that is in the shell has to be able to come over the top of the baffle and under the bottom of the next baffle, etc., as it passes across the tubes that are in the heat exchanger. When it is desired that a fluid pass through the shell with an extremely small pressure drop, these will usually be half-circle, 50% plates that provide rigidity and prevent the tubes from sagging.

Transverse baffles can help maintain greater turbulence for the shell-side fluid, resulting in a higher rate of heat transfer. The transverse baffles cause the liquid to flow through the shell at right angles to the axis of the tubes. This can cause considerable turbulence, even when a small quantity of liquid flows through the shell if the center-to-center distance between baffles, called baffle spacing, is sufficiently small. The baffles are held securely by means of baffle spacers, which consist of through-bolts screwed into the tube sheet and a number of smaller lengths of pipe that form shoulders between adjacent baffles.

Transverse baffles are drilled plates with heights that are generally 75% of the inside diameter of the shell. They may be arranged, as shown in Figure 3-3, for "up-and-down" flow or may be rotated 90° to provide "side-to-side" flow, the latter being desirable when a mixture of liquid and gas flows through the shell. Although other types of transverse baffles are sometimes used, such as the orifice baffle shown in Figure 3-4, they are not of general importance.

Impingement baffles are placed opposite the shell-side inlet nozzle. The flow into the shell hits the impingement baffle and is dispersed around the tubes, rather than impinging directly on the top tubes. This keeps the full force of the momentum of the flow from impinging on and eroding the top tubes.

Longitudinal baffles force the shell-side fluid to make more than one pass through an exchanger. With no longitudinal baffle, such as in Figure

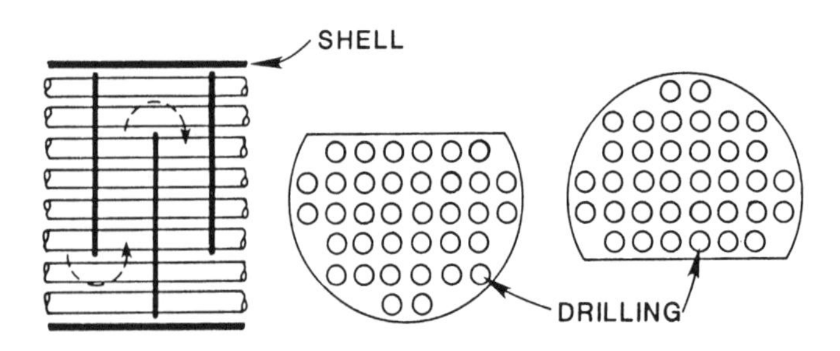

Figure 3-3. Transverse baffle detail.

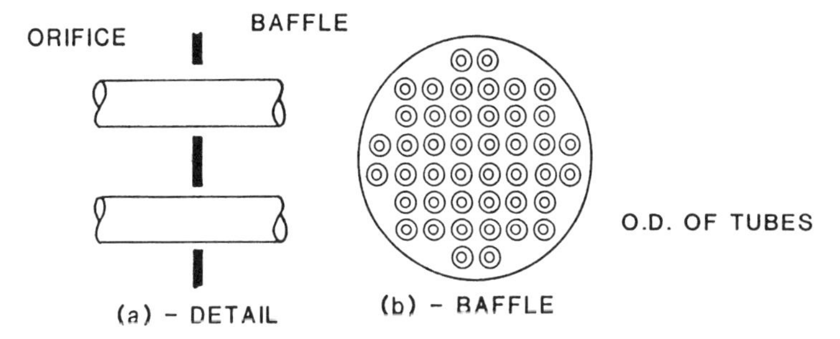

Figure 3-4. Orifice baffle.

3-1, the shell-side fluid makes one pass from inlet to outlet. With a longitudinal baffle, and with the nozzles placed 180° around the shell, the shell-side fluid would be forced to enter at the left, flow to the right to get around the baffle, and flow to the left to reach the exit nozzle. This would be required to approximate true counter-current flow, which was assumed in the heat transfer equations of Chapter 2.

Tubes

Heat-exchanger tubes should not be confused with steel pipe or other types of pipe that are extruded to steel pipe sizes. The outside diameter of heat-exchanger tubes is the actual outside diameter in inches within a very strict tolerance. Heat-exchanger tubes are available in a variety of metals including steel, copper, brass, 70-30 copper-nickel, aluminum bronze, aluminum, and stainless steel. They are obtainable in several wall thicknesses defined by the Birmingham Wire Gauge (B.W.G.), which is usually referred to as the B.W.G. or gauge of the tube. Table 2-1 shows tube outside diameter and B.W.G. gauge commonly used.

Tube Pitch

Tube holes cannot be drilled very close together, since this may structually weaken the tube sheet. The shortest distance between two adjacent tube holes is called the "clearance." Tubes are laid out in either square or triangular patterns as shown in Figure 3-5. The advantage of square pitch is that the tubes are accessible for external cleaning and cause a lower pressure drop when shell-side fluid flows perpendicularly to the tube axis. The tube pitch is the shortest center-to-center distance between adjacent tubes. The common pitches for square patterns are ¾-in. OD on

1-in. and 1-in. OD on 1¼-in. For triangular patterns these are ¾-in. OD on ¹⁵⁄₁₆-in., ¾-in. OD on 1-in., and 1-in. OD on 1¼-in.

In Figure 3-5c the square-pitch layout has been rotated 45°, yet it is essentially the same as Figure 3-5a. In Figure 3-5d a mechanically cleanable modification of triangular pitch is shown. If the tubes are spread wide enough, it is possible to allow the cleaning lanes indicated.

Shells

Shells up to 24-in. OD are fabricated from pipe using standard pipe nominal diameters. Standard pipe diameters and wall thicknesses are given in Volume 1, Tables 9-5 and 9-9 (1st edition: 9-4 and 9-7). Shells larger than 24-in. in diameter are fabricated by rolling steel plate.

The wall thickness of the pipe or plate used for the shell is normally determined from the American Society of Mechanical Engineers (ASME) Boiler and Pressure Vessel Code. TEMA standards also specify some minimum wall thicknesses for the shell.

Options

There are many different arrangements of the shells, tubes and baffles in heat exchangers. Figure 3-6 is a list of TEMA standard classifications for heat exchangers, which helps to describe the various options. These

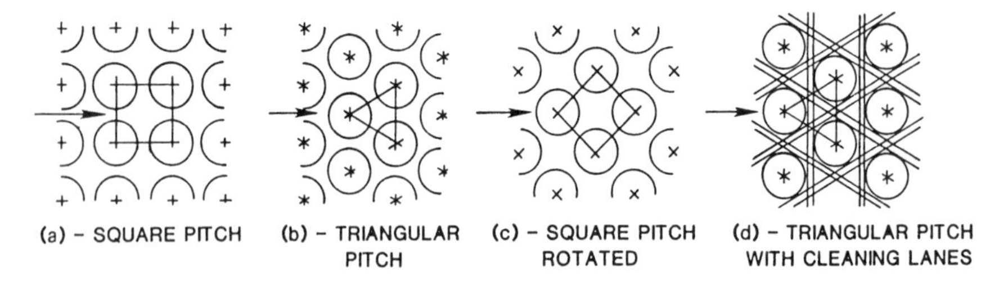

Figure 3-5. Common tube layouts for shell-and-tube heat exchangers.

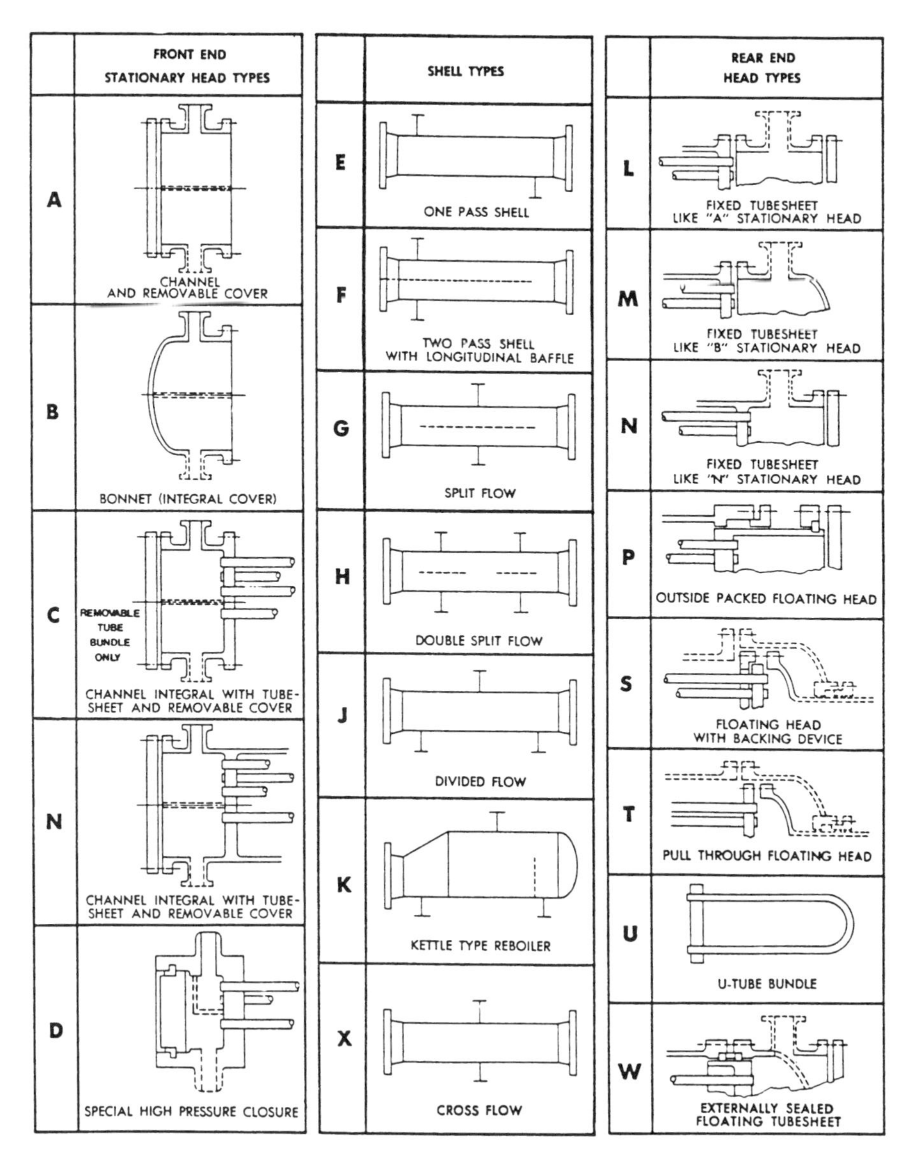

Figure 3-6. Heat exchanger nomenclature. (From Tubular Exchanger Manufacturers Association, © 1978.)

are best understood in conjunction with the example configurations given in Figures 3-7 through 3-9.

The first letter designates the front end of the heat exchanger, the second letter designates the shell type, or the middle of the heat exchanger, and the third letter designates the back, or the rear, of the heat exchanger.

An AES classification for a heat exchanger means that the heat exchanger has a channel and removable cover type front end. The cover can be unbolted to perform maintenance and the channel can be unbolted without pulling the tube sheet. The "E" designates a one-pass shell. The shell fluid comes in one end and goes out the other. The rear of the heat exchanger is an internal floating head. The head can move back and forth as the tubesheet expands and contracts.

1. Stationary Head—Channel
2. Stationary Head—Bonnet
3. Stationary Head Flange—Channel or Bonnet
4. Channel Cover
5. Stationary Head Nozzle
6. Stationary Tubesheet
7. Tubes
8. Shell
9. Shell Cover
10. Shell Flange—Stationary Head End
11. Shell Flange—Rear Head End
12. Shell Nozzle
13. Shell Cover Flange
14. Expansion Joint
15. Floating Tubesheet
16. Floating Head Cover
17. Floating Head Flange
18. Floating Head Backing Device
19. Split Shear Ring
20. Slip-on Backing Flange
21. Floating Head Cover—External
22. Floating Tubesheet Skirt
23. Packing Box
24. Packing
25. Packing Gland
26. Lantern Ring
27. Tierods and Spacers
28. Transverse Baffles or Support Plates
29. Impingement Plate
30. Longitudinal Baffle
31. Pass Partition
32. Vent Connection
33. Drain Connection
34. Instrument Connection
35. Support Saddle
36. Lifting Lug
37. Support Bracket
38. Weir
39. Liquid Level Connection

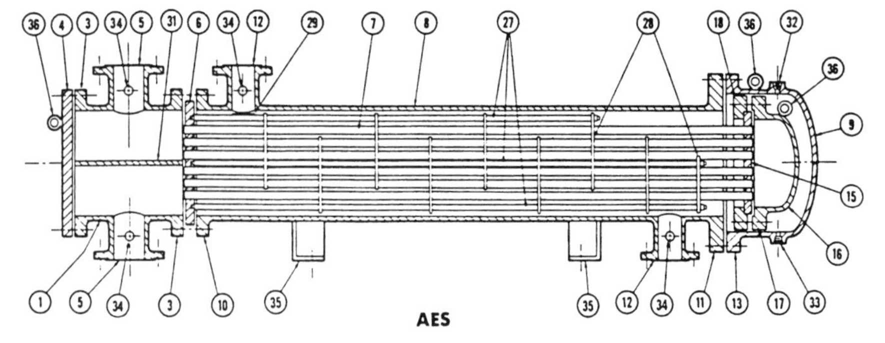

Figure 3-7. Heat exchanger components. (From Tubular Exchanger Manufacturers Association, © 1978.)

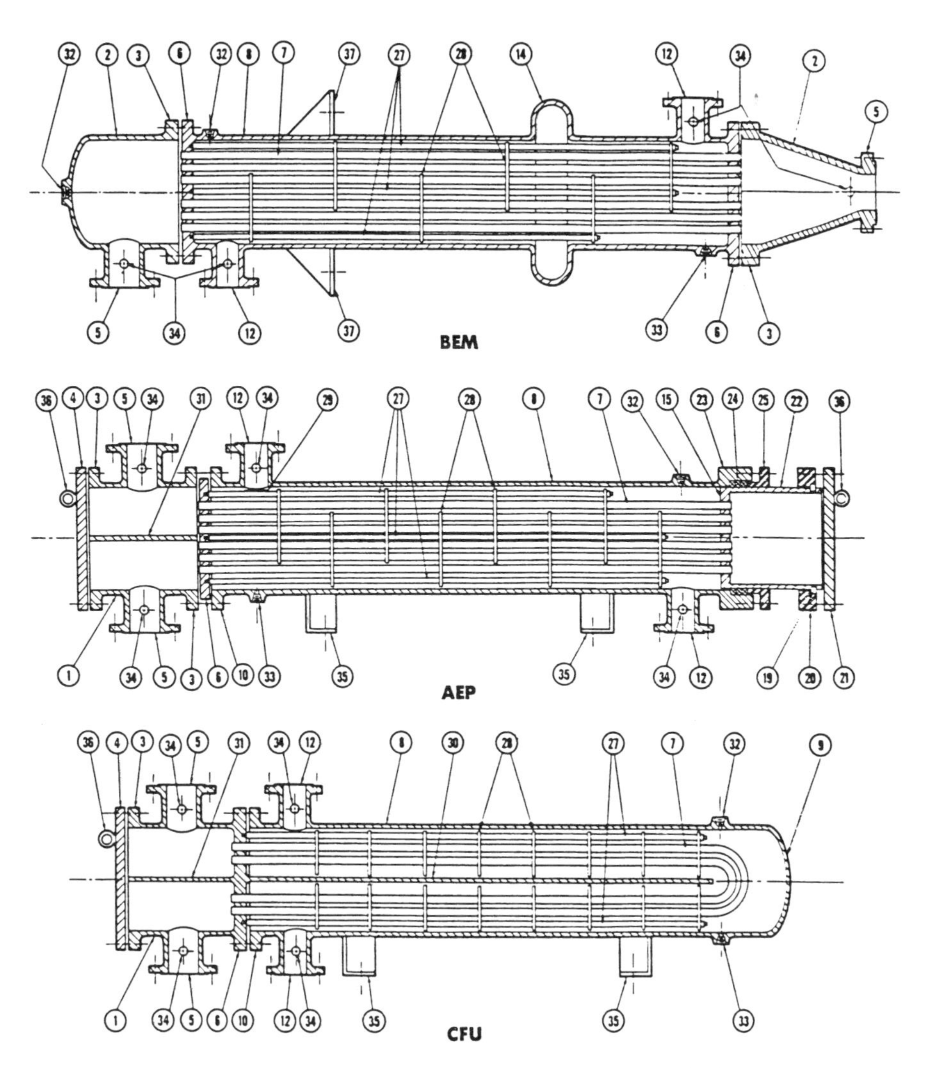

Figure 3-8. Heat exchanger components (continued). (From Tubular Exchanger Manufacturers Association, © 1978.)

If there were no removable cover on the front end of the exchanger, it would be designated "BES." The second nozzle and pass partition in the front end are discretionary depending upon the shell type. Types A and B bolt onto the shell. In type C, the head cannot be unbolted for maintenance.

The shell types are E, F, G, H, J, and K. E is a one-pass shell. The fluid comes in on one side and goes out the other side. F is a two-pass shell with a longitudinal plate in it. The fluid in the shell makes two passes.

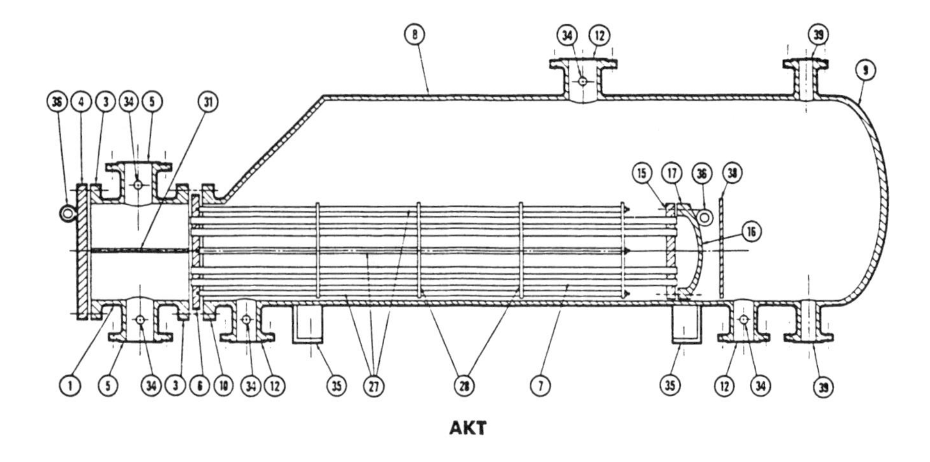

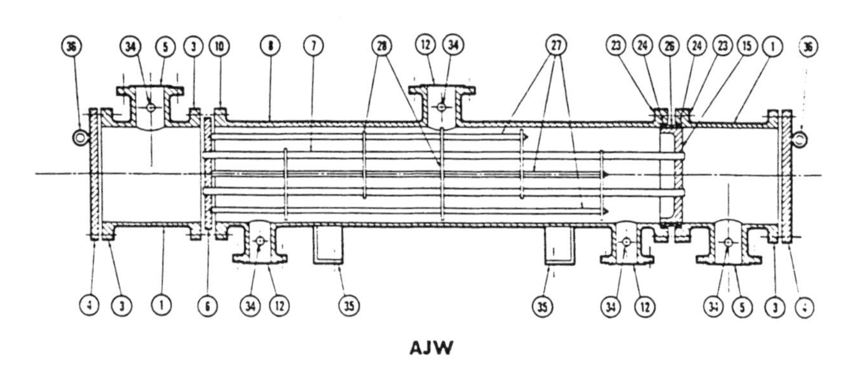

Figure 3-9. Heat exchanger components (continued). (From Tubular Exchanger Manufacturers Association, © 1978.)

G is a split flow. The fluid comes in and goes both ways around the longitudinal baffle and then exits. H is very rare; a double split flow. J is a divided flow. K is a kettle type reboiler, which is a special type and is best explained by looking at the example "AKT" in Figure 3-9. Kettle types are common where there is a boiling liquid or where gas is liberated from shell fluid as it is heated. The weir controls the liquid, making sure the tubes are always immersed in liquid. Gas that flashes from the liquid can exit the top nozzle.

Many of the rear end types allow the head to "float" as the tube bundle expands and contracts.

Type S is a floating head type. As the tubes heat up, they expand. As they expand, the floating head moves back and forth, but the pressure seal is not at the sliding joint. The pressure seal is at the fixed shell joint in the outer head, which contains the pressure. The floating head floats free inside the pressure vessel as the tubes move. Types P and W are floating heads where the movement of the head effects the seal between either the shell-side or tube-side fluid and atmosphere.

Type U indicates the tubes are in a U-shaped bundle and no special closure is needed for the rear end, other than the shell itself. The tubes are free to expand or contract. This is by far the least expensive type of floating head.

Classification

In addition to the type description code there is also a shorthand that is used for classifying heat exchangers. The first element of the shorthand is the nominal diameter, which is the inside diameter of the shell in inches, rounded off to the nearest integer. For kettle reboilers and chillers (remember the kettle has a narrow end and a fat end), the nominal diameter is the port diameter (the narrow end) followed by the shell diameter, each rounded off to the nearest integer.

The second element is the nominal tube length in inches. Tube length for straight tubes is taken as the actual overall length. For U-tubes the length is taken as the straight length from the end of the tube to the bend tangent.

The third element is a three-letter code indicating the type of front end, shell, and rear end, in that order. For example, a fixed tubesheet exchanger (the tubesheet does not expand and contract as it heats and cools) with removable channel and cover, single-pass shell, 23¼-in, inside diameter with tubes 16 ft long is denoted as SIZE 23-192 TYPE AEL.

A pull-through, floating head, kettle-type reboiler having stationary head integral with the tubesheet, 23-in. port diameter and 37-in. inside shell diameter with tubes 16 ft long is denoted as SIZE 23/37-192 TYPE CKT.

Selection of Types

In selecting an exchanger, one must know the advantages and disadvantages of each type. The three basic types of shell-and-tube exchangers are fixed tube sheet, floating head, and U-tube. Table 3-1 summarizes the comparison between these three exchangers.

Table 3-1
Heat Exchange Summary

Heat Exchanger Type	Advantages	Disadvantages	Remarks
Fixed Tube	• Least expensive • Fewer gaskets • Individual tubes are replaceable	• Cannot clean or inspect shell • Bundle is replaceable • Limited to temperature difference of 200°F without costly shell expansion joint	• For clean shell fluids and low differential temperature, most comonly used • Types A and L heads are most common • Types B and M heads are for large diameter and high pressure
Floating Head	• Ability to handle dirty fluids and high differential temperatures • Head and tubes can be mechanically cleaned • Individual tubes are replaceable	• Higher cost than fixed tube • Internal gaskets can leak	• Types T and W heads are least expensive but have possibilities of shell fluids leakage • Type P head is most expensive; type S is intermediate in cost
U-Tube	• Fairly low cost may be lower than fixed tube • Can handle thermal expansion • No internal gasketed joint • Bundle is replaceable	• Tube cannot be cleaned mechanically • Fewer tubes can be installed in a given diameter shell • Difficult to unbolt the flange from the end of U-tube • Individual tubes are not replaceable	• Commonly used for high differential temperature • Minimizes possibility for contamination of tube fluids with higher pressure shell fluids • Types B and U heads are the most common on U-tube

Placement of Fluid

The question always comes up of which fluid to put in the tubes and which fluid to put in the shell. Consider placing a fluid through the tubes when:

1. Special alloy materials are required for corrosion control and high temperatures.
2. Fluid is at high pressures.
3. Fluid contains vapors and non-condensable gases.
4. Fluid is scale forming.

Consider placing a fluid through the shell when:

1. Small pressure drops are desired.
2. Fluid is viscous.
3. Fluid is non-fouling.
4. Boiling service is desired.
5. Fluid has low film rate and is non-fouling (finned tubes can be used).

Placing the fluid through the tubes is a consideration when special alloy materials are needed for corrosion control, because the materials would be needed only on the tubes. If the corrosive material is in the shell, both the tubes and the shell would need to be protected with special alloy. If the fluid is at high pressure, it should be put in tubes because tubes can contain high pressure much more cheaply as they are much smaller in diameter than the shell. The low-pressure fluid would be in the shell. If the fluid contains vapor and non-condensable gases, heat transfer will be greater if it is placed in the tubes. If the fluid is scale forming it should be in the tubes, which can be reamed out.

Similarly, the fluid should be put in the shell when small pressure drops are desired. There is less of a pressure drop going through the shell than there is in going through the tubes. If the fluid is very viscous, pressure drop will be less and heat transfer improved if it is placed in the shell. It is harder to clean the shell than it is the tubes, so the non-fouling fluid should be put in the shell. If boiling service is desired, a kettle design should be used and the boiling fluid should be put in the shell, not inside the tubes. If the fluid has a low film rate and is non-fouling, it can be put in the shell, and finned tubes can be used to add to the U-value. In this case, the fluid would have to be a non-fouling fluid; otherwise, it would plug the fins.

TEMA Classes and Tube Materials

TEMA standards provide for two classes of shell and tube exchanger qualities. Class C is the less stringent and is typically used in onshore applications and where the temperature is above –20°F. Class R is normally used offshore and in cold temperature service. Table 3-2 shows the most important differences between a Class R and a Class C TEMA exchanger.

Table 3-3 is a tube materials selection chart (the common tube materials to use for different kinds of services). Standard tube lengths are either 20 ft or 40 ft.

Table 3-2
TEMA Classes

Major Features	R	C
Corrosion allowance	⅛-inch	1/16-inch
Tube diameters	¾, 1, 1¼, 1½	¼, ⅜, ½, ⅝, ¾, 1, 1¼, 1½
Square pitch cleaning lane	¼-inch	3/16-inch
Min. shell thickness	R greater than C	
Transverse baffle and support plate thickness	In some cases R greater than C	
Min. thickness longitudinal baffles	¼-inch	⅛-inch
Impingement protection required	$V > \frac{22}{\rho^{1/2}}$	$V > \frac{39}{\rho^{1/2}}$
Cross-over area for multi-pass floating heads or channels	1.3 times flow area through tubes of one pass	1.0 times flow area through tubes of one pass
Stress relieving of fabricated floating covers, or channels	Required	Only if required by ASME
Gaskets in contact with oil or gas	Metal jacketed or solid metal	Metal jacketed or solid material
Minimum gasket width	⅜-inch	¼-inch
Minimum tube sheet thickness	Tube diameter plus twice corrosion allowance	Factor less than tube diameter plus twice corrosion allowance
Edges of tubesheet holes	Chamfered	Deburred
Pressure gauge and thermowells on flange nozzles	Required for certain sizes	Must be specified
Min. bolt diameter	¾-inch	½-inch to ⅝-inch

Table 3-3
Tube Materials

Sweet Service Glycol, MEA, and Sulfinol	
Temperatures above	−20°F, A-214 ERW or A-179 (seamless)
	−50°F to −21°F, A-334 Grade 1
	−150°F to −51°F, A334 Grade 3
Sour and Low Temperature	
304 SS	
Brackish Water	
90/10 Cu-Ni	
70/30 Cu-Ni Use higher nickel content the more brackish the water	

Sizing

The required heat duty, film coefficients, conductivity, etc. for a shell-and-tube heat exchanger can be calculated using the procedures in Chapter 2. Approximate U-values are given in Table 2-8.

In the basic heat transfer equation it is necessary to use the log mean temperature difference. In Equation 2-4 it was assumed that the two fluids are flowing counter-current to each other. Depending upon the configuration of the exchanger, this may not be true. That is, the way in which the fluid flows through the exchanger affects LMTD. The correction factor is a function of the number of tube passes and the number of shell passes.

Figures 3-10 and 3-11 can be used to calculate a corrected LMTD from the formula.

$$\text{LMTD} = \left[\frac{\Delta T_1 - \Delta T_2}{\log_e (\Delta T_1 / \Delta T_2)}\right] F \tag{3-1}$$

where T_1 = hot fluid inlet temperature, °F
T_2 = hot fluid outlet temperature, °F
T_3 = cold fluid inlet temperature, °F
T_4 = cold fluid outlet temperature, °F
ΔT_1 = larger temperature difference, °F
ΔT_2 = smaller temperature difference, °F
LMTD = log mean temperature difference, °F
F = correction factor

(*text continued on page 64*)

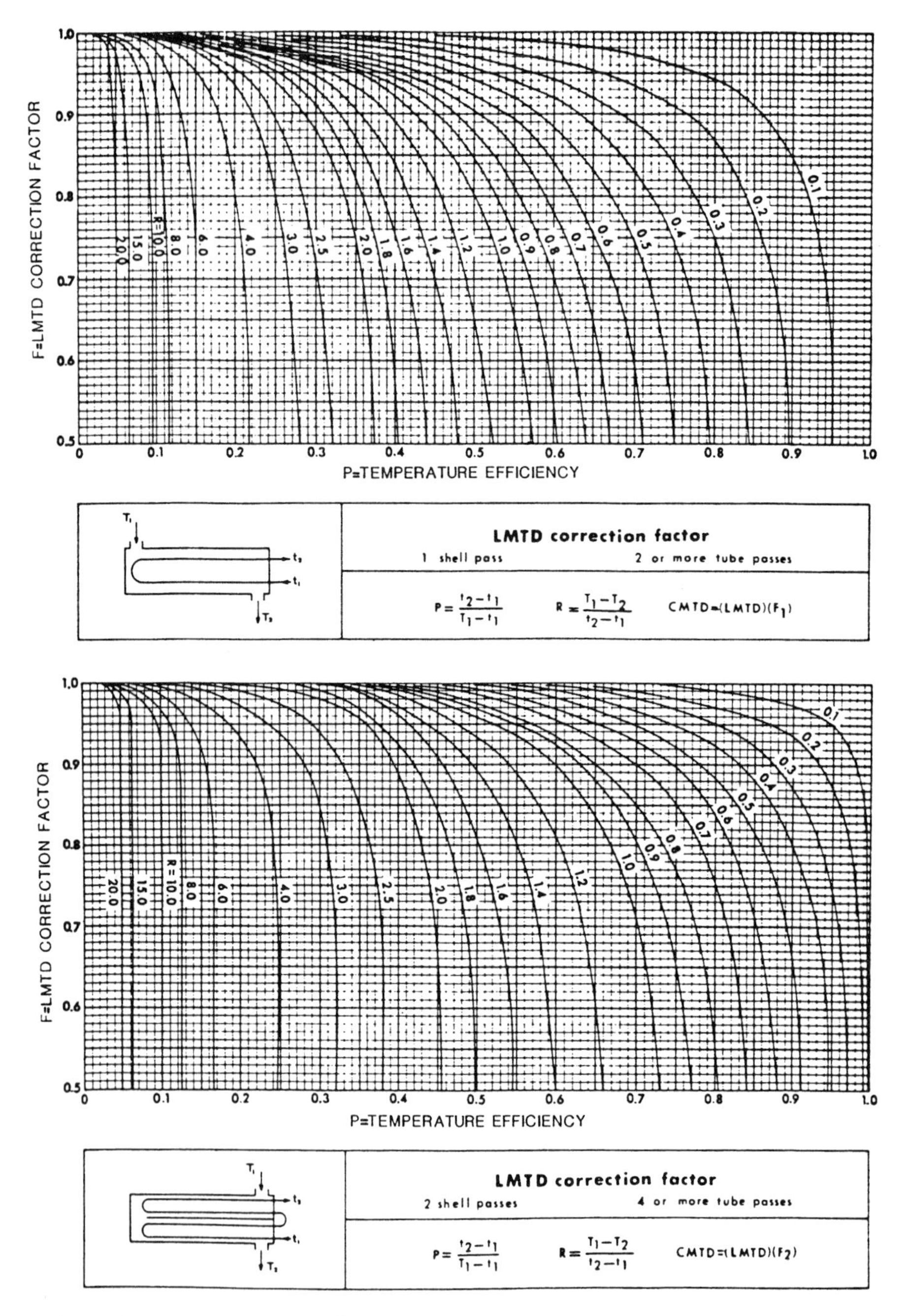

Figure 3-10. LMTD correction factors. (From Gas Processors Suppliers Association, *Engineering Data Book*, 9th Edition.)

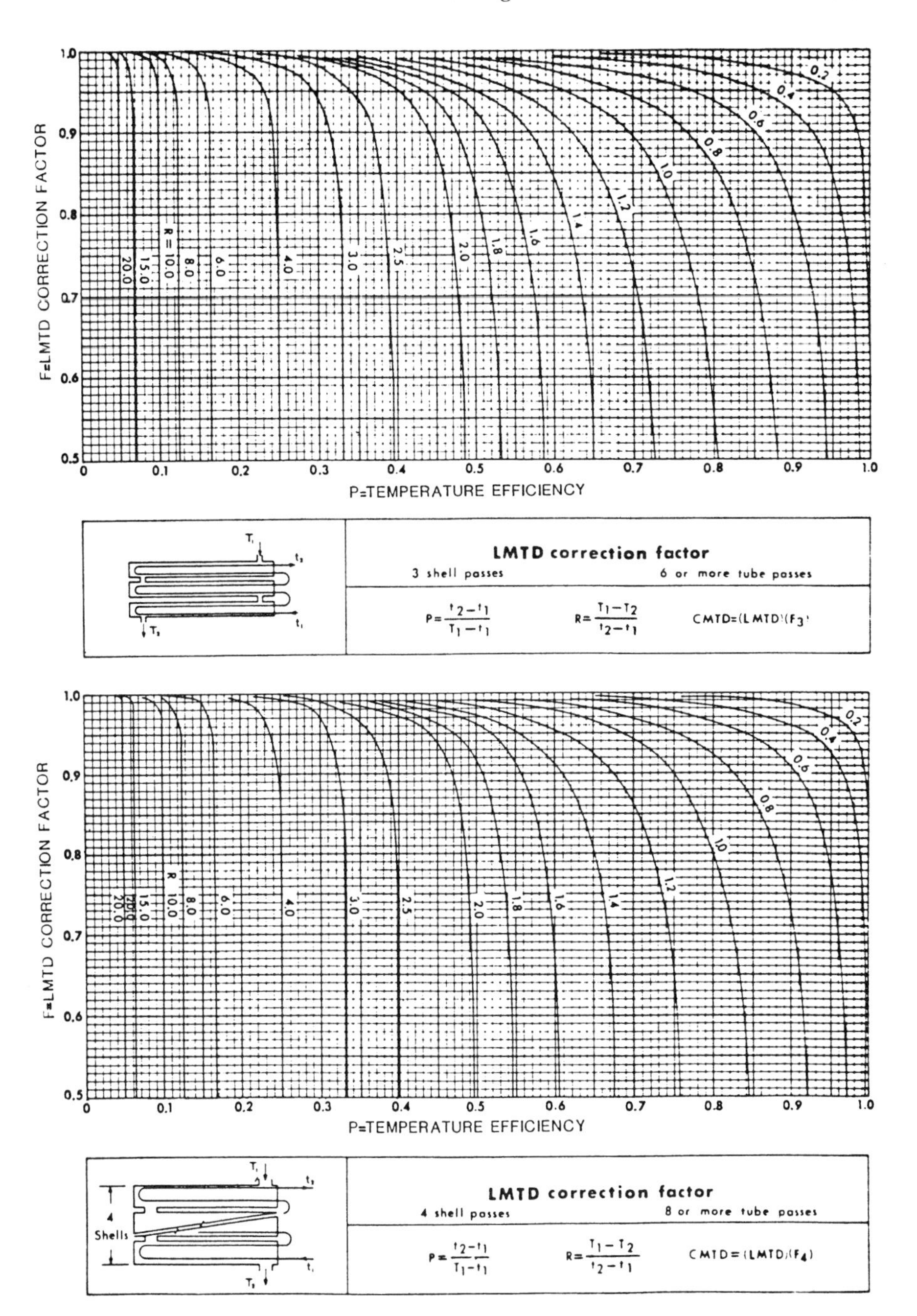

Figure 3-11. LMTD correction factors. (From Gas Processors Suppliers Association, *Engineering Data Book*, 9th Edition.)

(text continued from page 61)

To size a shell-and-tube exchanger, first the duty is calculated. Then it is determined which fluid will be in the shell and which in the tube, and a heat transfer coefficient assumed or calculated. A choice is made of the number of shell and tube passes to get a reasonable LMTD correction factor (F), and a corrected LMTD as calculated from Equation 3-1.

Next, a tube diameter and tube length are chosen. The number of tubes required is calculated by:

$$N = \frac{q}{U\,A'\,(LMTD)\,L} \tag{3-2}$$

where
- N = required number of tubes
- q = heat duty, Btu/hr
- U = overall heat transfer coefficient, Btu/hr-ft^2-°F
- LMTD = corrected log mean temperature difference, °F
- L = tube length, ft
- A′ = tube external surface area per foot of length, ft^2/ft (Table 2-1)

From Table 3-4 it is then possible to pick a shell diameter that can accommodate the number of tubes required. Please note that Equation 3-2 calculates the total number of tubes required and not the number of tubes per pass. Similarly, Table 3-4 lists the total number of tubes and not the number per pass. There are fewer total tubes in the same diameter exchanger for more passes of the tube fluid because of the need for partition plates. There are fewer tubes for floating head than fixed head designs because the heads and seals restrict the use of space. U-tubes have the lowest number of tubes because of the space required for the tightest radius bend in the U-tube bundle.

Once the number of tubes is determined, the flow velocity of fluid inside the tubes should be checked, using the criteria set forth for flow in pipes in Chapter 8 of Volume 1.

DOUBLE-PIPE EXCHANGERS

A double-pipe exchanger is made up of one pipe containing the tube fluid concentric with another pipe, which serves as the shell. The tube is often finned to give additional surface area. The double-pipe exchanger was developed to fit applications that are too small to economically apply the requirements of TEMA for shell and tube exchangers.

Double-pipe exchangers can be arranged as in Figure 3-12 such that two shells are joined at one end through a "return bonnet," which causes the shell-side fluid to flow in series through each of the two shells. In this configuration, the central tube is bent or welded into a "U" shape, with the U-bend inside the return bonnet. The principal advantage to this configuration is that a more compact exchanger can be designed, thus simplifying installation.

A variation of the U-tube exchanger is the hairpin style of exchanger. In the hairpin exchanger, multiple small tubes are bent into a "U" shape in place of the single central tube. This variation allows for more surface area to be provided in the exchanger than would be obtained with a single tube. U-tube exchangers may be designed with or without fins.

The advantages of double-pipe exchangers are that they are cheap and readily available, and because of the U-tube type of construction, thermal expansion is not a problem. Double-pipe exchangers are normally designed and built in accordance with the applicable requirements of TEMA. Thus, they can be applied to most services encountered in oil and gas production facilities as long as the required surface area can be fit into the physical configuration of the exchanger. Although they can be built in almost any size, double-pipe exchangers are most frequently used when the required surface area is 1,000 ft^2 or less.

PLATE-AND-FRAME EXCHANGERS

Plate-and-frame exchangers are an arrangement of gasketed, pressed metal plates aligned on carrying bars and secured between two covers by compression bolts. The pressed metal plates are corrugated in patterns to provide increased surface area, to direct the flow in specific directions, and to promote turbulence. The plates are gasketed such that each of the

(text continued on page 72)

Table 3-4
Heat Exchanger Tube Count

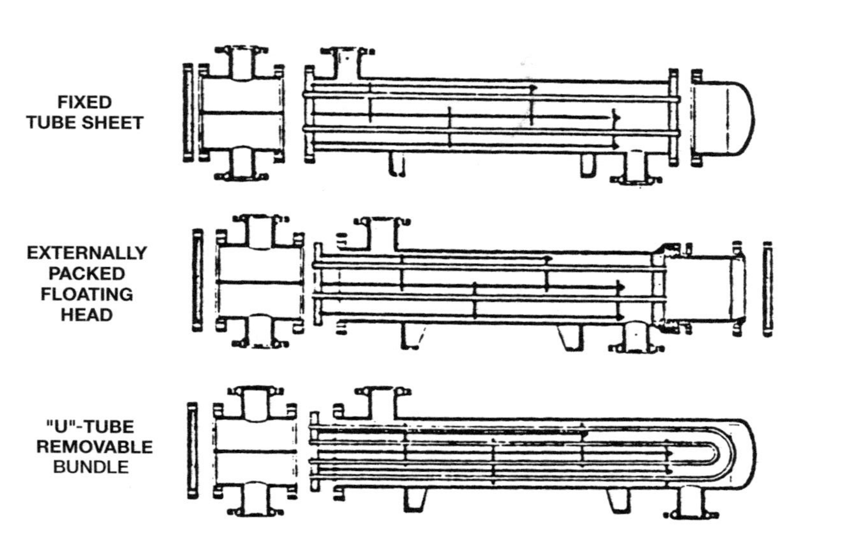

(table continued on next page)

Table 3-4 (Continued)
Heat Exchanger Tube Count

3/4″ O.D. Tubes on 15/16″ Δ Pitch

Shell I.D. (Inches)	Fixed Tube Sheet No. of Passes			Outside Packed Floating Head No. of Passes			"U"-Tube No. of Passes	
	1	2	4	1	2	4	2	4
5.047	19	14	12	10	10	4	3	2
6.065	25	20	16	19	18	12	7	4
7.981	52	48	36	38	36	28	16	12
10.02	85	76	68	70	66	56	28	26
12.00	126	114	100	109	98	92	56	40
13.25	147	140	128	130	126	112	57	52
15.25	206	196	176	187	176	160	83	74
17.25	268	252	234	241	236	220	110	102
19.25	335	326	302	313	298	276	145	134
21.25	416	397	376	384	368	344	180	170
23.25	499	480	460	469	449	430	220	210
25.00	576	558	530	544	529	500	253	244
27.00	675	661	632	643	616	600	307	290
29.00	790	773	736	744	732	704	360	342
31.00	806	875	858	859	835	812	415	402
33.00	1018	1011	976	973	959	926	477	458
35.00	1166	1137	1098	1118	1093	1054	538	520
37.00	1307	1277	1242	1253	1224	1184	609	592
39.00	1464	1425	1386	1392	1359	1318	683	662
42.00	1688	1669	1618	1616	1602	1552	800	776
45.00	1943	1912	1878	1870	1833	1800	927	900
48.00	2229	2189	2134	2145	2107	2060	1061	1032
51.00	2513	2489	2432	2411	2395	2344	1205	1178
54.00	2823	2792	2752	2733	2683	2642	1366	1334
60.00	3527	3477	3414	3400	3359	3294	1699	1668

3/4″ O.D. Tubes on 1″ Δ Pitch

Shell I.D. (Inches)	Fixed Tube Sheet No. of Passes			Outside Packed Floating Head No. of Passes			"U"-Tube No. of Passes	
	1	2	4	1	2	4	2	4
5.047	14	14	8	10	6	4	3	2
6.065	22	19	16	19	14	12	5	4
7.981	42	38	36	37	30	28	14	12
10.02	71	68	60	61	56	48	28	22
12.00	106	102	92	92	90	76	43	36
13.25	130	124	114	121	110	100	53	48
15.25	184	169	160	163	152	140	74	68
17.25	237	228	212	212	202	188	100	92
19.25	296	290	272	268	262	244	127	120
21.25	361	354	336	335	330	308	157	150
23.25	434	420	408	416	395	380	194	184
25.00	507	489	476	475	466	452	226	216
27.00	596	585	562	556	554	528	269	262
29.00	689	679	660	653	642	620	316	306
31.00	790	775	756	756	734	720	366	354
33.00	906	891	860	859	848	818	419	404
35.00	1031	1003	976	978	959	932	475	458
37.00	1152	1134	1090	1106	1081	1054	537	520
39.00	1273	1259	1222	1218	1208	1174	600	582
42.00	1485	1461	1434	1426	1399	1376	703	682
45.00	1721	1693	1650	1652	1620	1586	816	792
48.00	1968	1941	1902	1894	1861	1820	935	916
51.00	2221	2187	2134	2142	2101	2060	1061	1038
54.00	2502	2465	2414	2417	2379	2326	1198	1170
60.00	3099	3069	3010	2990	2957	2906	1496	1468

(*table continued on next page*)

Table 3-4 (Continued)
Heat Exchanger Tube Count

3/4″ O.D. Tubes on 1″ □ Pitch

Shell I.D. (Inches)	Fixed Tube Sheet			Outside Packed Floating Head			"U"-Tube	
	No. of Passes			No. of Passes			No. of Passes	
	1	2	4	1	2	4	2	4
5.047	12	12	12	12	6	4	3	2
6.065	21	16	16	16	16	12	4	4
7.981	37	34	32	32	28	24	12	10
10.02	61	60	52	52	52	52	22	20
12.00	97	88	88	81	78	76	34	34
13.25	112	112	112	97	94	88	45	44
15.25	156	148	148	140	132	124	64	60
17.25	208	196	188	188	178	172	88	84
19.25	250	249	244	241	224	216	112	108
21.25	316	307	296	296	280	276	138	134
23.25	378	370	370	356	344	332	170	166
25.00	442	432	428	414	406	392	200	194
27.00	518	509	496	482	476	468	236	230
29.00	602	596	580	570	562	548	277	272
31.00	686	676	676	658	640	640	320	312
33.00	782	768	768	742	732	732	362	360
35.00	896	868	868	846	831	820	418	406
37.00	1004	978	964	952	931	928	470	462
39.00	1102	1096	1076	1062	1045	1026	524	520
42.00	1283	1285	1270	1232	1222	1218	611	602
45.00	1484	1472	1456	1424	1415	1386	710	700
48.00	1701	1691	1670	1636	1634	1602	812	802
51.00	1928	1904	1888	1845	1832	1818	926	910
54.00	2154	2138	2106	2080	2066	2044	1042	1032
60.00	2683	2650	2636	2582	2566	2556	1298	1282

3/4″ O.D. Tubes on 1″ ◇ Pitch

Shell I.D. (Inches)	Fixed Tube Sheet			Outside Packed Floating Head			"U"-Tube	
	No. of Passes			No. of Passes			No. of Passes	
	1	2	4	1	2	4	2	4
5.047	12	10	8	12	10	8	2	2
6.065	21	18	16	16	12	8	5	4
7.981	37	32	28	32	28	24	12	10
10.02	61	54	48	52	46	40	21	18
12.00	97	90	84	81	74	68	33	32
13.25	113	108	104	97	92	84	43	40
15.25	156	146	136	140	134	128	62	58
17.25	208	196	184	188	178	168	87	82
19.25	256	244	236	241	228	216	109	104
21.25	314	299	294	300	286	272	136	130
23.25	379	363	352	359	343	328	167	160
25.00	448	432	416	421	404	392	195	190
27.00	522	504	486	489	472	456	234	226
29.00	603	583	568	575	556	540	275	266
31.00	688	667	654	660	639	624	313	304
33.00	788	770	756	749	728	708	360	350
35.00	897	873	850	849	826	804	409	398
37.00	1009	983	958	952	928	908	464	452
39.00	1118	1092	1066	1068	1041	1016	518	508
42.00	1298	1269	1250	1238	1216	1196	610	596
45.00	1500	1470	1440	1432	1407	1378	706	692
48.00	1714	1681	1650	1644	1611	1580	804	788
51.00	1939	1903	1868	1864	1837	1804	917	902
54.00	2173	2135	2098	2098	2062	2026	1036	1018
60.00	2692	2651	2612	2600	2560	2520	1292	1272

Table 3-4 (Continued)
Heat Exchanger Tube Count

1" O.D. Tubes on 1-1/4" △ Pitch

Shell I.D. (Inches)	Fixed Tube Sheet, No. of Passes			Outside Packed Floating Head, No. of Passes			"U"-Tube, No. of Passes	
	1	2	4	1	2	4	2	4
5.047	8	6	4	7	4	4	0	0
6.065	14	14	8	10	10	4	2	2
7.981	26	26	16	22	18	16	7	4
10.02	42	40	36	38	36	28	13	12
12.00	64	61	56	56	52	48	22	18
13.25	85	76	72	73	72	60	28	26
15.25	110	106	100	100	98	88	43	38
17.25	147	138	128	130	126	116	57	52
19.25	184	175	168	170	162	148	76	68
21.25	227	220	212	212	201	188	96	88
23.25	280	265	252	258	250	232	116	110
25.00	316	313	294	296	294	276	135	128
27.00	371	370	358	355	346	328	161	152
29.00	434	424	408	416	408	392	189	182
31.00	503	489	468	475	466	446	222	212
33.00	576	558	534	544	529	510	254	246
35.00	643	634	604	619	604	582	289	280
37.00	738	709	684	696	679	660	330	316
39.00	804	787	772	768	753	730	370	356
42.00	946	928	898	908	891	860	436	418
45.00	1087	1069	1042	1041	1017	990	505	490
48.00	1240	1230	1198	1189	1182	1152	578	562
51.00	1397	1389	1354	1348	1337	1300	661	642
54.00	1592	1561	1530	1531	1503	1462	748	726
60.00	1969	1945	1904	1906	1879	1842	933	914

1" O.D. Tubes on 1-1/4" □ Pitch

Shell I.D. (Inches)	Fixed Tube Sheet, No. of Passes			Outside Packed Floating Head, No. of Passes			"U"-Tube, No. of Passes	
	1	2	4	1	2	4	2	4
5.047	9	6	4	5	4	4	0	0
6.065	12	12	12	12	6	4	2	2
7.981	22	20	16	21	16	16	6	4
10.02	38	38	32	32	32	32	12	10
12.00	56	56	52	52	52	44	19	18
13.25	69	66	66	61	60	52	25	24
15.25	97	90	88	89	84	80	36	34
17.25	129	124	120	113	112	112	49	48
19.25	164	158	148	148	144	140	64	62
21.25	202	191	184	178	178	172	83	78
23.25	234	234	222	216	216	208	100	98
25.00	272	267	264	258	256	256	120	116
27.00	328	317	310	302	300	296	142	138
29.00	378	370	370	356	353	338	166	166
31.00	434	428	428	414	406	392	145	192
33.00	496	484	484	476	460	460	221	218
35.00	554	553	532	542	530	518	254	248
37.00	628	621	608	602	596	580	287	280
39.00	708	682	682	676	649	648	322	314
42.00	811	811	804	782	780	768	379	374
45.00	940	931	918	904	894	874	436	434
48.00	1076	1061	1040	1034	1027	1012	501	494
51.00	1218	1202	1192	1178	1155	1150	573	570
54.00	1370	1354	1350	1322	1307	1284	650	644
60.00	1701	1699	1684	1654	1640	1632	810	802

(table continued on next page)

Table 3-4 (Continued)
Heat Exchanger Tube Count

1" O.D. Tubes on 1-1/4" ◇ Pitch

Shell I.D. (Inches)	Fixed Tube Sheet No. of Passes			Outside Packed Floating Head No. of Passes			"U"-Tube No. of Passes	
	1	2	4	1	2	4	2	4
5.047	8	6	4	5	4	4	0	0
6.065	12	10	8	12	10	8	2	2
7.981	24	20	16	21	18	16	5	4
10.02	37	32	28	32	32	28	12	10
12.00	57	53	48	52	46	40	18	16
13.25	70	70	64	61	58	56	25	22
15.25	97	90	84	89	82	76	35	32
17.25	129	120	112	113	112	104	48	44
19.25	162	152	142	148	138	128	62	60
21.25	205	193	184	180	174	168	78	76
23.25	238	228	220	221	210	200	100	94
25.00	275	264	256	261	248	236	116	110
27.00	330	315	300	308	296	286	141	134
29.00	379	363	360	359	345	336	165	160
31.00	435	422	410	418	401	388	191	184
33.00	495	478	472	477	460	448	220	212
35.00	556	552	538	540	526	508	249	242
37.00	632	613	598	608	588	568	281	274
39.00	705	685	672	674	654	640	315	310
42.00	822	799	786	788	765	756	372	364
45.00	946	922	912	910	885	866	436	426
48.00	1079	1061	1052	1037	1018	1000	501	490
51.00	1220	1199	1176	1181	1160	1142	569	558
54.00	1389	1359	1330	1337	1307	1292	646	632
60.00	1714	1691	1664	1658	1626	1594	802	788

1-1/4" O.D. Tubes on 1-9/16" △ PITCH

Shell I.D. (Inches)	Fixed Tube Sheet No. of Passes			Outside Packed Floating Head No. of Passes			"U"-Tube No. of Passes	
	1	2	4	1	2	4	2	4
5.047	7	4	4	0	0	0	0	0
6.065	8	6	4	7	6	4	0	0
7.981	19	14	12	14	14	8	3	2
10.02	29	26	20	22	20	16	7	6
12.00	42	38	34	37	36	28	11	10
13.25	52	48	44	44	44	36	16	14
15.25	69	68	60	64	62	48	24	22
17.25	92	84	78	85	78	72	32	30
19.25	121	110	104	109	102	96	43	40
21.25	147	138	128	130	130	116	57	52
23.25	174	165	156	163	152	144	69	66
25.00	196	196	184	184	184	172	81	76
27.00	237	226	224	221	216	208	98	92
29.00	280	269	256	262	252	242	116	110
31.00	313	313	294	302	302	280	134	128
33.00	357	346	332	345	332	318	155	148
35.00	416	401	386	392	383	364	178	172
37.00	461	453	432	442	429	412	202	194
39.00	511	493	478	493	479	460	226	220
42.00	596	579	570	576	557	544	267	260
45.00	687	673	662	657	640	828	313	306
48.00	790	782	758	756	745	728	360	350
51.00	896	871	860	859	839	832	411	400
54.00	1008	994	968	964	959	940	465	454
60.00	1243	1243	1210	1199	1195	1170	580	570

Table 3-4 (Continued)
Heat Exchanger Tube Count

1-1/4" O.D. Tubes on 1-9/16" □ Pitch

Shell I.D. (Inches)	Fixed Tube Sheet No. of Passes			Outside Packed Floating Head No. of Passes			"U"-Tube No. of Passes	
	1	2	4	1	2	4	2	4
5.047	4	4	4	0	0	0	0	0
6.065	6	6	4	6	6	4	0	0
7.981	12	12	12	12	12	12	3	2
10.02	24	22	16	21	16	16	6	4
12.00	37	34	32	32	32	32	10	10
13.25	45	42	42	38	38	32	14	14
15.25	61	60	52	52	52	52	21	18
17.25	80	76	76	70	70	68	28	28
19.25	97	95	88	89	88	88	37	34
21.25	124	124	120	112	112	112	49	48
23.25	145	145	144	138	138	130	62	60
25.00	172	168	164	164	164	156	70	68
27.00	210	202	202	193	184	184	88	88
29.00	241	234	230	224	224	216	100	98
31.00	272	268	268	258	256	256	116	116
33.00	310	306	302	296	296	282	136	134
35.00	356	353	338	336	332	332	156	148
37.00	396	387	384	378	370	370	174	174
39.00	442	438	434	428	426	414	198	196
42.00	518	518	502	492	492	484	236	228
45.00	602	602	588	570	566	556	276	268
48.00	682	681	676	658	648	648	314	310
51.00	770	760	756	742	729	722	356	354
54.00	862	860	856	838	823	810	404	402
60.00	1084	1070	1054	1042	1034	1026	506	496

1-1/4" O.D. Tubes on 1-9/16" ◇ Pitch

Shell I.D. (Inches)	Fixed Tube Sheet No. of Passes			Outside Packed Floating Head No. of Passes			"U"-Tube No. of Passes	
	1	2	4	1	2	4	2	4
5.047	5	4	4	0	0	0	0	0
6.065	6	6	4	5	4	4	0	0
7.981	13	10	8	12	10	8	2	2
10.02	24	20	16	21	18	16	6	6
12.00	37	32	28	32	28	28	10	10
13.25	45	40	40	37	34	32	13	12
15.25	60	56	56	52	52	48	20	18
17.25	79	76	76	70	70	64	28	26
19.25	97	94	94	90	90	84	37	34
21.25	124	116	112	112	108	104	48	44
23.25	148	142	136	140	138	128	60	56
25.00	174	166	160	162	162	156	71	68
27.00	209	202	192	191	188	184	85	82
29.00	238	232	232	221	215	208	100	96
31.00	275	264	264	261	249	244	114	110
33.00	314	307	300	300	286	280	134	128
35.00	359	345	334	341	330	320	153	148
37.00	401	387	380	384	372	360	173	168
39.00	442	427	424	428	412	404	195	190
42.00	522	506	500	497	484	472	228	224
45.00	603	583	572	575	562	552	271	264
48.00	682	669	660	660	648	640	309	302
51.00	777	762	756	743	728	716	354	346
54.00	875	857	850	843	822	812	401	392
60.00	1088	1080	1058	1049	1029	1016	505	492

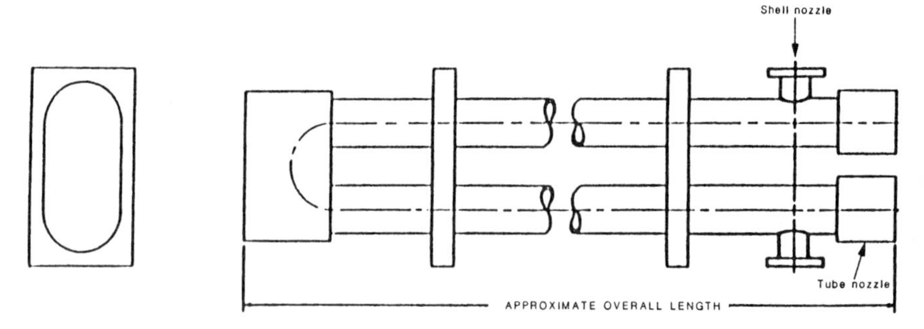

Figure 3-12. Double pipe exchanger.

(*text continued from page 65*)

two fluid streams flow in parallel between alternate pairs of plates. In addition to directing the flow patterns, the plate gasket keeps the fluids retained within the plate pack. Figure 3-13 shows a plate-and-frame exchanger.

Major advantages of plate-and-frame exchangers include the following: They have a low cost (especially for corrosive service), they are lighter and smaller than comparable shell-and-tube heat exchangers, full counter current flow and an LMTD correction factor are not required, and a close temperature approach is possible. Standard components allow simple stocking of spare parts, low maintenance, easy accessibility, and easy expansion by adding more plates. Metal plate-and-frame exchangers are particularly attractive for seawater and brackish water services. However, because of the design of plate-and-frame exchangers, wherein fluids are separated and retained across gasketed surfaces, they are limited to moderate temperature and pressure applications. In addition, some operators do not allow the use of plate-and-frame exchangers in hydrocarbon service or limit their use to pressures below 150 psig to 300 psig and temperatures less than 300°F. Plate-and-frame exchangers cannot be used for high viscosity liquids and slurry/suspended solids.

Because the plates are made of thin pressed metal, materials resistant to corrosive attack can be easily selected. Plates are standard and mass-produced. Specific applications are dealt with by changing plate arrangements. Stainless steels, monel, titanium, aluminum bronze, and other exotic metals

Figure 3-13. Plate-and-frame exchanger. (*Courtesy of Tranter, Inc.*)

may be used if desired. It is important to select the gasket materials to be compatible with the fluids and temperatures being handled.

AERIAL COOLERS

Aerial coolers are often used to cool a hot fluid to near ambient temperature. They are mechanically simple and flexible, and they eliminate the nuisance and cost of a cold source. In warm climates, aerial coolers may not be capable of providing as low a temperature as shell-and-tube exchangers, which use a cool medium. In aerial coolers the tube bundle is on the discharge or suction side of a fan, depending on whether the fan is blowing air across the tubes or sucking air through them. This type of exchanger can be used to cool a hot fluid to something near ambient temperature as in a compressor interstage cooler, or it can be used to heat the air as in a space heater.

When the tube bundle is on the discharge of the fan, the exchanger is referred to as "forced draft." When the tube bundle is on the suction of the fan it is referred to as an "induced draft" exchanger. Figure 3-14 shows a typical air cooled exchanger, and Figure 3-15 shows a detail of the headers and tube bundle. In Figure 3-15 the process fluid enters one of the nozzles on the fixed end and the pass partition plate forces it to flow through the tubes to the floating end (tie plate). Here it crosses over to the remainder of the tubes and flows back to the fixed end and out the other nozzle. Air is blown vertically across the finned section to cool the process fluid. Plugs are provided opposite each tube on both ends so that the tubes can be cleaned or individually plugged if they develop leaks. The tube bundle could also be mounted in a vertical plane, in which case air would be blown horizontally through the cooler.

Forced-air exchangers have tube lengths of 6 to 50 ft and tube diameters of ⅝ to 1½-in. The tubes have fins on them since air is non-fouling and it has a very low heat transfer efficiency. The fins increase efficiency by effectively adding surface area to the outside surface of the tubes. Some of the typical sizes of air cooled exchangers are shown in Table 3-5.

In a single aerial cooler there may be several fans and several tube bundles as shown in Figure 3-16, which defines bay width, tube length, and number of fans. Typically, on a compressor cooler there may be many tube bundles—one for cooling the gas after each stage, one for engine cooling water, one for lube oil, etc.

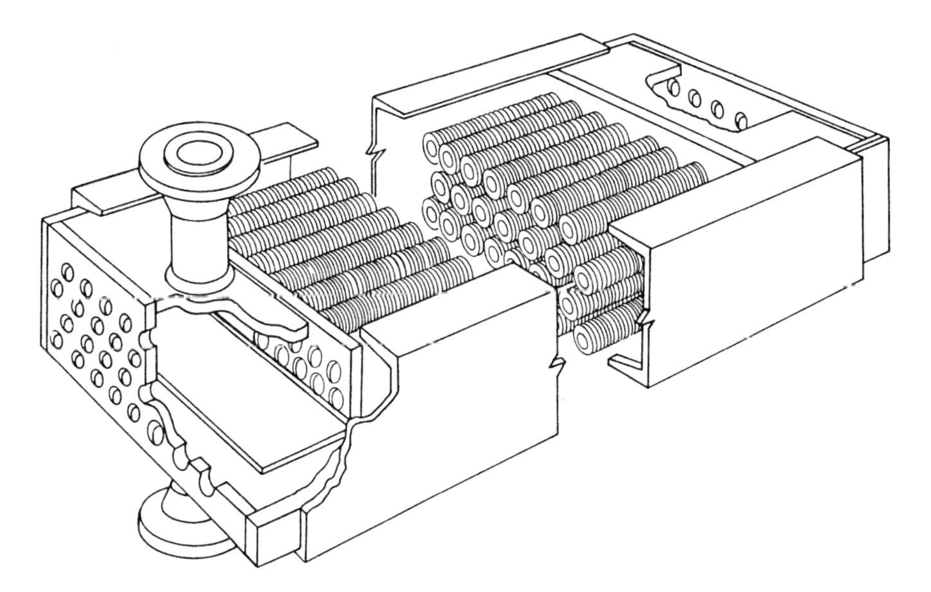

Figure 3-14. Aerial cooler.

Process outlet temperature in an aerial cooler can be controlled by louvers, fan variable speed drives, blade pitch or recirculation of process fluid. As the process flow rate and heat duties change, and as the temperature of the air changes from season to season and night to day, some adjustment must be made to assure adequate cooling while assuring that the process fluid is not over cooled. Too cool a gas temperature could lead to hydrates forming (Chapter 4) and developing ice plugs in the cooler. Too cool a lube oil temperature could lead to high viscosities, resulting in high pressure drops and inadequate lubrication.

Louvers are probably the most common type of temperature control device on aerial coolers. They may be either automatically adjusted by sensing the process temperature or manually adjusted. Blade pitch is probably second most common, and variable speed drive is third.

The procedure for calculating the number of tubes required for an aerial cooler is similar to that for a shell- and-tube exchanger. Table 3-6 shows approximate overall heat transfer coefficients. U_b should be used when the outside surface area of the bare tube (neglecting fins) is used in the heat

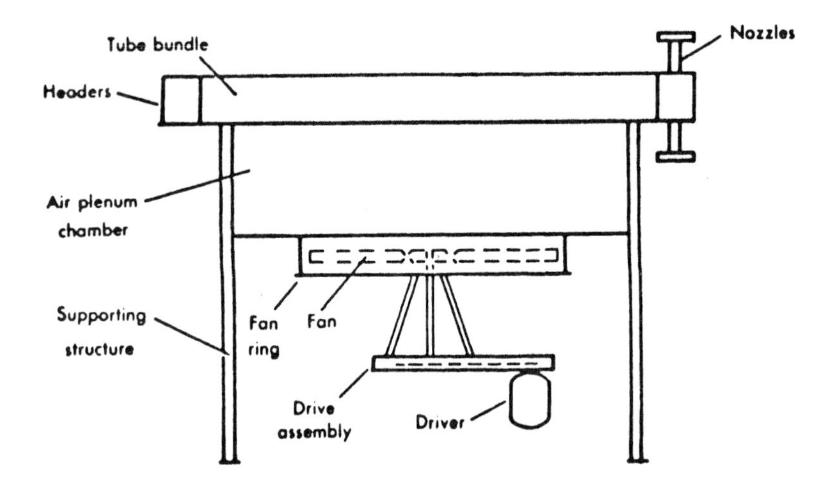

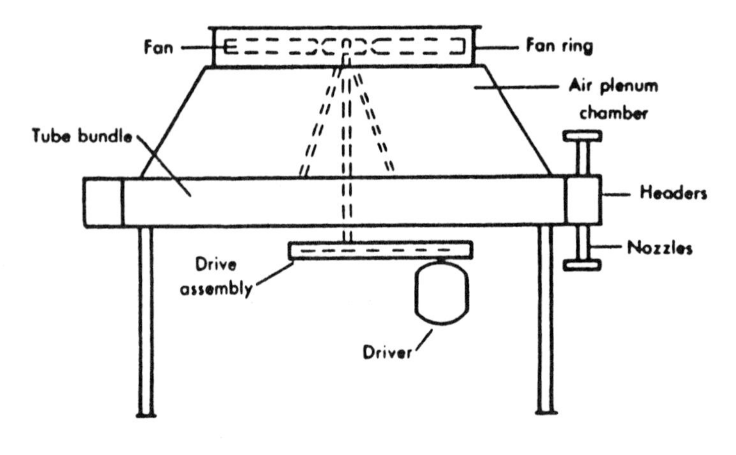

Figure 3-15. Side elevations of air coolers. (From Gas Processors Suppliers Association, *Engineering Data Book,* 9th Edition.)

transfer equation. U_x is used when the extended surface area including fins is used for the area term in the general heat transfer equation.

Figures 3-17 and 3-18 are LMTD correction charts. In using these figures the exit air temperature is needed. This can be approximated by:

Table 3-5
Typical Sizes of Air-Cooled Exchangers

Tube Lengths	— 6 ft to 50 ft
Tube Diameter	— ⅝ in. to 1½ in.
Fins	— ½ in. to 1 in. height 7 to 11 per in.
Depth	— 3 to 8 rows of fin tubes Triangular pitch with fins separated by ¹⁄₁₆ in. to ¼ in.
Bay Widths	— 4 ft to 30 ft
Fan Diameters	— 3 ft to 16 ft

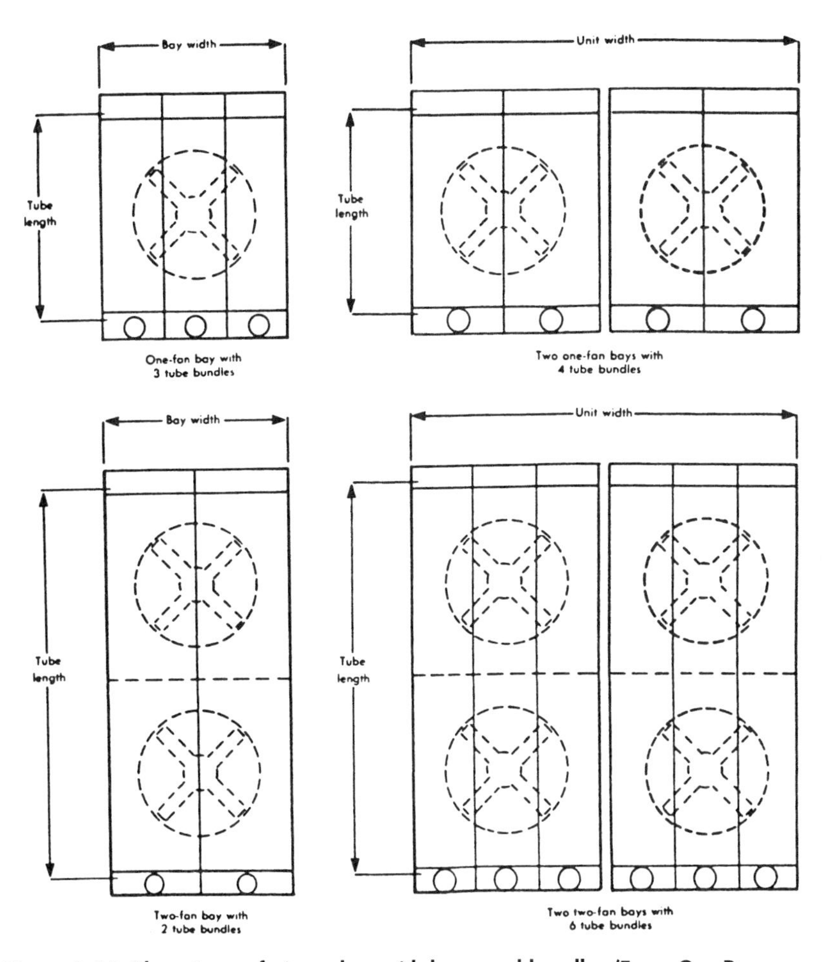

Figure 3-16. Plan views of air coolers with bays and bundles (From Gas Processors Suppliers Association, *Engineering Data Book,* 9th Edition.)

Table 3-6
Typical Overall Heat-Transfer Coefficients for Air Coolers

Service	Fintube ½ in. by 9		Fintube ⅝ in. by 10	
Water & Water Solutions				
	U_b	U_x	U_b	U_x
Engine jacket water (r_f = .001)	110	7.5	130	6.1
Process water (r_f = .002)	95	6.5	110	5.2
50-50 ethyl glycol-water (r_f = .001)	90	6.2	105	4.9
50-50 ethyl glycol-water (r_f = .002)	80	5.5	95	4.4
Hydrocarbon Liquid Coolers				
Viscosity C_p	U_b	U_x	U_b	U_x
0.2	85	5.9	100	4.7
0.5	75	5.2	90	4.2
1.0	65	4.5	75	3.5
2.5	45	3.1	55	2.6
4.0	30	2.1	35	1.6
6.0	20	1.4	25	1.2
10.0	10	0.7	13	0.6
Hydrocarbon Gas Coolers				
Temperature, °F	U_b	U_x	U_b	U_x
50	30	2.1	35	1.6
100	35	2.4	40	1.9
300	45	3.1	55	2.6
500	55	3.8	65	3.0
750	65	4.5	75	3.5
1000	75	5.2	90	4.2
Air and Flue-Gas Coolers				
Use one-half of value given for hydrocarbon gas coolers				
Steam Condensers				
(Atmospheric pressure & above)				
	U_b	U_x	U_b	U_x
Pure steam (r_f = 0.005)	125	8.6	145	6.8
Steam with non-condensables	60	4.1	70	3.3
HC Condensers				
Pressure, psig	U_b	U_x	U_b	U_x
0° range	85	5.9	100	4.7
10° range	80	5.5	95	4.4
25° range	75	5.2	90	4.2
60° range	65	4.5	75	3.5
100° & over range	60	4.1	70	3.3
Other Condensers				
	U_b	U_x	U_b	U_x
Ammonia	110	7.6	130	6.1
Freon 12	65	4.5	75	3.5

Note: U_b is overall rate based on bare tube area and U_x is overall rate based on extended surface.
Source: Gas Processors Suppliers Association, Engineering Data Book, *9th Edition.*

$$\Delta T_a = \left(\frac{U_x + 1}{10}\right)\left(\frac{T_1 + T_2}{2} - t_1\right) \quad (3\text{-}3)$$

where ΔT_a = air temperature rise, °F
U_x = overall heat transfer coefficient based on extended area, Btu/hr-ft^2-°F
T_1 = process fluid inlet temperature, °F
T_2 = process fluid outlet temperature, °F
t_i = ambient air temperature, °F

Table 3-7 gives the external area of fin tubes per square foot of bundle surface area. From this data the area of bundle surface area can be calculated from:

$$A = \frac{q}{U_x\,(LMTD)\,(APSF)} \quad (3\text{-}4)$$

where A = required area of bundle face, ft^2
q = heat duty, Btu/hr
U_x = overall heat transfer coefficient, Btu/hr-ft^2-°F
LMTD = log mean temperature difference corrected by Figures 3-17 and 3-18, °F
APSF = tube expanded area per square foot of bundle face

(*text continued on page 82*)

Table 3-7
External Area of Fin Tube Per Ft^2 of Bundle Surface Area (APSF) for 1-in. OD Tubes

	½ in. Height by 9 fins/in.		⅝ in. Height by 10 fins/in.	
Tube Pitch	2 in. Δ	2¼ in. Δ	2¼ in. Δ	2½ in. Δ
3 rows	68.4	60.6	89.1	80.4
4 rows	91.2	80.8	118.8	107.2
5 rows	114.0	101.0	148.5	134.0
6 rows	136.8	121.2	178.2	160.8

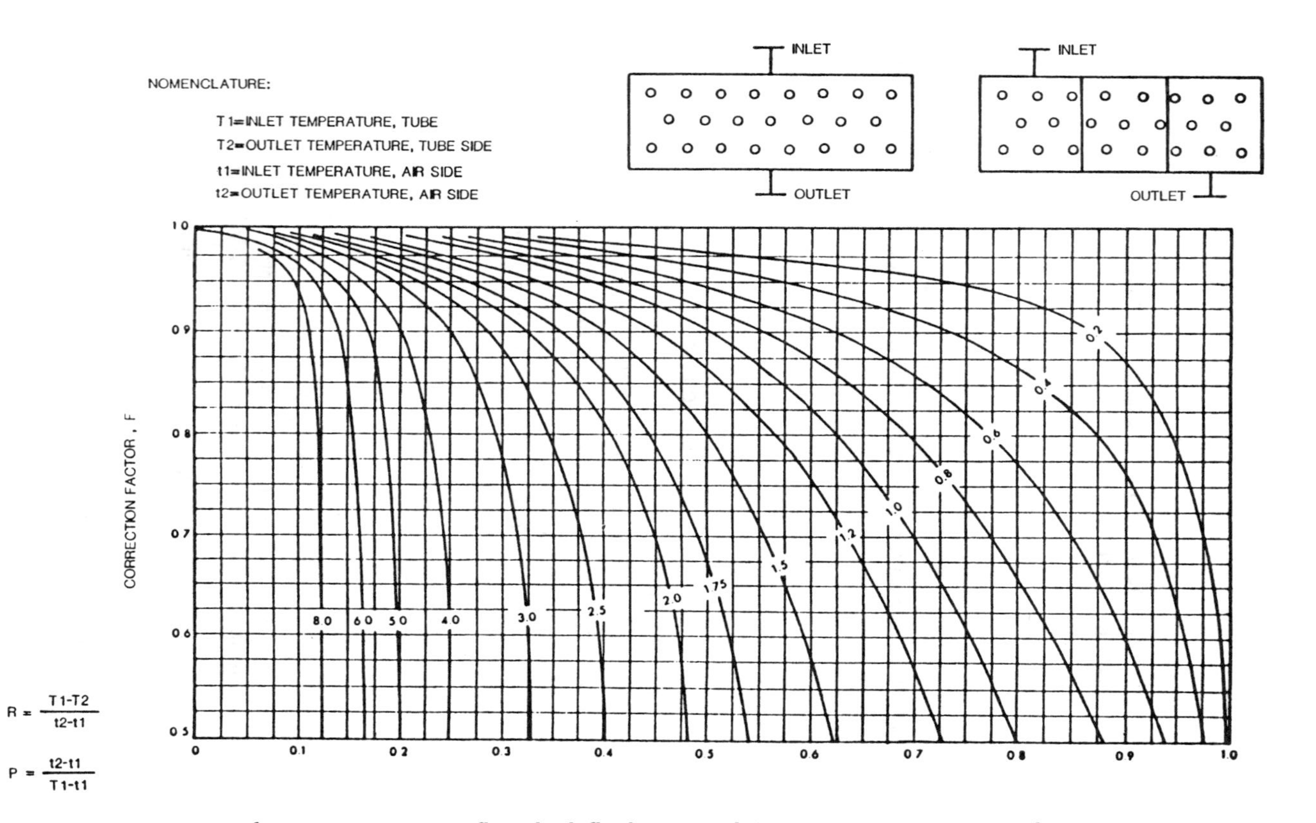

Figure 3-17. LMTD correction factors; 1-pass, cross-flow; both fluids unmixed. (From Gas Processors Suppliers Association, *Engineering Data Book*, 9th Edition.)

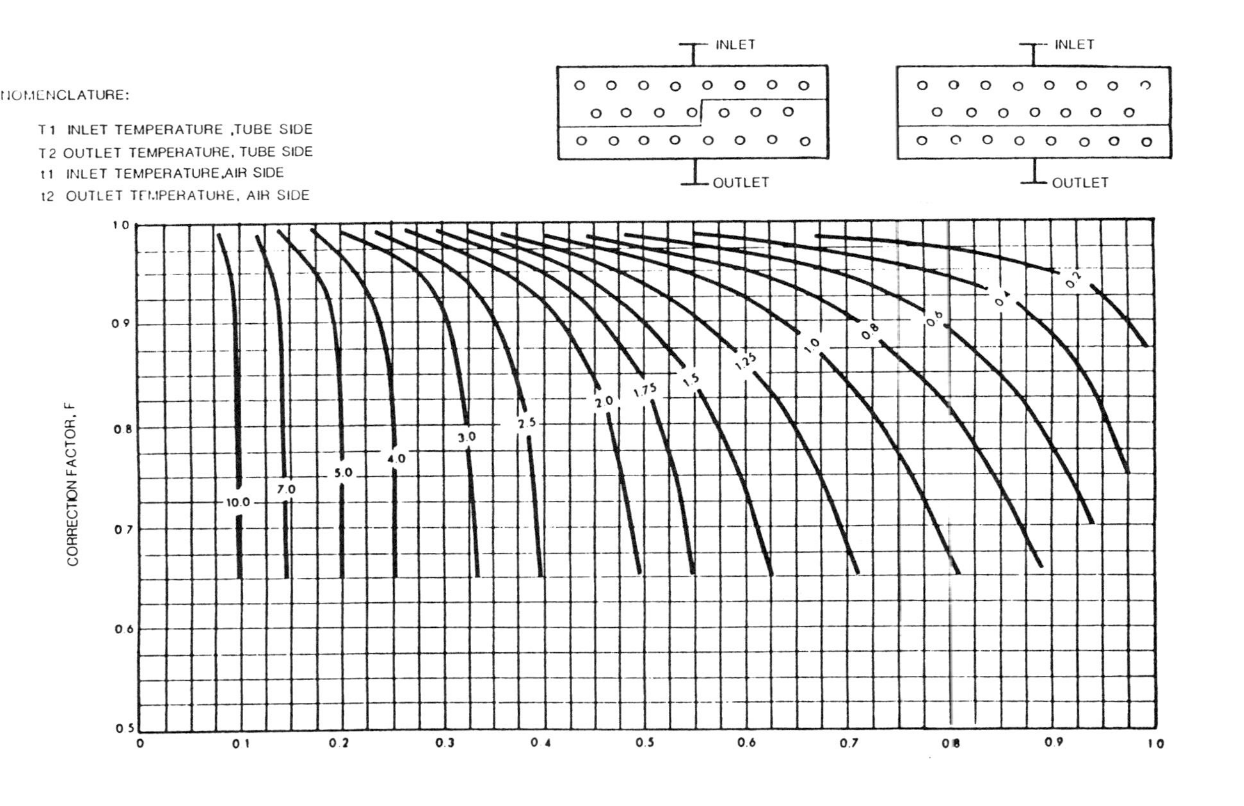

Figure 3-18. LMTD correction factors; 2-pass, cross-flow; both fluids unmixed. (From Gas Processors Suppliers Association, *Engineering Data Book*, 9th Edition.)

(*text continued from page 79*)

FIRED HEATER

Direct-fired combustion equipment is that in which the flame and/or products of combustion are used to achieve the desired result by radiation and convection. Common examples include rotary kilns and open-hearth furnaces. Indirect-fired combustion equipment is that in which the flame and products of combustion are separated from any contact with the principal material in the process by metallic or refractory walls. Examples are steam boilers, vaporizers, heat exchangers, and melting pots.

The heat exchangers previously discussed rely on convection and conduction for heat transfer. In a fired heater, such as shown in Figure 3-19, radiation plays a major role in heat transfer. The process fluid flows through tubes around a flame. These tubes receive most of the heat directly by radiation from the flame. A small amount of heat is also received by convection from the air between the tubes and flame. Heaters are not common in most field installations but are much more commonly used in plant situations in which competent operators routinely maintain and inspect the equipment.

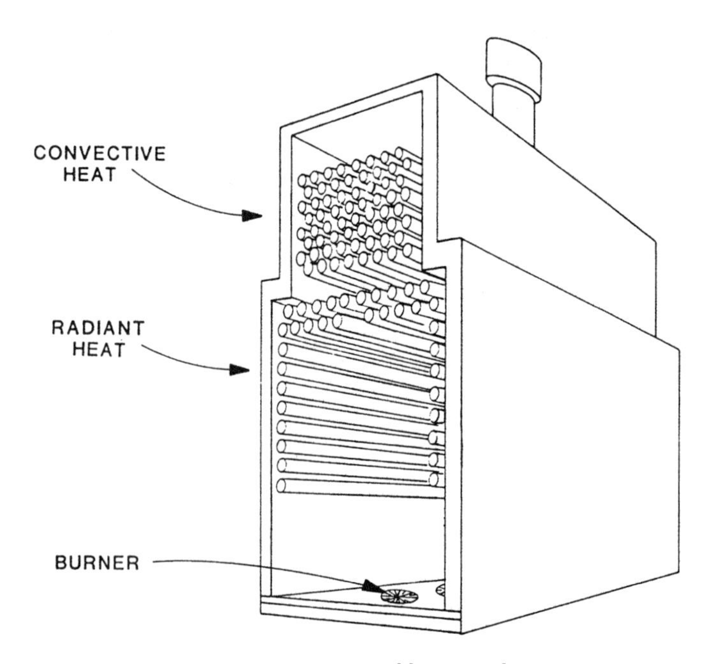

Figure 3-19. Furnace and heating elements.

For safety reasons, heaters are most often used to heat a heat medium system (water, steam, or heat transfer fluid) rather than to heat the gas or oil stream directly. If the fluid to be heated contains hydrocarbons, the heater can be located safely away from other equipment. If it catches on fire, damage can be limited.

The tubes that are around the flame get most of their heat energy from radiation. The tubes in the top of the chamber get their heat from convection as the hot exhaust gases rise up through the heater and heat the process fluid in the tubes. The principal classification of fired heaters relates to the orientation of the heating coil in the radiant section. The tube coils of vertical fired heaters are placed vertically along the walls of the combustion chamber. Firing also occurs vertically from the floor of the heater. All the tubes are subjected to radiant energy.

These heaters represent a low-cost, low-efficiency design that requires a minimum of plot area. Typical duties run from 0.5 to 200 MMBtu/hr. Six types of vertical-tube-fired heaters are shown in Figure 3-20.

The radiant section tube coils of horizontal fired heaters are arranged horizontally so as to line the sidewalls and the roof of the combustion chamber. In addition, there is a convection section of tube coils, which are positioned as a horizontal bank of tubes above the combustion chamber. Normally the tubes are fired vertically from the floor, but they can also be fired horizontally by side wall mounted burners located below the tube coil. This economical, high efficiency design currently represents the majority of new horizontal-tube-fired heater installations. Duties run from 5 to 250 MMBtu/hr. Six types of horizontal-tube-fired heaters are shown in Figure 3-21.

HEAT RECOVERY UNITS

In the interest of energy conversion, process heat can be obtained from a heat recovery unit in which heat is recovered from turbine or reciprocating engine exhaust. In a heat recovery unit, an exhaust gas flows over finned tubes carrying the fluid to be heated. The hot exhaust gas (900°F to 1,200°F) heats the fluid in the tubes in a manner similar to that in which air cools the fluid in an aerial cooler. It is also possible to recover heat from exhausts by routing the exhaust duct directly through a fluid bath. The latter option is relatively inefficient but easy to install and control.

(*text continued on page 86*)

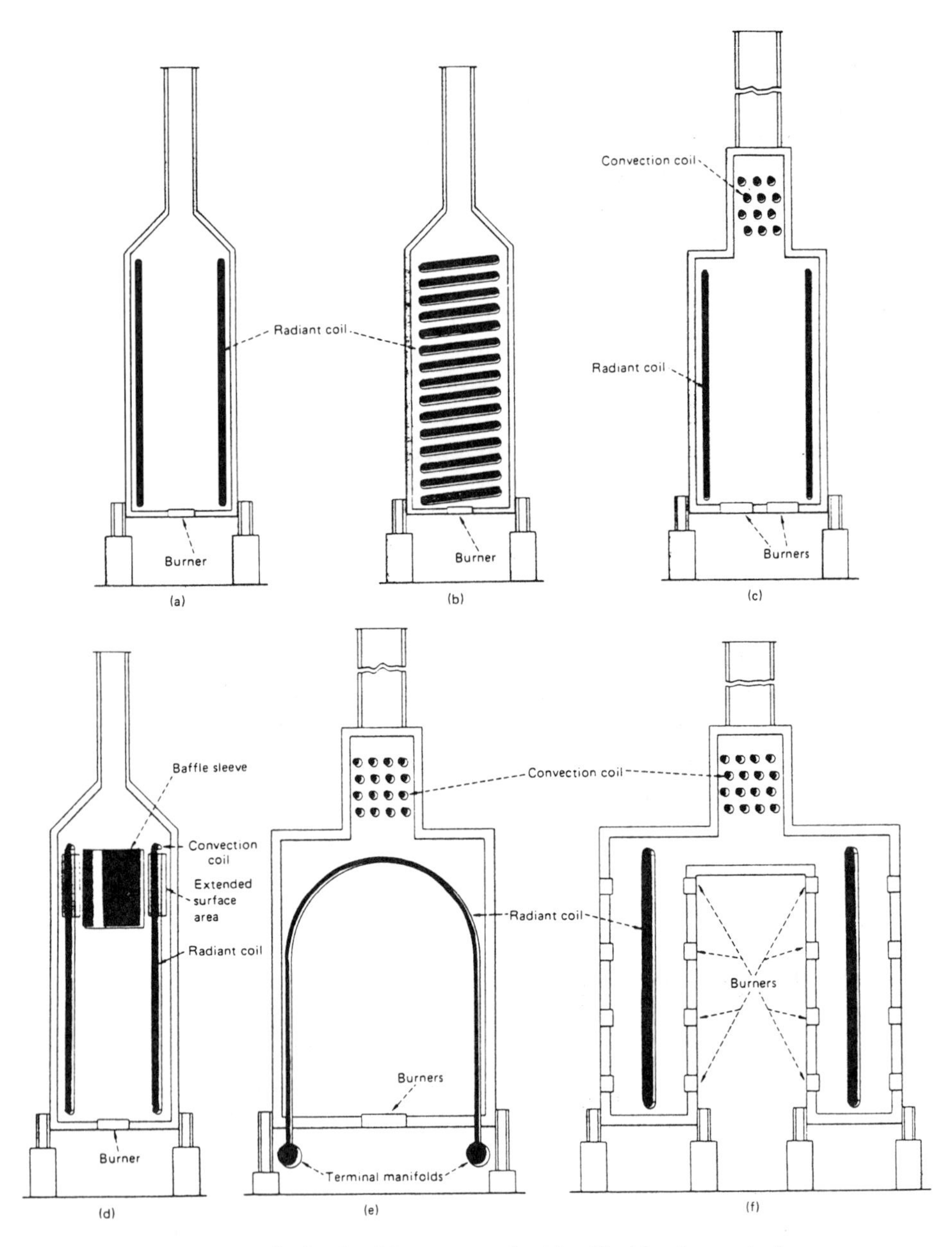

Figure 3-20. Vertical-tube-fired heaters can be identified by the vertical arrangement of the radiant-section coil. (a) Vertical-cylindrical; all radiant. (b) Vertical-cylindrical; helical coil. (c) Vertical-cylindrical, with cross-flow-convection section. (d) Vertical-cylindrical, with integral-convection section. (e) Arbor or wicket type. (f) Vertical-tube, single-row, double-fired. [From *Chem. Eng.*, 100–101 (June 19, 1978).]

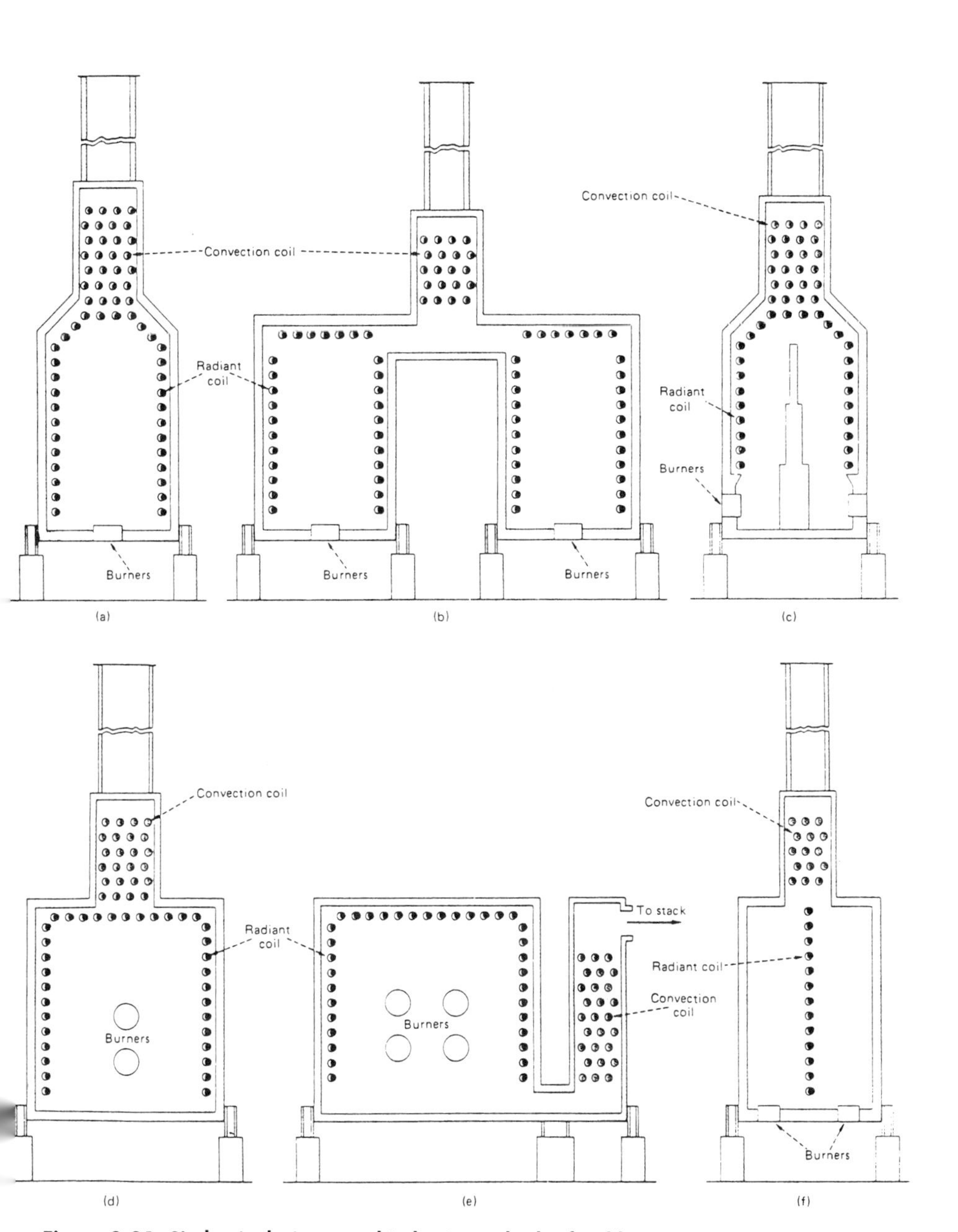

Figure 3-21. Six basic designs used in horizontal-tube-fired heaters. Radiant-section coil is horizontal. (a) Cabin. (b) Two-cell box. (c) Cabin with dividing bridgewall. (d) End-fired box. (e) End-fired box, with side-mounted convection section. (f) Horizontal-tube, single-row, double-fired. [From *Chem. Eng.*, 102–103 (June 19, 1978).]

(*text continued from page 83*)

Generally, the following design criteria should be provided to the manufacturers or vendors for sizing an exhaust heat recovery unit.

1. Total heat duty required to heat the fluid
2. Properties of the fluid to be heated
3. The outlet temperature of the heated fluid
4. Operational relationships between heat sources and users (e.g., which users continue to operate when sources shut down?)
5. Exhaust gas flow rates at anticipated ambient and at various loads from maximum to minimum
6. Exhaust gas temperature at anticipated ambient and at various loads
7. Maximum exhaust back pressure
8. Ambient temperature range

The design of heaters and waste heat recovery units is beyond the scope of this book. Sizing and design are best left to manufacturers. However, the concepts discussed in this chapter and in Chapter 2 can be used to verify the manufacturer's proposals.

HEAT EXCHANGER EXAMPLE PROBLEM

Design a seawater cooler to cool the total stream from the example field in its later stages of life from a flowing temperature of 175°F to a temperature of 100°F to allow further treating.

Given:

Inlet
100 MMscfd at 0.67 SG (from Table 2-10)
6,000 bopd at 0.77 SG
15 bbl water/MMscf
$T_1 = 175°F$
$P_1 = 1{,}000$ psig
Water vapor in gas = 60 lb/MMscf (See Chapter 4.)

Outlet
$T_2 = 100°F$
$P_2 = 990$ psig
Water vapor in gas = 28 lb/MMscf (See Chapter 4.)
Seawater $T_3 = 75°F$
Limit temperature rise to 10°F
Use 1-in. OD 10 BWG Tubes on 1¼-in. Pitch

Problem:

1. Calculate water flow rate in outlet and water vapor condensed.
2. Calculate heat duty.
3. Determine seawater circulation rate.
4. Pick a type of exchanger and number of tubes required.

Solution:

1. Calculate free water and water vapor flow rates.

Water flow rate in inlet:

Free water = (100 MMscfd)(15 bbl/MMscfd) = 1,500 bwpd

Water flow rate in outlet:

Free water = 1,500 bwpd

Water vapor condensed:

$$\frac{(60-28)\text{ lb}}{\text{MMscf}} \times \frac{100\text{ MMscf}}{\text{D}} = 3,200\text{ lb / d}$$

$$3,200\,\frac{\text{lb}}{\text{d}} \times \frac{\text{bbl}}{350\text{ lb}} = 9\text{ bwpd}$$

Water flow rate in outlet:

$$\begin{array}{r} 9\text{ bwpd} \\ \underline{+\ 1500\text{ bwpd}} \\ 1509\text{ bwpd} \end{array}$$

2. Calculate heat duty

a. Gas duty

$T_1 = 635°R$
$T_2 = 560°R$
$T_{av} = 597.5°R$
$P_C = 680$ psia (Table 2-10)
$P_R = P/P_C = 1.47$
$T_C = 375°R$ (Table 2-10)
$T_R = T_{av}/T_c = 1.59$
$q_g = 41.7\ (\Delta T)\ C_g Q_g$
$C_g = 2.64\ [29 \times S \times C + \Delta C_p]$
$C = 0.528$ Btu/lb°F (Figure 2-14)

$\Delta C_p = 1.6$ Btu/lb-mol °F (Figure 2-15)
$S = 0.67$ (Table 2-10)
$C_g = 2.64\,[(29)(0.67)(0.528) + 1.6]$
$C_g = 31.3$
$q_g = 41.7\,(100 - 175)(31.3)(100)$
$= -9{,}789{,}000$ Btu/hr

b. Condensate duty
$q_o = 14.6\,(SG)(\Delta T)C_oQ_o$
$C_o = 0.535$ Btu/lb °F (Figure 2-13)
$q_o = 14.6\,(0.77)(100 - 175)(0.535)(6000)$
$q_o = -2{,}707{,}000$ Btu/hr

c. Free water duty
$q_w = 14.6\,(\Delta T)Q_w$
$q_w = 14.6\,(100 - 175)(1509)$
$q_w = 1{,}652{,}000$ Btu/hr

d. Water latent heat duty

$$q_{lh} = W\,(\lambda)$$

$$W = 3{,}200\,\frac{\text{lb}}{\text{d}} \times \frac{1}{24} = 133\,\frac{\text{lb}}{\text{hr}}$$

$$\lambda = -996.3 \text{ Btu / lb (Table 2-6, 170°F)}$$

$$q_{lh} = (133)\,(-996.3) = -133{,}000 \text{ Btu / hr}$$

e. Total heat duty
$q = -9{,}789{,}000 - 2{,}707{,}000 - 1{,}652{,}000 - 133{,}000$
$q = -14{,}281{,}000$ Btu/hr

3. Water circulation rate
$q_w = 14.6\,(T_2 - T_1)\,Q_w$
$Q_w = q_w/14.6\,(T_2 - T_1)$

Limit ΔT for water to 10°F to limit scale.

$Q_w = 14.3 \times 10^6/14.6\,(10)$
$Q_w = 97{,}945$ bwpd $= 2{,}858$ gpm

4. Heat exchanger type and number of tubes
Choose TEMA R because of large size.
Select type AFL because of low temperature change and LMTD correction factor.

The water is corrosive and may deposit solids. Therefore, flow water through tubes and make the tubes 70/30 Cu/Ni. Flow the gas through the shell.
$U \simeq 90$ Btu/hr-ft^2-°F (Table 2-8)

Calculate LMTD:

$$T_1 = 175 \qquad T_2 = 100$$
$$T_3 = 75 \qquad T_4 = 85$$
$$\Delta T_1 = 175 - 85 = 90$$
$$\Delta T_2 = 100 - 75 = 25$$

$$\text{LMTD} = \frac{90 - 25}{\log_e\left(\dfrac{90}{25}\right)}$$

$$\text{LMTD} = 50.7°\text{F}$$

Correction factor (Figure 3-10):

$$P = \frac{85 - 75}{175 - 75} = 0.1$$

$$R = \frac{175 - 100}{85 - 75} = 7.5$$

$$F = 0.95$$

$$\text{LMTD} = (50.7)(0.95)$$
$$= 48.2°\text{F}$$

Calculate number of tubes:

$$N = \frac{q}{UA'\,(\text{LMTD})\,L}$$

Assume L = 40 ft

$$A = 0.2618\ \text{ft}^2\ /\ \text{ft (Table 2-1)}$$

$$N = \frac{14.3 \times 10^6}{(90)\,(0.2618)\,(48.2)\,(40)}$$

$$N = 315\ \text{tubes}$$

From Table 3-4 for 1-in. OD, 1¼-in. square pitch, fixed tube sheet, four passes, shell ID = 29 in.

Assume L = 20 ft
N = 629 tubes
Shell ID = 39 in.

Use 39-in. ID × 20 ft Lg w/682 1-in. OD, 10 BWG tubes 1¼-in., square pitch with four tube passes:

Size 39 - 240 Type AFL

Check the water velocity in tubes. From Volume 1:

$$V = 0.012\,\frac{Q_w}{d^2}$$

There are four passes. Thus, 682/4 tubes are used in each pass.

$$Q_w \text{ per tube} = \frac{(4)(97{,}945)}{682}$$

$$= 574 \text{ bwpd}$$

$$d = 0.732 \text{ in. (Table 2-1)}$$

$$V = \frac{(0.012)(574)}{(0.732)^2} = 12.9 \text{ ft/sec} \text{ which is less than 15 ft/sec.}$$

Comments About Example

Once a specific heat exchanger is chosen, the flow per tube is known, so it is possible to use the correlations of Chapter 2 to calculate a more precise overall heat transfer coefficient (U). An example of calculation of U is given in Chapter 5.

Note that more than 30% of the heat duty was required to cool the water and condensate. If the liquids had first been separated, a smaller exchanger and lower seawater flow rate could have been used. In most gas facilities, where cooling is required, the cooler is placed downstream of the first separator for this reason. Often an aerial cooler is used for this service.

In this example we selected a final outlet temperature of 100°F. This would be sufficiently low if the gas were only going to be compressed and dehydrated. For our case, we must also treat the gas for H_2S and CO_2 removal (Chapter 7). If we chose an amine unit, which we will in all likelihood, the heat of the reaction could heat the gas more than 10° to 20°F, making the next step, glycol dehydration, difficult (Chapter 8). In such a case, it may be better to cool the gas initially to a lower temperature so that it is still below 110°F at the glycol dehydrator. Often this is not possible, since cooling water is not available and ambient air conditions are in the 95°F to 100°F range. If this is so, it may be necessary to use an aerial cooler to cool the gas before treating, and another one to cool it before dehydration.

CHAPTER

4

Hydrates*

Resembling dirty ice, hydrates consist of a water lattice in which light hydrocarbon molecules are embedded. They are a loosely-linked crystalline chemical compound of hydrocarbon and water called cathrates, a term denoting compounds that may exist in stable form but do not result from true chemical combination of all the molecules involved. Hydrates normally form when a gas stream is cooled below its hydrate formation temperature. At high pressure these solids may form at temperatures well above 32°F. Hydrate formation is almost always undesirable because the crystals may cause plugging of flow lines, chokes, valves, and instrumentation; reduce line capacities; or cause physical damage. This is especially true in chokes and control valves where there are large pressure drops and small orifices. The pressure drops cause the temperature to decrease, and the small orifices are susceptible to plugging if hydrates form. Hydrate formation leading to flow restrictions is referred to as "freezing."

The two major conditions that promote hydrate formation are (1) the gas being at the appropriate temperature and pressure, and (2) the gas being at or below its water dew point with "free water" present. For any particular composition of gas at a given pressure there is a temperature below which hydrates will form and above which hydrates will not form. As the pressure increases, the hydrate formation temperature also increases. If there is no

*Reviewed for the 1999 edition by Dennis A. Crupper of Paragon Engineering Services, Inc.

free water, that is, liquid water, hydrates cannot form. Secondary conditions such as high gas velocities, agitation of any type, and the formation of a nucleation site may also help form hydrates. These secondary conditions are almost always present in the process piping stream.

Methods of preventing hydrate formation include adding heat to assure that the temperature is always above the hydrate formation temperature, lowering the hydrate formation temperature with chemical inhibition, or dehydrating the gas so that water vapor will not condense into free water. It is also feasible to design the process so that if hydrates form they can be melted before they plug equipment.

Before choosing a method of hydrate prevention or dehydration, the operating system should be optimized so as to minimize the necessary treating. Some general factors to consider include the following: (1) reduce pressure drops by minimizing line lengths and restrictions, (2) take required pressure drops at the warmest conditions possible, and (3) check the economics of insulating pipe in cold areas.

This chapter discusses the procedures used to calculate the temperature at which hydrates will form for a given pressure (or the pressure at which hydrates will form for a given temperature), the amount of dehydration required to assure that water vapor does not condense from a natural gas stream, and the amount of chemical inhibitor that must be added to lower the hydrate formation temperature. It also discusses the temperature drop that occurs as gas is expanded across a choke. This latter calculation is vital to the calculation of whether hydrates will form in a given stream.

The next chapter discusses the use of LTX units to melt the hydrates as they form, and the use of indirect fired heaters to keep the gas temperature above the hydrate formation temperature. Chapter 8 describes processes and equipment to dehydrate the gas and keep free water from forming.

DETERMINATION OF HYDRATE FORMATION TEMPERATURE OR PRESSURE

Knowledge of the temperature and pressure of a gas stream at the wellhead is important for determining whether hydrate formation can be expected when the gas is expanded into the flow lines. The temperature at the wellhead can change as the reservoir conditions or production rate changes over the producing life of the well. Thus, wells that initially flowed at conditions at which hydrate formation in downstream equipment was not expected may eventually require hydrate prevention, or vice versa.

If the composition of the stream is known, the hydrate temperature can be predicted using vapor-solid (hydrate) equilibrium constants. The basic equation for this prediction is:

$$\text{SUM}\left(\frac{y_n}{K_n}\right) = 1.0 \tag{4-1}$$

where Y_n = mol fraction of hydrocarbon component n in gas on a water-free basis
K_n = vapor-solid equilibrium constant for hydrocarbon component n

The vapor-solid equilibrium constant is determined experimentally and is defined as the ratio of the mol fraction of the hydrocarbon component in gas on a water-free basis to the mol fraction of the hydrocarbon component in the solid on a water-free basis. That is:

$$K_n = \left(\frac{y_n}{x_n}\right) \tag{4-2}$$

where x_n = mol fraction of hydrocarbon component in the solid on a water-free basis

Graphs giving the vapor-solid equilibrium constants at various temperatures and pressures are given in Figures 4-1 through 4-4. For nitrogen and components heavier than butane, the equilibrium constant is taken as infinity.

The steps for determining the hydrate temperature at a given system pressure are as follows:

1. Assume a hydrate formation temperature.
2. Determine K_n for each component.
3. Calculate $\frac{y_n}{K_n}$ for each component.
4. Sum the values of $\frac{y_n}{K_n}$.
5. Repeat steps 1–4 for additional assumed temperatures until the summation of $\frac{y_n}{K_n}$ is equal to 1.0

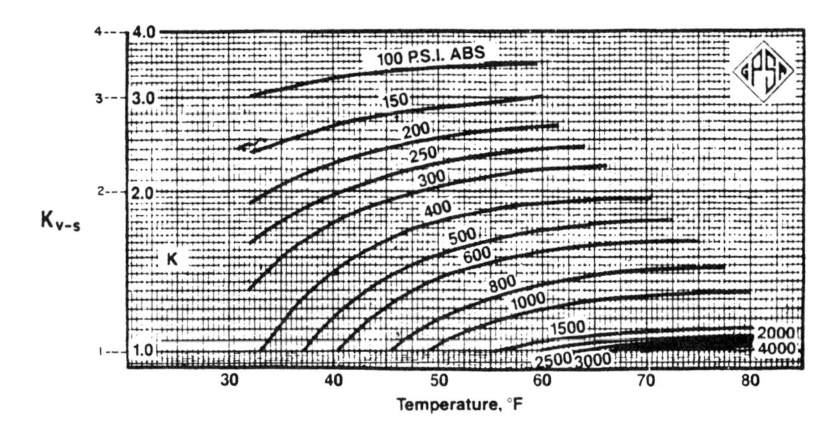

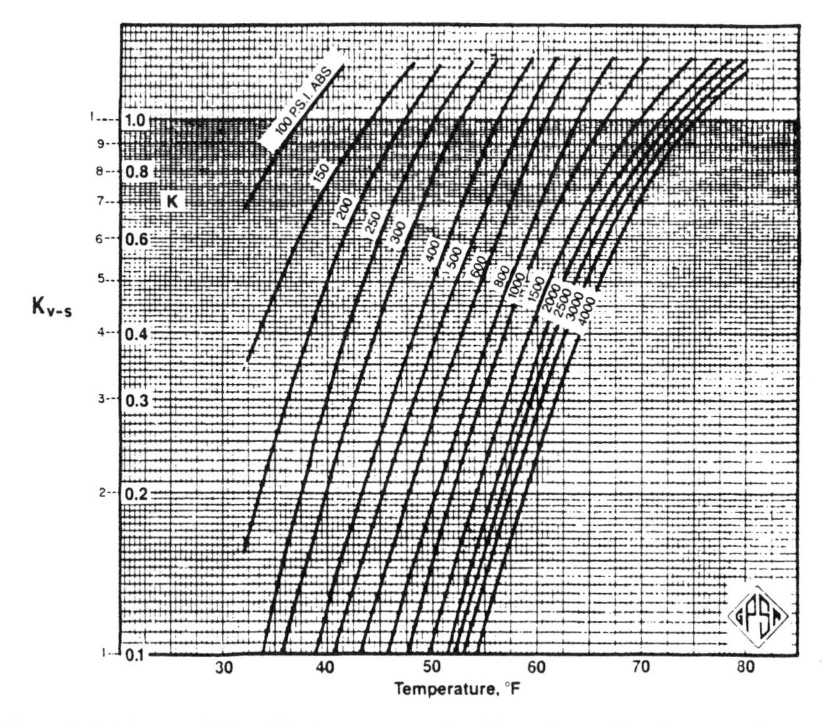

Figure 4-1. Vapor-solid equilibrium constant for (a) methane, (b) ethane, and n-butane. (From Gas Processors Suppliers Association, *Engineering Data Book.*)

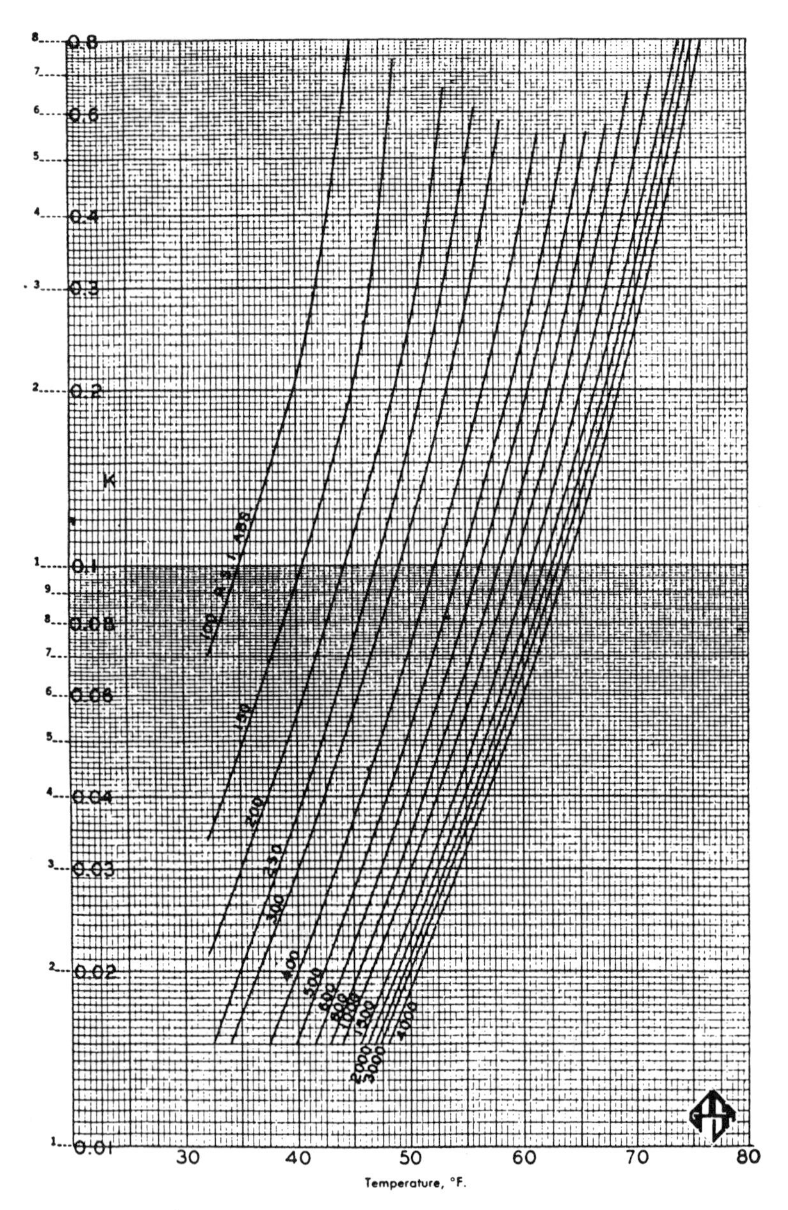

Figure 4-2. Vapor-solid equilibrium constant for propane. (From Gas Processors Suppliers Association, *Engineering Data Book,* 10th Edition.)

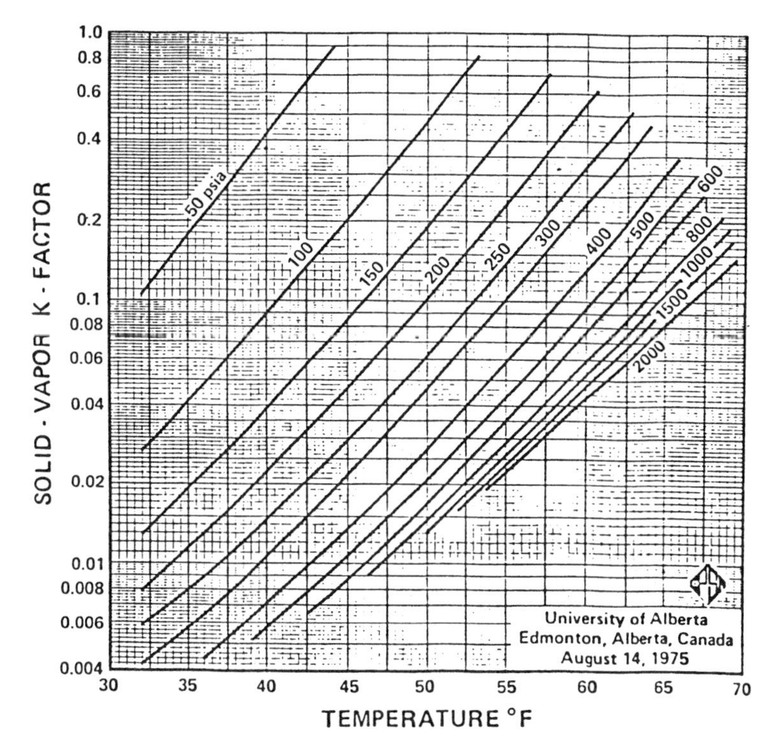

Figure 4-3. Vapor-solid equilibrium constants for isobutane. (From Gas Processors Suppliers Association, *Engineering Data Book,* 10th Edition.)

The presence of H_2S should not be overlooked in the determination of susceptibility of a gas to form hydrates. At concentrations of 30% or greater, hydrates will form in hydrocarbon gases at about the same temperature as in pure H_2S.

Table 4-1 is an example calculation of the temperature below which hydrates will form at the 4,000 psia flowing temperature for the example gas composition of Table 1-1. From this calculation, hydrates will form at temperatures below 74°F.

If the gas composition is not known, this procedure cannot be used to develop the hydrate formation point. Figure 4-5 gives approximate hydrate formation temperatures as a function of gas gravity and pressure. For example, for the 0.67 specific gravity gas of our example field (Table 2-10), Figure 4-5 predicts a hydrate formation temperature at 4,000 psia at 76°F.

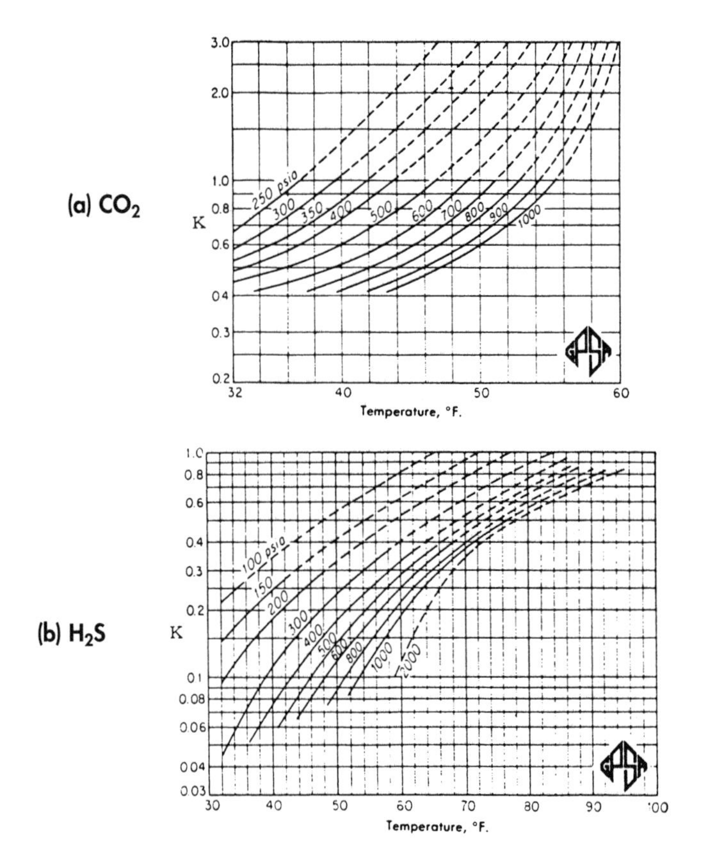

Figure 4-4. Vapor-solid equilibrium constants (a) for carbon dioxide, (b) for hydrogen sulfide. (From Gas Processors Suppliers Association, *Engineering Data Book,* 10th Edition.)

CONDENSATION OF WATER VAPOR

One method of assuring that hydrates do not form is to assure that the amount of water vapor in the gas is always less than the amount required to fully saturate the gas. Typically, but not always, the gas will be saturated with water in the reservoir. As the gas is cooled from reservoir temperature, the amount of water vapor contained in the gas will decrease. That is, water will condense.

The temperature at which water condenses from natural gas is called its dew point. If the gas is saturated with water vapor, it is by definition at its dew point. The amount of water vapor saturated in the gas is given

Table 4-1
Calculation of Temperature for Hydrate Formation at 4,000 psia

Component	y_n, Mole Fraction in Gas	At 70°F		At 80°F	
		K_n	y_n/K_n	K_n	y_n/g_n
Nitrogen	0.0144	Infinity	0.00	Infinity	0.00
Carbon dioxide	0.0403	Infinity	0.00	Infinity	0.00
Hydrogen sulfide	0.000019	0.3	0.00	0.5	0.00
Methane	0.8555	0.95	0.90	1.05	0.81
Ethane	0.0574	0.72	0.08	1.22	0.05
Propane	0.0179	0.25	0.07	Infinity	0.00
Isobutane	0.0041	0.15	0.03	0.6	0.01
n-Butane	0.0041	0.72	0.00	1.22	0.00
Pentane$^+$	0.0063	Infinity	0.00	Infinity	0.00
Total	1.0000		1.08		0.87

Interpolating linearly, Y/K = 1.0 at 74°F

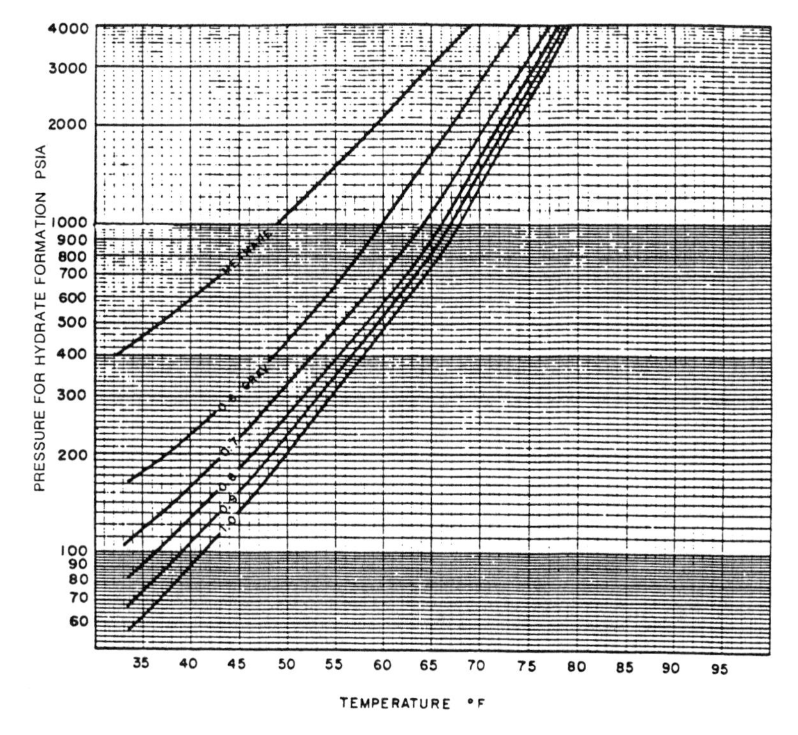

Figure 4-5. Approximate hydrate temperature formation. (*Courtesy of Smith Industries, Inc.*)

by Figure 4-6. The graph shows the water content in pounds of water per MMscf of saturated gas at any pressure and temperature. For example, at 150°F and 3,000 psi, saturated gas will contain approximately 105 lb of water vapor per MMscf of gas. If there is less water vapor, the gas is not saturated and its temperature can be reduced without water condensing. If the gas is saturated at a higher temperature and then cooled to 150°F, water will condense until there are only 105 lb of water vapor left in the gas. The dotted line crossing the family of curves shows the approximate temperature at which hydrates will probably form at any given pressure. Note the hydrates form more easily at higher pressures.

To keep water from condensing as the gas is processed, it is necessary to dehydrate the gas (that is, remove water vapor) until the amount of water vapor remaining in the gas is less than that required to fully saturate the gas at all conditions of temperature and pressure. Since the dehydrated gas will have a lower dew point, dehydration is sometimes called dew point depression. For example, if the amount of water vapor in the 3,000 psig gas stream referred to earlier were reduced from 105 lb/MMscf to 50 lb/MMscf, the dew point would be reduced from 150°F to 127°F. That is, its dew point will be depressed by 23°F.

Figure 4-6 contains an approximate hydrate formation line. This should be used with care, as the position of the line depends on the gas composition. It is better to calculate the hydrate formation temperature or use Figure 4-5 for approximation.

TEMPERATURE DROP DUE TO GAS EXPANSION

Choking, or expansion of gas from a high pressure to a lower pressure, is generally required for control of gas flow rates. Choking is achieved by the use of a choke or a control valve. The pressure drop causes a decrease in the gas temperature, thus hydrates can form at the choke or control valve. The best way to calculate the temperature drop is to use a simulation computer program. The program will perform a flash calculation, internally balancing enthalpy. It will calculate the temperature downstream of the choke, which assures that the enthalpy of the mixture of gas and liquid upstream of the choke equals the enthalpy of the new mixture of more gas and less liquid downstream of the choke.

For a single component fluid, such as methane, a Mollier diagram, such as Figure 4-7, can be used to calculate temperature drop directly.

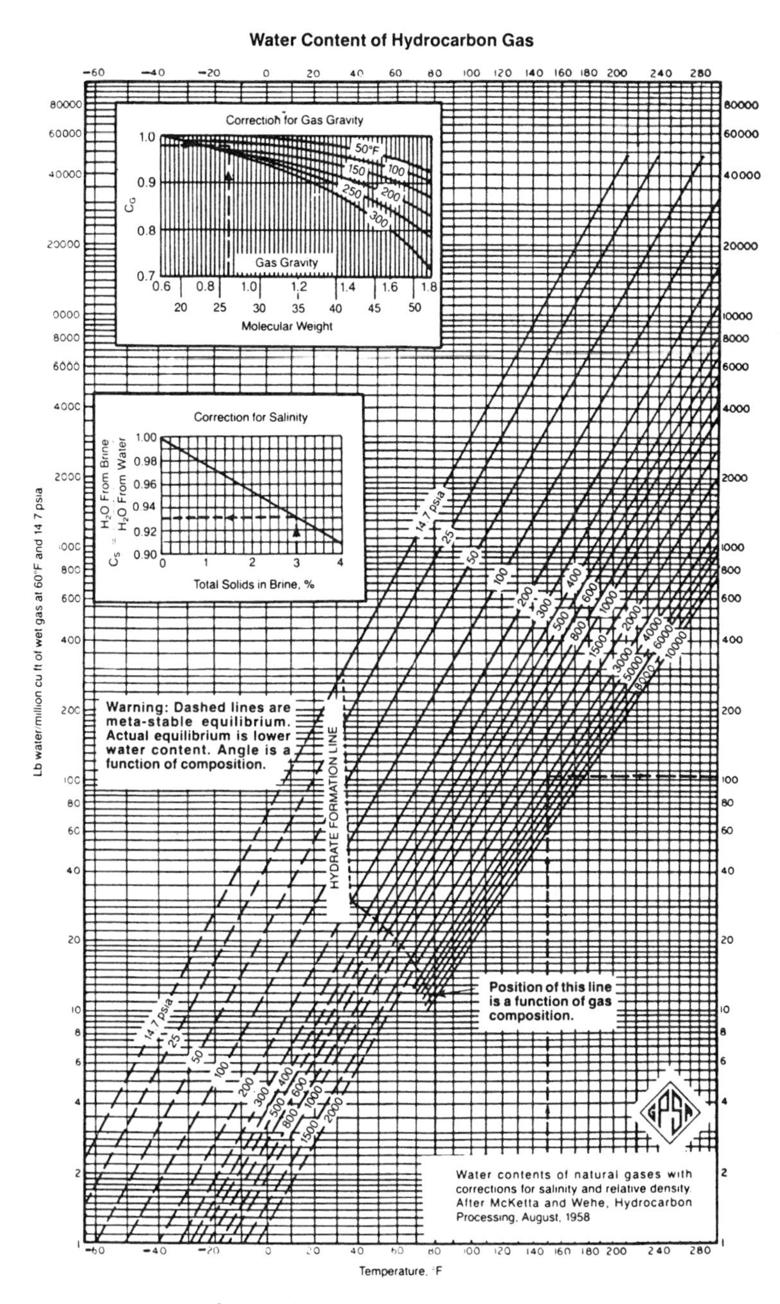

Figure 4-6. Dew point of natural gas. (From Gas Processors Suppliers Association, *Engineering Data Book,* 10th Edition.)

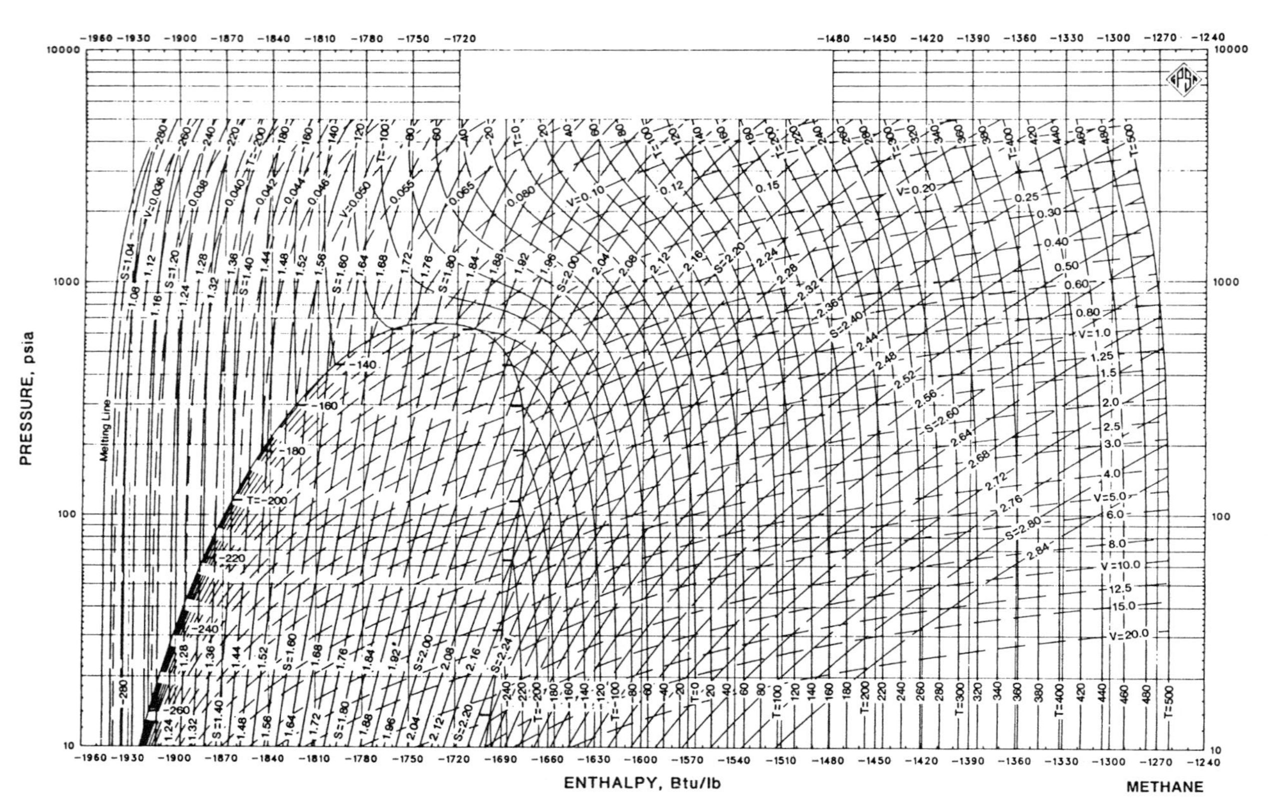

Figure 4-7. Typical Mollier diagram. (From Gas Processors Suppliers Association, *Engineering Data Book*, 10th Edition.)

Natural gas is not a single component and a Mollier diagram will probably not be available.

Figure 4-8 can be used to get a quick approximate solution for the temperature drop of a natural gas stream. For example, if the initial pressure is 4,000 psi and the final pressure is 1,000 psi, ΔP is 3,000 psi and the change in temperature is 80°F. This curve is based on a liquid concentration of 20 bbl/MMscf. The greater the amount of liquid in the gas the lower the temperature drop, that is, the higher the calculated final temperature. For each increment of 10 bbl/MMscf there is a correction of 5°F. For example, if there is no liquid, the final temperature is 10°F cooler (the temperature drop is 10°F more) than indicated by Figure 4-8.

Another technique that can be used to account for the presence of liquids is to assume that the water and oil in the stream pass through the choke with no phase change or loss of temperature. The gas is assumed to cool to a temperature given in Figure 4-8. The heat capacity of the liquids is then used to heat the gas to determine a new equilibrium temperature.

THERMODYNAMIC INHIBITORS

1. Chemicals can be injected into the production stream to depress the likelihood of significant hydrate formation.
2. A thermodynamic inhibitor alters the chemical potential of the hydrate phase such that the hydrate formation point is displaced to a lower temperature and/or a higher pressure.
3. Generally, an alcohol or one of the glycols—usually methanol, ethylene glycol (EG), or diethylene glycol (DEG)—is injected as an inhibitor. All may be recovered and recirculated, but the economics of methanol recovery will not be favorable in most cases.

The most common inhibitor in field gas situations, where the inhibitor will not be recovered and reused, is methanol. It is a relatively inexpensive inhibitor. Methanol is soluble in liquid hydrocarbons, about 0.5% by weight. If there is condensate in the stream, additional methanol is required because some of that methanol will dissolve in the condensate. Also, some of the methanol vaporizes and goes into the gas state.

Ethylene glycol is the most common recoverable inhibitor. It is less soluble in hydrocarbons and has less vaporization loss than methanol. This is common on the inlet to gas processing plants.

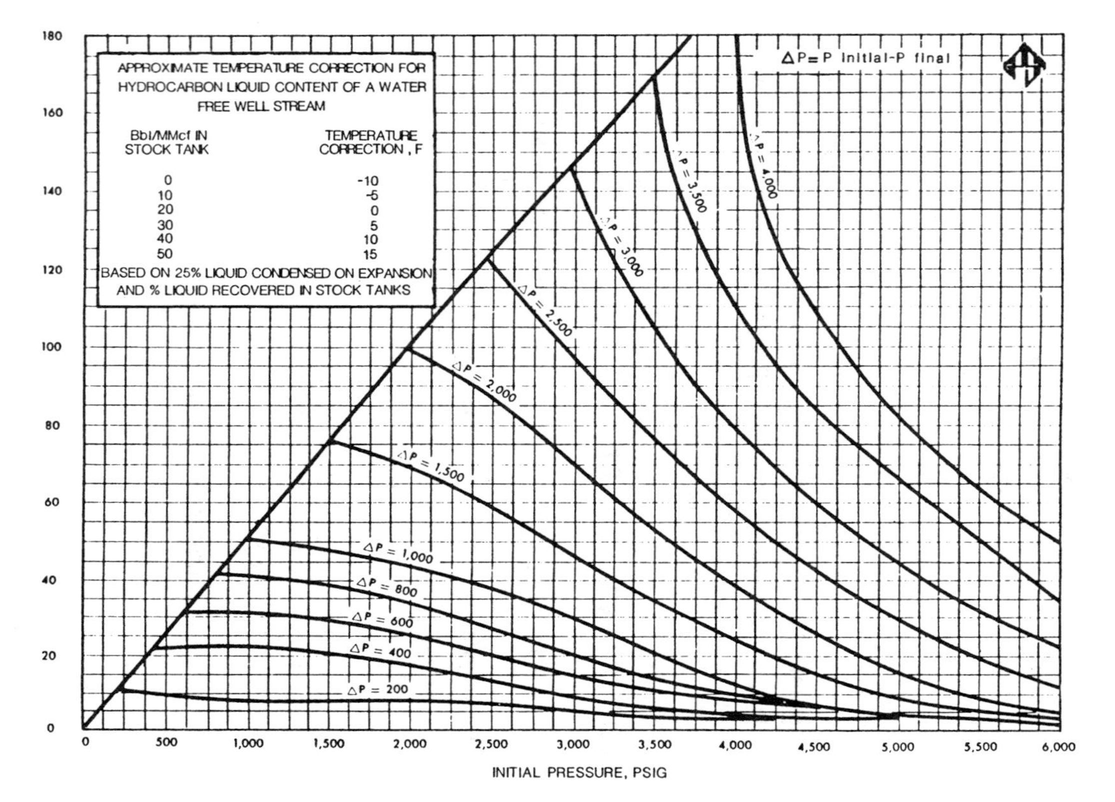

Figure 4-8. Temperature drop accompanying a given pressure drop for natural gas wellstream. (From Gas Processors Suppliers Association, *Engineering Data Book*, 9th Edition.)

The Hammerschmidt equation may be used in calculating the amount of inhibitor required in the water phase to lower the hydrate temperature:

$$\Delta T = \frac{KW}{100(MW) - (MW)W} \qquad (4\text{-}3)$$

where ΔT = depression of hydrate formation temperature, °F
MW = molecular weight of inhibitor (Table 4-2)
K = constant (Table 4-2)
W = weight percent of inhibitor in final water phase

The amount of inhibitor required to treat the free water, as given by Equation 4-3, plus the amount of inhibitor lost to the vapor phase and the amount that is soluble in the hydrocarbon liquid will be the total amount required. Figure 4-9 is a chart for determining the amount of methanol that will be lost to the vapor phase. Approximately 0.5% will be soluble in the hydrocarbon liquid.

The procedure for calculating methanol usage can best be explained by an example. Given a flowing temperature for one well of our example field of 65°F (as could occur with a remote well and subsea flow line), calculate the methanol required to prevent hydrates from forming. Assume that at the high flowing pressure there is no free water, but the gas is saturated.

First the amount of water that will be condensed will be determined from Figure 4-6, assuming the gas is saturated at reservoir conditions.

Table 4-2
Inhibitors and Constants

Inhibitor	MW	K
Methanol	32.04	2,335
Ethanol	46.07	2,335
Isopropanol	60.10	2,335
Ethylene glycol	62.07	2,200
Propylene glycol	76.10	3,590
Diethylene glycol	106.10	4,370

Water content (8,000 psig and 224°F) = 230 lb/MMscf
Water content (4,000 psig and 65°F) = 10 lb/MMscf
Water condensed = 220 lb/MMscf

From Table 4-1 the hydrate temperature is 74°F. The required dew-point depression then will be 74–65 = 9°F.

The concentration of methanol required in the liquid phase from Equation 4-3 is:

$$9 = \frac{2,335W}{100 \times 32 - 32W}$$

$$W = 11.0\%$$

Desired methanol in liquid phase = 0.11 × 220 lb/MMscf = 24.2 lb/MMscf

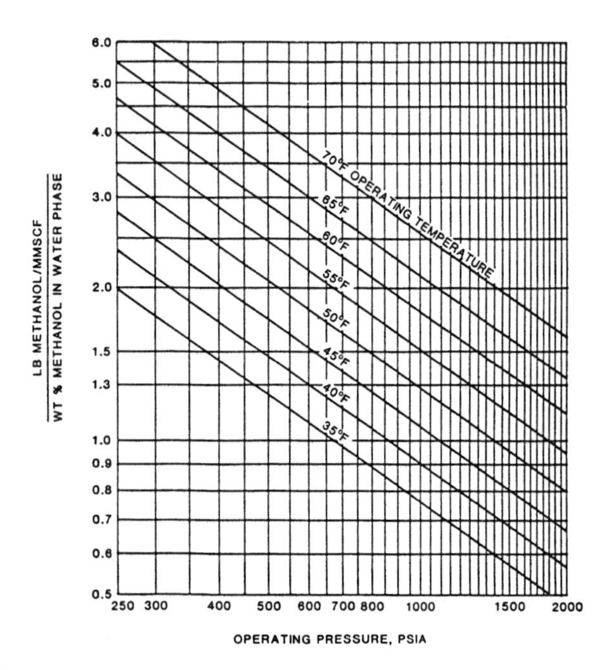

Figure 4-9. Ratio of methanol vapor to liquid composition vs. pressure at various temperatures.

From Figure 4-9 the methanol that will flash to the vapor at 4,000 psig and 65°F is:

$$\frac{\text{lb methanol / MMscf}}{\text{wt. \% methanol in water phase}} \cong 1.0$$

Methanol in vapor state = 1.0 × 11.0 = 11.0 lb/MMscf

Since a barrel of hydrocarbon weighs about 300 lb, the amount of methanol soluble in the hydrocarbon liquid phase will be:

0.005 × 300 lb/bbl × 60 bbl/MMscf = 90.0 lb/MMscf

Thus, the total amount of methanol required:

Water liquid phase	—	24.2 lb/MMscf
Vapor phase	—	11.0 lb/MMscf
Hydrocarbon liquid phase	—	90.0 lb/MMscf
		125.2 lb/MMscf

Note that the amount of methanol soluble in the condensate is crucial to determining the amount needed. Approximately 125 lb of methanol must be added so that approximately 25 lb will be dissolved in the water phase.

Since the specific gravity of methanol is 0.8, this is equivalent to:

$$\frac{125.2 \text{ lb/MMscf}}{0.8 \times 8.33 \text{ lb/gal}} = 18.8 \text{ gal/MMscf}$$

For this case it is impractical to lower the dew point with methanol. A more practical solution would be to separate the condensate first. At 1,000 psia the dew point is 68.4°F (see example in Chapter 5). Assuming a separator temperature of 75°F the amount of methanol needed to lower the dew point of gas to 65°F is 9.7 lb/MMscf. Using a surge factor of 1.4, the required injection rate is only 13.6 lb/MMscf or 2.0 gal/MMscf.

KINETIC INHIBITORS AND ANTI-AGGLOMERATORS

A kinetic inhibitor is a polymeric chemical that, when added to a production stream, will not change the hydrate formation temperature but will delay the growth of hydrate crystals. These chemicals are polymeric and include N-vinylpyrrolidone (5 ring), saccharides (6 ring), and N-vinylcaprolactam (7 ring).

An anti-agglomerator is an alkyl aromatic sulphonate, a quaternary ammonium salt, or an alkyl glycoside surfactant. When added to a production stream with a continuous oil phase, this chemical will minimize hydrate crystals from agglomerating or growing in size. The continuous oil phase provides a medium to transport the hydrate crystals through the piping system while crystal growth is delayed. This chemical and its physical reaction with hydrate crystals is not dependent upon the amount of subcooling and therefore has a wide range of pressure-temperature applications.

The use of kinetic inhibitors and/or anti-agglomerators in actual field operations is a new and evolving technology. These are various formulations of chemicals that can be used in a mixture of one or more kinetic inhibitors and/or anti-agglomerators. At the current time, to get an "optimum" mixture for a specific application it is necessary to set up a controlled bench test using the actual fluids to be inhibited and determine the resulting equilibrium phase line. As the mixture of chemicals is changed, a family of equilibrium phase lines will develop. This will result in an initial determination of a near "optimum" mixture of chemicals.

To determine the appropriate injection rate, a field test should first be performed at one of the industry-sponsored full-scale loop test facilities. The "optimum" mixture, its injection rate, and location of injection points will be a function of flow geometry, fluid properties, pressure-temperature relationships, etc., that will be encountered in the actual field application. The appropriate injection rate and location of injection points can be determined from this test by observing pressure increases, which indicate that hydrate plugs are forming.

Application of this "optimum mixture of chemicals" in the actual field installation would begin with the design injection rate of the mixture and by adding a sufficient concentration of a thermodynamic inhibitor to inhibit the free water that may exist in low spots in the piping system. After the system has reached an operating equilibrium, the volume of the thermodynamic inhibitor is then decreased in stages with a time period between each stage to ensure that no hydrate plug is formed. Some small amount of thermodynamic inhibitor will continue to be used to inhibit free water in low spots. In this manner the lowest chemical cost that will provide the necessary flow assurance is eventually reached.

CHAPTER

5

*LTX Units and Line Heaters**

Chapter 4 discussed the need to prevent hydrates, the techniques necessary to predict hydrate formation, and the use of chemical inhibitors. This chapter covers two types of equipment for handling hydrates.

Low-temperature exchange (LTX) units use the high flowing temperature of the well stream to melt the hydrates after they are formed. Since they operate at low temperatures, they also stabilize the condensate and recover more of the intermediate hydrocarbon components than would be recovered in a straight multistage flash separation process.

Indirect fired heaters (sometimes called line heaters) heat the gas stream before and/or after the choke so that the gas is maintained above the hydrate temperature. Indirect fired heaters can also be used to heat crude oil for treating, heat a hot fluid circulating medium (heat medium) that is used to provide process heat, etc.

There are many other types of heat exchanger devices that can be used to heat the gas above the hydrate temperature. These could include shell and tube heat exchangers, electrical immersion heaters, furnaces, etc. However, the most common equipment type used to heat a well stream is the indirect fired water bath heater.

---

*Reviewed for the 1999 edition by Kevin R. Mara of Paragon Engineering Services, Inc.

LTX UNITS

These units are designed to allow hydrates to form and to melt them with the heat of the incoming gas stream before they can plug downstream equipment. In addition, the low-temperature separation that occurs in an LTX unit results in stabilizing the liquids as discussed in Chapter 6. This results in an increase in liquids recovered and a corresponding decrease in the heating value of the gas over what would be the case with separation at normal temperatures.

Figure 5-1 shows a typical LTX process. The inlet gas stream is choked at the well to 2,000 to 3,000 psi or until the temperature declines to approximately 120°F, which is well above the hydrate formation temperature. The inlet stream next enters a coil in the bottom of the low temperature separator. The stream is then cooled to just above the hydrate formation temperature with the outlet gas coming off the low temperature separator. This assures the lowest possible temperature for the inlet stream when it enters the vessel after the choke. This choke is mounted in the vessel itself. When the pressure drop is taken, the temperature will

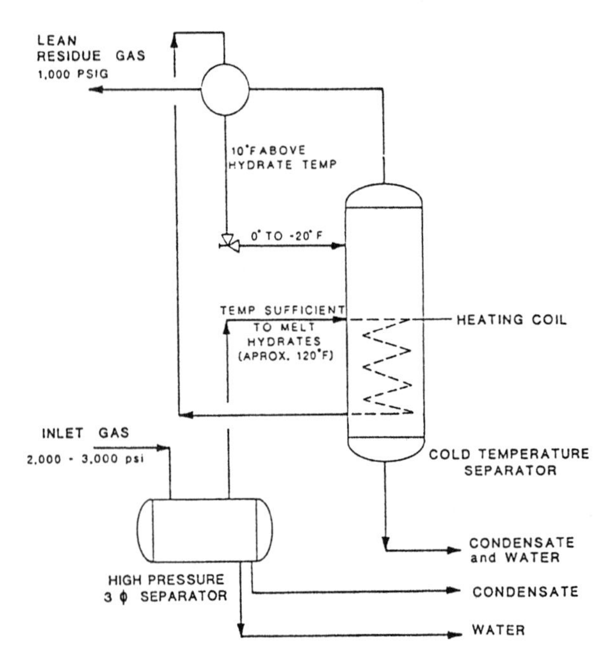

Figure 5-1. Typical low-temperature exchange unit (LTX).

decrease to well below the hydrate point. Hydrates form, but they fall into the bottom of the separator and are melted by the heating coil. The hydrates do not plug the choke because the choke is inside the separator.

The gas, condensate, and free water are then discharged from the vessel through backpressure and liquid dump valves. The gas leaving the separator is saturated with water vapor at the temperature and pressure of the top of the low temperature separator. If this temperature is low enough, the gas may be sufficiently dehydrated to meet sales specifications. Dehydration is discussed in greater detail in Chapter 8.

The low-temperature separator acts as a cold feed condensate stabilizer. A natural cold reflux action exists between the rising warmed gases liberated from the liquid phase and cold condensed liquid falling from the stream inlet. The lighter hydrocarbons rejoin the departing gas stream and the heavier components recondense and are drawn from the vessel as a stable stock tank product. This process is discussed in more detail in Chapter 6. The colder the temperature of the gas entering the separator downstream of the choke, the more intermediate hydrocarbons will be recovered as liquid. The hotter the gas in the heating coil, the less methane and ethane there will be in the condensate, and the lower its vapor pressure. In some cases, it may be necessary to heat the inlet gas stream upstream of the coil, or provide supplemental heating to the liquid to lower the vapor pressure of the liquid.

In summary, a colder separation temperature removes more liquid from the gas stream; adequate bottom heating melts the hydrates and revaporizes the lighter components so they may rejoin the sales gas instead of remaining in liquid form to be flashed off at lower pressure; and cold refluxing recondenses the heavy components that may also have been vaporized in the warming process and prevents their loss to the gas stream.

LTX units are not as popular as they once were. The process is difficult to control, as it is dependent on the well flowing-tubing pressure and flowing-tubing temperature. If it is being used to increase liquid recovery, as the flowing temperature and pressure change with time, controls have to be reset to assure that the inlet is cold enough and the coil hot enough. If the coil is not hot enough, it is possible to destabilize the condensate by increasing the fraction of light components in the liquid stream. This will lower the partial pressure of the intermediate components in the stock tank and more of them will flash to vapor. If the inlet stream is not cold enough, more of the intermediate components will be lost to the gas stream.

From a hydrate melting standpoint it is possible in the winter time to have too cold a liquid temperature and thus plug the liquid outlet of the low temperature separator. It is easier for field personnel to understand and operate a line heater for hydrate control and a multistage flash or condensate stabilizer system to maximize liquids recovery.

LINE HEATERS

As shown in Figure 5-2, the wellstream enters the first coil at its flowing-tubing temperature and pressure. Alternatively, it could be choked at the wellhead to a lower pressure, as long as its temperature remains above hydrate temperature.

There is typically a high-pressure coil of length L_1, which heats the wellstream to temperature, T_1. The wellstream at this point is at the same pressure as the inlet pressure, that is $P_1 = P_{in}$. The wellstream is choked and pressure drops to P_2. When the pressure drops there is a cooling effect and the wellstream temperature decreases to T_2. This temperature is usually below the hydrate temperature at P_2. Hydrates begin to form, but are melted as the wellstream is heated in the lower pressure coil of length L_2. This coil is long enough so that the outlet temperature is above the hydrate point at pressure, P_2. Typically, a safety factor of 10°F higher than the hydrate temperature is used to set T_{out}.

In fire tube type heaters, the coils are immersed in a bath of water. The water is heated by a fire tube that is in the bath below the coils. That is, the fire tube provides a heat flux that heats the water bath. The water bath

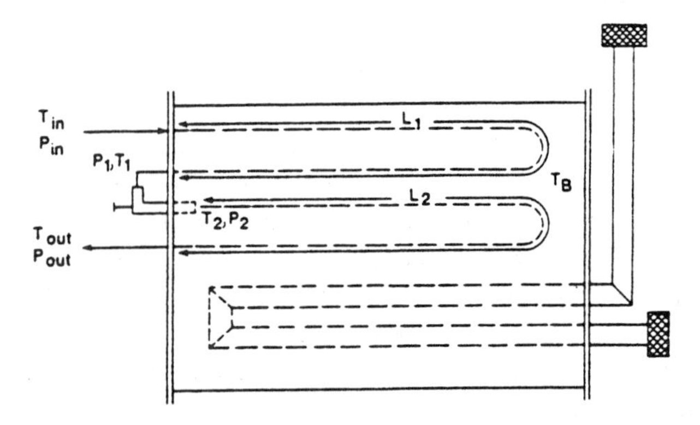

Figure 5-2. Schematic of line heater.

exchanges heat by convection and conduction to the process fluid. Instead of a fire tube, it is possible to use engine exhaust or electrical immersion heaters to heat the water bath. Fire tubes are by far the most common source of heat.

Since the bath fluid is normally water, it is desirable to limit the bath temperature to 190°F to 200°F to avoid evaporating the water. If higher bath temperatures are needed, glycol can be added to the water.

In order to adequately describe the size of a heater, the heat duty, the size of the fire tubes, the coil diameters and wall thicknesses, and the coil lengths must be specified. To determine the heat duty required, the maximum amounts of gas, water, and oil or condensate expected in the heater and the pressures and temperatures of the heater inlet and outlet must be known. Since the purpose of the heater is to prevent hydrates from forming downstream of the heater, the outlet temperature will depend on the hydrate formation temperature of the gas. The coil size of a heater depends on the volume of fluid flowing through the coil and the required heat duty.

Special operating conditions such as start up of a shut-in well must be considered in sizing the heater. The temperature and pressure conditions found in a shut-in well may require additional heater capacity over the steady state requirements. It may be necessary to temporarily install a heater until the flowing wellhead temperature increases as the hot reservoir fluids heat up the tubing, casing, and surrounding material.

It is perfectly acceptable for a line heater to have an L_1 equal to 0. In this case all the heat is added downstream of the choke. It is also possible to have L_2 equal to 0 and do all the heating before the choke. Most frequently it is found that it is better to do some of the heating before the choke, take the pressure drop, and do the rest of the heating at the lower temperature that exists downstream of the choke.

HEAT DUTY

To calculate the heat duty it must be remembered that the pressure drop through the choke is instantaneous. That is, no heat is absorbed or lost, but there is a temperature change. This is an adiabatic expansion of the gas with no change in enthalpy. Flow through the coils is a constant pressure process, except for the small amount of pressure drop due to friction. Thus, the change in enthalpy of the gas is equal to the heat absorbed.

The heat duty is best calculated with a process simulation program. This will account for phase changes as the fluid passes through the choke. It will balance the enthalpies and accurately predict the change in temperature across the choke. Heat duty should be checked for various combinations of inlet temperature, pressure, flow rate, and outlet temperature and pressure, so as to determine the most critical combination.

The heat duty can be approximately calculated using the techniques described in Chapter 2 once the required change in temperature is known. The change in temperature due to pressure drop through the choke can be approximated from Figure 4-8. The hydrate temperature can be calculated as described in Chapter 4, and the outlet wellstream temperature selected at approximately 10 °F above the hydrate temperature. The total temperature change for calculating gas, oil and water heat duties is then:

$$\Delta T = T_{out} - T_{in} + \Delta T \text{ (due to pressure drop)} \tag{5-1}$$

Recalling from Chapter 2, the general heat duty for multi-phase streams is expressed as:

$$q = q_g + q_o + q_w + q_l \tag{5-2}$$

where q = overall heat duty required, Btu/hr
q_g = gas heat duty, Btu/hr
q_o = oil heat duty, Btu/hr
q_w = water heat duty, Btu/hr
q_l = heat loss to the atmosphere, Btu/hr

The amount of heat required to heat the gas produced from the wellstream is calculated using the following equation:

$$q_g = 41.7\,(T_2 - T_1)C_gQ_g \tag{5-3}$$

where q_g = gas heat duty, Btu/hr
C_g = gas heat capacity, Btu/Mscf °F
Q_g = gas flow rate, MMscfd
T_1 = inlet temperature, °F
T_2 = outlet temperature, °F

The amount of heat required to heat any condensate or oil produced with the gas is calculated using the following equation:

$$q_o = 14.6\ SG(T_2 - T_1)CQ_o \tag{5-4}$$

where q_o = oil heat, Btu/hr
C = oil specific heat, Btu/lb °F (Figure 2-13)
Q_o = oil flow rate, bbl/day
SG = oil specific gravity
T_1 = inlet temperature, °F
T_2 = outlet temperature, °F

The amount of heat required to heat any free water produced with the gas is calculated using the following equation:

$$q_w = 14.6\ (T_2 - T_1)Q_w \tag{5-5}$$

where q_w = water heat duty, Btu/hr
Q_w = water flow rate, bbl/day
T_1 = inlet temperature, °F
T_2 = outlet temperature, °F

Heat loss varies greatly with weather conditions and is usually the greatest in heavy rain and extreme cold. As an approximation it can be assumed that the heat lost from the heater to the atmosphere is less than 10% of the process heat duty. Therefore:

$$q_l = 0.10\ (q_g + q_o + q_w) \tag{5-6}$$

The heat duty may have to be checked for various combinations of inlet temperature and pressure, flow rate, and outlet temperature and pressure to determine the most critical combinations.

FIRE-TUBE SIZE

The area of the fire tube is normally calculated based on a heat flux rate of 10,000 Btu/hr-ft^2. The fire-tube length can be determined from:

$$L = 3.8 \times 10^{-4}\ \frac{q}{d} \tag{5-7}$$

where L = fire tube length, ft
q = total heat duty, Btu/hr
d = fire tube diameter, in.

A burner must be chosen from the standard sizes in Table 2-12. For example, if the heat duty is calculated to be 2.3 MMBtu/hr, then a standard 2.5 MMBtu/hr fire tube should be selected.

Any combination of fire tube lengths and diameters that satisfies Equation 5-7 and is larger in diameter than those shown in Table 2-12 will be satisfactory. Manufacturers normally have standard diameters and lengths for different size fire tube ratings.

COIL SIZING

Choose Temperatures

In order to choose the coil length and diameter, a temperature must first be chosen upstream of the choke; the higher T_1, the longer the coil L_1 and the shorter the coil L_2. In Chapter 2 we showed that the greater the LMTD between the gas and the bath temperature, the greater the heat transfer per unit area, that is, the greater the LMTD, the smaller the coil surface area needed for the same heat transfer. The bath temperature is constant, and the gas will be coldest downstream of the choke. Therefore, the shortest total coil length ($L_1 + L_2$) will occur when L_1 is as small as possible (that is, T_1 is as low as possible).

Although the total coil length is always smaller when there is no upstream coil ($L_1 = 0$), the temperature could be so low at the outlet of the choke under these conditions that hydrates will form quickly and will partially plug the choke. In addition, the steel temperature in the choke body may become so cold that special steels are required. Therefore, some guidelines are necessary to choose T_1 for an economical design.

It is preferable to keep T_2 above 50°F to minimize plugging and above –20°F to avoid more costly steel. With this in mind the following guidelines have proven useful. For a water bath temperature of 190°F:

1. Set T_2 = 50°F. Solve for ΔT and calculate T_1. If T_1 is greater than 130°F, L_1 will become long. Consider going to the next step.
2. Set T_1 = 130°F. Solve for T_2. If T_2 is less than –20°F special steel will be needed. Consider lengthening L_1 instead and go to the next step.

3. Set $T_2 = -20°F$. Increase T_1 as needed.
4. If L_1 becomes too long, consider using glycol/water mixture or another heat medium liquid and raise the bath temperature above 190°F.

Choose Coil Diameter

Volume 1, Chapter 9 explains the criteria for choosing a diameter and wall thickness of pipe. This procedure can be applied to choosing a coil diameter in an indirect fired heater. Erosional flow criteria will almost always govern in choosing the diameter. Sometimes it is necessary to check for pressure drop in the coil. Typically, pressure drop will not be important since the whole purpose of the line heater is to allow a large pressure drop that must be taken. The allowable erosional velocity is given by:

$$V_e = \frac{c}{(\rho_m)^{1/2}} \tag{5-8}$$

where V_e = fluid erosional velocity, ft/sec
c = empirical constant (dimensionless); 125 for intermittent service, 100 for continuous service
ρ_m = fluid density at flowing temperature and pressure, lb/ft^3

The fluid density must be for the combined stream of oil and gas and should be calculated at the average gas temperature.

$$\rho_m = \frac{12{,}409(SG)P + 2.7RSP}{198.7P + RTZ} \tag{5-9}$$

where (SG) = specific gravity of liquid relative to water
P = operating pressure, psia
R = gas/liquid ratio, ft^3/bbl
S = specific gravity of gas at standard conditions
T = operating temperature, °R
Z = gas compressibility factor (from Volume 1, Chapter 3)

The required pipe internal diameter can be calculated based on the volumetric flow rate and a maximum velocity. The maximum velocity may be the erosional velocity or a limiting value based on noise or inability to use corrosion inhibitors. In gas lines it is recommended that the maximum allowable velocity would be 60 ft/sec, 50 ft/sec if CO_2 is present, or the erosional velocity, whichever is lower. (Please note that API Spec 12 K *Indirect Type Oil Field Heaters* uses 80 ft/sec as a limit.) In liquid lines a maximum velocity of 15 ft/sec should be used. A minimum velocity of 3 ft/sec should also be considered to keep liquids moving and to keep sand or other solids from settling and becoming a plugging problem.

The equation used for determining the pipe diameter is:

$$d = \left[\frac{\left(11.9 + \frac{Z\,R\,T}{16.7\,P} \right) Q_l}{1000\,V} \right]^{1/2} \tag{5-10}$$

where d = pipe inside diameter, in.
Z = gas compressibility factor
R = gas/liquid ratio, ft^3/bbl
T = operating temperature, °R
P = pressure, psia
Q_l = liquid flow rate, bbl/day
V = maximum allowable velocity, ft/sec

Choose Wall Thickness

Before choosing a wall thickness it is necessary to choose a pressure rating for the coil. Typically, the high-pressure coil (L_1) is rated for the shut-in pressure of the well, and the low-pressure coil (L_2) is rated for the maximum working pressure of the downstream equipment. There are many exceptions to this rule and reasons to deviate from it. If designing L_1 to withstand the well shut-in tubing pressure is too costly, it is common practice to design the coil above the normal operating pressure of the flow line and install a relief valve set at the maximum allowable operating pressure of the coil. If flow is accidentally shut-in by a hydrate plug or other blockage at the choke, L_1 could be subjected to total wellhead shut-in pressure unless it is protected by a relief valve.

The wall thickness of the coil can be chosen by using any number of recognized codes and standards. In the United States, the most commonly recognized are American National Standard Institute (ANSI) B31.3 and B31.8, or American Petroleum Institute (API) Specification 12 K. Volume 1 has the tables for ANSI B31.3 and ANSI B31.8. Table 5-1 illustrates the ratings from API Spec 12 K, which uses the calculation procedure from ANSI B31.3, but assumes no corrosion allowance.

After the minimum inside diameter and the required wall thickness, a coil diameter and wall thickness may be selected. Very often, the coil length downstream of the choke (L_2) is of a different diameter and wall thickness than the length upstream of the choke (L_1).

Coil Lengths

With the known temperatures on each end of the coil, the heat duty for each coil can be calculated from the heat transfer theory in Chapter 2. Since the bath is at a constant temperature, LMTD can be calculated as:

Table 5-1
Maximum Coil Working Pressure from API 12K

Nominal Pipe Size in.	Nominal Wall Thickness in.	Maximum Working Pressure* Grade B S = 20,000 psig	Maximum Working Pressure* Grade C S = 23,300 psig
1 XS	0.179	5,270	—
2 Std	0.154	2,380	—
2 XS	0.218	3,440	—
2 XXS	0.436	7,340	8,560
2½ Std	0.203	2,600	—
2½ XS	0.276	3,610	—
2½ XXS	0.552	7,770	9,050
2½	0.750	10,720	12,490
2½	0.875	12,530	14,600
3 Std	0.216	2,260	—
3 XS	0.300	3,200	—
3 XXS	0.600	6,820	7,940
4 Std	0.237	1,920	—
4 XS	0.337	2,770	—
4 XXS	0.674	5,860	6,830
6 Std	0.280	1,530	—
6 XS	0.432	2,400	—
6 XXS	0.864	5,030	5,860
8 Std	0.322	1,350	—
8 XS	0.500	2,120	—
8 XXS	0.875	3,830	4,460

**Maximum working pressure has been rounded up to the next higher unit of 10 psig. No corrosion allowance is assumed; same formula as ANSI B31.3*

$$\text{LMTD} = \frac{\Delta T_1 - \Delta T_2}{\log_e(\Delta T_1 / \Delta T_2)} \tag{5-11}$$

where ΔT_1 = temperature difference between coil inlet and bath
ΔT_2 = temperature difference between coil outlet and bath

The overall film coefficient, U, for the coil can be calculated or read from the charts and tables in Chapter 2. Since U, LMTD, q, and the diameter of the pipe are known, the length of the coil can be solved from the following equation:

$$L = \frac{12q}{\pi(\text{LMTD})Ud} \tag{5-12}$$

where d = coil outside diameter, in.

Equation 5-12 describes an overall length required for the coil. Since the shell length of the heater will probably be much less, several passes of the coil through the length of the shell may be required, as shown in Figure 5-3, to develop this length. For a given shell diameter there is a limit to the number of passes of coil. Therefore, the correct selection of coil length also requires determining the length of the shell and number of passes. As the shell length decreases, the number of passes increases, and a larger shell diameter is required.

For a given shell length the number of passes for each coil can be determined. Since the number of passes both upstream and downstream of the choke must be an even integer, actual T_1 and T_2 may differ slightly from that assumed in the calculation. The actual values of T_1 and T_2 can be calculated from actual coil lengths L_1 and L_2.

Once the total number of passes is known, the coil can be laid out geometrically assuming that the center-line minimum radius of bends is 1½ times nominal pipe size. The required shell diameter is then determined.

Other selections of shell length, number of passes and required diameters can then be made to obtain an optimum solution.

STANDARD SIZE LINE HEATERS

The previous procedure is very helpful for reviewing existing designs or proposals from vendors. In most situations, however, it will be eco-

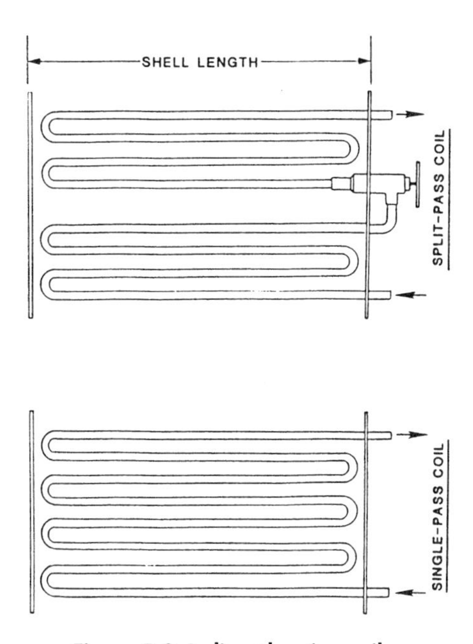

Figure 5-3. Indirect heating coils.

nomically advantageous to select a standard size line heater. Figure 5-4 shows dimensions of standard line heaters available from one manufacturer. It is possible to mix and match coil sizes. In Figure 5-4 a standard 250 MBtu/hr line heater is available with either eight 2-in. XH coils or eight 2-in. XXH coils. For a given situation, it may be necessary to use six 3-in. XXH coils instead.

Other Uses for Indirect Fired Heaters

The previous discussion focused on the use of indirect fired heaters as line heaters to provide the necessary heat to avoid hydrate formation at wellstream chokes. Indirect fired heaters have many other potential uses in production facilities. For example, indirect fired heaters can be used to provide heat to emulsions prior to treating, as reboilers on distillation towers, and to heat liquids that are circulated to several heat users. The sizing of indirect fired heaters for these uses relies on the same principles and techniques discussed for wellstream line heaters.

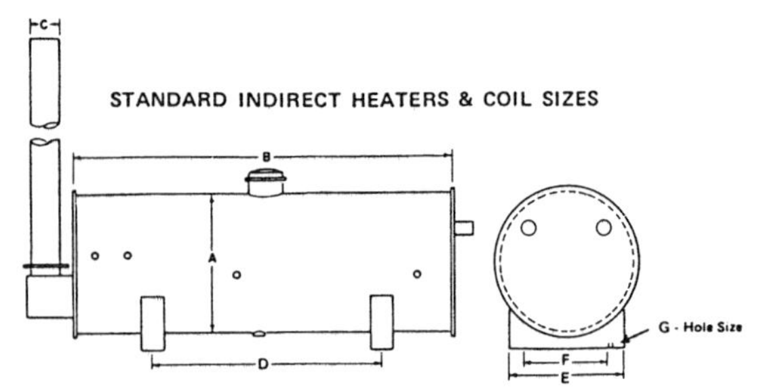

NOMINAL DIMENSIONAL DATA

Heater	A	B	C	D	E	F	G
BTU HR	Ft. In.	Ft. In.	Ft. In.	Ft. In.	Ft. In.	Ft. In.	In. only
250,000	2'-0"	7'-6"	0'-8"	5'-6"	1'-0"	1'-5"	11/16*
500,000	2'-6"	10'-0"	0'-10"	6'-0"	1'-9"	1'-5"	11/16"
750,000	3'-0"	12'-0"	0'-12"	6'-0"	2'-2"	1'-10"	11/16"
1,000,000	3'-6"	14'-4"	1'-2"	11'-0"	3'-0"	2'-4"	3/4"
1,500,000	4'-0"	17'-6"	1'-4"	12'-6"	3'-6"	3'-0"	3/4"
2,000,000	5'-0"	20'-0"	1'-8"	12'-6"	4'-4"	3'-0"	7/8"

SPECIFICATIONS

Heater Furnace Input BTU/HR	Shell Size O.D. x Lgt.	Std. No. & Size Tubes	Coil W.P. PSI	Std. Mean Coil Area Sq. Ft.	Approx. Coil Lin. Ft.	Water Fill Vol: Bbls.	Shipping Weight Pounds
250,000	24" x 7'6"	8-2"XH	3440	29.5	54	2.9	1,400
250,000	24" x 7'6"	8-2"XXH	7340	26.5	54	2.9	1,610
500,000	30" x 10'0"	8-2"XH	3440	42.6	76	6.0	2,210
500,000	30" x 10'0"	8-2"XXH	7340	38.3	76	6.0	2,510
750,000	36" x 12'0"	10-2"XH	3440	64.4	114	10.5	2,875
750,000	36" x 12'0"	10-2"XXH	7340	58.0	114	10.5	3,325
750,000	36" x 12'0"	6-3"XH	3200	59.4	70.9	10.3	3,030
750,000	36" x 12'0"	6-3"XXH	6820	53.8	70.9	10.3	3,615
1,000,000	42" x 14'4"	12-2"XH	3440	93.4	166	17.9	4,060
1,000,000	42" x 14'4"	12-2"XXH	7340	85.9	166	17.9	4,725
1,000,000	42" x 14'4"	8-3"XH	3200	94.8	113.2	17.5	4,390
1,000,000	42" x 14'4"	8-3"XXH	6820	85.9	113.2	17.5	5,335
1,500,000	48" x 17'6"	14-2"XH	3440	134.0	237	28.7	5,650
1,500,000	48" x 17'6"	14-2"XXH	7340	120.5	237	28.7	6,600
1,500,000	48" x 17'6"	10-3"XH	3200	145.0	173.1	28.0	6,235
1,500,000	48" x 17'6"	10-3"XXH	6820	131.4	173.1	28.0	7,675
2,000,000	60" x 20'0"	16-2"XH	3440	175.7	311	51.8	10,110
2,000,000	60" x 20'0"	16-2"XXH	7340	158.0	311	51.8	11,360
2,000,000	60" x 20'0"	10-3"XH	3200	165.9	198.1	51.2	10,580
2,000,000	60" x 20'0"	10-3"XXH	6820	150.4	198.1	51.2	12,240

*Subject to change without notice. Other combinations are available.

Figure 5-4. Dimensions of standard line heaters. (From Smith Industries, Inc.)

LINE HEATER DESIGN EXAMPLE PROBLEM

Design a line heater for each of the 10 wells that make up the total 100 MMscfd field rate. That is, each well flows at 10 MMscfd.

Determine:

1. Temperature for hydrate formation at 1,000 psia.
2. Heat duty for a single pass coil downstream of choke.

3. Coil length for a 3-in. XX coil.
   a. Calculate LMTD
   b. Calculate U
   c. Choose the coil length
4. Fire tube area required and heater size (shell diameter, shell length, fire tube rating, coil length and number of passes).

Solution:

1. Determine the temperature for hydrate formation at 1,000 psia
   a. From equilibrium values

	Mole Fraction	K_{v-s} Values at 1,000 psia 50°F	70°F
N_2	0.0144	—	—
CO_2	0.0403	0.60	—
H_2S	0.000019	0.07	0.38
C_1	0.8555	1.04	1.26
C_2	0.0574	0.145	1.25
C_3	0.0179	0.03	0.70
iC_4	0.0041	0.013	0.21
nC_4	0.0041	0.145	1.25
iC_5	0.0020	—	—
nC_5	0.0013	—	—
C_6	0.0015	—	—
C_7^+	0.0015	—	—
	1.0000	$\text{Sum}\left(\frac{Y}{K_{v-s}}\right) = 2.226$	$\text{Sum}\left(\frac{Y}{K_{v-s}}\right) = 0.773$

Interpolating linearly, $\text{Sum}\left(\frac{Y}{K_{v-s}}\right) = 1.0$ at 66.9°F

   b. From Figure 4-5
      Specific gravity from Table 2-10 is 0.67
      At 0.6 gravity hydrate temperature is 60°F
      At 0.7 gravity hydrate temperature is 64°F
      By interpolation, hydrate temperature at S = 0.67 is 62.8°F

2. Determine the process heat duty
   Temperature at outlet of heater should be about 5 to 15°F above hydrate temperature. Choose temperature at heater outlet as 75°F.

a. Temperature drop through choke

Flow tubing pressure, psig	4,000
Heater inlet pressure, psig	1,000
Pressure drop through choke, psia	3,000
Temperature drop from Figure 4-8	79°F

Gas has 60.0 bbl/MMscf condensate for which temperature correction is 20°F.

Corrected temperature drop = 79 – 20 = 59°F
Heater inlet temperature = 120 – 59 = 61°F

b. Gas duty

Flowing pressure, P, psia	1,015
P_c, psia (Table 2-10)	680
Reduced pressure, $P_R = P/P_c$	1.49
Heater inlet temperature, °F	61
Heater outlet temperature, °F	75
Average temperature, °F	68
Average temperature, °R	528
T_c, °R (Table 2-10)	375
$T_R = T/T_c$	1.41

$$q_g = 41.7\ (\Delta T)\ C_g\ Q_g$$

where q_g = gas heat duty, Btu/hr
$\Delta T = T_{out} - T_{in}$

Since flow through coil is a constant pressure process, we have:

$\Delta T = T_{out} - T_{in} = 75 - 61 = 14°F$
Q_g = gas flow, MMscfd = 10
C_g = gas heat capacity, Btu/Mscf-°F

Calculation of C_g

$$C_g = 2.64\ (29 \times S \times C + \Delta C_p)$$

where C = gas specific heat, Btu/lb°F

From Figure 2-14, C at 68°F is = 0.50

ΔC_p from Figure 2-15, (at $T_R = 1.41$, $P_R = 1.49$) = 2.6

$$S = 0.67$$

$$C_g = 2.64\ (29 \times .67 \times .50 + 2.6)$$

$$= 32.51 \text{ Btu/Mscf-°F}$$

$$q_g = 41.7 \times 14 \times 32.51 \times 10$$

$$= \underline{190 \text{ M Btu/hr}}$$

c. Oil duty:

$$q_o = 14.6\ SG\ (T_2 - T_1)\ C_o\ Q_o$$

where $SG = 0.77$

$$Q_o = \text{Oil flow rate, bpd}$$

$$= 60.0 \frac{\text{bbl}}{\text{MMscf}} \times 10 \text{MMscfd}$$

$$= 600 \text{ bpd}$$

$$T_2 - T_1 = 75 - 61 = 14°F$$

$$C_o = \text{oil specific heat, } \frac{\text{Btu}}{\text{lb°F}}$$

From Figure 2-13 at 68°F (52.3°API)

$$C_o = 0.48$$

$$q_o = 14.6 \times 0.77 \times 14 \times 0.48 \times 600$$

$$= \underline{46 \text{ MBtu/hr}}$$

d. Water duty:

$$q_w = 14.6\ (T_2 - T_1) Q_w$$

Gas is saturated with water at 8,000 psig and 224°F. From Figure 4-6, we have:

lb water/MMscf of wet gas at reservoir conditions (8,000 psig and 224°F)	= 260
lb water/MMscf of wet gas at 1,000 psig and 75 °F	= 28
Water to be heated, lb/MMscf	232

Water quantity = Q_w

$$= 232 \frac{\text{lb}}{\text{MMscf}} \times \frac{10\text{ MMscfd}}{62.4\text{ lb}/\text{ft}^3} \times \frac{7.48\text{ gal}}{\text{ft}^3} \times \frac{\text{bbl}}{42\text{ gal}}$$

$$= 6.6\text{ bpd}$$

$$q_w = 14.6 \times (75 - 61) \times 6.6$$

$$= 1.3\text{ MBtu/hr}$$

e. Total process duty

$$q_p = q_g + q_o + q_w$$
$$q_p = 190 + 46 + 1.3 = 237\text{ MBtu/hr}$$

3. Calculation of coil length

a. Calculate LMTD

Temperature of bath is 190°F

$$\Delta T_1 = 190 - 61 = 129$$

$$\Delta T_2 = 190 - 75 = 115$$

$$\text{LMTD} = \frac{14}{\log_e\left(\frac{129}{115}\right)} = 122°\text{F}$$

b. Calculate U

$$\frac{1}{U} = \frac{1}{h_o} + R_o + \frac{L}{k} + R_i + \frac{A_o}{h_i A_i}$$

Use $R_o + R_i = 0.003$ hr-ft^2-°F/Btu

$$h_o = 116\left[\frac{k^3\,C\,\rho^2\beta\,\Delta T}{\mu\,d_o}\right]^{0.25}$$

$k = 0.39$ (Figure 2-10)
$C = 1$ Btu/lb-°F (Figure 2-10)
$\rho = 60.35$ lb/ft^3 (Table 2-6, 1/0.01657)
$\beta = 0.0024$ 1/°F (Table 2-5)
$\mu = 0.32$ cp (Figure 2-10)
$\Delta T = 122$°F
$d_o = 3.5$ in.

$$h_o = 116\left[\frac{(0.39)^3(1)(60.35)^2(0.0024)(122)}{(0.32)(3.5)}\right]^{0.25}$$

$h_o = 318$ Btu/hr ft^2°F

For 3-in. XX Pipe A-106-B

$L = 0.60$ in. $= 0.05$ ft (Table 2-2)

$k = 30$ Btu/hr-ft-°F (Table 2-3)

$A_o = 0.916$ ft^2/ft (Table 2-2)

$A_i = 0.602$ ft^2/ft (Table 2-2)

$$h_i = \frac{0.022k}{D}\left(\frac{DG}{\mu_e}\right)^{0.8}\left(\frac{C\mu_e}{k}\right)^{0.4}\left(\frac{\mu_e}{\mu_{ew}}\right)^{0.16}$$

$D = 2.30$ in. $= 0.192$ ft (Table 2-2)

$$A = \frac{\pi D^2}{4} = 0.0289\ \text{ft}^2$$

$k = 0.017$ Btu/hr-ft-°F $\times$ 1.25 (Figure 2-5)

$= 0.021$ Btu/hr-ft-°F

$$\text{Gas flow} = 10\ \text{MMscf} \times \frac{\text{D}}{24\ \text{hr}} \times \frac{\text{lb/mol}}{379\ \text{SCF}} \times \frac{19.4\ \text{lb}}{\text{lb/mol}}$$

$$= 21{,}328\ \text{lb/hr}$$

$$\text{Oil flow} = \frac{600\ \text{bbl}}{\text{D}} \times \frac{\text{D}}{24\ \text{hr}} \times \frac{350\ \text{lb}}{\text{bbl}} \times 0.77 = 6{,}738\ \text{lb/hr}$$

$$\text{Water flow} = \frac{6.6\ \text{bbl}}{\text{D}} \times \frac{\text{D}}{24\ \text{hr}} \times \frac{350\ \text{lb}}{\text{bbl}} = 96\ \text{lb/hr}$$

$$G = \frac{21{,}328 + 6{,}738 + 96}{0.0289}$$

$$= 974{,}460\ \text{lb/hr-ft}^2$$

$$C = \left(\frac{32.51\ \text{Btu}}{\text{Mscf}^\circ\text{F}} \times \frac{0.379\ \text{Mscf}}{\text{lb/mol}} \times \frac{\text{lb/mol}}{19.4\ \text{lb}} \times \frac{21{,}328\ \text{lb}}{28{,}162\ \text{lb}}\right)$$

$C = 0.60$ Btu/lb-°F
$\mu_e = 0.0134$ cp (at 68°F, Volume 1) $\times\ 2.4 = 0.0322$ lb/hr-ft
$\mu_{ew} = 0.0142$ cp (at 129°F, Volume 1) $\times\ 2.4 = 0.0341$ lb/hr-ft

$$h_i = \frac{0.022\,(0.021)}{0.192}\left(\frac{(0.192)\,(974{,}460)}{(0.0322)}\right)^{0.8}\left(\frac{(0.6)\,(0.0322)}{(0.021)}\right)^{0.4}\left(\frac{0.0322}{0.0341}\right)^{0.16}$$

$h_i = 595$ Btu/hr-ft^2-°F

$$\frac{1}{U} = \frac{1}{318} + 0.003 + \frac{0.05}{30} + \frac{0.9163}{(595)\,0.6021}$$

$U = 96.4$ Btu/hr-ft^2-°F

Estimated U from Figure 2-11

$U = 106$ Btu/hr-ft^2-°F

Use $U = 96.4$ Btu/hr-ft^2-°F

c. Calculate Coil Length

$$L = \frac{12\,q}{\pi\,(LMTD)\,U\,d}$$

$q = 237$ MBtu/hr
$LMTD = 122$°F
$U = 96.4$ Btu/hr ft^2°F
$d = 3.5$ in.

$$L = \frac{(12)(237{,}000)}{\pi\,(122)(96.4)(3.5)}$$

$L = 22.0$ ft

4. Calculate fire tube area required
   For heat transfer to water use 10,000 Btu/hr-ft^2 flux rate:

$$A = \frac{237{,}000 \text{ Btu / hr}}{10{,}000 \text{ Btu / hr ft}^2} = 23.7 \text{ ft}^2$$

Estimate shell size:
Assuming a 10-ft shell, then four passes of 3-in. XXH are required. This will require a 30-in. OD shell for the coils and fire tube.

5. Summary of line heater size

Heater duty	250 MBtu/hr
Coil size	3-in. XXH
Minimum coil length	22.0 ft
Minimum fire tube area	23.7 ft^2
Shell size	30-in. OD × 10-ft F/F

CHAPTER

6

Condensate Stabilization*

The liquids that are separated from the gas stream in the first separator may be flowed directly to a tank or may be "stabilized" in some fashion. As was discussed in Chapter 2 of Volume 1, these liquids contain a large percentage of methane and ethane, which will flash to gas in the tank. This lowers the partial pressure of all other components in the tank and increases their tendency to flash to vapors. The process of increasing the amount of intermediate (C_3 to C_5) and heavy (C_6 +) components in the liquid phase is called "stabilization." In a gas field this process is called condensate stabilization and in an oil field it is called crude stabilization.

In almost all cases the molecules have a higher value as liquid than as gas. Crude oil streams typically contain a low percentage of intermediate components. Thus, it is not normally economically attractive to consider other alternatives to multistage separation to stabilize the crude. In addition, the requirement to treat the oil at high temperature is more important than stabilizing the liquid and may require the flashing of both intermediate and heavy components to the gas stream.

---

*Reviewed for the 1999 edition by Conrad F. Anderson of Paragon Engineering Services, Inc.

Gas condensate, on the other hand, may contain a relatively high percentage of intermediate components and can be easily separated from entrained water due to its lower viscosity and greater density difference with water. Thus, some sort of condensate stabilization should be considered for each gas well production facility.

PARTIAL PRESSURES

As pointed out in Volume 1, the fraction of any onc component that flashes to gas at any stage in a process is a function of the temperature, pressure, and composition of the fluid at that stage. For a given temperature this tendency to flash can be visualized by the partial pressure of the component in the gas phase that is in equilibrium with the liquid. Partial pressure is defined as:

$$\text{Partial Pressure}_n = \frac{\text{Moles}_n}{\text{Sum of Moles}} \times \text{Gas Pressure}$$

The partial pressure at a given pressure and temperature is lower when there are more moles of other components in the gas phase. The lower the partial pressure the greater the tendency of the component to flash to gas. Thus, the higher the fraction of light components in the inlet fluid to any separator, the lower the partial pressure of intermediate components in the gas phase of the separator, and the greater the number of intermediate component molecules that flash to gas.

MULTISTAGE SEPARATION

Figure 6-1 shows a multistage separation process. By removing molecules of the light components in the first separator they are not available to flash to gas from the liquid in the second separator, and the partial pressure of intermediate components in the second separator is higher than it would have been if the first separator did not exist. The second separator serves the same function of increasing the partial pressure of the intermediate components in the third separator and so forth.

The simplest form of condensate stabilization is to install a low-pressure separator downstream of an initial high-pressure separator. Unless the gas well produces at low pressure (less than 500 psi) and the gas contains very little condensate (less than 100 bpd), the additional expendi-

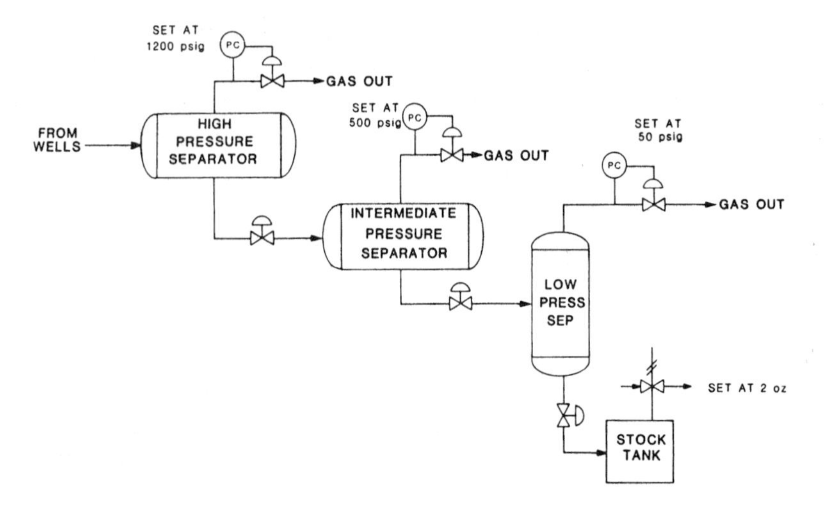

Figure 6-1. Multistage separation process.

ture for this stage of separation is almost always economical when balanced against increased liquid production. If vapor recovery from the tank is required by environmental regulations, the flash separator will significantly reduce the horsepower required. If vapor recovery is not required, the gas from the flash separator may be economically feasible to be recovered and recompressed for sales even if it is not feasible to recover stock tank vapors.

MULTIPLE FLASHES AT CONSTANT PRESSURE AND INCREASING TEMPERATURE

It is possible to stabilize a liquid at a constant pressure by successively flashing it at increasing temperatures as shown in Figure 6-2. At each successive stage the partial pressure of the intermediate components is higher than it could have been at that temperature if some of the lighter components had not been removed by the previous stage. It would be very costly to arrange a process as shown in Figure 6-2, and this is never done. Instead, the same effect is obtained in a tall, vertical pressure vessel with a cold temperature at the top and a hot temperature at the bottom. This is called a "condensate stabilizer."

Figure 6-3 shows a condensate stabilizer system. The well stream flows to a high pressure, three-phase separator. Liquids containing a high fraction of light ends are cooled and enter the stabilizer tower at approxi-

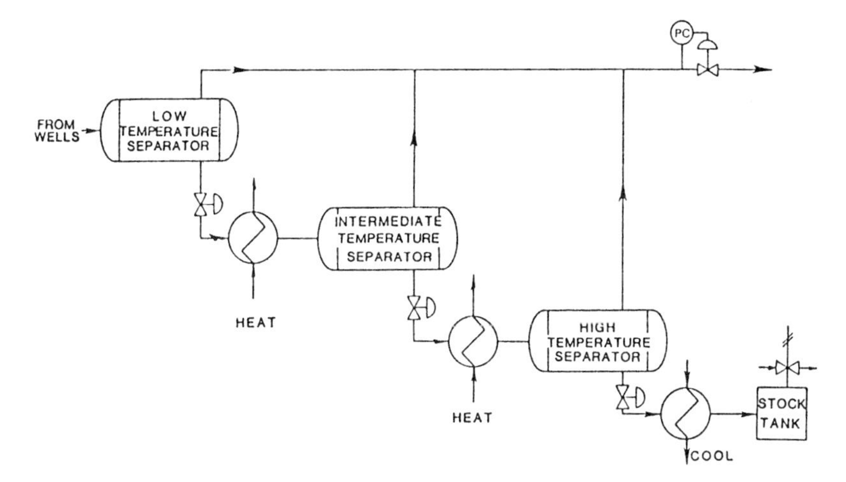

Figure 6-2. Multiple flashes at constant pressure and increasing temperature.

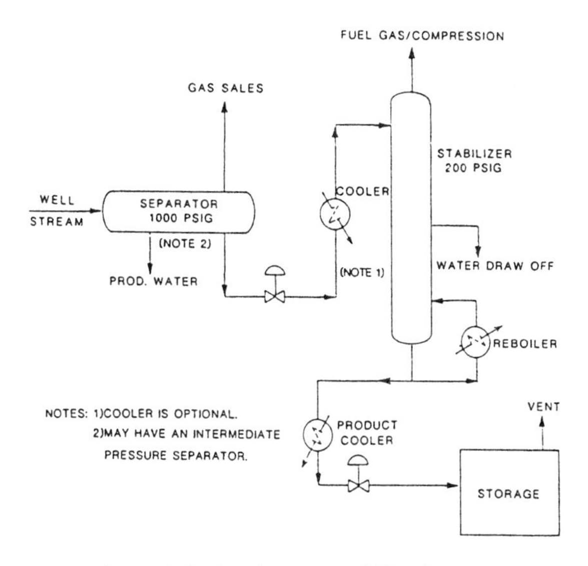

Figure 6-3. Condensate stabilization system.

mately 200 psi. In the tower the liquid falls downward in a process that results in many flashes at ever-increasing temperatures. At the bottom of the tower, some of the liquids are cycled to a reboiler where they receive heat to provide the necessary bottoms temperature (200°F to 400°F). The reboiler could be either a direct-fired bath, an indirect-fired bath, or a heat medium exchanger.

The liquids leaving the bottom of the tower have undergone a series of stage flashes at ever-increasing temperatures, driving off the light components, which exit the top of the tower. These liquids must be cooled to a sufficiently low temperature to keep vapors from flashing to atmosphere in the storage tank.

COLD FEED DISTILLATION TOWER

Figure 6-4 shows the cold feed distillation tower of Figure 6-3. The inlet stream enters the top of the tower. It is heated by the hot gases bubbling up through it as it falls from tray to tray through the downcomers. A flash occurs on each tray so that the liquid is in near-equilibrium with the gas above it at the tower pressure and the temperature of that particular tray.

As the liquid falls, it becomes leaner and leaner in light ends, and richer and richer in heavy ends. At the bottom of the tower some of the liquid is circulated through a reboiler to add heat to the tower. As the gas goes up from tray to tray, more and more of the heavy ends get stripped out of the gas at each tray and the gas becomes richer and richer in the light

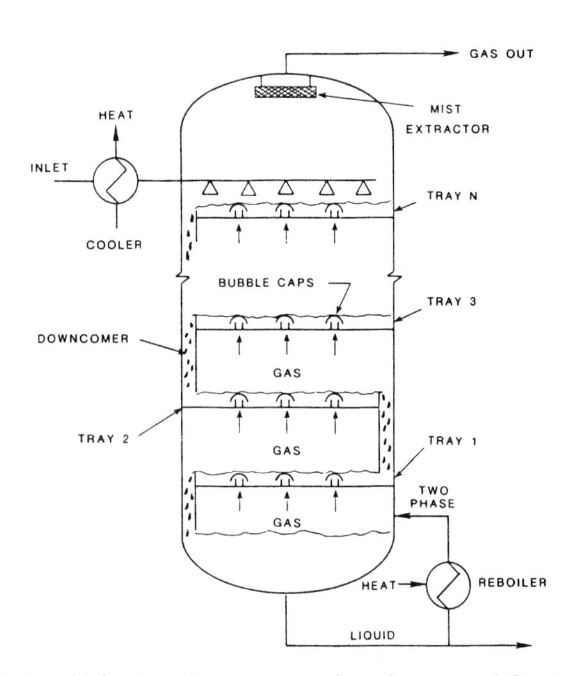

Figure 6-4. Cold-feed distillation tower of condensate stabilization system.

ends and leaner and leaner in the heavy ends (just the opposite of the liquid). The gas exits the top of the tower.

The lower the temperature of the inlet liquid, the lower the fraction of intermediate components that flash to vapor on the top trays and the greater the recovery of these components in the liquid bottoms. However, the colder the feed, the more heat is required from the reboiler to remove light components from the liquid bottoms. If too many light components remain in the liquid, the vapor pressure limitations for the liquid may be exceeded. Light components may also encourage flashing of intermediate components (by lowering their partial pressure) in the storage tank. There is a balance between the amount of inlet cooling and the amount of reboiling required.

Typically, the liquid out the bottom of the tower must meet a specified vapor pressure. The tower must be designed to maximize the molecules of intermediate components in the liquid without exceeding the vapor pressure specification. This is accomplished by driving the maximum number of molecules of methane and ethane out of the liquid and keeping as much of the heavier ends as possible from going out with the gas.

Given inlet composition, pressure, and temperature, a tower temperature and the number of trays that produce a liquid with a specified vapor pressure can be chosen as follows:

1. Assume an initial split of components in the inlet that yields the desired vapor pressure. That is, assume a split of each component between the tower overhead (gas) and bottoms (liquid). There are various rules of thumb that can be used to estimate this split in order to give a desired vapor pressure. Once the split is made, both the assumed composition of the liquid and the assumed composition of the gas are known.
2. Calculate the temperature required at the base of the tower to develop this liquid. This is the temperature at the bubble point for the tower pressure and for the assumed outlet composition. Since the composition and pressure are known, the temperature at its bubble point can be calculated.
3. Calculate the composition of the gas in equilibrium with the liquid. The composition, pressure, and temperature of the liquid are known, and the composition of the gas that is in equilibrium with this liquid can be calculated.
4. Calculate the composition of the inlet liquid falling from Tray 1. Since the composition of the bottom liquid and gas in equilibrium

with the liquid is known, the composition of the feed to this tray is also known. This is the composition of the liquid falling from Tray 1.

5. Calculate the temperature of Tray 1. From an enthalpy balance, the temperature of the liquid falling from Tray 1, and thus the temperature of the flash on Tray 1, can be calculated. The composition is known, the enthalpy can be calculated. Enthalpy must be maintained, so the enthalpy of the liquid of known composition falling fom Tray 1 must equal the sum of the enthalpies of the liquid and gas flashing from it at known temperature.
6. This procedure can then be carried on up the tower to Tray N, which establishes the temperature of the inlet and the gas outlet composition.
7. From the composition of the inlet and gas outlet the liquid outlet composition can be calculated and compared to that assumed in step 1.
8. The temperature or number of trays can then be varied until the calculated outlet liquid composition equals the assumed composition, *and* the vapor pressure of the liquid is equal to or less than that assumed. If the vapor pressure of the liquid is too high, the bottoms temperature must be increased.

DISTILLATION TOWER WITH REFLUX

Figure 6-5 shows a stabilizer with reflux. The well fluid is heated with the bottoms product and injected into the tower, below the top, where the temperature in the tower is equal to the temperature of the feed. This minimizes the amount of flashing. In the tower, the action is the same as in a cold-feed stabilizer or any other distillation tower. As the liquid falls

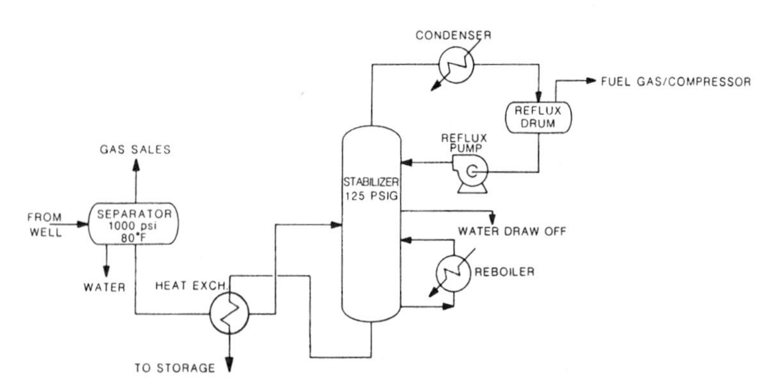

Figure 6-5. Stabilizer with reflux and feed/bottoms heat exchanger.

through the tower, it goes from tray to tray, and gets increasingly richer in the heavy components and increasingly leaner in the light components. The stabilized liquid is cooled in the heat exchanger by the feed stream before flowing to the stock tank.

At the top of the tower any intermediate components going out with the gas are condensed, separated, pumped back to the tower, and sprayed down on the top tray. This liquid is called "reflux," and the two-phase separator that separates it from the gas is called a "reflux tank" or "reflux drum." The reflux performs the same function as the cold feed in a cold-feed stabilizer. Cold liquids strip out the intermediate components from the gas as the gas rises.

The heat required at the reboiler depends upon the amount of cooling done in the condenser. The colder the condenser, the purer the product and the larger the percentage of the intermediate components that will be recovered in the separator and kept from going out with the gas. The hotter the bottoms, the greater the percentage of light components will be boiled out of the bottoms liquid and the lower the vapor pressure of the bottoms liquid.

A condensate stabilizer with reflux will recover more intermediate components from the gas than a cold-feed stabilizer. However, it requires more equipment to purchase, install, and operate. This additional cost must be justified by the net benefit of the incremental liquid recovery, less the cost of natural gas shrinkage and loss of heating value, over that obtained from a cold-feed stabilizer.

CONDENSATE STABILIZER DESIGN

It can be seen from the previous description that the design of both a cold-feed stabilizer and a stabilizer with reflux is a rather complex and involved procedure. Distillation computer simulations are available that can be used to optimize the design of any stabilizer if the properties of the feed stream and desired vapor pressure of the bottoms product are known. Cases should be run of both a cold-feed stabilizer and one with reflux before a selection is made. Because of the large number of calculations required, it is not advisable to use hand calculation techniques to design a distillation process. There is too much opportunity for computational error.

Normally, the crude or condensate sales contract will specify a maximum Reid Vapor Pressure (RVP). This pressure is measured according to a specific ASTM testing procedure. A sample is placed in an evacuated container such that the ratio of the vapor volume to the liquid volume is 4 to 1. The sample is then immersed in a 100°F liquid bath. The absolute pressure then measured is the RVP of the mixture.

Since a portion of the liquid was vaporized to the vapor space, the liquid will have lost some of its lighter components. This effectively changes the composition of the liquid and yields a slightly lower vapor pressure than the true vapor pressure of the liquid at 100°F. Figure 6-6 can be used to estimate true vapor pressure at any temperature from a known RVP.

The inherent error between true vapor pressure and RVP means that a stabilizer designed to produce a bottoms liquid with a true vapor pressure equal to the specified RVP will be conservatively designed. The vapor pressures of various hydrocarbon components at 100°F are given in Table 6-1.

The bottoms temperature of the tower can be approximated if the desired vapor pressure of the liquid is known. The vapor pressure of a mixture is given by:

$$VP = \text{Sum}\ [VP_n \times MF_n] \tag{6-1}$$

where VP = vapor pressure of mixture, psia
VP_n = vapor pressure of component n, psia
MF_n = mole fraction of component n in liquid

To estimate the desired composition of the bottom liquid, the vapor pressures of the different components at 100°F can be assumed to be a measure of the volatility of the component. Thus, if a split of n-C_4 is assumed, the mole fraction of each component in the liquid can be estimated from:

$$L_n = \frac{F_n\ (\text{n-}C_4\ \text{split})}{RV_n} \tag{6-2}$$

$$MF_n = \frac{L_n}{\text{SUM}\ (L_n)} \tag{6-3}$$

(equation continued on page 140)

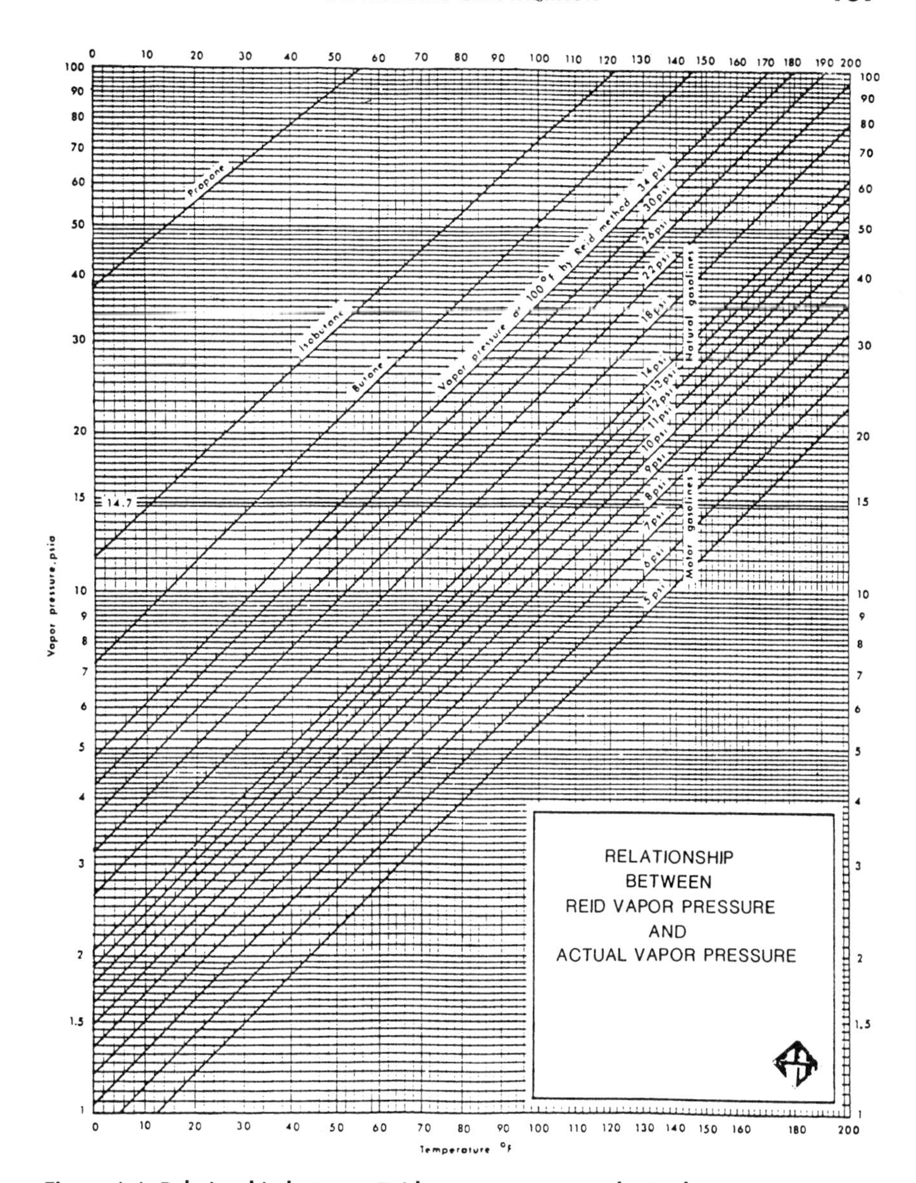

Figure 6-6. Relationship between Reid vapor pressure and actual vapor pressure. (From Gas Processors Suppliers Association, *Engineering Data Book*, 9th Edition.)

Table 6-1
Vapor Pressure and Relative Volatility of Various Components

Component	Vapor Pressure At 100°F psia	Relative Volatility
C_1	5000	96.9
C_2	800	15.5
C_3	190	3.68
$i\text{-}C_4$	72.2	1.40
$n\text{-}C_4$	51.6	1.00
$i\text{-}C_5$	20.4	0.40
$n\text{-}C_5$	15.6	0.30
C_6	5.0	0.10
C_7^+	$\simeq$0.1	0.00
CO_2	—	Infinite
N_2	—	Infinite
H_2S	394	7.64

(equation continued from page 138)

where F_n = total number of moles of component n in feed
L_n = total number of moles of component n in the bottom liquid
($n\text{-}C_4$ split) = assumed moles of component $n\text{-}C_4$ in bottom liquid divided by moles of $n\text{-}C_4$ in feed
RV_n = relative volatility of component n from Table 6-1

To determine the compositon of the bottom liquid, assume a split of $n\text{-}C_4$ and compute MF_n from Equations 6-2 and 6-3. The vapor pressure can then be computed from Equation 6-1. If the vapor pressure is higher than the desired RVP choose a lower number for the $n\text{-}C_4$ split. If the calculated vapor pressure is lower than the desired RVP, choose a higher number for the $n\text{-}C_4$ split. Iterate until the calculated vapor pressure equals the desired RVP.

The bottoms temperature can then be determined by calculating the bubble point of the liquid described by the previous iteration at the chosen operating pressure in the tower. This is done by choosing a temperature, determining equilibrium constants from Chapter 3, Volume 1, and computing:

$$C = \text{Sum}\ (L_n \times K_n) \qquad (6\text{-}4)$$

If C is greater than 1.0, the assumed temperature is too high. If C is lower than 1.0, the assumed temperature is too low. By iteration a temperature can be determined where C = 1.0.

Typically, bottoms temperatures will range from 200–400°F depending on operating pressure, bottoms composition, and vapor pressure requirements. Temperatures should be kept to a minimum to decrease the heat requirements, limit salt build-up, and prevent corrosion problems.

When stabilizer operating pressures are kept below 200 psig the reboiler temperatures will normally be below 300°F. A water glycol heating medium can then be used to provide the heat. Higher stabilizer operating pressures require the use of steam- or hydrocarbon-based heating mediums. Operating at higher pressures, however, decreases the flashing of the feed on entering the column, which decreases the amount of feed cooling required. In general, a crude stabilizer should be designed to operate between 100 and 200 psig.

TRAYS AND PACKING

The number of actual equilibrium stages determines the number of flashes that will occur. The more stages, the more complete the split, but the taller and more costly the tower. Most condensate stabilizers will normally contain approximately five theoretical stages. In a refluxed tower, the section above the feed is known as the rectification section, while the section below the feed is known as the stripping section. The rectification section normally contains about two equilibrium stages above the feed, and the stripping section normally contains three equilibrium stages.

Theoretical stages within a tower are provided by actual stage devices (typically either trays or packings). The actual diameter and height of the tower can be derived using manufacturer's data for the particular device. The height of the tower is a function of the number of theoretical stages and of the efficiency of the actual stages. The diameter of the tower is a function of the hydraulic capacity of the actual stages.

Trays

For most trays, liquid flows across an "active area" of the tray and then into a "downcomer" to the next tray below, etc. Inlet and/or outlet weirs control the liquid distribution across the tray. Vapor flows up the tower and passes through the tray active area, bubbling up through (and thus contacting) the liquid flowing across the tray. The vapor distribution is controlled by (1) perforations in the tray deck (sieve trays), (2) bubble caps (bubble cap trays), or (3) valves (valve trays).

Trays operate within a hydraulic envelope. At excessively high vapor rates, liquid is carried upward from one tray to the next (essentially back-mixing the liquid phase in the tower). For valve trays and sieve trays, a capacity limit can be reached at low vapor rates when liquid falls through the tray floor rather than being forced across the active area into the downcomers. Because the liquid does not flow across the trays, it misses contact with the vapor, and the separation efficiency drops dramatically.

Trays are generally divided into four categories: (1) sieve trays, (2) valve trays, (3) bubble cap trays, and (4) high capacity/high efficiency trays.

Sieve Trays

Sieve trays are the least expensive tray option. In sieve trays, vapor flowing up through the tower contacts the liquid by passing through small perforations in the tray floor (Figure 6-7b). Sieve trays rely on vapor velocity to exclude liquid from falling through the perforations in the tray floor. If the vapor velocity is much lower than design, liquid will begin to flow through the perforations rather than into the downcomer. This condition is known as weeping. Where weeping is severe, the equilibrium efficiency will be very low. For this reason, sieve trays have a very small turndown ratio.

Valve Trays

Valve trays are essentially modified sieve trays. Like sieve trays, holes are punched in the tray floor. However, these holes are much larger than those in sieve trays. Each of these holes is fitted with a device called a "valve." Vapor flowing up through the tower contacts the liquid by passing through valves in the tray floor (Figure 6-7c). Valves can be fixed or moving. Fixed valves are permanently open and operate as deflector plates for the vapor coming up through the holes in the tray floor. For moving valves, vapor passing through the tray floor lifts the valves and contacts the liquid. Moving valves come in a variety of designs, depending on the manufacturer and the application. At low vapor rates, valves will close, helping to keep liquid from falling through the holes in the deck. At sufficiently low vapor rates, a valve tray will begin to weep. That is, some liquid will leak through the valves rather than flowing to the tray downcomers. At very low vapor rates, it is possible that all the liquid will fall through the valves and no liquid will reach the downcomers. This severe weeping is known as "dumping." At this point, the efficiency of the tray is nearly zero.

Bubble Cap Trays

In bubble cap trays, vapor flowing up through the tower contacts the liquid by passing through bubble caps (Figure 6-7a). Each bubble cap assembly consists of a riser and a cap. The vapor rising through the column passes up through the riser in the tray floor and then is turned downward to bubble into the liquid surrounding the cap. Because of their design, bubble cap trays cannot weep. However bubble cap trays are also more expensive and have a lower capacity/higher pressure drop than valve trays or sieve trays.

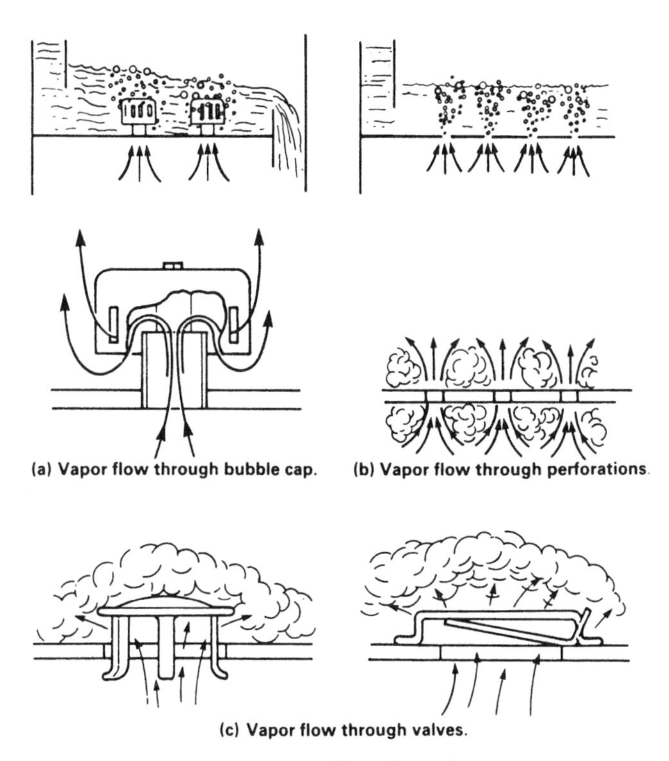

Figure 6-7. Vapor flow through trays.

High Capacity/High Efficiency Trays

High capacity/high efficiency trays have valves or sieve holes or both. They typically achieve higher efficiencies and capacities by taking advantage of the active area under the downcomer. At this time, each of the major vendors has its own version of these trays, and the designs are proprietary.

Bubble Cap Trays vs. Valve Trays

At low vapor rates, valve trays will weep. Bubble cap trays cannot weep (unless they are damaged). For this reason, it is generally assumed that bubble cap trays have nearly an infinite turndown ratio. This is true in absorption processes (e.g., glycol dehydration), in which it is more important to contact the vapor with liquid than the liquid with vapor. However, this is not true of distillation processes (e.g., stabilization), in which it is more important to contact the liquid with the vapor.

As vapor rates decrease, the tray activity also decreases. There eventually comes a point at which some of the active devices (valves or bubble caps) become inactive. Liquid passing these inactive devices gets very little contact with vapor. At very low vapor rates, the vapor activity will concentrate only in certain sections of the tray (or, in the limit, one bubble cap or one valve). At this point, it is possible that liquid may flow across the entire active area without ever contacting a significant amount of vapor. This will result in very low tray efficiencies for a distillation process. Nothing can be done with a bubble cap tray to compensate for this.

However, a valve tray can be designed with heavy valves and light valves. At high vapor rates, all the valves will be open. As the vapor rate decreases, the valves will begin to close. With light and heavy valves on the tray, the heavy valves will close first, and some or all of the light valves will remain open. If the light valves are properly distributed over the active area, even though the tray activity is diminished at low vapor rates, what activity remains will be distributed across the tray. All liquid flowing across the tray will contact some vapor, and mass transfer will continue. Of course, even with weighted valves, if the vapor rate is reduced enough, the tray will weep and eventually become inoperable. However, with a properly designed valve tray this point may be reached after the loss in efficiency of a comparable bubble cap tray. So, in distillation applications, valve trays can have a greater vapor turndown ratio than bubble cap trays.

Tray Efficiency and Tower Height

In condensate stabilizers, trays generally have 70% equilibrium stage efficiency. That is, 1.4 actual trays are required to provide one theoretical stage. The spacing between trays is a function of the spray height and the downcomer backup (the height of clear liquid established in the downcomer). The tray spacing will typically range from 20 to 30 in. (with 24 in. being the most common), depending on the specific design and the internal vapor and liquid traffic. The tray spacing may increase at higher operating pressures (greater than 165 psia) because of the difficulty in disengaging vapor from liquid on both the active areas and in the downcomers.

Packing

Packing typically comes in two types: random and structured.

Liquid distribution in a packed bed is a function of the internal vapor/liquid traffic, the type of packing employed, and the quality of the liquid distributors mounted above the packed bed. Vapor distribution is controlled by the internal vapor/liquid traffic, by the type of packing employed, and by the quality of the vapor distributors located below the packed beds.

Packing material can be plastic, metal, or ceramic. Packing efficiencies can be expressed as HETP (height equivalent to a theoretical plate).

Random Packing

A bed of random packing typically consists of a bed support (typically a gas injection support plate) upon which pieces of packing material are randomly arranged (they are usually poured or dumped onto this support plate). Bed limiters, or hold-downs, are sometimes set above random beds to prevent the pieces of packing from migrating or entraining upward. Random packing comes in a variety of shapes and sizes. For a given shape (design) of packing, small sizes have higher efficiencies and lower capacities than large sizes.

Figure 6-8 shows a variety of random packing designs. An early design is known as a Raschig ring. Raschig rings are short sections of tubing and are low-capacity, low-efficiency, high-pressure drop devices. Today's industry standard is the slotted metal (Pall) ring. A packed bed made of 1-in. slotted metal rings will have a higher mass transfer efficiency and a higher capacity than will a bed of 1-in. Raschig rings. The

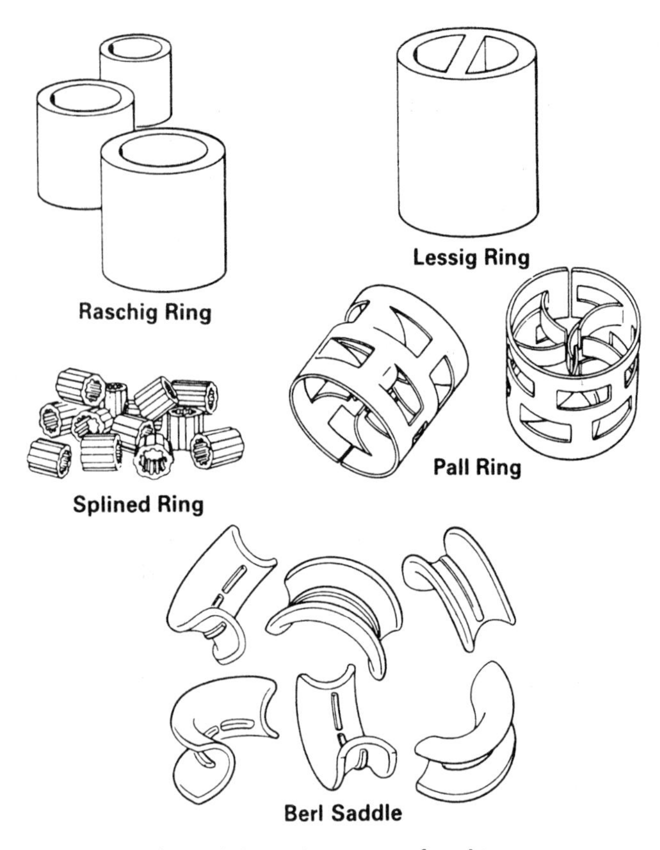

Figure 6-8. Various types of packing.

HETP for a 2-in. slotted metal ring in a condensate stabilizer is about 36 in. This is slightly more than a typical tray design, which would require 34 in. (1.4 trays × 24-in. tray spacing) for one theoretical plate or stage.

Structured Packing

A bed of structured packing consists of a bed support upon which elements of structured packing are placed. Beds of structured packing typically have lower pressure drops than beds of random packing of comparable mass transfer efficiency. Structured packing elements are composed of grids (metal or plastic) or woven mesh (metal or plastic) or of thin vertical crimped sheets (metal, plastic, or ceramic) stacked parallel to each other. Figure 6-9 shows examples of the vertical crimped sheet style of structured packing.

Figure 6-9. Structured packing can offer better mass transfer than trays. (*Courtesy of Koch Engineering Co., Inc.*)

The grid types of structured packing have very high capacities and very low efficiencies, and are typically used for heat transfer or for vapor scrubbing. The wire mesh and the crimped sheet types of structured packing typically have lower capacities and higher efficiencies than the grid type.

Trays or Packing

There is no umbrella answer. The choice is dictated by project scope (new tower or retrofit), current economics, operating pressures, anticipated operating flexibility, and physical properties.

Distillation Service

For distillation services, as in condensate stabilization, tray design is well understood, and many engineers are more comfortable with trays than with packing. In the past, bubble cap trays were the standard. However, they are not commonly used in this service anymore. Sieve trays are inexpensive but offer a very narrow operating range when compared with valve trays. Although valve trays offer a wider operating range than sieve trays, they have moving parts and so may require more maintenance. High capacity/high efficiency trays can be more expensive than standard valve trays. However, high capacity/high efficiency trays require smaller diameter towers, so they can offer significant savings in the overall cost of the distillation tower. The high capacity/high efficiency tray can also be an ideal candidate for tower retrofits in which increased throughputs are required for existing towers.

Random packing has traditionally been used in small diameter (<20 in.) towers. This is because it is easier and less expensive to pack these small diameter towers. However, random packed beds are prone to channeling and have poor turndown characteristics when compared with trays. For these reasons, trays were preferred for tower diameters greater than 20 in. In recent years an improved understanding of the impact of high pressure on packing performance has been gained. Improved vapor and liquid distributor designs and modified bed heights have made the application of packing to large-diameter, high-pressure distillation towers more common. A properly designed packed bed system (packing, liquid distributors, vapor distributors) can be an excellent choice for debottlenecking existing distillation towers.

Stripping Service

For stripping service, as in a glycol or amine contactor (see Chapters 7 and 8), bubble cap trays are the most common. In recent years, there has been a growing movement toward crimped sheet structured packing. Improved vapor and liquid distributor design in conjunction with structured packing can lead to smaller-diameter and shorter stripping towers than can be obtained with trays.

CONDENSATE STABILIZER AS A GAS PROCESSING PLANT

A gas-processing plant, as described in Chapter 9, is designed to recover ethane, propane, butane, and other natural gas liquids from the gas stream. A condensate stabilizer also recovers some portion of these liquids. The colder the temperature of the gas leaving the overhead condenser in a reflux stabilizer, or the colder the feed stream in a cold-feed stabilizer, and the higher the pressure in the tower, the greater the recovery of these components as liquids. Indeed, any stabilization process that leads to recovery of more molecules in the final liquid product is removing those molecules from the gas stream. In this sense, a stabilizer may be considered as a simple form of a gas-processing plant.

It is difficult to determine the point at which a condensate stabilizer becomes a gas plant. Typically, if the liquid product is sold as a condensate, the device would be considered a condensate stabilizer. If the product is sold as a mixed natural gas liquid stream (NGL) or is fractionated into its various components, the same process would be considered a gas plant. The least volatile NGL stream has an RVP between 10 and 14 and has sufficient light hydrocarbons such that 25% of the total volume is vaporized at 140°F.

LTX UNIT AS A CONDENSATE STABILIZER

It should be clear from the description of LTX units in Chapter 5 that the lower pressure separator in an LTX unit is a simple form of cold-feed condensate stabilizer. In the cold, upper portion of the separator some of the intermediate hydrocarbon components condense. In the hot, lower portion some of the lighter components flash.

An LTX unit is not a very efficient stabilizer because the absence of trays or packing keeps the two phases from approaching equilibrium at the various temperatures that exist in the vessel. In addition, it is difficult to control the process. Typically, for a 100-psi to 200-psi operating pressure, a 300°F to 400°F bottoms temperature is required to stabilize completely the condensate. The heating coil in an LTX separator is more likely to be in the range of 125°F to 175°F, and thus complete stabilization will not occur even if the flash were capable of reaching equilibrium.

There may be some additional recovery from an LTX unit than would be realized from a straight two-stage flash separation process, but this increment is normally small and may not justify the increased equipment cost and operating complexity associated with an LTX unit.

CHAPTER

7

*Acid Gas Treating**

In addition to heavy hydrocarbons and water vapor, natural gas often contains other contaminants that may have to be removed. Carbon dioxide (CO_2), hydrogen sulfide (H_2S), and other sulfur compounds such as mercaptans are compounds that may require complete or partial removal for acceptance by a gas purchaser. These compounds are known as "acid gases." H_2S combined with water forms a weak form of sulfuric acid, while CO_2 and water forms carbonic acid, thus the term "acid gas."

Natural gas with H_2S or other sulfur compounds present is called "sour gas," while gas with only CO_2 is called "sweet." Both H_2S and CO_2 are undesirable, as they cause corrosion and reduce the heating value and thus the sales value of the gas. In addition, H_2S may be lethal in very small quantities. Table 7-1 shows physiological effects of H_2S concentrations in air.

At 0.13 ppm by volume, H_2S can be sensed by smell. At 4.6 ppm the smell is quite noticeable. As the concentration increases beyond 200 ppm, the sense of smell fatigues, and the gas can no longer be detected by odor. Thus, H_2S cannot always be detected by smell. Even if H_2S cannot be smelled, it is possible that there is sufficient H_2S present to be life threat-

---

*Reviewed for the 1999 edition by K. S. Chiou of Paragon Engineering Services, Inc.

Table 7-1
Physiological Effects of H_2S Concentrations in Air

Concentrations in Air				
Percent by Volume	Parts per Million by Volume	Grains per 100 scf*	Milligrams per m³*	Physiological Effects
0.00013	0.13	0.008	0.18	Obvious and unpleasant odor generally perceptible at 0.13 ppm and quite noticeable at 4.6 ppm. As the concentration increases, the sense of smell fatigues and the gas can no longer be detected by odor.
0.001	10	0.63	14.41	Acceptable ceiling concentration permitted by federal OSHA standards.
0.005	50	3.15	72.07	Acceptable maximum peak above the OSHA acceptable ceiling concentrations permitted once for 10 minutes per 8-hour shift, if no other measurable exposure occurs.
0.01	100	6.30	144.14	Coughing, eye irritation, loss of sense of smell after 3 to 15 minutes. Altered respiration, pain in eyes, and drowsiness after 15 to 30 minutes, followed by throat irritation after one hour. Prolonged exposure results in a gradual increase in the severity of these symptoms.
0.02	200	12.59	288.06	Kills sense of smell rapidly, burns eyes and throat.
0.05	500	31.49	720.49	Dizziness, loss of sense of reasoning and balance. Breathing problems in a few minutes. Victims need prompt artificial resuscitation.
0.07	700	44.08	1008.55	Unconscious quickly. Breathing will stop and deaths will result if not rescued promptly. Artificial resuscitation is needed.
0.10+	1000+	62.98	1440.98+	Unconsciousness at once. Permanent brain damage or death may result unless rescued promptly and given artificial resuscitation.

**Based on 1% hydrogen sulfide = 629.77 gr/100 scf at 14.696 psia and 59°F, or 101.325 kPa and 15°C.*

ening. At 500 ppm, H_2S can no longer be smelled, but breathing problems and then death can be expected within minutes. At concentrations above 700 to 1,000 ppm, death can be immediate and without warning. Generally, a concentration of 100 ppm H_2S or more in a process stream is cause for concern and the taking of proper operating precautions.

Gas sales contracts for natural gases will limit the concentration of acid compounds. In the United States, typically, gas sales contracts will permit up to 2 to 3% carbon dioxide and ¼ grain per 100 scf (approximately 4 ppm) of hydrogen sulfide. The actual requirement for any sales contract may vary, depending upon negotiations between seller and purchaser.

Next to sales contract specifications, corrosion protection ranks highest among the reasons for the removal of acid gases. The partial pressure of the acid gases may be used as a measure to determine whether treatment is required. The partial pressure of a gas is defined as the total pressure of the system times the mole % of the gaseous component. Where CO_2 is present with free water, a partial pressure of 30 psia or greater would indicate that CO_2 corrosion should be expected. If CO_2 is not removed, inhibition and special metallurgy may be required. Below 15 psia, CO_2 corrosion is not normally a problem, although inhibition may be required.

H_2S may cause hydrogen embrittlement in certain metals. Figures 7-1 and 7-2 show the H_2S concentration at which the National Association of Corrosion Engineers (NACE) recommends special metallurgy to guard against H_2S corrosion.

In the sulfide stress cracking region, appropriate metallurgy is required in line piping, pressure vessels, etc. There is a listing of acceptable steels in the NACE standard. Steels with a hardness of less than 22 Rockwell C hardness should be used in areas where sulfide-stress cracking is a problem.

The concentration of H_2S required for sulfide-stress cracking in a multiphase gas/liquid system (Figure 7-2) is somewhat higher than in pure gas streams (Figure 7-1). The liquid acts as an inhibitor.

This chapter discusses the different processes that are commonly used in field gas treating of acid gases and presents a method that can be used to select from among the various processes. Design procedures for determining critical sizing parameters for iron sponge and amine systems are presented, as these are the most common field gas treating processes currently employed and they are not proprietary in nature.

(*text continued on page 156*)

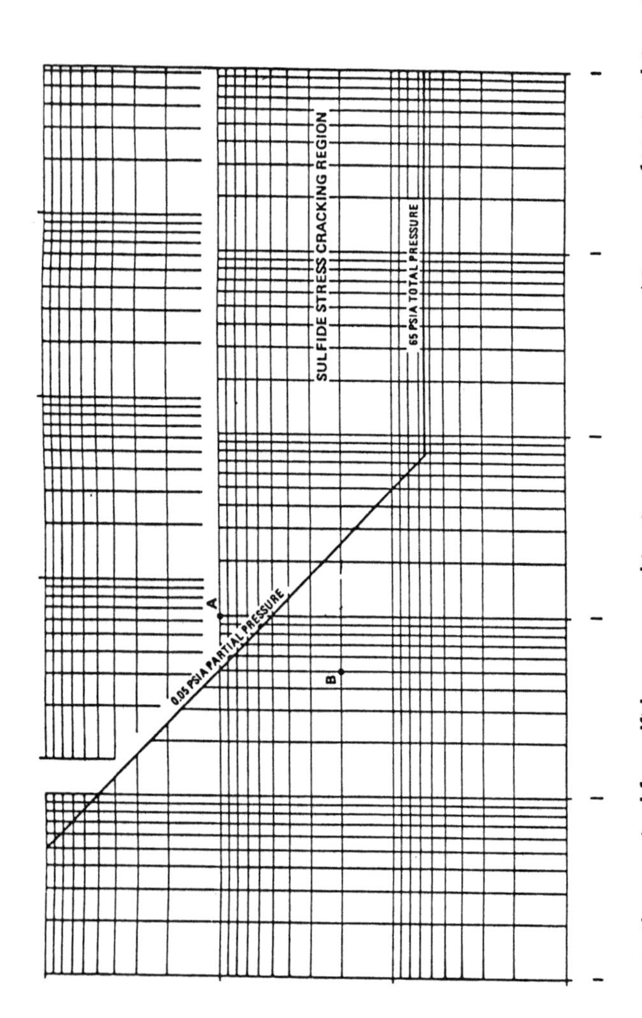

Figure 7-1. H_2S concentration required for sulfide-stress cracking in a pure gas system. *(Courtesy of National Association of Corrosion Engineers.)*

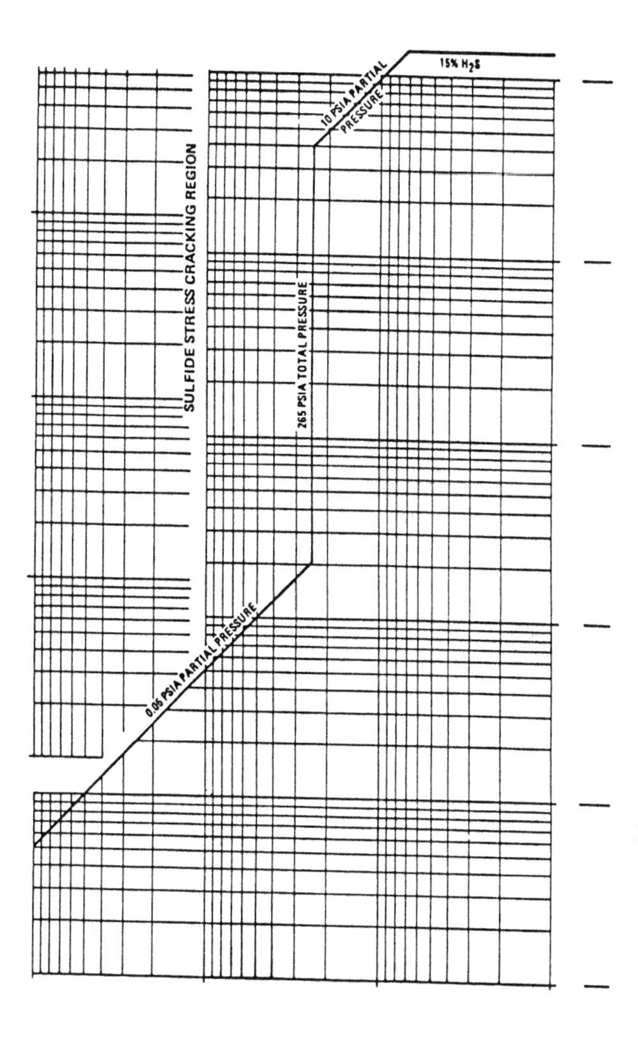

Figure 7-2. H_2S concentration required for sulfide-stress cracking in a multiphase gas/liquid system. *(Courtesy of National Association of Corrosion Engineers.)*

(text continued from page 153)

GAS SWEETENING PROCESSES

Numerous processes have been developed for gas sweetening based on a variety of chemical and physical principles. These processes can be categorized by the principles used in the process to separate the acid gas and the natural gases as follows:

Process	Licensor
1. Solid bed absorption	
Iron Sponge	
SulfaTreat®	The SulfaTreat Company
Zinc Oxide	
Molecular Sieves	Union Carbide Corporation
2. Chemical solvents	
Monoethanol amine (MEA)	
Diethanol amine (DEA)	
Methyldiethanol amine (MDEA)	
Diglycol amine (DGA)	
Diisopropanol amine (DIPA)	
Hot potassium carbonate	
Proprietary carbonate systems	
3. Physical solvents	
Fluor Flexsorb®	Fluor Daniel Corporation
Shell Sulfinol®	
Selexol®	Norton Co., Chemical Process Products
Rectisol®	Lurg, Kohle & Mineraloltechnik GmbH & Linde A.G.
4. Direct conversion of H_2S to sulfur	
Claus	
LOCAT	ARI Technologies
Stretford	Ralph M. Parsons Co.
IFP	Institute Français du Petrole
Sulfa-check	Exxon Chemical Co.
5. Sulfide scavenger	
6. Distillation	
Amine-aldehyde condensates	
7. Gas permeation	

The list, although not complete, does represent many of the commonly available commercial processes. New proprietary processes are being developed. The design engineer is cautioned to consult with vendors and experts in acid gas treating before making a selection for any large plant.

Solid Bed Absorption

A fixed bed of solid particles can be used to remove acid gases either through chemical reactions or ionic bonding. Typically, in solid bed absorption processes the gas stream must flow through a fixed bed of solid particles that remove the acid gases and hold them in the bed. When the bed is saturated with acid gases, the vessel must be removed from service and the bed regenerated or replaced. Since the bed must be removed from service to be regenerated, some spare capacity must be provided. There are three commonly used processes under this category: the iron oxide process, the zinc oxide process, and the molecular sieve process.

Iron Sponge

The iron sponge process uses the chemical reaction of ferric oxide with H_2S to sweeten gas streams. This process is applied to gases with low H_2S concentrations (300 ppm) operating at low to moderate pressures (50–500 psig). Carbon dioxide is not removed by this process.

The reaction of H_2S and ferric oxide produces water and ferric sulfide as follows:

$$2Fe_2O_3 + 6H_2S \rightarrow 2Fe_2S_3 + 6H_20 \quad (7\text{-}1)$$

The reaction requires the presence of slightly alkaline water and a temperature below 110°F. If the gas does not contain sufficient water vapor, water may need to be injected into the inlet gas stream. Additionally, bed alkalinity should be checked daily. A pH level of 8–10 should be maintained through the injection of caustic soda with the water.

The ferric oxide is impregnated on wood chips, which produces a solid bed with a large ferric oxide surface area. Several grades of treated wood chips are available, based on iron oxide content. The most common grades are 6.5-, 9.0-, 15.0-, and 20-lb iron oxide/bushel. The chips are contained in a vessel, and sour gas flows through the bed and reacts with the ferric oxide. Figure 7-3 shows a typical vessel for the iron sponge process.

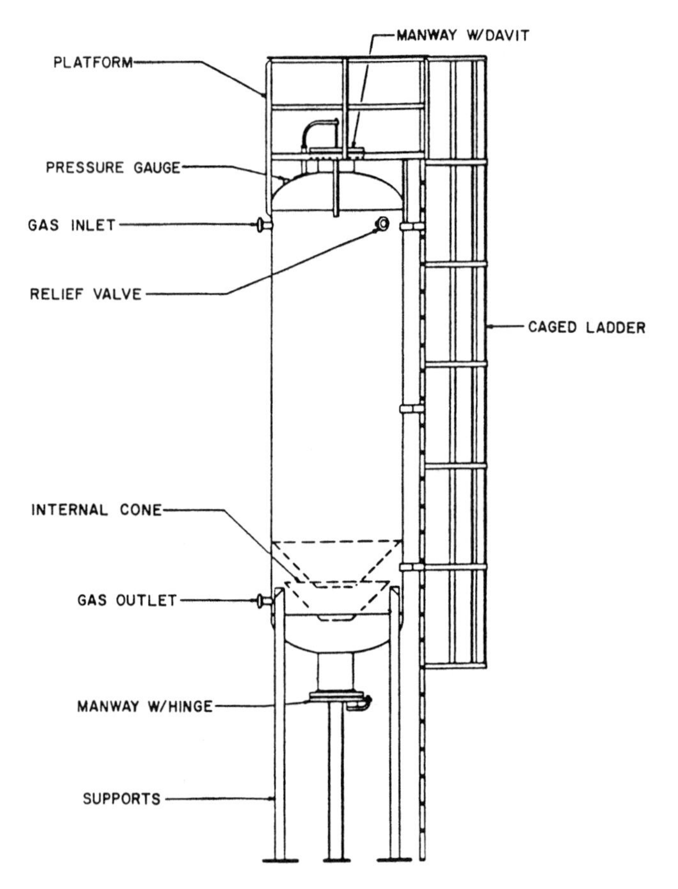

Figure 7-3. Iron oxide acid treating unit.

The ferric sulfide can be oxidized with air to produce sulfur and regenerate the ferric oxide. The reaction for ferric oxide regeneration is as follows:

$$2Fe_2S_3 + 3O_2 \rightarrow 2Fe_2O_3 + 6S \tag{7-2}$$

$$S_2 + 2O_2 \rightarrow 2SO_2 \tag{7-3}$$

The regeneration step must be performed with great care as the reaction with oxygen is exothermic (that is, gives off heat). Air must be introduced slowly so the heat of reaction can be dissipated. If air is introduced quickly the heat of reaction may ignite the bed.

Some of the elemental sulfur produced in the regeneration step remains in the bed. After several cycles this sulfur will cake over the ferric oxide, decreasing the reactivity of the bed. Typically, after 10 cycles the bed must be removed from the vessel and replaced with a new bed.

In some designs the iron sponge may be operated with continuous regeneration by injecting a small amount of air into the sour gas feed. The air regenerates ferric sulfide while H_2S is being removed by ferric oxide. This process is not as effective at regenerating the bed as the batch process. It requires a higher pressure air stream, and if not properly controlled may create an explosive mixture of air and gas.

Hydrocarbon liquids in the gas tend to coat the iron sponge media, inhibiting the reactions. The use of an adequately designed gas scrubber or filter separator upstream of the iron sponge unit will minimize the amount of liquids that condense on the bed. Sometimes the process can be arranged so that the scrubber operates at a lower temperature or higher pressure than the iron sponge unit, so that there is no possibility of hydrocarbon liquids condensing in the iron sponge unit.

Due to the difficulty of controlling the regeneration step, the eventual coating of the bed with elemental sulfur, the low cost of iron sponge material, and the possibility of hydrocarbon liquids coating the bed, iron sponge units are normally operated in the batch mode. The spent bed is removed from the unit and trucked to a disposal site. It is replaced with a new bed and the unit put back in service. The spent bed will react with the oxygen in air as shown in Equations 7-2 and 7-3 unless it is kept moist. In areas where iron sponge units are installed, service companies exist that can replace iron sponge beds and properly dispose of the waste material.

SulfaTreat®

SulfaTreat® process is similar to the iron sponge process. It uses a patented proprietary mixture of ferric oxide and triferric oxide to react with H_2O to sweeten gas streams. In SulfaTreat® process the iron oxides are supported on the surface of an inert, inorganic substrate forming a granular material, while in the iron sponge process the ferric oxide is impregnated on wood chips. The SulfaTreat® starting material and the spent product are safe and stable. The spent product can be recycled or disposed in a landfill.

Two vessels arranged in series, a lead/lag arrangement, will allow the SulfaTreat® material to be used more efficiently with no interruption in unit service and greater process reliability. The first vessel, the "lead" unit, acts as the "working" unit to remove all the H_2S at the beginning of a treatment period with its outlet H_2S increasing over time. The exit gas from the first vessel can go to the second vessel, the "lag" unit, for further polishing or bypass the second vessel as though the first vessel is operating in a single vessel arrangement. The second vessel is to be placed in operation as the lag unit to polish the H_2S remaining in the gas when the lead unit outlet H_2S starts to approach the specification.

Once the lead unit inlet and outlet concentrations are equal, the SulfaTreat® material is considered spent or exhausted. Then, the gas flow is directed to the second vessel, which becomes the lead unit. The spent material is removed from the first vessel and the fresh SulfaTreat® material reloaded to be placed into operation as the lag unit without gas flow interruption. Removal of the spent SulfaTreat® material and the reload of the fresh material could be conveniently scheduled using the change-out "window" available without exceeding maximum outlet H_2S concentrations when operating in this lead/lag mode.

Zinc Oxide

The zinc oxide process is similar to the iron sponge process. It uses a solid bed of granular zinc oxide to react with the H_2S to form water and zinc sulfide:

$$ZnO + H_2S \rightarrow ZnS + H_2O \quad (7\text{-}4)$$

The rate of reaction is controlled by the diffusion process, as the sulfide ion must first diffuse to the surface of the zinc oxide to react. High temperature (>250°F) increases the diffusion rate and is normally used to promote the reaction rate.

Zinc oxide is usually contained in long, thin beds to lessen the chances of channeling. Pressure drop through the beds is low. Bed life is a function of gas H_2S content and can vary from 6 months to in excess of 10 years. The spent catalyst is discharged by gravity flow and contains up to 20 weight percent of sulfur.

The process has seen decreasing use due to increasing disposal problems with the spent catalyst, which is classified as a heavy metal salt.

Molecular Sieves

The molecular sieve process uses synthetically manufactured solid crystalline zeolite in a dry bed to remove gas impurities. The crystalline structure of the solids provides a very porous solid material with all the pores exactly the same size. Within the pores the crystal structure creates a large number of localized polar charges called active sites. Polar gas molecules, such as H_2S and water, that enter the pores form weak ionic bonds at the active sites. Nonpolar molecules such as paraffin hydrocarbons will not bond to the active sites. Thus, molecular sieve units will "dehydrate" the gas (remove water vapor) as well as sweeten it.

Molecular sieves are available with a variety of pore sizes. A molecu lar sieve should be selected with a pore size that will admit H_2S and water while preventing heavy hydrocarbons and aromatic compounds from entering the pores. However, carbon dioxide molecules are about the same size as H_2S molecules and present problems. Even though the CO_2 is non-polar and will not bond to the active sites, the CO_2 will enter the pores. Small quantities of CO_2 will become trapped in the pores. In this way small portions of CO_2 are removed. More importantly, CO_2 will obstruct the access of H_2S and water to active sites and decrease the effectiveness of the pores. Beds must be sized to remove all water and to provide for interference from other molecules in order to remove all H_2S.

The absorption process usually occurs at moderate pressure. Ionic bonds tend to achieve an optimum performance near 450 psig, but the process can be used for a wide range of pressures. The molecular sieve bed is regenerated by flowing hot sweet gas through the bed. Typical regeneration temperatures are in the range of 300–400°F.

Molecular sieve beds do not suffer any chemical degradation and can be regenerated indefinitely. Care should be taken to minimize mechanical damage to the solid crystals as this may decrease the bed's effectiveness. The main causes of mechanical damage are sudden pressure and/or temperature changes when switching from absorption to regeneration cycles.

Molecular sieves for acid gas treatment are generally limited to small gas streams operating at moderate pressures. Due to these operating limitations, molecular sieve units have seen limited use for gas sweetening operations. They are generally used for polishing applications following one of the other processes and for dehydration of sweet gas streams where very low water vapor concentrations are required. Techniques for sizing molecular sieve units are discussed in Chapter 8.

Chemical Solvents

Chemical solvent processes use an aqueous solution of a weak base to chemically react with and absorb the acid gases in the natural gas stream.

The absorption occurs as a result of the driving force of the partial pressure from the gas to the liquid. The reactions involved are reversible by changing the system temperature or pressure, or both. Therefore, the aqueous base solution can be regenerated and thus circulated in a continuous cycle. The majority of chemical solvent processes use either an amine or carbonate solution.

Amine Processes

Several processes are available that use the basic action of various amines. These amines can be categorized as primary, secondary, or tertiary according to the number of organic groups bonded to the central nitrogen atom.

Primary amines are stronger bases than secondary amines, which are stronger than tertiary amines. Amines with stronger base properties will be more reactive toward CO_2 and H_2S gases and will form stronger chemical bonds.

A typical amine system is shown in Figure 7-4. The sour gas enters the system through an inlet separator to remove any entrained water or hydrocarbon liquids. Then the gas enters the bottom of the amine absorber and flows counter-current to the amine solution. The absorber can be either a trayed or packed tower. Conventional packing is usually used for 20-in. or smaller diameter towers, and trays or structured packing for larger towers. An optional outlet separator may be included to recover entrained amines from the sweet gas.

The amine solution leaves the bottom of the absorber carrying with it the acid gases. This solution containing the CO_2 and H_2S is referred to as the rich amine. From the absorber the rich amine is flashed to a flash tank to remove almost all the dissolved hydrocarbon gases and entrained hydrocarbon condensates. A small percentage of the acid gases will also flash to the vapor phase in this vessel. From the flash tank the rich amine proceeds to the rich/lean amine exchanger. This exchanger recovers some of the sensible heat from the lean amine stream to decrease the heat duty on the amine reboiler. The heated rich amine then enters the amine stripping tower where heat from the reboiler breaks the bonds between the amines and acid gases. The acid gases are removed overhead and lean amine is removed from the bottom of the stripper.

The hot lean amine proceeds to the rich/lean amine exchanger and then to additional coolers to lower its temperature to no less than 10°F above the inlet gas temperature. This prevents hydrocarbons from con-

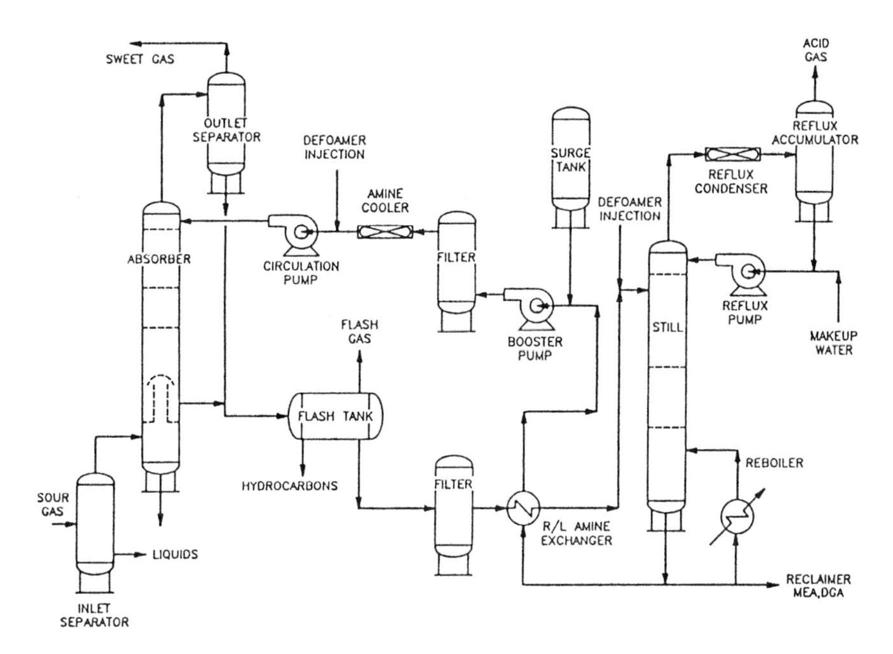

Figure 7-4. Amine system for gas sweetening.

densing in the amine solution when the amine contacts the sour gas. The cooled lean amine is then pumped up to the absorber pressure and enters the top of the absorber. As the amine solution flows down the absorber it absorbs the acid gases. The rich amine is then removed at the bottom of the tower and the cycle is repeated.

Of the following amine systems that are discussed, diethanol amine (DEA) is the most common. Even though a DEA system may not be as efficient as some of the other chemical solvents, it may be less expensive to install because standard packaged systems are readily available. In addition, it may be less expensive to operate and maintain because field personnel are likely to be more familiar with it.

Monoethanolamine Systems. Monoethanolamine (MEA) is a primary amine that can meet nominal pipeline specifications for removing both H_2S and CO_2. MEA is a stable compound and in the absence of other chemicals suffers no degradation or decomposition at temperatures up to its normal boiling point. MEA reacts with CO_2 and H_2S as follows:

$$2(RNH_2) + H_2S \underset{\text{high temp.}}{\overset{\text{low temp.}}{\rightleftharpoons}} (RNH_3)_2S \tag{7-5}$$

$$(RNH_3)_2S + H_2S \underset{\text{high temp.}}{\overset{\text{low temp.}}{\rightleftharpoons}} 2(RNH_3)HS \tag{7-6}$$

$$2(RNH_2) + CO_2 \underset{\text{high temp.}}{\overset{\text{low temp.}}{\rightleftharpoons}} RNHCOONH_3R \tag{7-7}$$

These reactions are reversible by changing the system temperature. MEA also reacts with carbonyl sulfide (COS) and carbon disulfide (CS_2) to form heat-stable salts that cannot be regenerated. At temperatures above 245°F a side reaction with CO_2 exists that produces oxazolidone-2, a heat-stable salt, and consumes MEA from the process.

The reactions with CO_2 and H_2S shown are reversed in the stripping column by heating the rich MEA to approximately 245°F at 10 psig. The acid gases evolve into the vapor and are removed from the still overhead. Thus, the MEA is regenerated.

The normal regeneration temperature in the still will not regenerate heat-stable salts or oxazolidone-2. Therefore, a reclaimer is usually included to remove these contaminants. A side stream of from 1 to 3% of the MEA circulation is drawn from the bottom of the stripping column. This stream is then heated to boil the water and MEA overhead while the heat-stable salts and oxazolidone-2 are retained in the reclaimer. The reclaimer is periodically shut in and the collected contaminants are cleaned out and removed from the system. However, any MEA bonded to them is also lost.

MEA is usually circulated in a solution of 15–20% MEA by weight in water. From operating experience the solution loading should be between 0.3–0.4 moles of acid gas removed per mole of MEA. Both the solution strength and the solution loading are limited to avoid excessive corrosion. The higher the concentration of H_2S relative to CO_2, the higher the amine concentration and allowable loading. This is due to the reaction of H_2S and iron (Fe) to form iron sulfide (Fe_2S_3), which forms a protective barrier on the steel surface.

The acid gases in the rich amine are extremely corrosive. The corrosion commonly shows up on areas of carbon steel that have been stressed, such as heat-affected zones near welds, in areas of high acid-gas concentration, or at a hot gas-liquid interface. Therefore, stress-relieving all equipment after manufacturing is necessary to reduce corrosion, and special metallurgy in specific areas such as the still overhead or the reboiler tubes may be required.

MEA systems foam rather easily resulting in excessive amine carry-over from the absorber. Foaming can be caused by a number of foreign materials such as condensed hydrocarbons, degradation products, solids such as carbon or iron sulfide, excess corrosion inhibitor, valve grease, etc. Solids can be removed with cartridge filters. Hydrocarbon liquids are usually removed in the flash tank. Degradation products are removed in a reclaimer as previously described.

Storage tanks and surge vessels for MEA must have inert blanket-gas systems. Sweet natural gas or nitrogen can be used as the blanket gas. This is required because MEA will oxidize when exposed to the oxygen in air.

As the smallest of the ethanolamine compounds, MEA has a relatively high vapor pressure. Thus, MEA losses of 1 to 3 lb/MMscf are common.

In summation, MEA systems can efficiently sweeten sour gas to pipeline specifications; however, great care in designing the system is required to limit equipment corrosion and MEA losses.

Diethanolamine Systems. Diethanolamine (DEA) is a secondary amine that has in recent years replaced MEA as the most common chemical solvent. As a secondary amine, DEA is a weaker base than MEA, and therefore DEA systems do not typically suffer the same corrosion problems. In addition, DEA has lower vapor loss, requires less heat for regeneration per mole of acid gas removed, and does not require a reclaimer. DEA reacts with H_2S and CO_2 as follows:

$$2R_2NH + H_2S \underset{\text{high temp.}}{\overset{\text{low temp.}}{\rightleftharpoons}} (R_2NH_2)_2S \tag{7-8}$$

$$(R_2NH_2)_2S + H_2S \underset{\text{high temp.}}{\overset{\text{low temp.}}{\rightleftharpoons}} 2(R_2NH_2)HS \tag{7-9}$$

$$2R_2NH + CO_2 \underset{\text{high temp.}}{\overset{\text{low temp.}}{\rightleftharpoons}} R_2NCOONH_2R_2 \tag{7-10}$$

These reactions are reversible. DEA reacts with carbonyl sulfide (COS) and carbon disulfide (CS_2) to form compounds that can be regenerated in the stripping column. Therefore, COS and CS_2 are removed without a loss of DEA. Typically, DEA systems include a carbon filter but do not include a reclaimer.

The stoichiometry of the reactions of DEA and MEA with CO_2 and H_2S is the same. The molecular weight of DEA is 105, compared to 61 for MEA. The combination of molecular weights and reaction stoichiometry means that approximately 1.7 lb of DEA must be circulated to react with the same amount of acid gas as 1.0 lb of MEA. However, because of its lower corrosivity, the solution strength of DEA ranges up to 35% by weight compared to only 20% for MEA. Loadings for DEA systems range to 0.65 mole of acid gas per mole of DEA compared to a maximum of 0.4 mole of acid gas per mole of MEA. The result of this is that the circulation rate of a DEA solution is slightly less than for a comparable MEA system.

The vapor pressure of DEA is approximately 1/30th of the vapor pressure of MEA; therefore, amine losses as low as ¼–½ lb/MMscf can be expected.

Diglycolamine Systems. The Fluor Econamine® process uses diglycolamine (DGA) to sweeten natural gas. The active DGA reagent is 2-(2-amino-ethoxy) ethanol, which is a primary amine. The reactions of DGA with acid gases are the same as for MEA. Degradation products from reactions with COS and CS_2 can be regenerated in a reclaimer.

DGA systems typically circulate a solution of 50–70% DGA by weight in water. At these solution strengths and a loading of up to 0.3 mole of acid gas per mole of DGA, corrosion in DGA systems is slightly less than in MEA systems, and the advantages of a DGA system are that the low vapor pressure decreases amine losses, and the high solution strength decreases circulation rates and heat required.

Diisopropanolamine Systems. Diisopropanolamine (DIPA) is a secondary amine used in the Shell ADIP® process to sweeten natural gas. DIPA systems are similar to MEA systems but offer the following advantages: carbonyl sulfide (COS) can be removed and regenerated easily and the system is generally noncorrosive and requires less heat input.

One feature of this process is that at low pressures DIPA will preferentially remove H_2S. As pressure increases the selectivity of the process decreases. The DIPA removes increasing amounts of CO_2 as well as the H_2S. Therefore, this system can be used either selectively to remove H_2S or to remove both CO_2 and H_2S.

Hot Potassium Carbonate Process

The hot potassium carbonate (K_2CO_3) process uses hot potassium carbonate to remove both CO_2 and H_2S. It works best on a gas with CO_2 partial pressures in the range of 30–90 psi. The main reactions involved in this process are:

$$K_2CO_3 + CO_2 + H_2O \underset{\text{low } CO_2 \text{ press.}}{\overset{\text{high } CO_2 \text{ press.}}{\rightleftharpoons}} 2KHCO_3 \tag{7-11}$$

$$K_2CO_3 + H_2S \underset{\text{low } H_2S \text{ press.}}{\overset{\text{high } H_2S \text{ press.}}{\rightleftharpoons}} KHS + KHCO_3 \tag{7-12}$$

It can be seen from Equation 7-12 that H_2S alone cannot be removed unless there is sufficient CO_2 present to provide $KHCO_3$, which is needed to regenerate potassium carbonate. Since these equations are driven by partial pressures, it is difficult to treat H_2S to the very low requirements usually demanded (¼ grain per 100 scf). Thus, final polishing to H_2S treatment may be required.

The reactions are reversible based on the partial pressures of the acid gases. Potassium carbonate also reacts reversibly with COS and CS_2.

Figure 7-5 shows a typical hot carbonate system for gas sweetening. The sour gas enters the bottom of the absorber and flows counter-current to the potassium carbonate. The sweet gas then exits the top of the absorber. The absorber is typically operated at 230°F; therefore, a sour/sweet gas exchanger may be included to recover sensible heat and decrease the system heat requirements.

The acid-rich potassium carbonate solution from the bottom of the absorber is flashed to a flash drum, where much of the acid gas is removed. The solution then proceeds to the stripping column, which operates at approximately 245°F and near-atmospheric pressure. The low pressure, combined with a small amount of heat input, drives off the remaining acid gases. The lean potassium carbonate from the stripper is pumped back to the absorber. The lean solution may or may not be cooled slightly before entering the absorber. The heat of reaction from the absorption of the acid gases causes a slight temperature rise in the absorber.

The solution concentration for a potassium carbonate system is limited by the solubility of the potassium bicarbonate ($KHCO_3$) in the rich

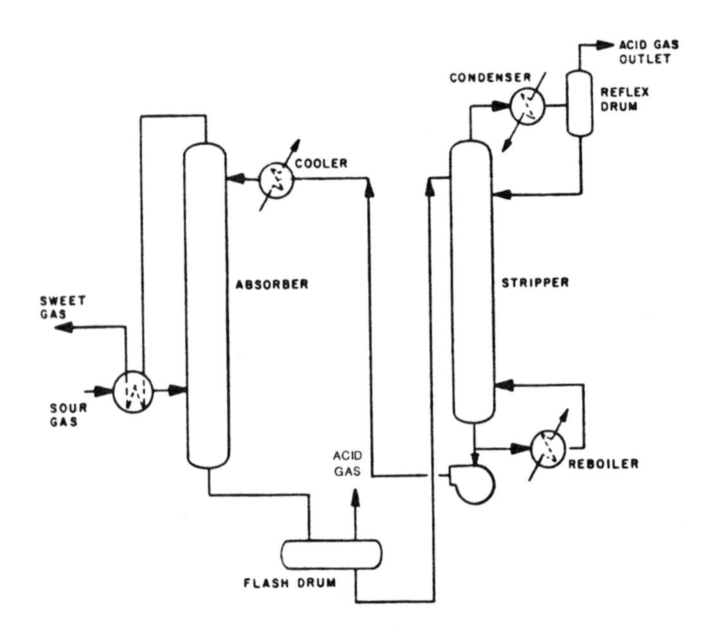

Figure 7-5. Hot carbonate system for gas sweetening.

stream. The high temperature of the system increases the solubility of $KHCO_3$, but the reaction with CO_2 produces two moles of $KHCO_3$ per mole of K_2CO_3 reacted. For this reason the $KHCO_3$ in the rich stream limits the lean solution K_2CO_3 concentration to 20–35% by weight.

The entire system is operated at high temperatures to increase the solubility of potassium carbonate. Therefore, the designer must be careful to avoid dead spots in the system where the solution could cool and precipitate solids. If solids do precipitate, the system may suffer from plugging, erosion, or foaming.

The hot potassium carbonate solutions are extremely corrosive. All carbon steel must be stress-relieved to limit corrosion. A variety of corrosion inhibitors are available to decrease corrosion.

Proprietary Carbonate Systems

Several proprietary processes have been developed based on the hot carbonate system with an activator or catalyst. These activators increase the performance of the hot PC system by increasing the reaction rates both in the absorber and the stripper. In general, these processes also decrease corrosion in the system. The following are some of the proprietary processes for hot potassium carbonate:

Benfield: Several activators
Girdler: Alkanolamine activators
Catacarb: Alkanolamine and/or borate activators
Giammarco-Vetrocoke: Arsenic and other activators

Physical Solvent Processes

These processes are based on the solubility of the H_2S and/or CO_2 within the solvent, instead of on chemical reactions between the acid gas and the solvent. Solubility depends first and foremost on partial pressure and secondarily on temperature. Higher acid-gas partial pressures and lower temperatures increase the solubility of H_2S and CO_2 in the solvent and thus decrease the acid-gas components.

Various organic solvents are used to absorb the acid gases. Regeneration of the solvent is accomplished by flashing to lower pressures and/or stripping with solvent vapor or inert gas. Some solvents can be regenerated by flashing only and require no heat. Other solvents require stripping and some heat, but typically the heat requirements are small compared to chemical solvents.

Physical solvent processes have a high affinity for heavy hydrocarbons. If the natural gas stream is rich in C_{3+} hydrocarbons, then the use of a physical solvent process may result in a significant loss of the heavier molecular weight hydrocarbons. These hydrocarbons are lost because they are released from the solvent with the acid gases and cannot be economically recovered.

Under the following circumstances physical solvent processes should be considered for gas sweetening:

1. The partial pressure of the acid gases in the feed is 50 psi or higher.
2. The concentration of heavy hydrocarbons in the feed is low. That is, the gas stream is lean in propane-plus.
3. Only bulk removal of acid gases is required.
4. Selective H_2S removal is required.

A physical solvent process is shown in Figure 7-6. The sour gas contacts the solvent using counter-current flow in the absorber. Rich solvent from the absorber bottom is flashed in stages to a pressure near atmospheric. This causes the acid-gas partial pressures to decrease; the acid gases evolve to the vapor phase and are removed. The regenerated solvent is then pumped back to the absorber.

The example in Figure 7-6 is a simple one in that flashing is sufficient to regenerate the solvent. Some solvents require a stripping column just prior to the circulation pump.

Most physical solvent processes are proprietary and are licensed by the company that developed the process.

Fluor Solvent Process®

This process uses propylene carbonate as a physical solvent to remove CO_2 and H_2S. Propylene carbonate also removes C_{2+} hydrocarbons, COS, SO_2, CS_2, and H_2O from the natural gas stream. Thus, in one step the natural gas can be sweetened and dehydrated to pipeline quality. In general, this process is used for bulk removal of CO_2 and is not used to treat to less than 3% CO_2, as may be required for pipeline quality gas. The system requires special design features, larger absorbers, and higher circulation rates to obtain pipeline quality and usually is not economically applicable for these outlet requirements.

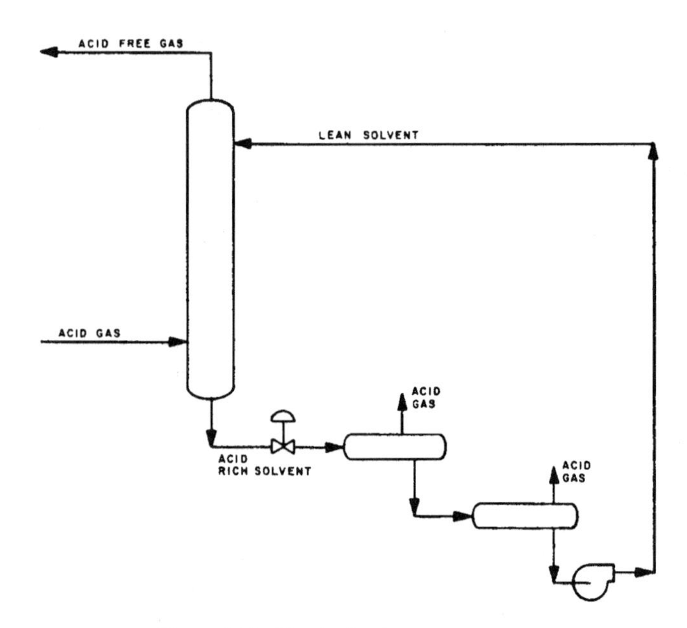

Figure 7-6. Physical solvent process.

Propylene carbonate has the following characteristics, which make it suitable as a solvent for acid gas treating:

1. High degree of solubility for CO_2 and other gases.
2. Low heat of solution for CO_2.
3. Low vapor pressure at operating temperature.
4. Low solubility for light hydrocarbons (C_1, C_2).
5. Chemically nonreactive toward all natural gas components.
6. Low viscosity.
7. Noncorrosive toward common metals.

These characteristics combine to yield a system that has low heat and pumping requirements, is relatively noncorrosive, and suffers only minimal solvent losses (less than 1 lb/MMscf).

Solvent temperatures below ambient are usually used to increase the solubility of acid gas components and therefore decrease circulation rates.

Sulfinol® Process

Licensed by Shell the Sulfinol® process combines the properties of a physical and a chemical solvent. The Sulfinol® solution consists of a mixture of sulfolane (tetrahydrothiophene 1-1 dioxide), which is a physical solvent, diisopropanolamine (DIPA), and water. DIPA is a chemical solvent that was discussed under the amines.

The physical solvent sulfolane provides the system with bulk removal capacity. Sulfolane is an excellent solvent of sulfur compounds such as H_2S, COS, and CS_2. Aromatic and heavy hydrocarbons and CO_2 are soluble in sulfolane to a lesser degree. The relative amounts of DIPA and sulfolane are adjusted for each gas stream to custom fit each application. Sulfinol® is usually used for streams with an H_2S to CO_2 ratio greater than 1:1 or where it is not necessary to remove the CO_2 to the same levels as is required for H_2S removal. The physical solvent allows much greater solution loadings of acid gas than for pure amine-based systems. Typically, a Sulfinol® solution of 40% sulfolane, 40% DIPA and 20% water can remove 1.5 moles of acid gas per mole of Sulfinol® solution.

The chemical solvent DIPA acts as secondary treatment to remove H_2S and CO_2. The DIPA allows pipeline quality residual levels of acid gas to be achieved easily. A stripper is required to reverse the reactions of the DIPA with CO_2 and H_2S. This adds to the cost and complexity of the system compared to other physical solvents, but the heat requirements are much lower than for amine systems.

A reclaimer is also required to remove oxazolidones produced in a side reaction of DIPA and CO_2.

Selexol® Process

Developed by Allied Chemical Company, this process is selective toward removing sulfur compounds. Levels of CO_2 can be reduced by approximately 85%. This process may be used economically when there are high acid-gas partial pressures and the absence of heavy ends in the gas, but it will not normally meet pipeline gas requirements. This process also removes water to less than 7 lb/MMscf. DIPA can be added to the solution to remove CO_2 down to pipeline specifications. This system then functions much like the Sulfinol® process discussed earlier. The addition of DIPA will increase the stripper heat duty; however, this duty is relatively low.

Rectisol® Process

The German Lurgi Company and Linde A. G. developed the Rectisol® process to use methanol to sweeten natural gas. Due to the high vapor pressure of methanol this process is usually operated at temperatures of −30 to −100°F. It has been applied to the purification of gas for LNG plants and in coal gasification plants, but is not used commonly to treat natural gas streams.

Direct Conversion of H_2S to Sulfur

The chemical and solvent processes previously discussed remove acid gases from the gas stream but result in a release of H_2S and CO_2 when the solvent is regenerated. The release of H_2S to the atmosphere may be limited by environmental regulations. The acid gases could be routed to an incinerator or flare, which would convert the H_2S to SO_2. The allowable rate of SO_2 release to the atmosphere may also be limited by environmental regulations. For example, currently the Texas Air Control Board generally limits H_2S emissions to 4 lb/hr (17.5 tons/year) and SO_2 emissions to 25 tons/year. There are many specific restrictions on these limits, and the allowable limits are revised periodically. In any case, environmental regulations severely restrict the amount of H_2S that can be vented or flared in the regeneration cycle.

Direct conversion processes use chemical reactions to oxidize H_2S and produce elemental sulfur. These processes are generally based either on the reaction of H_2S and O_2 or H_2S and SO_2. Both reactions yield water and elemental sulfur. These processes are licensed and involve specialized catalysts and/or solvents. A direct conversion process can be used directly on the produced gas stream. Where large flow rates are encountered, it is more common to contact the produced gas stream with a chemical or physical solvent and use a direct conversion process on the acid gas liberated in the regeneration step.

Claus® Process

This process is used to treat gas streams containing high concentrations of H_2S. The chemistry of the units involves partial oxidation of hydrogen sulfide to sulfur dioxide and the catalytically promoted reaction of H_2S and SO_2 to produce elemental sulfur. The reactions are staged and are as follows:

$$H_2S + \tfrac{3}{2}\,O_2 \rightarrow SO_2 + H_2O \qquad \text{Thermal Stage} \qquad (7\text{-}13)$$

$$SO_2 + 2H_2S \rightarrow 3S + 2H_2O \qquad \text{Thermal and Catalytic Stage} \qquad (7\text{-}14)$$

Figure 7-7 shows a simplified process flow diagram of the Claus® process. The first stage of the process converts H_2S to sulfur dioxide and

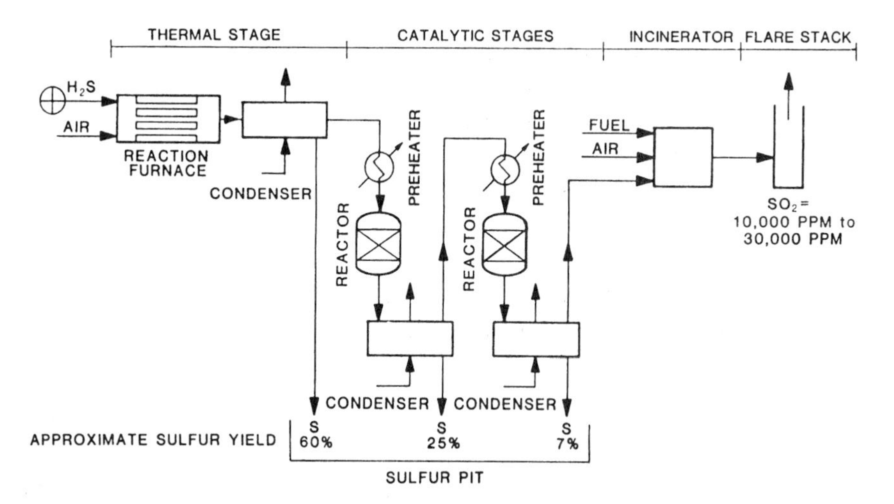

Figure 7-7. Two-stage Claus process plant.

to sulfur by burning the acid-gas stream with air in the reaction furnace. This stage provides SO_2 for the next catalytic phase of the reaction. Multiple catalytic stages are provided to achieve a more complete conversion of the H_2S. Condensors are provided after each stage to condense the sulfur vapor and separate it from the main stream. Conversion efficiencies of 94–95% can be attained with two catalytic stages, while up to 97% conversion can be attained with three catalytic stages. The effluent gas is incinerated or sent to another treating unit for "tail-gas treating" before it is exhausted to atmosphere.

There are many processes used in tail-gas treating. The Sulfreen® and the Cold Bed Absorption® (CBA) processes use two parallel reactors in a cycle, where one reactor operates below the sulfur dew point to absorb the sulfur while the second is regenerated with heat to recover molten sulfur. Even though sulfur recoveries with the additional reactors are normally 99–99.5% of the inlet stream to the Claus unit, incineration of the outlet gas may still be required.

The SCOTT® process uses an amine to remove the H_2S. The acid gas off the amine still is recycled back to the Claus plant. Other types of processes oxidize the sulfur compounds to SO_2 and then convert the SO_2 to a secondary product such as ammonium thiosulfate, a fertilizer. These plants can remove more than 99.5% of the sulfur in the inlet stream to the Claus plant and may eliminate the need for incineration. Costs of achieving this removal are high.

LOCAT® Process

The LOCAT® process is a liquid phase oxidation process based on a dilute solution of a proprietary, organically chelated iron in water that converts the hydrogen sulfide to water and elemental sulfur. The process is not reactive to CO_2. A small portion of the chelating agent degrades in some side reactions and is lost with the precipitated sulfur. Normally, sulfur is separated by gravity, centrifuging, or melting.

Figure 7-8 represents a process flow diagram of the LOCAT® process. The H_2S is contacted with the reagent in an absorber; it reacts with the dissolved iron to form elemental sulfur. The reactions involved are the following:

$$H_2S + 2Fe^{+++} \rightarrow 2H^+ + S + 2Fe^{++} \tag{7-15}$$

$$1/2\ O_2 + H_2O + 2Fe^{++} \rightarrow 2(OH)^- + 2Fe^{+++} \tag{7-16}$$

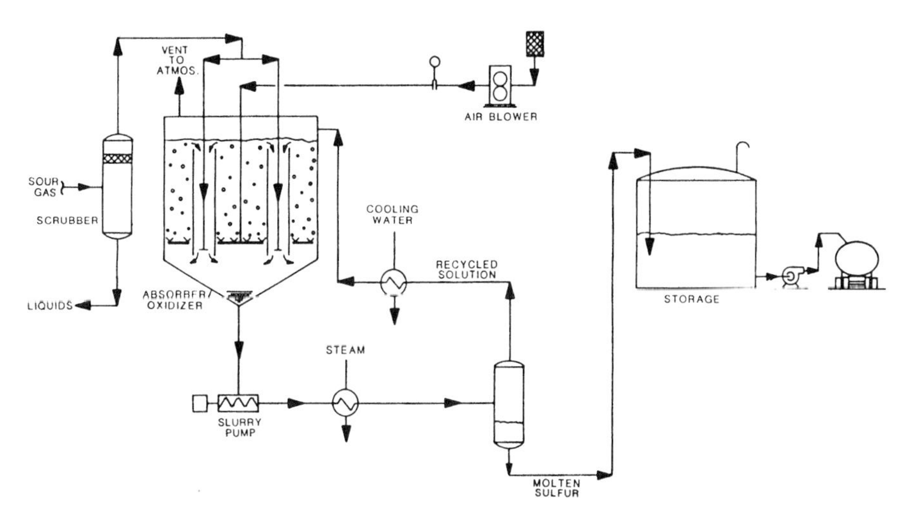

Figure 7-8. LOCAT® process.

The iron, now in a reduced ferrous form, is not consumed; instead, it is continuously regenerated by bubbling air through the solution. The sulfur precipitates out of the solution and is removed from the reactor with a portion of the reagent. The sulfur slurry is pumped to a melter requiring a small amount of heat and then to a sulfur separator where the reagent in the vapor phase is recovered, condensed, and recycled back to the reactor.

LOCAT® units can be used for tail-gas clean-up from chemical or physical solvent processes. They can also be used directly as a gas sweetening unit by separating the absorber/oxidizer into two vessels. The regenerated solution is pumped to a high-pressure absorber to contact the gas. A light slurry of rich solution comes off the bottom of the absorber and flows to an atmospheric oxidizer tank where it is regenerated. A dense slurry is pumped off the base of the oxidizer to the melter and sulfur separator.

Stretford® Process

An example of a process using O_2 to oxidize H_2S is the Stretford® process, which is licensed by the British Gas Corporation. In this process the gas stream is washed with an aqueous solution of sodium carbonate, sodium vanadate, and anthraquinone disulfonic acid. Figure 7-9 shows a simplified process diagram of the process.

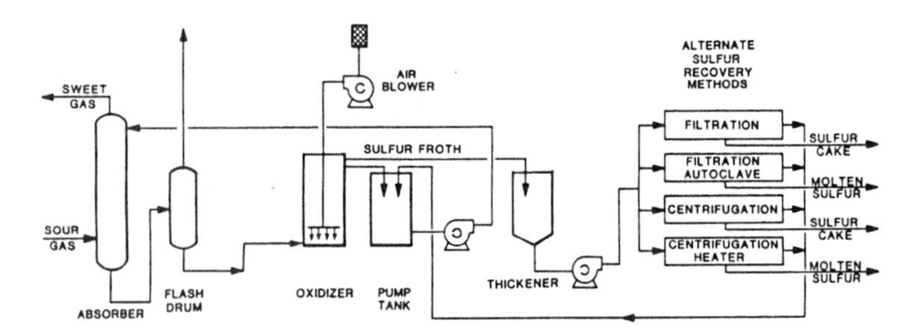

Figure 7-9. Stretford® process.

Oxidized solution is delivered from the pumping tank to the top of the absorber tower, where it contacts the gas stream in a counter-current flow. The reduced solution flows from the contactor to the solution flash drum. Hydrocarbon gases that have been dissolved in the solution are flashed and the solution flows to the base of the oxidizer vessel. Air is blown into the oxidizer, and the solution, now re-oxidized, flows to the pumping tank.

The sulfur is carried to the top of the oxidizer by a froth created by the aeration of the solution and passes into the thickener. The function of the thickener is to increase the weight percent of sulfur, which is pumped to one of the alternate sulfur recovery methods.

IFP Process

The Institute Francais du Petrole has developed a process for reacting H_2S with SO_2 to produce water and sulfur. The overall reaction is $2H_2S + SO_2 \rightarrow 2H_2O + 3S$. Figure 7-10 is a simplified diagram of the process. This process involves mixing the H_2S and SO_2 gases and then contacting them with a liquid catalyst in a packed tower. Elemental sulfur is recovered in the bottom of the tower. A portion of this must be burned to produce the SO_2 required to remove the H_2S. The most important variable is the ratio of H_2S to SO_2 in the feed. This is controlled by analyzer equipment to maintain the system performance.

Sulfa-Check®

Sulfa-Check® process uses sodium nitrite ($NaNO_2$) in aqueous solution to oxidize H_2S to sulfur. This process was developed and patented by NL Treating Chemicals and is now a product of Exxon Energy Chem-

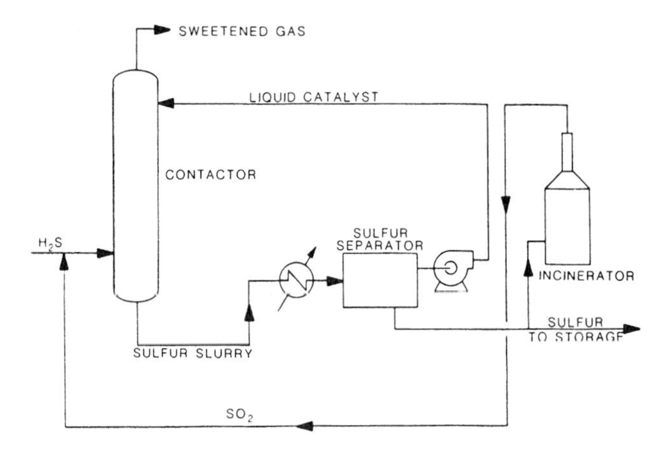

Figure 7-10. IFP process.

icals. It will generate NO_x in presence of CO_2 and O_2. Therefore, local air quality emission standard should be consulted. This process is most suited for small gas streams, generally 0.1 to 10 MMscfd and containing 100 ppm to < 1% H_2S.

Sulfa-Check® gas sweetening process is generally carried out in a contact tower. The sour gas flows into the bottom of the tower and through a sparging system to disperse the gas throughout the chemical solution. The maximum linear gas velocity should be < 0.12 ft/sec. The sweetened gas exits the contact tower at the top and goes to a gas/liquid separator to catch any liquids that may be carried over. An inverted U with a syphon breaker on top should be designed into the gas inlet line to prevent the liquid from being siphoned back. When the chemical is spent, the system is shut down to remove the spent chemical and recharged with a fresh solution to resume the operation.

Sulfide Scavengers

Sour gas sweetening may also be carried out continuously in the flowline by continuous injection of H_2S scavengers, such as amine-aldehyde condensates. Contact time between the scavenger and the sour gas is the most critical factor in the design of the scavenger treatment process. Contact times shorter than 30 sec can be accommodated with faster reacting and higher volatility formulations. The amine-aldehyde condensates process is best suited for wet gas streams of 0.5–15 MMscfd containing less than 100 ppm H_2S.

The advantages of amine-aldehyde condensates are water (or oil) soluble reaction products, lower operating temperatures, low corrosiveness to steel, and no reactivity with hydrocarbons.

Distillation

The Ryan-Holmes distillation process uses cryogenic distillation to remove acid gases from a gas stream. This process is applied to remove CO_2 for LPG separation or where it is desired to produce CO_2 at high pressure for reservoir injection. This complicated process is beyond the scope of this book.

Gas Permeation

Gas permeation is based on the mass transfer principles of gas diffusion through a permeable membrane. In its most basic form, a membrane separation system consists of a vessel divided by a single flat membrane into a high- and a low-pressure section. Feed entering the high-pressure side selectively loses the fast-permeating components to the low-pressure side. Flat plate designs are not used commercially, as they do not have enough surface area. In the hollow-fiber design, the separation modules contain anywhere from 10,000 to 100,000 capillaries, each less than 1 mm diameter, bound to a tube sheet surrounded by a metal shell. Feed gas is introduced into either the shell or the tube side. Where gas permeability rates of the components are close, or where high product purity is required, the membrane modules can be arranged in series or streams recycled.

The driving force for the separation is differential pressure. CO_2 tends to diffuse quickly through membranes and thus can be removed from the bulk gas stream. The low pressure side of the membrane that is rich in CO_2 is normally operated at 10 to 20% of the feed pressure.

It is difficult to remove H_2S to pipeline quality with a membrane system. Membrane systems have effectively been used as a first step to remove the CO_2 and most of the H_2S. An iron sponge or other H_2S treating process is then used to remove the remainder of the H_2S.

Membranes will also remove some of the water vapor. Depending upon the stream properties, a membrane designed to treat CO_2 to pipeline specifications may also reduce water vapor to less than 7 lb/MMscf. Often, however, it is necessary to dehydrate the gas downstream of the membrane to attain final pipeline water vapor requirements.

Membranes are a relatively new technology. They are an attractive economic alternative for treating CO_2 from small streams (up to 10 MMscfd). With time they may become common on even larger streams.

PROCESS SELECTION

Each of the previous treating processes has advantages relative to the others for certain applications; therefore, in selection of the appropriate process, the following facts should be considered:

1. The type of acid contaminants present in the gas stream.
2. The concentrations of each contaminant and degree of removal desired.
3. The volume of gas to be treated and temperature and pressure at which the gas is available.
4. The feasibility of recovering sulfur.
5. The desirability of selectively removing one or more of the contaminants without removing the others.
6. The presence and amount of heavy hydrocarbons and aromatics in the gas.

Figures 7-11 to 7-14 can be used as screening tools to make an initial selection of potential process choices. These graphs are not meant to supplant engineering judgment. New processes are continuously being developed. Modifications to existing proprietary products will change their range of applicability and relative cost. The graphs do enable a first choice of several potential candidates that could be investigated to determine which is the most economical for a given set of conditions.

To select a process, determine flow rate, temperature, pressure, concentration of the acid gases in the inlet gas, and allowed concentration of acid gases in the outlet stream. With this information, calculate the partial pressure of the acid gas components.

$$PP_i = X_i P_t \quad (7\text{-}17)$$

where PP_i = partial pressure of component i, psia
P_t = systems pressure, psia
X_i = mole fraction of component i

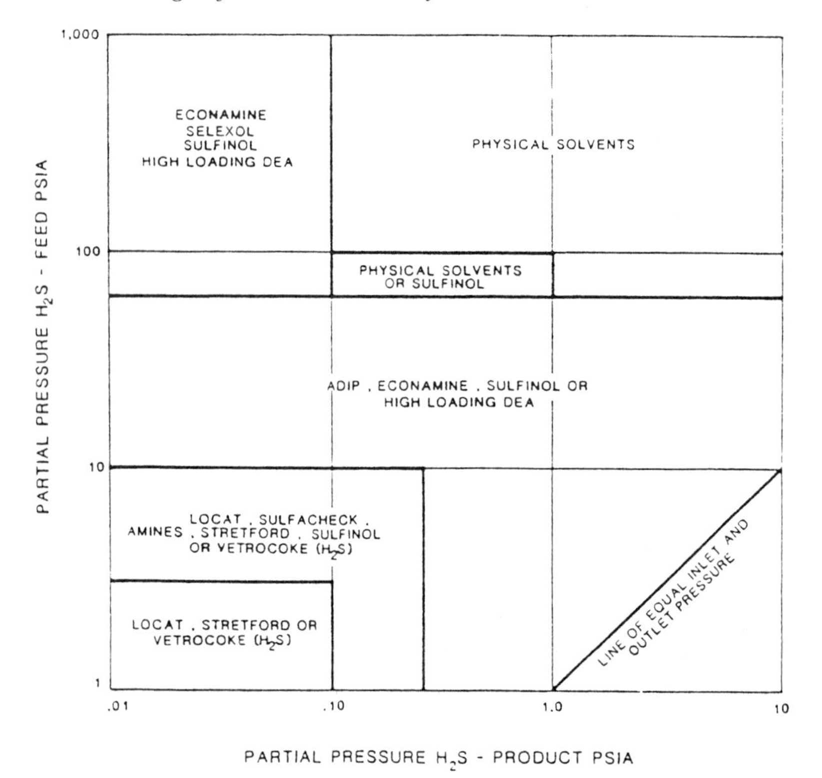

Figure 7-11. H_2S removal, no CO_2.

Next, determine if one of the four following situations is required and use the appropriate guide:

Removal of H_2S with no CO_2 present (Figure 7-11)
Removal of H_2S and CO_2 (Figure 7-12)
Removal of CO_2 with no H_2S present (Figure 7-13)
Selective removal of H_2S with CO_2 present (Figure 7-14)

DESIGN PROCEDURES FOR IRON-SPONGE UNITS

The iron-sponge process generally uses a single vessel to contain the hydrated ferric oxide wood shavings. A drawing of an iron-sponge unit showing typical provisions for internal and external design requirements was presented in Figure 7-3.

The inlet gas line should have taps for gas sampling, temperature measurement, pressure measurement, and for an injection nozzle for

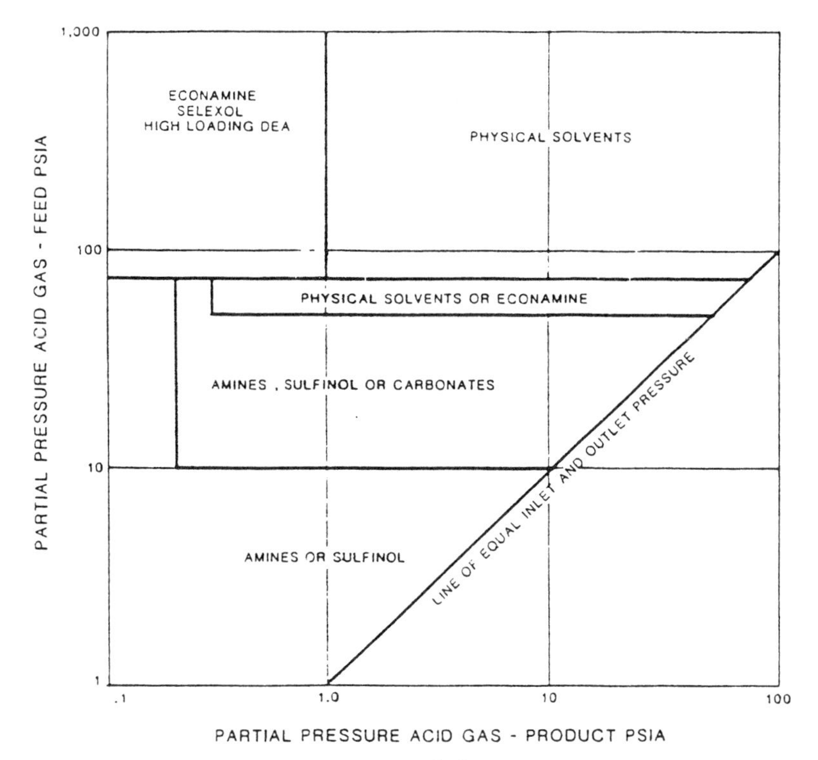

Figure 7-12. Removal of H_2S and CO_2.

methanol, water, or inhibitors. The gas is carried into the top section of the vessel in a distributor and discharged upward. This causes the gas to reverse flow downward and provides for more uniform flow through the bed, minimizing the potential for channeling.

Supporting the hydrated ferric oxide chips is a combination of a perforated, heavy metal support plate and a coarse support packing material. This material may consist of scrap pipe thread protectors and 2–3-in. sections of small diameter pipe. This provides support for the bed, while offering some protection against detrimental pressure surges.

Gas exits the vessel at the bottom through the vessel sidewall. This arrangement minimizes entrainment of fines. Additionally, a cone strainer should be included in the exit line. This line should also have a pressure tap and sample test tap.

The vessel is generally constructed of carbon steel that has been heat treated. Control of metal hardness is required because of the potential of sulfide-stress cracking. The iron-sponge vessel is either internally coated or clad with stainless steel.

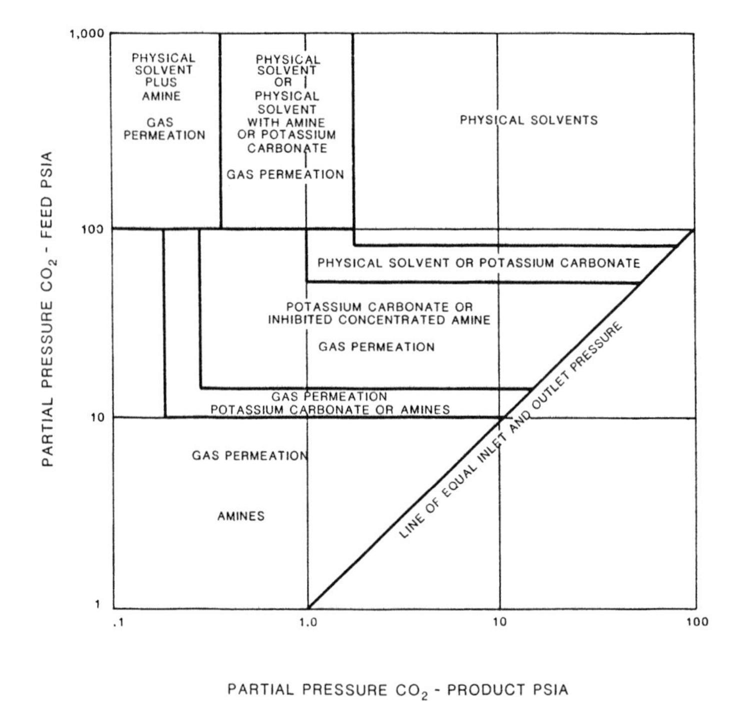

Figure 7-13. CO_2 removal, no H_2S.

The superficial gas velocity (that is, gas flow rate divided by vessel cross-sectional area) through the iron-sponge bed is normally limited to a maximum of 10 ft/min at actual flow conditions to promote proper contact with the bed and to guard against excessive pressure drop. Thus, the vessel minimum diameter is given by:

$$d_{min}^2 = 360 \frac{Q_g T Z}{P} \tag{7-18}$$

where d_{min} = minimum required vessel diameter, in.
Q_g = gas flow rate, MMscfd
T = operating temperature, °R
Z = compressibility factor
P = operating pressure, psia

A maximum rate of deposition of 15 grains of $H_2S/min/ft^2$ of bed cross-sectional area is also recommended to allow for the dissipation of

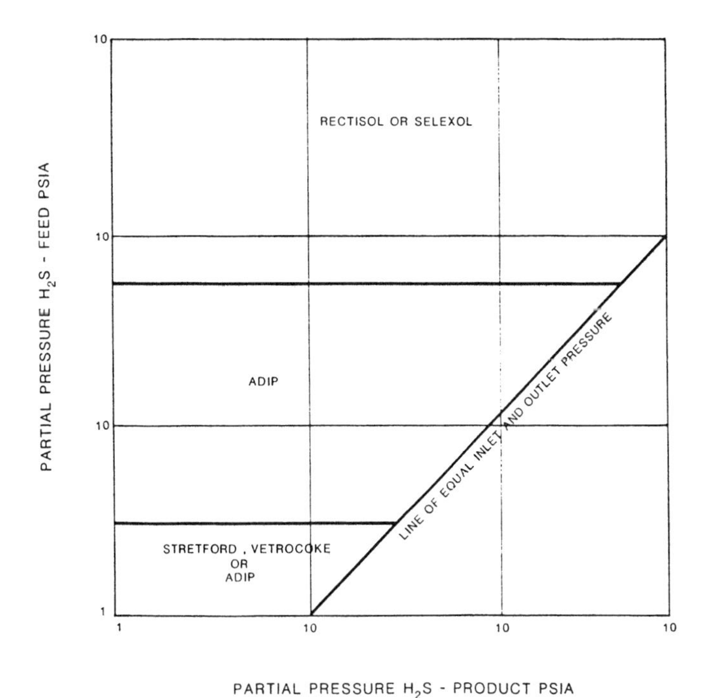

Figure 7-14. Selective removal of H_2S in presence of CO_2.

the heat of reaction. This requirement also establishes a minimum required diameter, which is given by:

$$d^2_{min} = 5.34 \times 10^6 \, Q_g \, (MF) \tag{7-19}$$

where d_{min} = minimum required vessel diameter, in.
Q_g = gas flow rate, MMscfd
MF = mole fraction of H_2S

The larger of the diameters calculated by Equation 7-18 or 7-19 will set the minimum vessel diameter. Any choice of diameter equal to or larger than this diameter will be an acceptable choice.

At very low superficial gas velocities (less than 2 ft/min) channeling of the gas through the bed may occur. Thus, it is preferred to limit the vessel diameter to:

$$d^2_{max} = 1{,}800 \frac{Q_g T Z}{P} \tag{7-20}$$

where d_{max} = maximum recommended vessel diameter, in.

A contact time of 60 seconds is considered a minimum in choosing a bed volume. A larger volume may be considered, as it will extend the bed life and thus extend the cycle time between bed change outs. Assuming a minimum contact time of 60 seconds, any combination of vessel diameter and bed height that satisfies the following is acceptable:

$$d^2H \geq 3{,}600 \frac{Q_g T Z}{P} \tag{7-21}$$

where d = vessel diameter, in.
H = bed height, ft

In selecting acceptable combinations, the bed height should be at least 10 ft for H_2S removal and 20 ft for mercaptan removal. This height will produce sufficient pressure drop to assure proper flow distribution over the entire cross-section. Thus, the correct vessel size will be one that has a bed height of at least 10 ft (20 ft if mercaptans must be removed) and a vessel diameter between d_{min} and d_{max}.

Iron sponge is normally sold in the U.S. by the bushel. The volume in bushels can be determined from the following equation once the bed dimensions of diameter and height are known:

$$Bu = 4.4 \times 10^{-3} d^2 H \tag{7-22}$$

where Bu = volume, bushels

The amount of iron oxide that is impregnated on the wood chips is normally specified in units of pounds of iron oxide (Fe_2O_3) per bushel. Common grades are 9, 15 or 20 lb Fe_2O_3/bushel.

Bed life for the iron sponge between change outs is determined from:

$$t_c = 3.14 \times 10^{-8} \frac{Fe\, d^2\, H\, e}{Q_g MF} \tag{7-23}$$

where t_c = cycle time, days
Fe = iron-sponge content, lb Fe_2O_3/bushel
e = efficiency (0.65 to 0.8)
MF = mole fraction H_2S

The iron-sponge material is normally specified to have a size distribution with 0% retained on 16 mesh, 80% between 30 and 60 mesh, and 100% retained on 325 mesh. It is purchased with a moisture content of 20% by weight and buffering to meet a flood pH of 10. Because it is necessary to maintain a moist alkaline condition, provisions should be included in the design to add water and caustic.

DESIGN PROCEDURES FOR AMINE SYSTEMS

The types of equipment and thc methods for designing the equipment are similar for both MEA and DEA systems. For other amine systems such as SNPA-DEA, Fluor Econamine (DGA), and Shell ADIP (DIPA) the licensee should be contacted for detailed design information.

Amine Absorber

Amine absorbers use counter-current flow through a trayed or packed tower to provide intimate mixing between the amine solution and the sour gas. Typically, small diameter towers use stainless steel packing, while larger towers use stainless steel trays. For systems using the recommended solution concentrations and loadings, a tower with 20 to 24 actual trays is normal. Variations in solution concentrations and loadings may require further investigation to determine the number of trays.

In a trayed absorber the amine falls from one tray to the one below in the same manner as the liquid in a condensate stabilizer (Chapter 6, Figure 6-4). It flows across the tray and over a weir before flowing into the next downcomer. The gas bubbles up through the liquid and creates a froth that must be separated from the gas before it reaches the underside of the next tray. For preliminary design, a tray spacing of 24 in. and a minimum diameter capable of separating 150 to 200 micron droplets (using the equations developed in Volume 1 for gas capacity of a vertical separator) can be assumed. The size of packed towers must be obtained from manufacturer's published literature.

Commonly, amine absorbers include an integral gas scrubber section in the bottom of the tower. This scrubber would be the same diameter as required for the tower. The gas entering the tower would have to pass through a mist eliminator and then a chimney tray. The purpose of this scrubber is to remove entrained water and hydrocarbon liquids from the gas to protect the amine solution from contamination.

Alternately, a separate scrubber vessel can be provided so that the tower height can be decreased. This vessel should be designed in accordance with the procedures in Volume 1 for design of two-phase separators.

For MEA systems with a large gas flow rate, a scrubber should be considered for the outlet sweet gas. The vapor pressure of MEA is such that the separator may be helpful in reducing MEA losses in the overhead sweet gas. DEA systems do not require this scrubber because the vapor pressure of DEA is very low.

Amine Circulation Rates

The circulation rates for amine systems can be determined from the acid gas flow rates by selecting a solution concentration and an acid gas loading.

The following equations can be used:

$$L_{MEA} = \frac{112\,(Q_g)\,(MF)}{(c)\,(\rho)\,(A_L)} \tag{7-24}$$

$$L_{DEA} = \frac{192\,(Q_g)\,(MF)}{(c)\,(\rho)\,(A_L)} \tag{7-25}$$

where L_{MEA} = MEA circulation rate, gpm
L_{DEA} = DEA circulation rate, gpm
Q_g = gas flow rate, MMscfd
MF = total acid-gas fraction in inlet gas, moles acid gas/mole inlet gas
c = amine weight fraction, lb amine/lb solution
ρ = solution density, lb/gal at 60°F
A_L = acid-gas loading, mole acid-gas/mole amine

For design, the following solution strengths and loadings are recommended to provide an effective system without an excess of corrosion:

MEA: c = 20 wt. %
A_L = 0.33 mole acid gas/mole MEA
DEA: c = 35 wt. %
A_L = 0.5 mole acid gas/mole DEA

For the recommended concentrations the densities at 60°F are:

20% MEA = 8.41 lb/gal = 0.028 mole MEA/gal
35 % DEA = 8.71 lb/gal = 0.029 mole DEA/gal

Using these design limits, Equations 7-24 and 7-25 can be simplified to:

$$L_{MEA} = 201\,(Q_g)\,(MF) \tag{7-26}$$

$$L_{DEA} = 126\,(Q_g)\,(MF) \tag{7-27}$$

The circulation rate determined with these equations should be increased by 10–15% to supply an excess of amine.

Flash Drum

The rich amine solution from the absorber is flashed to a separator to remove any hydrocarbons. A small percentage of acid gases will also flash when the pressure is reduced. The dissolved hydrocarbons should flash to the vapor phase and be removed. However, a small amount of hydrocarbon liquid may begin to collect in this separator. Therefore, a provision should be made to remove these liquid hydrocarbons.

Typically the flash tanks are designed for 2 to 3 minutes of retention time for the amine solution while operating half full.

Amine Reboiler

The reboiler provides the heat input to an amine stripper, which reverses the chemical reactions and drives off the acid gases. Amine reboilers may be either a kettle reboiler (see Chapter 3) or an indirect fired heater (see Chapter 5).

The heat duty of amine reboilers varies with the system design. The higher the reboiler duty, the higher the overhead condenser duty, the higher the reflux ratio, and thus the lower the number of trays required. The lower the reboiler duty, the lower the reflux ratio will be and the more trays the tower must have.

Typically for a stripper with 20 trays, the reboiler duties will be as follows:

MEA system—1,000 to 1,200 Btu/gal lean solution
DEA system—900 to 1,000 Btu/gal lean solution

For design, reboiler temperatures in a stripper operating at 10 psig can be assumed to be 245°F for 20% MEA and 250°F for 35% DEA.

Amine Stripper

Amine strippers use heat and steam to reverse the chemical reactions with CO_2 and H_2S. The steam acts as a stripping gas to remove the CO_2 and H_2S from the liquid solution and to carry these gases to the overhead. To promote mixing of the solution and the steam, the stripper is a trayed or packed tower with packing normally used for small diameter columns.

The typical stripper consists of a tower operating at 10–20 psig with 20 trays, a reboiler, and an overhead condenser. The rich amine feed is introduced on the third or fourth tray from the top. The lean amine is removed at the bottom of the stripper and acid gases are removed from the top.

Liquid flow rates are greatest near the bottom tray of the tower where the liquid from the bottom tray must provide the lean-amine flow rate from the tower plus enough water to provide the steam generated by the reboiler. The lean-amine circulation rate is known, and from the reboiler duty, pressure, and temperature, the amount of steam generated and thus the amount of water can be calculated.

The vapor flow rate within the tower must be studied at both ends of the stripper. The higher of these vapor rates should be used to size the tower for vapor. At the bottom of the tower the vapor rate equals the amount of steam generated in the reboiler. Near the top of the tower the vapor rate equals the steam rate overhead plus the acid-gas rate. The steam overhead can be calculated from the steam generated in the reboiler by subtracting the amount of steam condensed by raising the lean amine from its inlet temperature to the reboiler temperature and the amount of steam condensed by vaporizing the acid gases.

For most field gas units it is not necessary to specify a stripper size. Vendors have standard design amine circulation packages for a given amine circulation rate, acid-gas loading, and reboiler. These concepts can be used in a preliminary check of the vendor's design. However, for detailed design and specification of large units, a process simulation computer model should be used.

Overhead Condenser and Reflux Accumulator

Amine-stripper overhead condensers are typically air-cooled, fin-fan exchangers. Their duty can be determined from the concepts in Chapter 3 as required to cool the overhead gases and condense the overhead steam to water. Thc inlet temperature to the cooler can be found using the partial pressure of the overhead steam to determine the temperature from steam tables. The cooler outlet temperature is typically 130 to 145°F depending on the ambient temperature.

The reflux accumulator is a separator used to separate the acid gases from the condensed water. The water is accumulated and pumped back to the top of the stripper as reflux. With the vapor and liquid rates known, the accumulator can be sized using the procedures in Volume 1 for two-phase separators.

Rich/Lean Amine Exchanger

Rich/lean amine exchangers are usually shell-and-tube exchangers with the corrosive rich amine flowing through the tubes. The purpose of these exchangers is to reduce the reboiler duty by recovering some of the sensible heat from the lean amine.

The flow rates and inlet temperatures are typically known. Therefore, the outlet temperatures and duty can be determined by assuming an approach temperature for one outlet. The closer the approach temperature selected, the greater the duty and heat recovered, but the larger and more costly the exchanger. For design, an approach temperature of about 30°F provides an economic design balancing the cost of the rich/lean exchanger and the reboiler. The reboiler duties recommended above assume a 30°F approach.

Amine Cooler

The amine cooler is typically an air-cooled, fin-fan cooler, which lowers the lean amine temperature before it enters the absorber. The lean amine entering the absorber should be approximately 10°F warmer than the sour gas entering the absorber. Lower amine temperatures may cause the gas to cool in the absorber and thus condense hydrocarbon liquids. Higher temperatures would increase the amine vapor pressure and thus increase amine losses to the gas. The duty for the cooler can be calculated from the lean-amine flow rate, the lean-amine temperature leaving the rich/lean exchanger and the sour-gas inlet temperature.

Amine Solution Purification

Due to side reactions and/or degradation, a variety of contaminants will begin to accumulate in an amine system. The method of removing these depends on the amine involved.

When MEA is used in the presence of COS and CS_2, they react to form heat-stable salts. Therefore, MEA systems usually include a reclaimer. The reclaimer is a kettle-type reboiler operating on a small side stream of lean solution. The temperature in the reclaimer is maintained such that the water and MEA boil to the overhead and are piped back to the stripper. The heat-stable salts remain in the reclaimer until the reclaimer is full. Then the reclaimer is shut-in and dumped to a waste disposal. Thus, the impurities are removed but the MEA bonded to the salts is also lost.

For DEA systems a reclaimer is not required because the reactions with COS and CS_2 are reversed in the stripper. The small amount of degradation products from CO_2 can be removed by a carbon filter on a side stream of lean solution.

Materials of Construction

Amine systems are extremely corrosive due to the acid-gas concentrations and the high temperatures. It is important that all carbon steel exposed to the amine be stress-relieved after the completion of welding on the particular piece. A system fabricated from stress-relieved carbon steel for DEA solutions, as recommended, will not suffer excessive corrosion. For MEA systems, corrosion-resistant metals (304 SS) should be used in the following areas:

1. Absorber trays or packing
2. Stripper trays or packing
3. Rich/Lean amine exchanger tubes
4. Any part of the reboiler tube bundle that may be exposed to the vapor phase
5. Reclaimer tube bundle
6. Pressure-reduction valve and pipe leading to the flash tank
7. Pipe from the rich/lean exchange to the stripper inlet

EXAMPLE PROBLEMS

Example No. 7-1: Iron-Sponge Unit

Given: $Q_g = 2$ MMscfd
$SG = 0.6$
$H_2S = 19$ ppm
$P = 1{,}200$ psig
$T = 100°F$

Problem: Design an Iron-Sponge Unit

Solution:

1. Minimum diameter for gas velocity:

$$d_{min}^2 = 360 \frac{Q_g \, T \, Z}{P}$$

$$Z = 0.85 \text{ (From Volume 1, Figure 3-2)}$$

$$d_{min}^2 = 360 \frac{(2)(560)(0.85)}{1{,}215}$$

$$d_{min} = 16.8 \text{ in.}$$

2. Minimum diameter for deposition:

$$d_{min}^2 = 5.34 \times 10^6 \, Q_g \, (MF)$$

$$d_{min}^2 = 5.34 \times 10^6 (2) \left(\frac{19}{1{,}000{,}000} \right)$$

$$d_{min} = 14.2 \text{ in. (Does not govern)}$$

3. Minimum diameter to prevent channeling:

$$d_{max}^2 = 1{,}800 \frac{Q_g \, T \, Z}{P}$$

$$d_{max} = 37.6 \text{ in.}$$

Therefore, any diameter from 16.8 in. to 37.6 in. is acceptable.

4. Choose a cycle time for one month:

$$t_c = 3.14 \times 10^{-8} \frac{Fe \, d^2 \, H \, e}{Q_g \, MF}$$

Assume Fe = 9 lb/Bu and rearrange:

$$d^2 H = \frac{(30)(2)(19/1{,}000{,}000)}{(3.14 \times 10^{-8})(9)(0.65)}$$

$$d^2 H = 6{,}206$$

d in.	H ft
18	19.2
20	15.5
22	12.8
24	10.8
30	6.9
36	4.8

An acceptable choice is a 30-in. OD vessel. The wall thickness can be calculated from Chapter 12 and a value of bed height determined. However, since t_c and e are arbitrary, a 10-ft bed seems appropriate.

5. Calculate volume of iron sponge to purchase:

$$Bu = 4.4 \times 10^{-3}\, d^2H$$
$$Bu = (4.4 \times 10^{-3})(30)^2(10)$$
$$Bu = 39.6 \text{ bushels}$$

Example No. 7-2: Specify Major Parameters for DEA

Given:

$Q_g = 100$ MMscfd
$S = 0.67$
$P = 1{,}000$ psig
$T = 100°F$
CO_2 inlet = 4.03%
outlet = 2%
H_2S inlet = 19 ppm = 0.0019%
outlet = ¼ grain/Mscf = 4 ppm
C_D (contactor) = 0.689 (From Volume 1, Chapter 4)

Required:

1. Show that a DEA unit is an acceptable process selection.
2. Determine DEA circulation rate using 35 wt. % DEA and an acid-gas loading of 0.50 mole acid gas/mole DEA.
3. Determine preliminary diameter and height for DEA contact tower.
4. Calculate approximate reboiler duty with 250°F reboiler temperature.

Solution:

1. Process selection
   Total acid gas inlet = 4.03 + 0.0019
   = 4.032%
   $P_{acid,\ in} = 1{,}015 \times (4.032/100) = 40.9$ psia
   Total acid gas outlet = 2.0%
   $P_{acid,\ out} = 1{,}015 \times (2.0/100) = 20.3$ psia
   From Figure 7-12 for removing CO_2 and H_2S, possible processes are amines, Sulfinol, or carbonates.
   The most common selection for this application is a DEA unit.

2. DEA circulation rate

$$L_{DEA} = \frac{192\, Q_g\, MF}{c\, \rho\, A_L}$$

$\rho = 8.71$ lb/gal
$c = 0.35$ lb/lb
$A_L = 0.50$ mole/mole
$Q_g = 100$ MMscfd
$MF = 4.032\% = 0.04032$

Note: In order to meet the H_2S outlet, virtually all the CO_2 must be removed, as DEA is not selective for H_2S.

$$L_{DEA} = \frac{192\,(100)\,(0.04032)}{(0.35)\,(8.71)\,(0.50)}$$

$$= 508 \text{ gpm}$$

3. Tower size

From Volume 1:

$$d^2 = 5{,}040\, \frac{T\, Z\, Q_g}{P} \left[\left(\frac{\rho_g}{\rho_l - \rho_g} \right) \frac{C_D}{d_m} \right]^{1/2}$$

C_D = 0.689 (given)
d_m = 150 (assumed)
T = 560°R
P = 1015 psia
Q_g = 100 MMscfd
Z = 0.84 (Volume 1)

$$\rho_g = \frac{(2.7)(0.67)(1{,}015)}{(560)(0.84)} = 3.90 \text{ lb/ft}^3 \text{ (Volume 1)}$$

$$\rho_1 = 8.71 \text{ lb/gal} = 65.1 \text{ lb/ft}^3$$

$$d^2 = 5{,}040 \frac{(560)(0.84)(100)}{1{,}015} \left[\left(\frac{3.90}{65.1 - 3.90} \right) \frac{0.689}{150} \right]^{1/2}$$

$$d = 63.2 \text{ in.}$$

Use 72-in. ID tower w/24 trays.

4. Determine reboiler duty:

Using 1,000 Btu/gal lean solution

q = 1,000(508)(60 min/hr)
q = 30.5 MMBtu/hr

CHAPTER

8

Gas Dehydration*

Gas dehydration is the process of removing water vapor from a gas stream to lower the temperature at which water will condense from the stream. This temperature is called the "dew point" of the gas. Most gas sales contracts specify a maximum value for the amount of water vapor allowable in the gas. Typical values are 7 lb/MMscf in the Southern U.S., 4 lb/MMscf in the Northern U.S. and 2 to 4 lb/MMscf in Canada. These values correspond to dew points of approximately 32°F for 7 lb/ MMscf, 20°F for 4 lb MMscf, and 0°F for 2 lb/MMscf in a 1,000 psi gas line.

Dehydration to dew points below the temperature to which the gas will be subjected will prevent hydrate formation and corrosion from condensed water. The latter consideration is especially important in gas streams containing CO_2 or H_2S where the acid gas components will form an acid with the condensed water.

The capacity of a gas stream for holding water vapor is reduced as the stream is compressed or cooled. Thus, water can be removed from the gas stream by compressing or cooling the stream. However, the gas stream is still saturated with water so that further reduction in temperature or increase in pressure can result in water condensation.

This chapter discusses the design of liquid glycol and solid bed dehydration systems that are the most common methods of dehydration used

*Reviewed for the 1999 edition by Lindsey S. Stinson of Paragon Engineering Services, Inc.

for natural gas. In producing operations gas is most often dehydrated by contact with triethylene glycol. Solid bed adsorption units are used where very low dew points are required, such as on the inlet stream to a cryogenic gas plant where water contents of less than 0.05 lb/MMscf may be necessary.

WATER CONTENT DETERMINATION

The first step in evaluating and/or designing a gas dehydration system is to determine the water content of the gas. The water content of a gas is dependent upon gas composition, temperature, and pressure. For sweet natural gases containing over 70% methane and small amounts of "heavy ends," the McKetta-Wehe pressure-temperature correlation, as shown in Figure 8-1, may be used. As an example, assume it is desired to determine the water content for a natural gas with a molecular weight of 26 that is in equilibrium with a 3% brine at a pressure of 3,000 psia and a temperature of 150°F. From Figure 8-1 at a temperature of 150°F and pressure of 3,000 psia there is 104 lb of water per MMscf of wet gas. The correction for salinity is 0.93 and for molecular weight is 0.98. Therefore, the total water content is $104 \times 0.93 \times 0.98 = 94.8$ lb/MMscf.

A correction for acid gas should be made when the gas stream contains more than 5% CO_2 and/or H_2S. Figures 8-2 and 8-3 may be used to determine the water content of a gas containing less than 40% total concentration of acid gas. As an example, assume the example gas from the previous paragraph contains 15% H_2S. The water content of the hydrocarbon gas is 94.8 lb/MMscf. From Figure 8-3, the water content of H_2S is 400 lb/MMscf. The effective water content of the stream is equal to $(0.85)(94.8) + (0.15)(400)$ or 141 lb/MMscf.

GLYCOL DEHYDRATION

By far the most common process for dehydrating natural gas is to contact the gas with a hygroscopic liquid such as one of the glycols. This is an absorption process, where the water vapor in the gas stream becomes dissolved in a relatively pure glycol liquid solvent stream. Glycol dehydration is relatively inexpensive, as the water can be easily "boiled" out of the glycol by the addition of heat. This step is called "regeneration" or "reconcentration" and enables the glycol to be recovered for reuse in absorbing additional water with minimal loss of glycol.

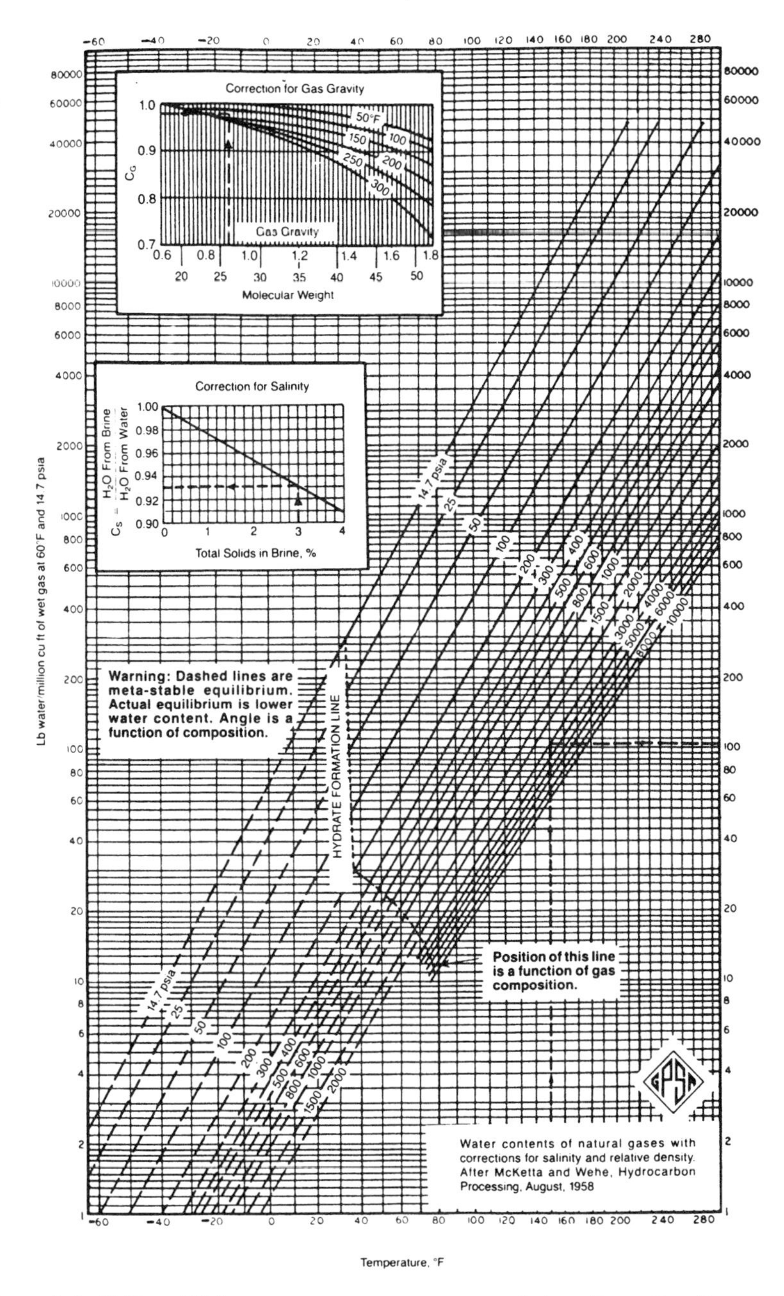

Figure 8-1. McKetta-Wehe pressure-temperature correlation. (From Gas Processors Suppliers Association, *Engineering Data Book,* 10th Edition.)

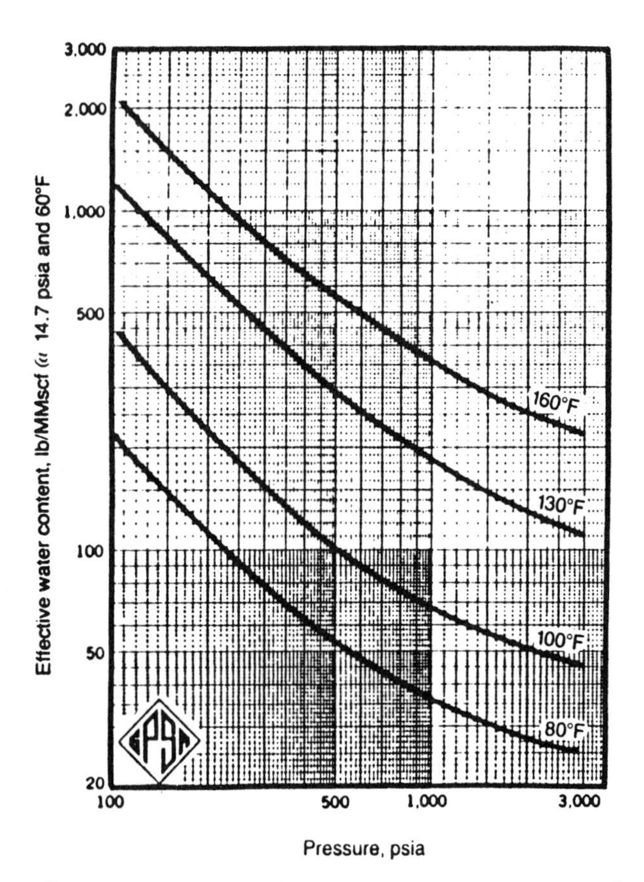

Figure 8-2. Effective water content of CO_2. (From Gas Processors Suppliers Association, *Engineering Data Book,* 10th Edition.)

Process Description

Most glycol dehydration processes are continuous. That is, gas and glycol flow continuously through a vessel (the "contactor" or "absorber") where they come in contact and the glycol absorbs the water. The glycol flows from the contactor to a "reboiler" (sometimes called "reconcentrator" or "regenerator") where the water is removed or "stripped" from the glycol and is then pumped back to the contactor to complete the cycle.

Figure 8-4 shows a typical trayed contactor in which the gas and liquid are in counter-current flow. The wet gas enters the bottom of the contactor and contacts the "richest" glycol (glycol containing water in solution)

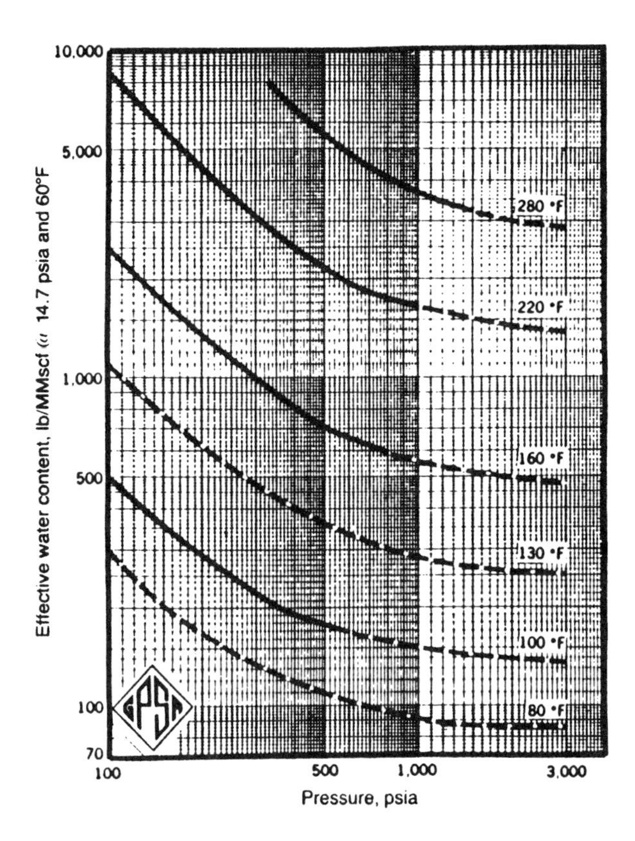

Figure 8-3. Effective water content of H_2S. (From Gas Processors Suppliers Association, *Engineering Data Book,* 10th Edition.)

just before the glycol leaves the column. The gas encounters leaner and leaner glycol (that is, glycol containing less and less water in solution), as it rises through the contactor. At each successive tray the leaner glycol is able to absorb additional amounts of water vapor from the gas. The counter-current flow in the contactor makes it possible for the gas to transfer a significant amount of water to the glycol and still approach equilibrium with the leanest glycol concentration.

The contactor works in the same manner as a condensate stabilizer tower described in Chapter 6. As the glycol falls from tray to tray it becomes richer and richer in water. As the gas rises it becomes leaner and leaner in water vapor. Glycol contactors will typically have between 6 and 12 trays, depending upon the water dew point required. To obtain a 7 lb/MMscf specification, 6 to 8 trays are common.

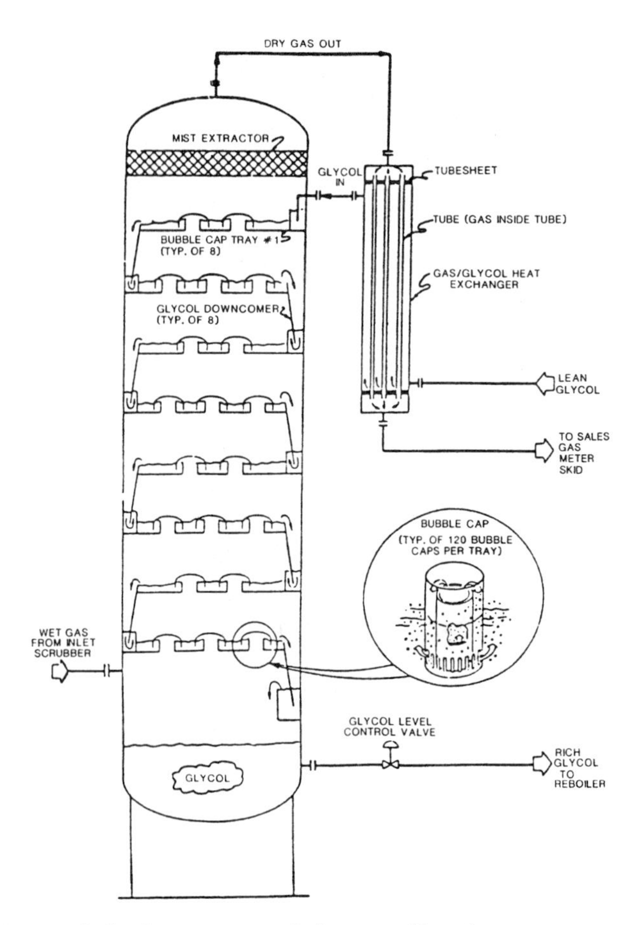

Figure 8-4. Typical glycol contactor in which gas and liquid are in counter-current flow.

As with a condensate stabilizer, glycol contactors may have bubble cap trays as shown in Figure 8-4, or they may have valve trays, perforated trays, regular packing or structured packing. Contactors that are 12¾ in. and less in diameter usually use regular packing, while larger contactors usually use bubble cap trays to provide adequate contact at gas flow rates much lower than design. Structured packing is becoming more common for very large contactors.

It is possible to inject glycol in a gas line and have it absorb the water vapor in co-current flow. Such a process is not as efficient as countercurrent flow, since the best that can occur is that the gas reaches near equilibrium with the *rich* glycol as opposed to reaching near equilibrium with the *lean* glycol as occurs in counter-current flow. Partial co-current flow can be used to reduce the height of the glycol contactor by eliminating the need for some of the bottom trays.

The glycol will absorb heavy hydrocarbon liquids present in the gas stream. Thus, before the gas enters the contactor it should pass through a separate inlet gas scrubber to remove liquid and solid impurities that may carry over from upstream vessels or condense in lines leading from the vessels. The inlet scrubber should be located as close as possible to the contactor.

On larger streams filter separators (Volume 1, Chapter 4) are used as inlet scrubbers to further reduce glycol contamination and thus increase the life of the glycol charge. Due to their cost, filter separators are not normally used on streams less than approximately 50 MMscfd. Often on these smaller units a section in the bottom of the contactor is used as a vertical inlet scrubber as shown in Figure 8-5.

Dry gas from the top of the gas/glycol contactor flows through an external gas/glycol heat exchanger. This cools the incoming dry glycol to increase its absorption capacity and decrease its tendency to flash in the contactor and be lost to the dry gas. In some systems, the gas passes over a glycol cooling coil inside the contactor instead of the external gas/glycol heat exchanger.

The glycol reconcentration system is shown in Figure 8-6. The rich or "wet" glycol from the base of the contactor passes through a reflux condensor to the glycol/glycol preheater where the rich glycol is heated by the hot lean glycol to approximately 170°F to 200°F. After heating, the glycol flows to a low pressure separator operating at 35 to 50 psig, where the entrained gas and any liquid hydrocarbons present are removed. The glycol/condensate separator is a standard three-phase vessel designed for at least 15–30 minutes retention time and may be either horizontal or vertical. It is important to heat the glycol before flowing to this vessel to reduce its viscosity and encourage easier separation of condensate and gas.

The gas from the glycol/condensate separator can be used for fuel gas. In many small field gas packaged units this gas is routed directly to fire tubes in the reboiler, and provides the heat for reconcentrating the glycol. This separator is sometimes referred to as a gas/glycol separator or "pump gas" separator.

The wet glycol from the separator flows through a sock filter to remove solids and a charcoal filter to absorb small amounts of hydrocarbons that may build up in the circulating glycol. Sock filters are normally designed for the removal of 5-micron solids. On units larger than 10 gpm

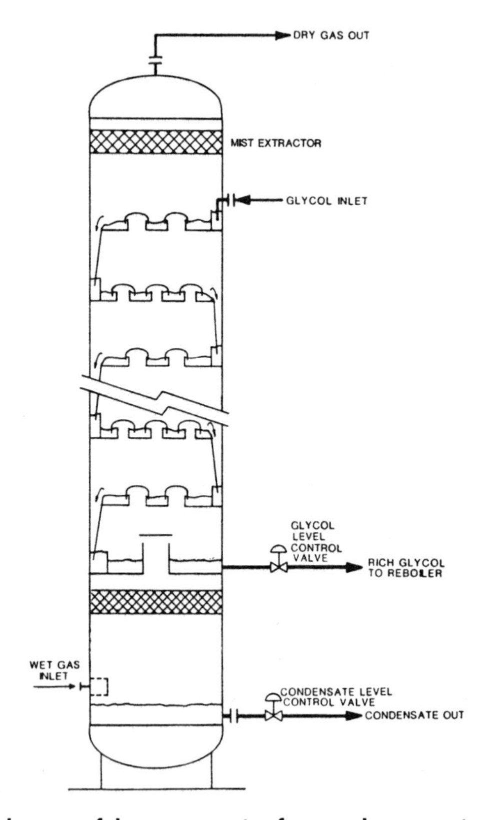

Figure 8-5. The bottom of the contactor is often used as a vertical inlet scrubber.

it is common to route only a sidestream of 10 to 50% of total glycol flow through the charcoal filter. The filters help minimize foaming and sludge build-up in the reconcentrator.

The glycol then flows through the glycol/glycol heat exchanger to the still column mounted on the reconcentrator, which operates at essentially atmospheric pressure. As the glycol falls through the packing in the still column, it is heated by the vapors being boiled off the liquids in the reboiler. The still works in the same manner as a condensate stabilizer. The falling liquid gets hotter and hotter. The gas flashing from this liquid is mostly water vapor with a small amount of glycol. Thus, as the liquid falls through the packing it becomes leaner and leaner in water. Before the vapors leave the still, they encounter the reflux condenser. The cold rich glycol from the contactor cools them, condensing the glycol vapors and approximately 25 to 50% of the rising water vapor. The result is a

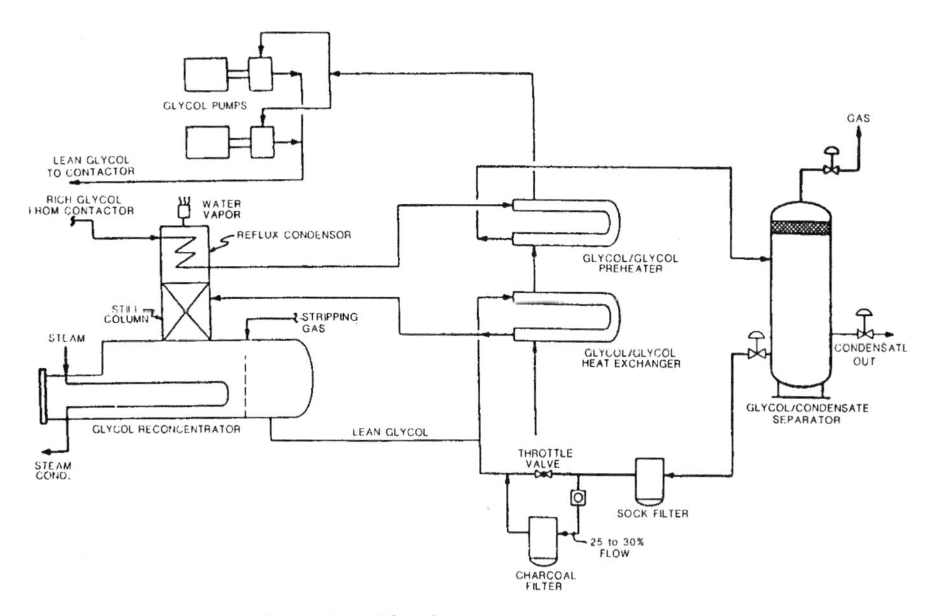

Figure 8-6. Glycol reconcentration system.

reflux liquid stream, which reduces the glycol losses to atmosphere to almost zero. The water vapor exiting the top of the still contains a small amount of volatile hydrocarbons and is normally vented to atmosphere at a safe location. If necessary, the water vapor can be condensed in an aerial cooler and routed to the produced water treating system to eliminate any potential atmospheric hydrocarbon emission.

Since there is a large difference between the boiling point of triethylene glycol (546°F) and water (212°F), the still column can be relatively short (10 to 12 ft of packing). The glycol liquid in the reboiler is heated to 340°F to 400°F to provide the heat necessary for the still column to operate. Higher temperatures would vaporize more water, but may degrade the glycol.

If a very lean glycol is required, it may be necessary to use stripping gas. A small amount of wet natural gas can be taken from the fuel stream or contactor inlet stream and injected into the reboiler. The stripping gas can be taken from the fuel stream or the contactor inlet stream and injected into the reboiler. The "leaness" of the gas depends on the purity of the wet glycol and the number of stages below the reconcentrator. The stripping gas is saturated with water at the inlet temperature and pressure conditions, but adsorbs water at the reboiler conditions of atmospheric pressure and high temperatures. The gas will adsorb the water from the glycol by lowering the partial pressure of the water vapor in the reboiler. Stripping gas exits in the still column with the water vapor. If necessary, the gas can be recovered by condensing the water and routing the gas to a vapor recovery compressor.

The lean glycol flows from the reboiler to a surge tank which could be constructed as an integral part of the reboiler as in Figure 8-6. The surge tank must be large enough to allow for thermal expansion of the glycol and to allow for reasonable time between additions of glycol. A well designed and operated unit will have glycol losses to the dry gas from the contactor and the water vapor from the still of between 0.01 and 0.05 gal/MMscf of gas processed.

The lean glycol from the atmospheric surge tank is then pumped back to the contactor to complete the cycle. Depending upon the pump design, the lean glycol must be cooled by the heat exchangers to less than 200°F to 250°F before reaching the pumps. There are many variations to the basic glycol process described above. For higher "wet" gas flow rates greater than 500 MMscfd, the "cold finger" condenser process as shown in Figure 8-7 is often attractive. A cold finger condenser tube bundle with cold rich gas from the contactor is inserted either into the vapor space at the reboiler or into a separate separator. This creates a "cold finger" in the vapor space. The hydrocarbon liquid and vapor phases along with the glycol/water phase are separated in a three-phase separator. The lean glycol from the bottom of the condenser is cooled, pumped, cooled again, and fed to the contactor.

Choice of Glycol

The commonly available glycols and their uses are:

1. Ethylene glycol—High vapor equilibrium with gas so tend to lose to gas phase in contactor. Use as hydrate inhibitor where it can be recovered from gas by separation at temperatures below 50°F.
2. Diethylene glycol—High vapor pressure leads to high losses in contactor. Low decomposition temperature requires low reconcentrator temperature (315°F to 340°F) and thus cannot get pure enough for most applications.
3. Triethylene glycol—Most common. Reconcentrate at 340°F to 400°F for high purity. At contactor temperatures in excess of 120°F tends to have high vapor losses to gas. Dew point depressions up to 150°F are possible with stripping gas.

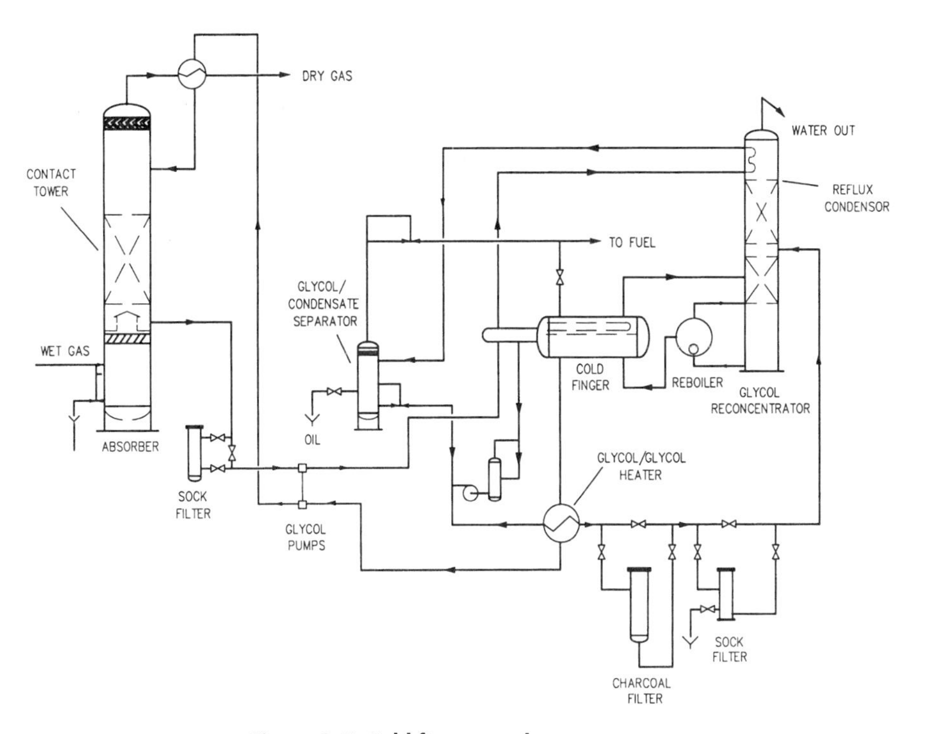

Figure 8-7. Cold finger condenser process.

4. Tetraethylene glycol—More expensive than triethylene but less losses at high gas contact temperatures. Reconcentrate at 400°F to 430°F.

Almost all field gas dehydration units use triethylene glycol for the reasons indicated. Normally when field personnel refer to "glycol" they mean triethylene glycol and we will use that convention in the remainder of this chapter.

Design Considerations

Inlet Gas Temperature

At constant pressure, the water content of the inlet gas increases as the inlet gas temperature increases. For example, at 1,000 psia and 80°F gas holds about 34 lb/MMscf, while at 1,000 psia and 120°F it will hold about 104 lb/MMscf. At the higher temperature, the glycol will have to remove over three times as much water to meet a pipeline specification of 7 lb/MMscf.

An increase in gas temperature may result in an increase in the required diameter of the contact tower. As was shown in separator sizing (Volume 1, Chapter 4), an increase in temperature increases the actual gas velocity, which in turn increases the diameter of the vessel.

Inlet gas temperatures above 120°F result in high triethylene glycol losses. At higher gas temperatures tetraethylene glycol can be used, but it is more common to cool the gas below 120°F before entering the contactor. The more the gas is cooled, while staying above the hydrate formation temperature, the smaller the glycol unit required.

The minimum inlet gas temperature is normally above the hydrate formation temperature and should always be above 50°F. Below 50°F glycol becomes too viscous. Below 60°F to 70°F glycol can form a stable emulsion with liquid hydrocarbons in the gas and cause foaming in the contactor.

There is an economic trade-off between the heat exchanger system used to cool the gas and the size of the glycol unit. A larger cooler provides for a smaller glycol unit, and vice versa. Typically, triethylene glycol units are designed to operate with inlet gas temperatures between 80°F and 110°F.

Contactor Pressure

Contactor pressures have little effect on the glycol absorption process as long as the pressures remain below 3,000 psig. At a constant temperature the water content of the inlet gas decreases with increasing pressure, thus less water must be removed if the gas is dehydrated at a higher pressure. In addition, a smaller contactor can be used at high pressure as the actual velocity of the gas is lower, which decreases the required diameter of the contactor.

At lower pressure less wall thickness is required to contain the pressure in a given diameter contactor, therefore, an economic trade-off exists between operating presssure and contactor cost. Typically, dehydration pressures of 500 to 1,200 psi are most economical.

Number of Contactor Trays

The glycol and the gas do not reach equilibrium on each tray. A tray efficiency of 25% is commonly used for design. That is, if one theoretical equilibrium tray is needed, four actual trays are specified. In bubble cap towers, tray spacing is normally 24 in.

The more trays the greater the dew-point depression for a constant glycol circulation rate and lean glycol concentration. By specifying more trays, fuel savings can be realized because the heat duty of the reboiler is directly related to the glycol circulation rate. Figure 8-8 shows how the number of trays can have a much greater effect on dew-point depression than the circulation rate.

The additional investment for a taller contactor is often easily justified by the resultant fuel savings. Most contactors designed for 7 lb/MMscf gas are sized for 6 to 8 trays.

Lean Glycol Temperatures

The temperature of the lean glycol entering the contactor has an effect on the gas dew-point depression and should be held low to minimize required circulation rate. High glycol losses to the gas exiting the contac-

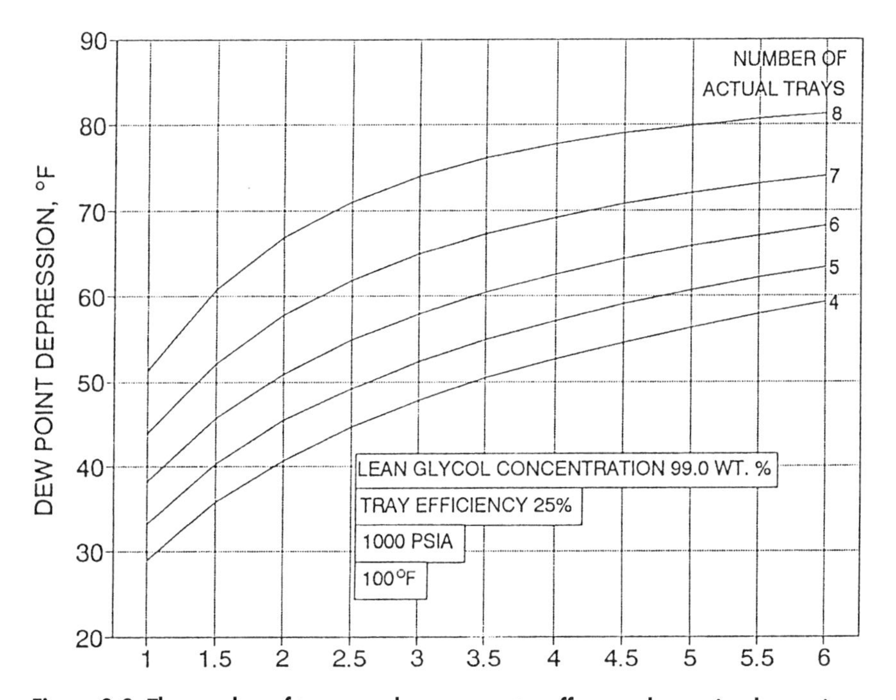

Figure 8-8. The number of trays can have a greater effect on dew-point depression than the circulation rate.

tor may occur when the lean glycol temperature gets too hot. On the other hand, the lean glycol temperature should be kept slightly above the contactor gas temperature to prevent hydrocarbon condensation in the contactor and subsequent foaming of the glycol. Most designs call for a lean glycol temperature 10°F hotter than the gas exiting the contactor.

Glycol Concentration

The higher the concentration of the lean glycol the greater the dew-point depression for a given glycol circulation rate and number of trays.

Figure 8-9 shows the equilibrium water dew point at different temperatures for gases in contact with various concentrations of glycol. At 100°F contact temperature there is an equilibrium water dew point of 25°F for 98% glycol and 10°F for 99% glycol. Actual dew points of gas leaving the contactor will be 10°F to 20°F higher than equilibrium.

Figure 8-10 shows that increasing the lean glycol concentration can have a much greater effect on dew-point depression than increasing the circulation rate. To obtain a 70°F dew-point depression a circulation rate

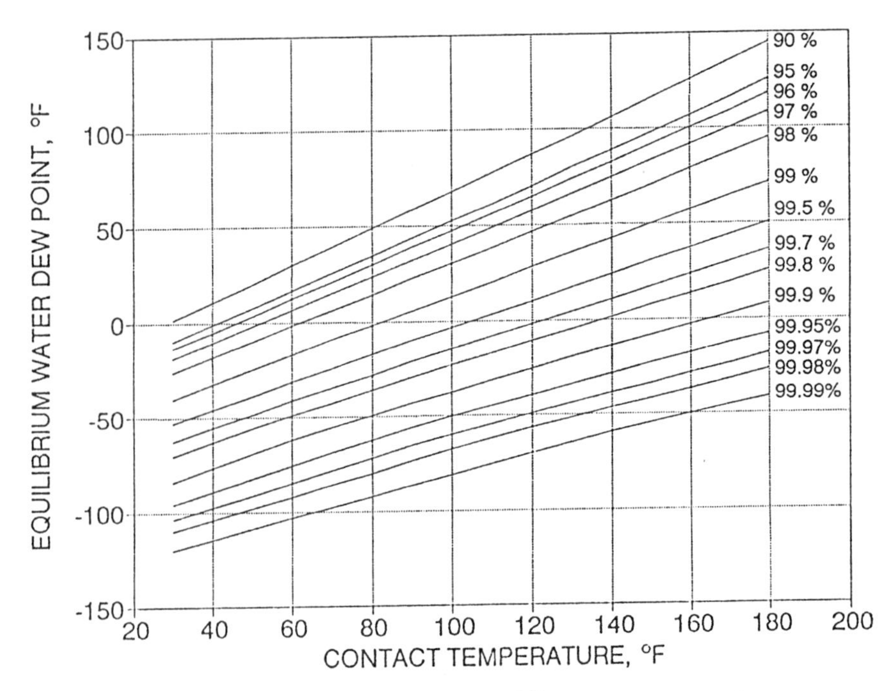

Figure 8-9. Equilibrium water dew points at different temperatures for gases.

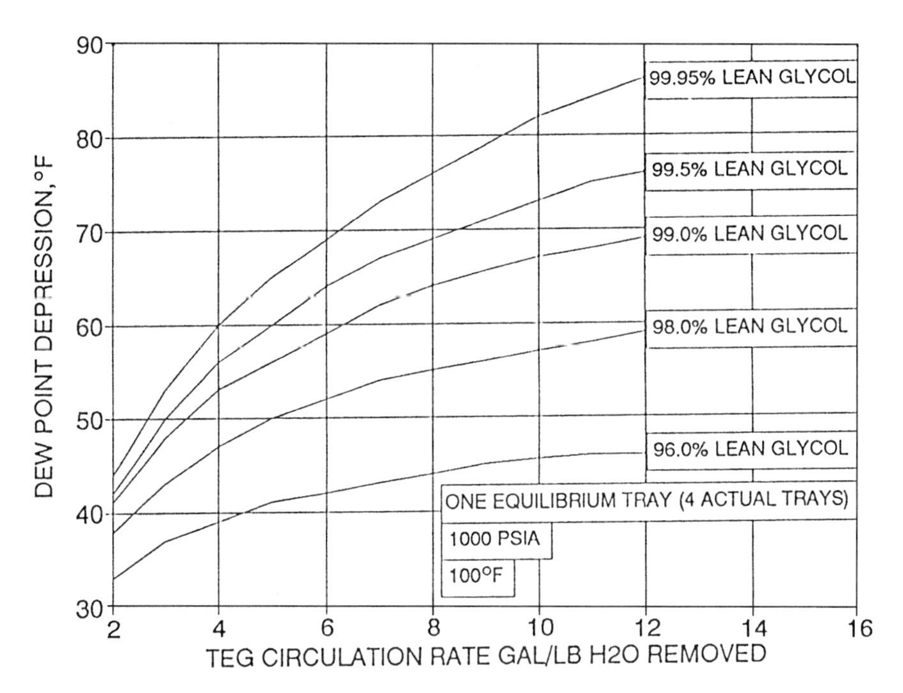

Figure 8-10. Increasing glycol concentration has a greater effect on dew-point depression than increasing the circulation rate.

of 6.2 gal/lb at 99.95%, 8.2 gal/lb at 99.5% or in excess of 12 gal/lb at 99% is required.

The lean glycol concentration is determined by the temperature of the reboiler, the gas stripping rate, and the pressure of the reboiler. Glycol concentrations between 98 and 99% are common for most field gas units.

Glycol Reboiler Temperature

The reboiler temperature controls the concentration of the water in the lean glycol. The higher the temperature the higher the concentration, as shown in Figure 8-11. Reboiler temperatures for triethylene glycol are limited to 400°F, which limits the maximum lean glycol concentration without stripping gas. It is good practice to limit reboiler temperatures to between 370°F and 390°F to minimize degradation of the glycol. This effectively limits the lean glycol concentration to between 98.5% and 98.9%.

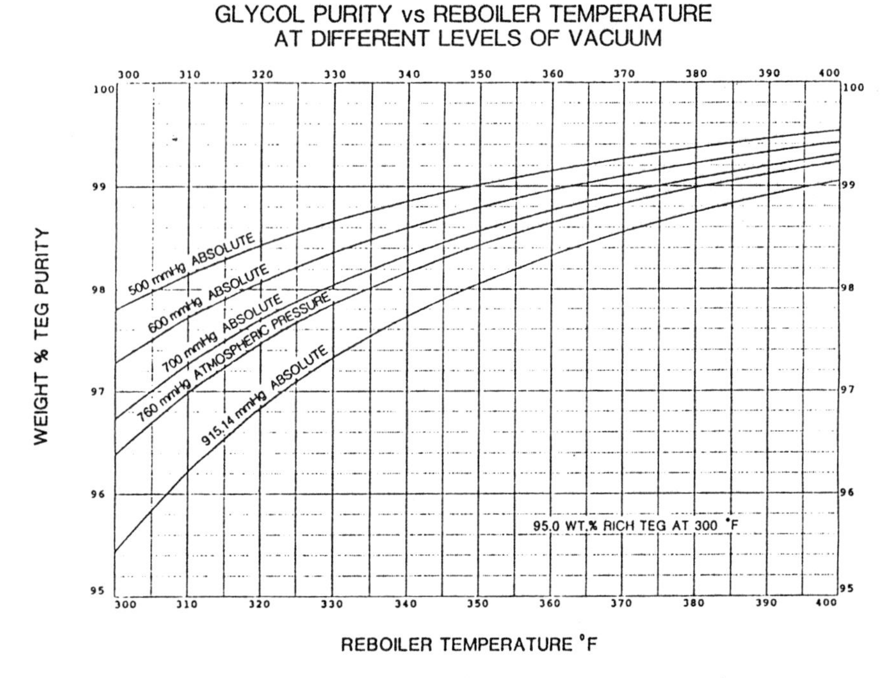

Figure 8-11. The higher the temperature the greater the lean glycol concentration.

When higher lean glycol concentrations are required, stripping gas can be added to the reboiler, or the reboiler and still column can be operated at a vacuum.

Reboiler Pressure

Pressures above atmospheric in the reboiler can significantly reduce lean glycol concentration and dehydration efficiency. The still column should be adequately vented and the packing replaced periodically to prevent excess back pressure on the reboiler.

At pressures below atmospheric the boiling temperature of the rich glycol/water mixture decreases, and a greater lean glycol concentration is possible at the same reboiler temperature. Reboilers are rarely operated at a vacuum in field gas installations, because of the added complexity and the fact that any air leaks will result in glycol degradation. In addition, it is normally less expensive to use stripping gas. However, if lean glycol concentrations in the range of 99.5% are required, consider using a reboiler pressure of 500 mm Hg absolute (approximately 10 psia) as well as using stripping gas. Sometimes the addition of a vacuum will help extend the range of an existing glycol system.

Figure 8-11 can be used to estimate the effect of vacuum on lean glycol concentration.

Stripping Gas

The lean glycol concentration leaving the reboiler can be lowered by contacting the glycol with stripping gas. Often, wet gas that is saturated with water vapor at ambient temperature and 25 to 100 psig is used. At 25 psig and 100°F this gas is saturated with 1,500 lb/MMscf of water vapor. At atmospheric pressure and the temperatures in the reboiler the gas can absorb over 100,000 lb/MMscf.

In most situations the additional fuel gas required to heat the reboiler to increase lean glycol concentration is less than the stripping gas required for the same effect. Thus, it is normally desirable to use stripping gas only to increase lean glycol concentration above 98.5 to 98.9%, which can be reached with normal reboiler temperatures and normal back pressure on the still column. If the glycol circulation rate must be increased above design on an existing unit and the reboiler cannot reach desired temperature, it is often possible to use stripping gas to achieve the desired lean glycol concentration.

Figure 8-12 shows the effects on the glycol purity of stripping gas flow rate for various reboiler temperatures, assuming the gas is injected directly into the reboiler. Greater purities are possible if stripping gas contacts the lean glycol in a column containing one or more stages of packing before entering the reboiler.

Glycol Circulation Rate

When the number of absorber trays and lean glycol concentration are fixed, the dew-point depression of a saturated gas is a function of the glycol circulation rate. The more glycol that comes in contact with the gas, the more water vapor is stripped out of the gas. Whereas the glycol concentration mainly affects the dew point of the dry gas, the glycol rate controls the total amount of water that can be removed. The minimum circulation rate to assure good glycol-gas contact is about two gallons of glycol for each pound of water to be removed. Seven gallons of glycol per pound of water removed is about the maximum rate. Most standard

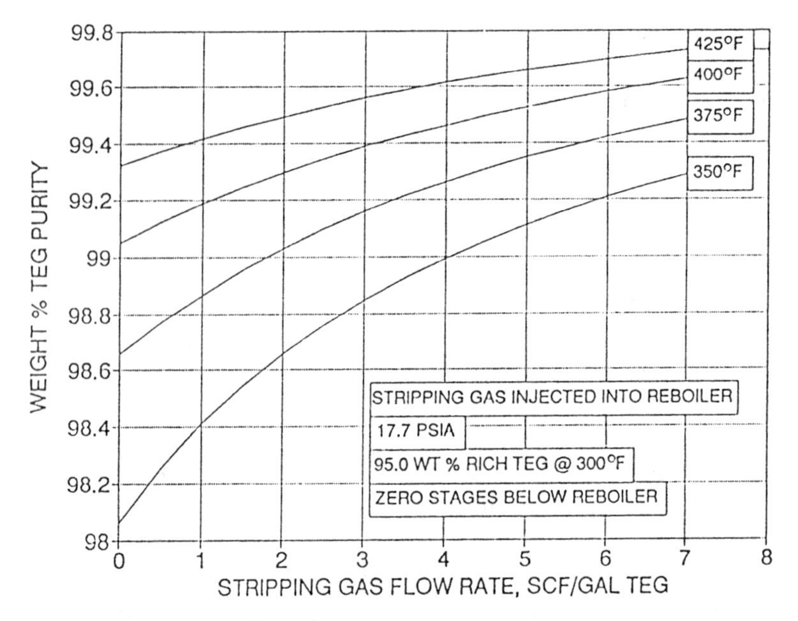

Figure 8-12. Effect of stripping gas on glycol concentration.

dehydrators are designed for approximately three gallons of glycol per pound of water removed.

An excessive circulation rate may overload the reboiler and prevent good glycol regeneration. The heat required by the reboiler is directly proportional to the circulation rate. Thus, an *increase* in circulation rate may decrease reboiler temperature, decreasing lean glycol concentration, and actually *decrease* the amount of water that is removed by the glycol from the gas. Only if the reboiler temperature remains constant will an increase in circulation rate lower the dew point of the gas.

Stripping Still Temperature

A higher temperature in the top of the still column can increase glycol losses due to excessive vaporization. The boiling point of water is 212°F and the boiling point of TEG is 546°F. The recommended temperature in the top of the still column is approximately 225°F. When the temperature exceeds 250°F the glycol vaporization losses may become substantial. The still top temperature can be lowered by increasing the amount of glycol flowing through the reflux coil.

If the temperature in the top of the still column gets too low, too much water can be condensed and increase the reboiler heat load. Too much cool glycol circulation in the reflux coil can sometimes lower the still top temperature below 220°F. Thus, most reflux coils have a bypass to allow manual or automatic control of the stripping still temperature.

Stripping gas will have the effect of requiring reduced top still temperature to produce the same reflux rate.

System Sizing

Glycol system sizing involves specifying the correct contactor diameter and number of trays, which establishes its overall height; selecting a glycol circulation rate and lean glycol concentration; and calculating the reboiler heat duty. As previously explained, the number of trays, glycol circulation rate and lean glycol concentration are all interrelated. For example, the greater the number of trays the lower the circulation rate or lean glycol concentration required. Figures 8-13, 8-16, and 8-17 can be used to relate these three parameters.

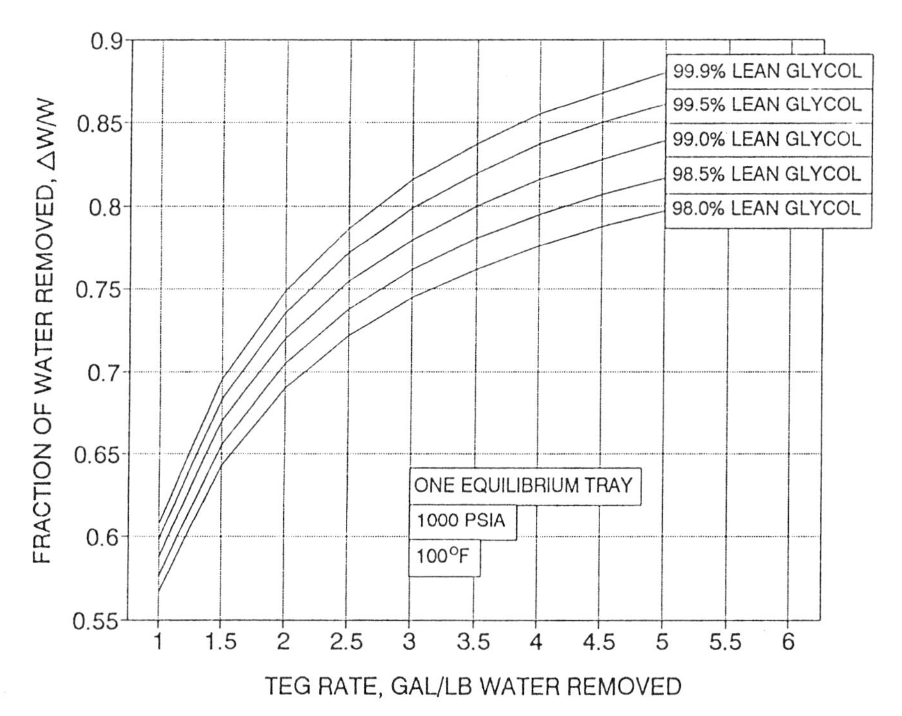

Figure 8-13. Glycol concentration vs. glycol circulation when n = 1 theoretical tray.

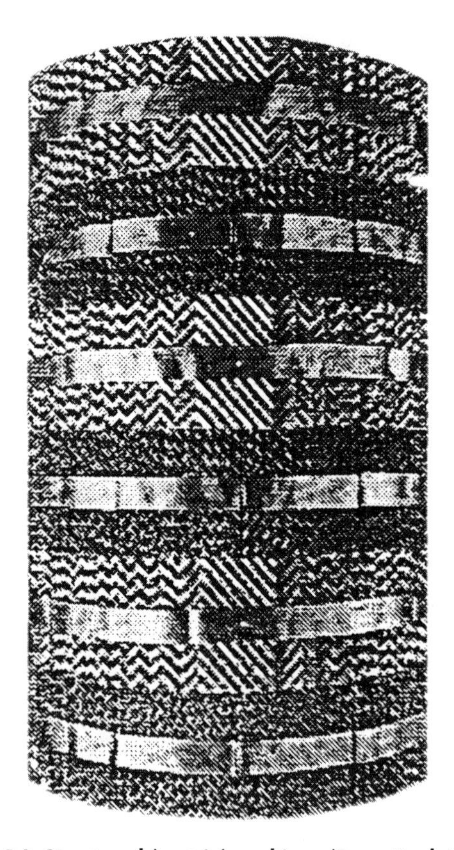

Figure 8-14. Structured (matrix) packing. ***(From Koch Industries.)***

Contactor Sizing

Bubble cap contactors are the most common. The minimum diameter can be determined using the equation derived for gas separation in vertical separators (Volume 1, Chapter 4). This is:

$$d^2 = 5{,}040\,\frac{TZQ_g}{P}\left[\left(\frac{\rho_g}{\rho_l - \rho_g}\right)\frac{C_D}{d_m}\right]^{1/2} \qquad (8\text{-}1)$$

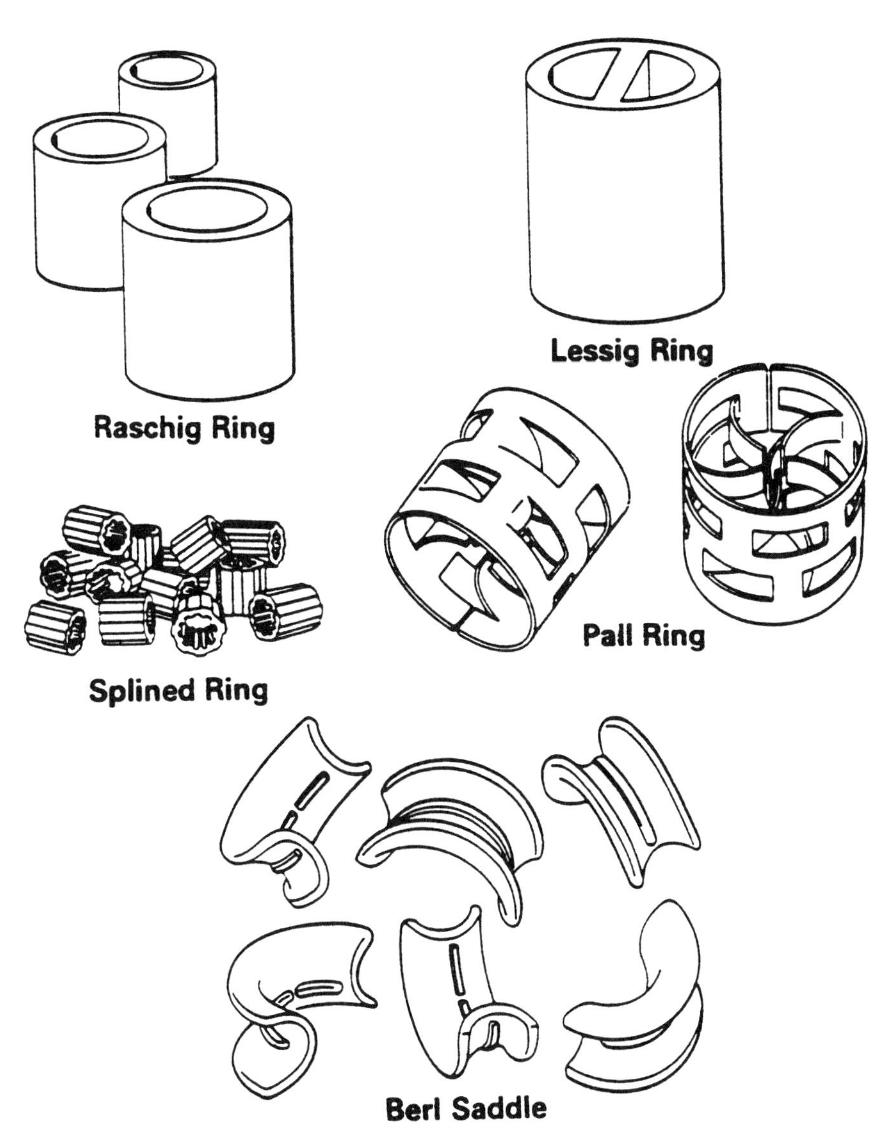

Figure 8-15. Various types of packing. *(Courtesy: McGraw-Hill Book Company.)*

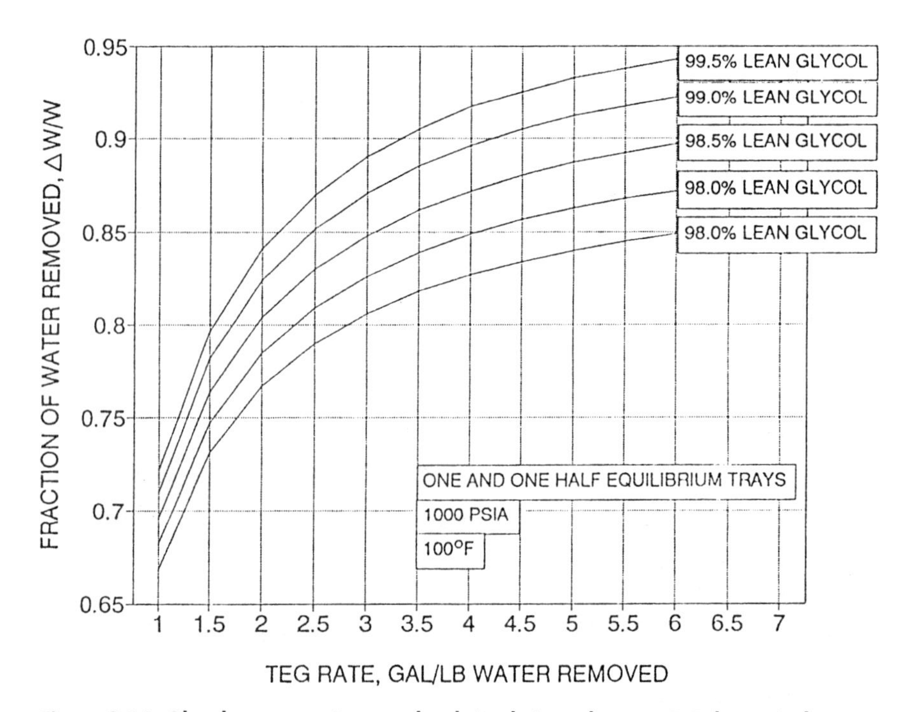

Figure 8-16. Glycol concentration vs. glycol circulation when n = 1.5 theoretical trays.

where d = column inside diameter, in.
d_m = drop size, micron
T = contactor operating temperature, °R
Q_g = design gas rate, MMscfd
P = contactor operating presssure, psia
C_D = drag coefficent
ρ_g = gas density, lb/ft^3
ρ_g = 2.7 SP/TZ (Volume 1, Chapter 3)
ρ_1 = density of glycol, lb/ft^3
Z = compressibility factor (Volume 1, Chapter 3)
S = specific gravity of gas relative to air

Reasonable choices of contactor diameter are obtained when the contactor is sized to separate 120–150 micron droplets of glycol in the gas. The density of glycol can be estimated as 70 lb/ft^3.

The diameter of packed towers may differ depending upon parameters developed by the packing manufacturers and random packing. Conventional packing will require approximately the same diameter as bubble

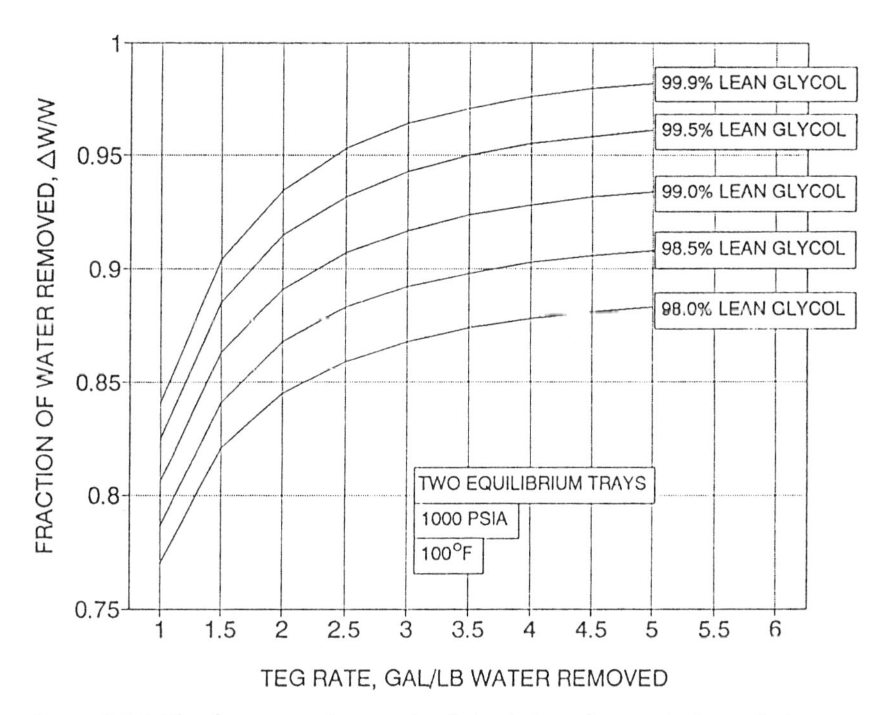

Figure 8-17. Glycol concentration vs. glycol circulation when n = 2 theoretical trays.

cap towers. Structured packing can handle higher gas flow rates than bubble cap trays in the same diameter contactor. (See Table 8-1.)

Conventional and random packing will require approximately the same diameter as bubble caps. Structured packing can handle higher gas flow rates than bubble caps in the same diameter contactor while requiring half the height. The height per equivalent theoretical tray normally ranges from 8 ft for low dewpoints to 4 ft for moderate dewpoints. Adequate mist eliminator and glycol distribution is needed for high gas flow rates.

Reboiler Heat Duty

The reboiler heat duty can be calculated using the techniques in Chapter 2, by sizing the reflux coil and heat exchangers and calculating the temperature at which the wet glycol enters the still. The reboiler duty is then the sum of the sensible heat required to raise the wet glycol to reboiler temperature, the heat required to vaporize the water in the glycol, the heat required for the reflux (which is estimated at 25 to 50% of the heat required to vaporize the water in the glycol) and losses to atmosphere.

Table 8-1
Example Contactor Sizes for Dehydrating 50 MMscfd at 1,000 psig and 100°F

Tower Internals	Tower Diameter (inch)	Tray/Packing Height (feet)
A. Tray		
Bubble Cap	48	16
B. Structural Packing (Figures 6-9, 8-14)		
B1-300	36	8
B1-100	30	8
Flexipac #1	42	6
Flexipac #2	30	8
C. Random Packing (Figure 8-15)		
2″ Pall Ring	48	16

In sizing the various heat exchangers it is common to assume a 10°F loss of rich glycol temperature in the reflux coil, a desired temperature of 175°F to 200°F for the rich glycol after the preheater and a rich glycol temperature after the glycol/glycol heat exchanger of 275°F to 300°F It is necessary to make sure that the lean glycol temperature to the pumps does not exceed 200°F for glycol powered pumps and 250°F for plunger pumps. The temperature of lean glycol after the glycol/gas exhanger should be approximately 10°F above the temperature of the gas in the contactor. The water vapor boiled from the rich glycol plus the reflux water vapor must be cooled from approximately 320°F to 220°F by the reflux coil.

Exchanger heat transfer factors, "U," can be approximated as 10 to 12 Btu/hr-ft^2-°F for glycol/glycol exchangers, 45 Btu/hr-ft^2-°F for the gas/glycol exchanger, and 100 Btu/hr-ft^2-°F for the reflux coil. The specific heat of triethylene glycol is given in Figure 8-15.

Table 8-2 can be used for an initial approximation of reboiler duties. If the reboiler is heated with a fire tube, the fire tube should be sized for a maximum flux rate of 8,000 Btu/hr-ft^2.

Glycol Powered Pumps

The process flow schematic in Figure 8-6 shows electric motor driven glycol pumps. On smaller systems it is common to use glycol powered pumps. These pumps use the energy contained in the rich (wet) glycol to

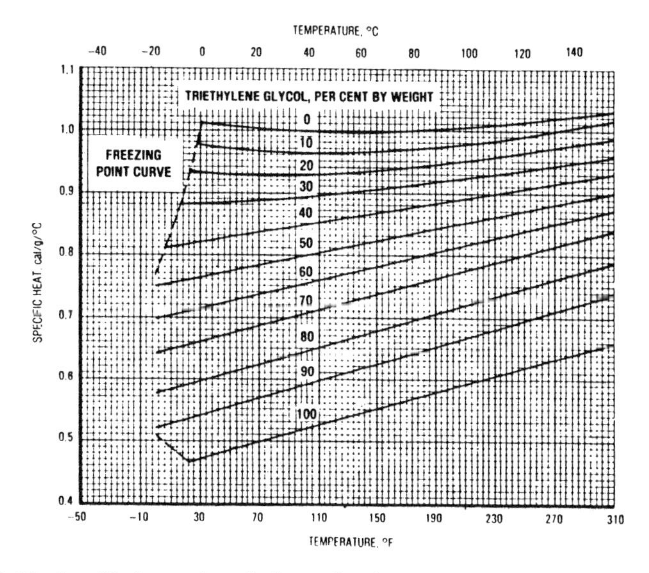

Figure 8-18. Specific heat of triethylene glycol. (*Courtesy of Union Carbide, Gas Treating Chemicals.*)

Table 8-2
Approximate Reboiler Heat Duty

Design Gallons of Glycol Circulated /lb H_2O Removed	Reboiler Heat Duty Btu/Gal of Glycol Circulated
2.0	1066
2.5	943
3.0	862
3.5	805
4.0	762
4.5	729
5.0	701
5.5	680
6.0	659

Size at 150% of above to allow for start-up, increased circulation, fouling.

pump the lean (dry) glycol to the contactor. The action of this type pump is shown in Figures 8-19 and 8-20. With the main piston moving to the left (Figure 8-19), dry glycol is drawn into the left cylinder and discharged from the one at the right. Wet glycol is drawn into the right cylinder and discharged from the left cylinder. As the piston completes

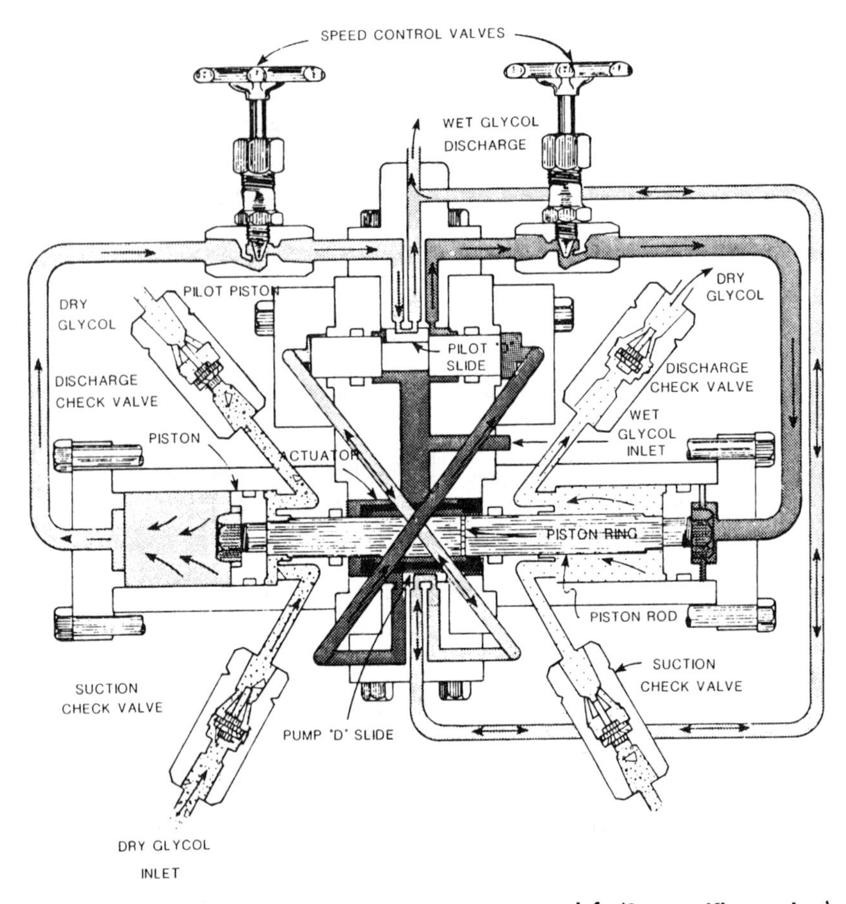

Figure 8-19. Glycol-powered pump—piston moving to left. (Source: Kimray, Inc.)

its movement to the left, it moves the "D" slide to the position shown in Figure 8-19. This reverses the pilot slide position, which reverses the action of the piston.

Even though the wet glycol drops in pressure from contactor pressure to condensate/separator pressure, it has enough energy to pump the dry glycol from atmospheric pressure to contactor pressure. This is because it contains more water and gas in solution, but also because gas from the contactor flows out with the wet glycol. There is no level control valve on the contactor when using a glycol powered pump. Sufficient contactor gas is automatically drawn into the wet glycol line to power the pump at the rate set by the speed control valves. This gas, as well as the approximately 1 scf/gal gas in solution in the glycol, is separated in the conden-

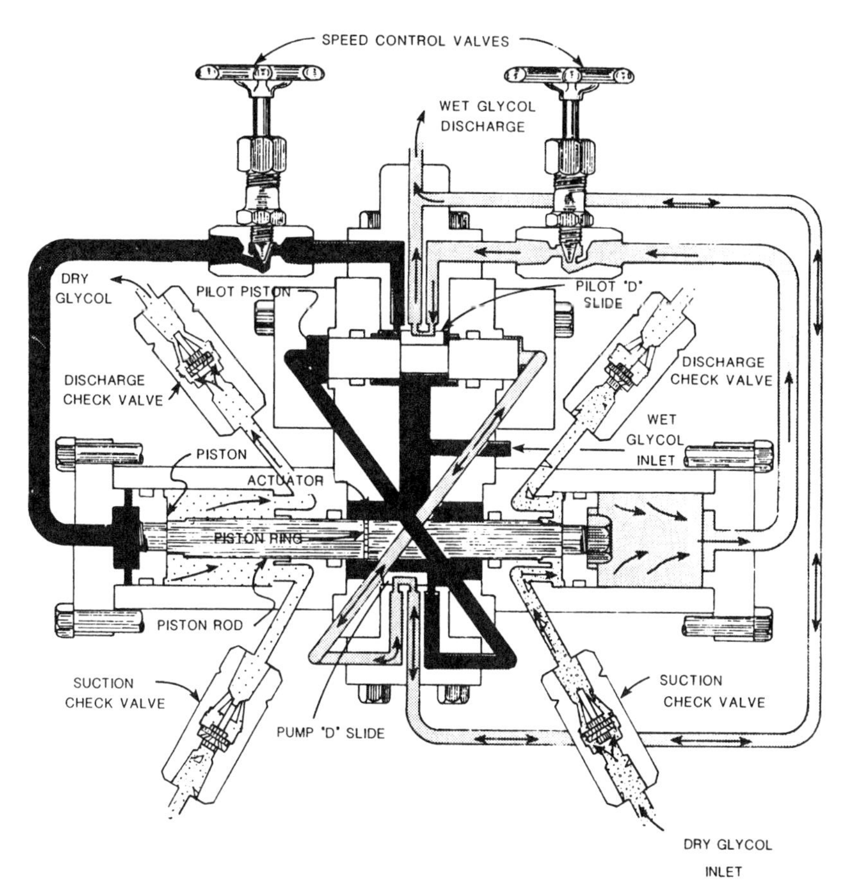

Figure 8-20. Glycol-powered pump—piston moving to right. (Source: Kimray, Inc.)

sate separator giving it an alternative designation as "pump gas separator." The free gas consumption of the glycol pump is given in Table 8-3.

The pump gas can be used to fuel the reboiler. The amount of pump gas is normally close to balancing the reboiler fuel gas requirements. The pump gas can also be routed to the facility fuel gas system or to a low-pressure system for compression and sales. If it is not recovered in one of these ways and is just vented locally, the cost of using this type of pump can be very high.

Glycol powered pumps are inexpensive and easy to repair in the field. They have many moving parts and because of their slamming reciprocating motion require constant attention. One spare pump should always be installed.

Table 8-3
Gas Consumption by Glycol Powered Pump

Contactor Operating Pressure psig	Pump Gas Consumption sc f/gal
300	1.7
400	2.3
500	2.8
600	3.4
700	3.9
800	4.5
900	5.0
1000	5.6
1100	6.1
1200	6.7
1300	7.2
1400	7.9
1500	8.3

EXAMPLE 8-1: GLYCOL DEHYDRATION

Given: Gas 98 MMscfd at 0.67 SG
Sat w/water at 1,000 psig, 100°F
Dehydrate to 7 lb/MMscf
Use triethylene glycol
No stripping gas,
98.5% TEG
C_D (contactor) = 0.852 (calculated from Volume 1, Chapter 4)
T_{CR} = 376°R P_{CR} = 669 psia

Problem:

1. Calculate contactor diameter.
2. Determine glycol circulation rate and estimate reboiler duty.
3. Calculate duties for gas/glycol exchanger and glycol/glycol exchangers.

Calculate Contactor Diameter

$$d^2 = 5{,}040 \frac{T\,Z\,Q_g}{P}\left[\left(\frac{\rho_g}{\rho_l - \rho_g}\right)\frac{C_D}{d_m}\right]^{1/2}$$

$d_m = 125$ microns
$T = 100 + 460 = 560°R$
$P = 1{,}015$ psia
$Q_g = 98$ MMscfd
$T_r = 560/376 = 1.49$
$P_r = 1{,}015/669 = 1.52$
$Z = 0.865$

$$\rho_g = 2.7 \frac{(0.67)(1{,}015)}{(560)(0.865)} = 3.79 \text{ lb/ft}^3$$

$\rho_l = 70 \text{ lb/ft}^3$
$C_D = 0.852$ (given)

$$d^2 = 5{,}040 \frac{(560)(0.865)(98)}{(1{,}015}\left[\left(\frac{3.79}{70 - 3.79}\right)\frac{0.852}{125}\right]^{1/2}$$

$d = 68.2$ in.

Use 72-in. ID contactor.

Determine Glycol Circulation Rate and Reboiler Duty

$W_i = 63$ lb/MMscf
$W_o = 7$ lb/MMscf
$\Delta W = 63 - 7 = 56$ lb/MMscf
$\Delta W/W_i = 56/63 = .889$

Assume 8 actual trays or 2 theoretical trays. From Figure 8-17 the glycol circulation rate is 2.8 gal TEG/lb H_20. Size for 3.0 gal/lb.

$$Q = \frac{3.0 \text{ gal}}{\text{lb}} \times \frac{56 \text{ lb}}{\text{MMscf}} \times \frac{98 \text{ MMscf}}{\text{D}} \times \frac{\text{D}}{24 \text{ hr}} \times \frac{\text{hr}}{60 \text{ min}}$$

$Q = 11.4$ gpm TEG

Estimate reboiler duty:

$$q = 862 \text{ Btu/gal (from Table 8-2)}$$

$$q = \frac{862 \text{ Btu}}{\text{gal}} \times \frac{11.4 \text{ gal}}{\text{min}} \times \frac{60 \text{ min}}{\text{hr}}$$

$$q = 590 \text{ MBtu/hr}$$

Use a 750 MBtu/hr reboiler to allow for startup heat loads.

Calculate Duties of Heat Exchangers

Rich TEG from contactor	T = 100°F (given)
Rich TEG from reflux	T = 110°F (assume 10°F increase in reflux coil)
Rich TEG to sep	T = 200°F (assume for good design)
Rich TEG to still	T = 300°F (assume for good design)
Lean TEG from reboiler	T = 353°F (Figure 8-11)
Lean TEG to pumps (max)	T = 210°F (given by manufacturer)
Lean TEG to contactor	T = 110°F (10°F above contactor temperature)

- Glycol/glycol preheater (rich side, duty)

Rich TEG $T_1 = 110°F$
$T_2 = 200°F$

Lean glycol composition

$$W_{TEG} = 0.985 \times \frac{70 \text{ lb}}{\text{ft}^3} \times \frac{\text{ft}^3}{7.48 \text{ gal}}$$

$$W_{TEG} = 9.22 \text{ lb TEG / gal of lean glycol}$$

$$W_{H_2O} = (0.015) \times \frac{70 \text{ lb}}{\text{ft}^3} \times \frac{\text{ft}^3}{7.48 \text{ gal}}$$

$$W_{H_2O} = 0.140 \text{ lb } H_2O \text{ / gal of lean glycol}$$

Rich glycol composition

$$W_{TEG} = 9.22 \text{ lb TEG/gal of lean glycol}$$

$$W_{H_2O} = 0.140 + \frac{1 \text{ lb } H_2O}{3.0 \text{ gal of lean glycol}}$$

$$W_{H_2O} = 0.473 \text{ lb } H_2O/\text{gal of lean glycol}$$

$$\text{Wt. concentration TEG} = \frac{9.22}{9.22 + 0.473} = 95.1\%$$

Rich glycol flow rate (W_{rich})

$$W_{rich} = (9.22 + 0.473)\frac{\text{lb}}{\text{gal}} \times 11.4\frac{\text{gal}}{\text{min}} \times 60\frac{\text{min}}{\text{hr}}$$

$$W_{rich} = 6{,}630 \text{ lb/hr}$$

Rich glycol heat duty (q_{rich})

$$C_p\ (95.1\%\ \text{TEG}) = 0.56 \text{ at } 110°\text{F (From Figure 8-18)}$$
$$= 0.63 \text{ at } 200°\text{F}$$

$$C_{p,av} = 0.60 \text{ Btu/lb°F}$$

$$q_{rich} = \frac{6{,}630 \text{ lb}}{\text{hr}} \times \frac{0.6 \text{ Btu}}{\text{lb °F}} \times (200 - 110°\text{F})$$

$$q_{rich} = 358 \text{ MBtu/hr}$$

- Glycol/glycol exchanger

$$\text{Rich } T_1 = 200$$
$$T_2 = 300$$
$$\text{Lean } T_3 = 353$$
$$T_4 = ?$$

Rich glycol heat duty

$$C_p \text{ (95.1\% TEG)} = 0.63 \text{ at } 200°F \text{ (From Figure 8-18)}$$
$$= 0.70 \text{ at } 300°F$$

$$C_{p,av} = 0.67 \text{ Btu/lb °F}$$

$$q_{rich} = \frac{6,630 \text{ lb}}{\text{hr}} \times \frac{0.67 \text{ Btu}}{\text{lb °F}} \times (300 - 200) \text{ °F}$$

$$q_{rich} = 444 \text{ MBtu/hr}$$

Lean glycol flow rate (W_{lean})

$$W_{lean} = \frac{11.4 \text{ gal}}{\text{min}} \times \frac{70 \text{ lb}}{\text{ft}^3} \times \frac{\text{ft}^3}{7.48 \text{ gal}} \times \frac{60 \text{ min}}{\text{hr}}$$

$$W_{lean} = 6,401 \text{ lb/hr}$$

Calculation of T_4

Assume T = 250°F

$$T_{av} \cong (353 + 250)/2 = 302°F$$

$$C_{p,av} \text{ (98.5\% TEG)} = 0.67 \text{ Btu/lb°F (From Figure 8-18)}$$

$$q_{lean} = W_{lean} C_p (T_4 - T_3)$$

$$q_{lean} = -q_{rich}$$

$$T_4 = T_3 - \frac{q_{rich}}{W_{lean} C_p}$$

$$T_4 = 353 - \frac{444,000}{(6,401)(0.67)}$$

$$T_4 = 249°F$$

- Glycol/glycol preheater—calculate lean side

Temperature

$$\text{Lean } T_4 = 249°F$$
$$T_5 = ?$$

Assume $T_5 = 175°F$

$T_{av} \cong (249 + 175)/2 = 212$

$C_{p,av}$ (98.5% TEG) = 0.61 Btu/lb°F (From Figure 8-18)

$$q_{lean} = W_{lean}\, C_p\, (T_4 - T_5)$$

$$q_{lean} = -\, q_{rich}$$

$$T_4 = T_3 - \frac{q_{rich}}{W_{lean}\ C_p}$$

$$T_4 = 353 - \frac{444{,}000}{(6{,}401)(0.67)}$$

$T_5 = 157°F$ (This is less than the maximum to the pumps.)

- Gas/glycol exchanger duty

$$\text{Lean } T_1 = 157°F$$
$$T_2 = 110°F$$

C_p (98.5% TEG) = 0.57 at 157°F (From Figure 8-18)
= 0.53 at 110°F

$C_{p,av} = 0.55$ Btu/lb°F

$$q_{lean} = (6{,}401)(0.55)(110 - 157)$$

$$q_{lean} = -165 \text{ MBtu/hr}$$

- Summary of exchangers

Glycol/glycol exchanger

Rich $T_i = 200°F$, $T_o = 300°F$

Lean $T_i = 353°F$, $T_o = 249°F$

Duty q = 444 MBtu/hr

Glycol/glycol preheater

Rich $T_i = 110°F$, $T_o = 200°F$

Lean $T_i = 249°F$, $T_o = 157°F$

Duty q = 358 MBtu/hr

Gas/glycol exchanger

Lean $T_i = 157°F$, $T_o = 110°F$

Duty q = 165 MBtu/hr

It is possible to recover more heat from the lean glycol and reduce the lean glycol temperature to the pumps to 180°F to 200°F by making the glycol/glycol exchanger larger, but this is not required in this case. This would also increase the temperature of the rich glycol flowing on the still to more than 300°F and would decrease the reboiler heat duty.

SOLID BED DEHYDRATION

Solid bed dehydration systems work on the principle of adsorption. Adsorption involves a form of adhesion between the surface of the solid desiccant and the water vapor in the gas. The water forms an extremely thin film that is held to the desiccant surface by forces of attraction, but there is no chemical reaction. The desiccant is a solid, granulated drying or dehydrating medium with an extremely large effective surface area per unit weight because of a multitude of microscopic pores and capillary openings. A typical desiccant might have as much as 4 million square feet of surface area per pound.

The initial cost for a solid bed dehydration unit generally exceeds that of a glycol unit. However, the dry bed has the advantage of producing very low dew points, which are required for cryogenic gas plants (see Chapter 9), and is adaptable to very large changes in flow rates. A dry bed can handle high contact temperatures. Disadvantages are that it is a batch process, there is a relatively high pressure drop through the system, and the desiccants are sensitive to poisoning with liquids or other impurities in the gas.

Process Description

Multiple desiccant beds are used in cyclic operation to dry the gas on a continuous basis. The number and arrangement of the desiccant beds may vary from two towers, adsorbing alternately, to many towers. Three separate functions or cycles must alternately be performed in each dehydrator. They are an adsorbing or gas drying cycle, a heating or regeneration cycle, and a cooling cycle.

Figure 8-21 is a flow diagram for a typical two-tower solid desiccant dehydration unit. The essential components of any solid desiccant dehydration system are:

1. Inlet gas separator.
2. Two or more adsorption towers (contactors) filled with a solid desiccant.
3. A high-temperature heater to provide hot regeneration gas to reactivate the desiccant in the towers.
4. A regeneration gas cooler to condense water from the hot regeneration gas.
5. A regeneration gas separator to remove the condensed water from the regeneration gas.
6. Piping, manifolds, switching valves and controls to direct and control the flow of gases according to the process requirements.

In the drying cycle, the wet inlet gas first passes through an inlet separator where free liquids, entrained mist, and solid particles are removed. This is a very important part of the system because free liquids can damage or destroy the desiccant bed and solids may plug it. If the adsorption

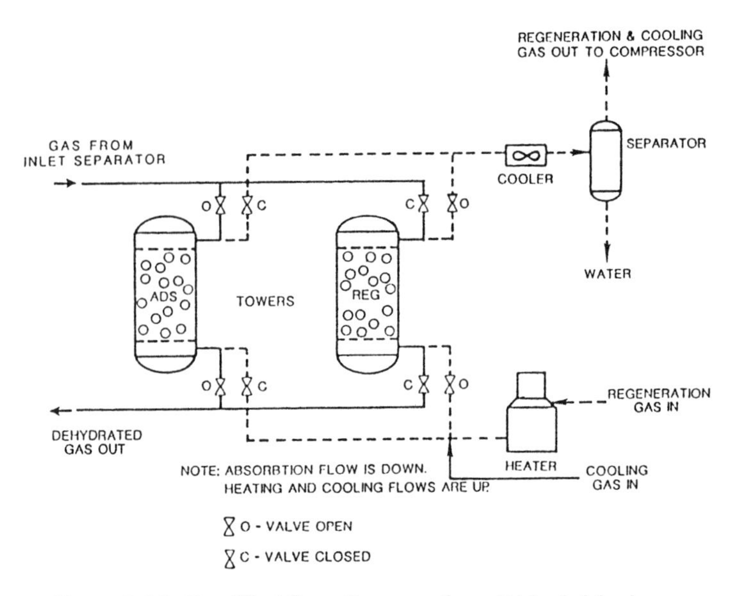

Figure 8-21. Simplified flow diagram of a solid bed dehydrator.

unit is downstream from an amine unit, glycol unit or compressors, a filter separator is preferred.

In the adsorption cycle, the wet inlet gas flows downward through the tower. The adsorbable components are adsorbed at rates dependent on their chemical nature, the size of their molecules, and the size of the pores. The water molecules are adsorbed first in the top layers of the desiccant bed. Dry hydrocarbon gases are adsorbed throughout the bed. As the upper layers of desiccant become saturated with water, the water in the wet gas stream begins displacing the previously adsorbed hydrocarbons in the lower desiccant layers. Liquid hydrocarbons will also be absorbed and will fill pore spaces that would otherwise be available for water molecules.

For each component in the inlet gas stream, there will be a section of bed depth, from top to bottom, where the desiccant is saturated with that component and where the desiccant below is just starting to adsorb that component. The depth of bed from saturation to initial adsorption is known as the mass transfer zone. This is simply a zone or section of the bed where a component is transferring its mass from the gas stream to the surface of the desiccant.

As the flow of gas continues, the mass transfer zones move downward through the bed and water displaces the previously adsorbed gases until finally the entire bed is saturated with water vapor. If the entire bed becomes completely saturated with water vapor, the outlet gas is just as wet as the inlet gas. Obviously, the towers must be switched from the adsorption cycle to the regeneration cycle (heating and cooling) before the desiccant bed is completely saturated with water.

At any given time, at least one of the towers will be adsorbing while the other towers will be in the process of being heated or cooled to regenerate the desiccant. When a tower is switched to the regeneration cycle some wet gas (that is, the inlet gas downstream of the inlet gas separator) is heated to temperatures of 450°F to 600°F in the high-temperature heater and routed to the tower to remove the previously adsorbed water. As the temperature within the tower is increased, the water captured within the pores of the desiccant turns to steam and is absorbed by the natural gas. This gas leaves the top of the tower and is cooled by the regeneration gas cooler. When the gas is cooled the saturation level of water vapor is lowered significantly and water is condensed. The water is separated in the regeneration gas separator and the cool, saturated regeneration gas is recycled to be dehydrated. This can be done by operating the dehydration tower at a lower pressure than the tower being regenerated or by recompressing the regeneration gas.

Once the bed has been "dried" in this manner, it is necessary to flow cool gas through the tower to return it to normal operating temperatures (about 100°F to 120°F) before placing it back in service to dehydrate gas. The cooling gas could either be wet gas or gas that has already been dehydrated. If wet gas is used, it must be dehydrated after being used as cooling gas. A hot tower will not sufficiently dehydrate the gas.

The switching of the beds is controlled by a time controller that performs switching operations at specified times in the cycle. The length of the different phases can vary considerably. Longer cycle times will require larger beds, but will increase the bed life. A typical two-bed cycle might have an eight-hour adsorption period with six hours of heating and two hours of cooling for regeneration. Adsorption units with three beds typically have one bed being regenerated, one fresh bed adsorbing, and one bed in the middle of the drying cycle.

Internal or external insulation for the adsorbers may be used. The main purpose of internal insulation is to reduce the total regeneration gas requirements and costs. Internal insulation eliminates the need to heat and cool the steel walls of the adsorber vessel. Normally, a castable refractory lining is used for internal insulation. The refractory must be applied and properly cured to prevent liner cracks. Liner cracks will permit some of the wet gas to bypass the desiccant bed. Only a small amount of wet, bypassed gas is needed to cause freezeups in cryogenic plants. Ledges installed every few feet along the vessel wall can help eliminate this problem.

Design Considerations

Temperature

Adsorption plant operation is very sensitive to the temperature of the incoming gas. Generally, the adsorption efficiency decreases as the temperature increases.

The temperature of the regeneration gas that commingles with the incoming wet gas ahead of the dehydrators is also important. If the temperature of these two gas streams differs more than 15°F to 20°F, liquid water and hydrocarbons will condense as the hotter gas stream cools. The condensed liquids can shorten the solid desiccant life.

The temperature of the hot gas entering and leaving a desiccant tower during the heating cycle affects both the plant efficiency and the desiccant life. To assure good removal of the water and other contaminants from the bed, a high regeneration gas temperature is needed. The maximum hot gas temperature depends on the type of contaminants and the "holding power" or affinity of the dessicant for the contaminants. A temperature of 450°F to 600°F is normally used.

The desiccant bed temperature attained during the cooling cycle is important. If wet gas is used to cool the desiccant, the cooling cycle should be terminated when the desiccant bed reaches a temperature of approximately 215°F. Additional cooling may cause water to be adsorbed from the wet gas stream and presaturate or preload the desiccant bed before the next adsorption cycle begins. If dry gas is used for cooling, the desiccant bed should be cooled within 10°F–20°F of the incoming gas temperature during the adsorption cycle, thereby maximizing the adsorption capacity of the bed.

The temperature of the regeneration gas in the regeneration gas scrubber should be low enough to condense and remove the water and hydrocarbons from the regeneration gas without causing hydrate problems.

Pressure

Generally, the adsorption capacity of a dry bed unit decreases as the pressure is lowered. If the dehydrators are operated well below the design pressure, the desiccant will have to work harder to remove the water and to maintain the desired effluent dew point. With the same volume of incoming gas, the increased gas velocity, occurring at the lower pressures, could also affect the effluent moisture content and damage the desiccant.

Cycle Time

Most adsorbers operate on a fixed drying cycle time and, frequently, the cycle time is set for the worst conditions. However, the adsorbent capacity is not a fixed value; it declines with usage. For the first few months of operation, a new desiccant has a very high capacity for water removal. If a moisture analyzer is used on the effluent gas, a much longer initial drying cycle can be achieved. As the desiccant ages, the cycle time will be automatically shortened. This will save regeneration fuel costs and improve the desiccant life.

Gas Velocities

Generally, as the gas velocity during the drying cycle decreases, the ability of the desiccant to dehydrate the gas increases. At lower actual velocities, drier effluent gases will be obtained. Consequently, it would seem desirable to operate at minimum velocities to fully use the desiccant.

However, low velocities require towers with large cross-sectional areas to handle a given gas flow, and allow the wet gas to channel through the desiccant bed and not be properly dehydrated. In selecting the design velocity therefore, a compromise must be made between the tower diameter and the maximum use of the desiccant. Figure 8-22 shows a maximum design velocity. Smaller velocities may be required due to pressure drop considerations.

The minimum vessel internal diameter for a specified superficial velocity is given by:

$$d^2 = 3{,}600 \frac{Q_g \, T \, Z}{V_m \, P} \qquad (8\text{-}2)$$

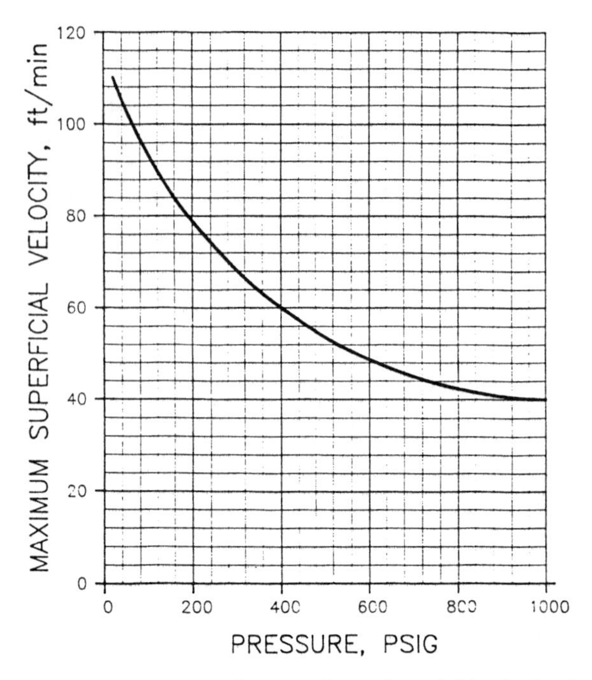

Figure 8-22. Maximum design velocity for solid bed adsorbers.

where d = vessel internal diameter, in.
Q_g = gas flow rate, MMscfd
T = gas temperature, °R
Z = compressibility factor
V_m = gas superficial velocity, ft/min
P = gas pressure, psia

Bed Height to Diameter Ratio

In its simplest form, an adsorber is normally a cylindrical tower filled with a solid desiccant. The depth of the desiccant may vary from a few feet to 30 ft or more. The vessel diameter may be from a few inches to 10 or 15 ft. A bed height to diameter (L/D) ratio of higher than 2.5 is desirable. Ratios as low as 1:1 are sometimes used; however, poor gas dehydration, caused by non-uniform flow, channeling and an inadequate contact time between the wet gas and the desiccant sometimes result.

Pressure Drop

Towers are sized for a design pressure drop of about 5 psi through the desiccant. The pressure drop can be estimated by:

$$\frac{\Delta P}{L} = B\mu\, V_m + C\rho\, V_m^2 \tag{8-3}$$

where ΔP = pressure drop, psi
L = length of bed, ft
μ = gas viscosity, cp
ρ = gas density, lb/ft^3
V_m = gas superficial velocity, ft/min

B and C are constants given by:

Particle Type	B	C
⅛-in. bead	0.0560	0.0000889
⅛-in. extrudate	0.0722	0.000124
1⁄16-in. bead	0.152	0.000136
1⁄16-in. extrudate	0.238	0.000210

Pressure drops of greater than approximately 8 psi are not recommended.

Moisture Content of Inlet Gas

An important variable that determines the size of a given desiccant bed is the relative saturation of the inlet gas. This variable is the driving force that affects the transfer of water to the adsorbent. If saturated gas (100% relative humidity) is being dried, higher useful capacities can be expected for most desiccants than when drying partially saturated gases. However, in most field gas dehydration installations the inlet gas is saturated with water vapor and this is not a variable that must be considered.

Desiccant Selection

No desiccant is perfect or best for all applications. In some applications the desiccant choice is determined primarily by economics. Sometimes the process conditions control the desiccant choice. Many times the

Table 8-4
Properties of Solid Desiccants

Desiccant	Bulk Density (lb/ft³)	Specific Heat (Btu/lb/°F)	Normal Sizes Used	Design Adsorptive Capacity (WT%)	Regeneration Temperature (°F)
Activated Alumina	51	0.24	¼ in.–8 mesh	7	350–600
Mobil SOR Beads	49	0.25	4–8 mesh	6	300–500
Fluorite	50	0.24	4–8 mesh	4–5	350+
Alumina Gel (H-151)	52	0.24	⅛–¼ inch	7	350–850
Silica Gel	45	0.22	4–8 mesh	7	350
Molecular Sieves (4A)	45	0.25	⅛ inch	14	450–550

Source: API.

desiccants are interchangeable and the equipment designed for one desiccant can often be operated effectively with another product. Table 8-4 illustrates the most common desiccants used for gas dehydration and some conservative parameters to use for initial design. Desiccant suppliers' information should be consulted for detail design.

All desiccants exhibit a decrease in capacity (design loading) with increase in temperature. Molecular sieves tend to be the less severely affected and aluminas the most affected by temperature.

Aluminas and molecular sieves act as a catalyst with H_2S to form COS. When the bed is regenerated, sulfur remains and plugs up the spaces. Liquid hydrocarbons also present a plugging problem to all desiccants, but molecular sieves are less susceptible to contamination with liquid hydrocarbons.

Silica gels will shatter in the presence of free water and are chemically attacked by many corrosion inhibitors. The chemical attack permanently destroys the silica gels. The other desiccants are not as sensitive to free water and are not chemically attacked by most corrosion inhibitors. However, unless the regeneration temperature is high enough to desorb the inhibitor, the inhibitor may adhere to the desiccants and possibly cause coking.

The alumina gels, activated aluminas, and molecular sieves are all chemically attacked by strong mineral acids and their adsorptive capacity will quickly decline. Special acid resistant molecular sieve desiccants are available.

EXAMPLE 8-2: DRY DESICCANT DESIGN

The detailed design of solid bed dehydrators should be left to experts. The general "rules of thumb" presented in this chapter can be used for preliminary design as shown in the following example:

Design Basis

Feed rate	50 MMscfd
Molecular weight of gas	17.4
Gas density	1.70 lb/ft^3
Operating temperature	110°F
Operating pressure	600 psia
Inlet dew point	100°F (equivalent to 90 lb of H_2O/MMcf)
Desired outlet dew point	1 ppm H_2O

Water Adsorbed

For this example, an 8-hour on-stream cycle with 6 hours of regeneration and cooling will be assumed. On this basis, the amount of water to be adsorbed per cycle is:

$$8/24 \times 50 \text{ MMcf} \times 90 \text{ lb/MMcf} = 1{,}500 \text{ lb } H_2O\text{/cycle}$$

Loading

Because of the relative cost, use Sorbeads as the desiccant and design on the basis of 6% loading. Sorbeads weigh approximately 49 lb/ft^3 (bulk density). The required weight and volume of desiccant per bed would be:

$$\frac{1{,}500 \text{ lb } H_2O}{0.06 \text{ lb } H_2O/\text{lb desiccant}} = 25{,}000 \text{ lb desiccant per bed}$$

$$\frac{25{,}000 \text{ lb desiccant / bed}}{49 \text{ lb desiccant/ft}^3} = 510 \text{ ft}^3 \text{ per bed}$$

Tower Sizing

Recommended maximum superficial velocity at 600 psia is about 55 ft/min. From Equation 8-2, assuming Z = 1.0:

$$d^2 = 3{,}600\,\frac{(50)(570)(1.0)}{(55)(600)}$$

$$d = 55.7\text{ in.} = 4.65\text{ ft}$$

The bed height is:

$$L = \frac{510\text{ ft}^2}{\pi\,(4.65)^2 / 4\text{ ft}^2} = 30.0\text{ ft}$$

The pressure drop from Equation 8-3, assuming ⅛-in. bead and $\mu = 0.01$ cp, is:

$$\Delta P = [(0.056)(0.01)(55) + (0.00009)(1.70)(55)^2]\;30$$

$$\Delta P = 14.8\text{ psi}$$

This is higher than the recommended 8 psi. Choose a diameter of 5 ft 6 in.

$$V = 39.2\text{ ft/min}$$
$$L = 21.5\text{ ft}$$
$$\Delta P = 5.5\text{ psi}$$

Leaving 6 ft above and below the bed, the total tower length including space to remove the desiccant and refill would be about 28 ft. This yields an L/D of 28/5.5 = 5.0.

Regeneration Heat Requirement

Assume the bed (and tower) is heated to 350°F. The average temperature will be (350 + 110)°F/2 = 230°F. The approximate weight of the 5 ft 6 in. ID × 28 ft × 700 psig tower is 53,000 lb including the shell, heads, nozzles and supports for the desiccant.

The heating and cooling requirement can be calculated using

$$Q = mCp\Delta t$$

Heating Requirement/Cycle

Desiccant	25,000 lb × (350°F – 110°F) × (0.25)	= 1,500,000 Btu
Tower	53,000 lb × (350°F – 110°F) × (0.12)*	= 1,520,000 Btu
Desorb water	1,500 lb × 1,100 Btu/lb**	= 1,650,000 Btu
	1,500 lb × (230°F – 110°F) × (1.0)***	= 200,000 Btu
		4,870,000 Btu
+ 10% for heat losses, etc.		490,000 Btu
		5,360,000 Btu/cycle

*Specific heat of steel.

**The number "1100 Btu/lb" is the heat of water desorption, a value supplied by the desiccant manufacturer.

***The majority of the water will desorb at the average temperature. This heat requirement represents the sensible heat required to raise the temperature of the water to the desorption temperature.

Cooling Requirement/Cycle

Desiccant	25,000 lb × (350°F – 110°F) × (0.25)	= 1,500,000 Btu
Tower	53,000 lb × (350°F – 110°F) × (0.12)	= 1,520,000 Btu
		3,020,000 Btu
+10% for non-uniform cooling, etc.		300,000 Btu
		3,320,000 Btu/cycle

These methods for calculating the heating and cooling requirements are conservative estimates assuming that the insulation is on the outside of the towers. The requirements will be less if the towers are insulated internally.

Regeneration Gas Heater

Assume the inlet temperature of regeneration gas is 400°F. In the beginning the initial outlet temperature of the bed will be the bed temperature of 110°F; at the end of the heating cycle, the outlet temperature will be the design value of 350°F. So the average outlet temperature is (350 + 110)/2 or 230°F. Then the volume of gas required for heating will be

$$V_{heating} = \frac{5,360,000 \text{ Btu/cycle}}{(400 - 230)\ °\text{F}\ (0.64)^{**} \text{Btu/lb/°F}} = 49,400 \text{ lb/cycle}$$

The regeneration gas heater load Q_H is then:

$$Q_H = 49{,}400\ (400 - 110)(0.62)^* = 8{,}900{,}000\ \text{Btu/cycle}$$

For design, add 25% for heat losses and non-uniform flow. Assuming a 3-hour heating cycle, the regenerator gas heater must be sized for

$$Q_H = 8{,}900{,}000 \times 1.25/3 = 3{,}710{,}000\ \text{Btu/hr}$$

Regeneration Gas Cooler

The regeneration gas cooling load is calculated using the assumption that all of the desorbed water is condensed during a half hour of the 3-hour cycle. The regeneration gas cooler load Q_c would be:

Regeneration gas	49,400 (230 – 110)(0.61)/3	= 1,205,000 Btu/hr
Water	1,500 (1,157 – 78)*/0.5	= 3,237,000 Btu/hr
		= 4,442,000 Btu/hr
	+ 10% =	444,000 Btu/hr
		4,886,000 Btu/hr

Cooling Cycle

For the cooling cycle the initial outlet temperature is 350°F and at the end of the cooling cycle, it is approximately 110°F. So the average outlet temperature is (350 + 110)/2 = 230°F. Assuming the cooling gas is at 110°F, the volume of gas required for cooling will be

$$V_{cooling} = \frac{3{,}320{,}000\ \text{Btu/cycle}}{(230 - 110)\ °\text{F}\ (0.59)^{**}\ \text{Btu/lb/°F}} = 46{,}900\ \text{lb/cycle}$$

---

*Steam tables.
**Specific heat of gas at the average temperature.

CHAPTER

9

Gas Processing*

The term "gas processing" is used to refer to the removing of ethane, propane, butane, and heavier components from a gas stream. They may be fractionated and sold as "pure" components, or they may be combined and sold as a natural gas liquids mix, or NGL.

The first step in a gas processing plant is to separate the components that are to be recovered from the gas into an NGL stream. It may then be desirable to fractionate the NGL stream into various liquefied petroleum gas (LPG) components of ethane, propane, iso-butane, or normal-butane. The LPG products are defined by their vapor pressure and must meet certain criteria as shown in Table 9-1. The unfractionated natural gas liquids product (NGL) is defined by the properties in Table 9-2. NGL is made up principally of pentanes and heavier hydrocarbons although it may contain some butanes and very small amounts of propane. It cannot contain heavy components that boil at more than 375°F.

In most instances gas processing plants are installed because it is more economical to extract and sell the liquid products even though this lowers the heating value of gas. The value of the increased volume of liquids sales may be significantly higher than the loss in gas sales revenue because of a decrease in heating value of the gas.

---

*Reviewed for the 1999 edition by Douglas L. Erwin of Paragon Engineering Services, Inc.

Table 9-1
GPA Liquefied Petroleum Gas Specifications
(from GPA Standard 2140-86)

Product Characteristics	Product Designation: Commercial Propane	Commercial Butane	Commercial B-P Mixtures	Propane HD-5	Test Methods
Composition	Predominantly propane and/or propylene.	Predominantly butanes and/or butylenes.	Predominantly mixtures of butanes and/or butylenes with propane and/or propylene.	Not less than 90 liquid volume percent propane; not more than 5 liquid volume percent propylene.	ASTM D-2163-82
Vapor pressure at 100°F,* psig, max.	208*	70*	208*	208*	ASTM D-1267-84
Volatile residue:					
temperature at 95% evaporation, deg. F, max.	–37*	36*	36*	–37*	ASTM D-1837-81
or					
butane and heavier, liquid volume percent max.	2.5	—	—	2.5	ASTM D-2163-82
pentane and heavier, liquid volume percent max.	—	2.0	2.0	—	ASTM D-2163-82
Residual matter:					
residue on evaporation of 100 ml,	0.05ml	—	—	0.05ml	ASTM D-2158-80
max. oil stain observation	pass (1)	—	—	pass (1)	ASTM D-2158-80
Corrosion, copper strip, max.	No. 1	No. 1	No. 1	No. 1	ASTM D-1838-84
Total sulfur, ppmw	185	140	140	123	ASTM D-2784-80I
Moisture content	pass	—	—	pass	GPA Propane Dryness Test (Cobalt Bromide) *or* D-2713-81
Free water content	—	none	none	—	—

(1) An acceptable product shall not yield a persistent oil ring when 0.3 ml of solvent residue mixture is added to a filter paper in 0.1 increments and examined in daylight after 2 minutes as described in ASTM D-2158.

**Metric Equivalents   208 psig = 1434 kPa gauge   70 psig = 483 kPa gauge   – 37°F = –38.3°C   36°F = 2.2°C   100°F = 37.8°C*

Courtesy: Gas Processing Suppliers Association, Tenth Edition, Engineering Data Book

Table 9-2
GPA Natural Gasoline Specifications

Product Characteristic	Specification	Test Method
Reid Vapor Pressure	10–34 pounds	ASTM D-323
Percentage evaporated at 140°F	25–85	ASTM D-216
Percentage evaporated at 275 °F	Not less than 90	ASTM D-216
End point	Not more than 375°F	ASTM D-216
Corrosion	Not more than classification 1	ASTM D-130 (modified)
Color	Not less than plus 25 (Saybolt)	ASTM D-156
Reactive Sulfur	Negative, "sweet"	GPA 1138

In addition to the above general specifications, natural gasoline shall be divided into 24 possible grades on the basis of Reid vapor pressure and percentage evaporated at 140°F. Each grade shall have a range in vapor pressure of four pounds, and a range in the percentage evaporated at 140°F of 15%. The maximum Reid vapor pressure of the various grades shall be 14, 18, 22, 26, 30, and 34 pounds respectively. The minimum percentage evaporated at 140°F shall be 25, 40, 55, and 70 respectively. Each grade shall be designated by its maximum vapor pressure and its minimum percentage evaporated at 140°F, as shown in the following:

Grades of Natural Gasoline
Percentage Evaporated at 140°F

Reid Vapor Pressure	25–40	40–55	55–70	70–85
34–30	Grade 34-25	Grade 34-40	Grade 34-55	Grade 34-70
30–26	Grade 30-25	Grade 30-40	Grade 30-55	Grade 30-70
26–22	Grade 26-25	Grade 26-40	Grade 26-55	Grade 26-70
22–18	Grade 22-25	Grade 22-40	Grade 22-55	Grade 22-70
18–14	Grade 18-25	Grade 18-40	Grade 18-55	Grade 18-70
14–10	Grade 14-25	Grade 14-40	Grade 14-55	Grade 14-70

**Courtesy: Gas Processing Suppliers Association, Tenth Edition,* Engineering Data Book

In deciding whether it is economical to remove liquids from a natural gas stream, it is necessary to evaluate the decrease in gas value after extraction of the liquid. Table 9-3 shows the break-even value for various liquids. Below these values the molecules will be more valuable as gas.

The difference between the actual sales price of the liquid and the break-even price of the liquid in Table 9-3 provides the income to pay out the capital cost, fuel cost, and other operating and maintenance expenses necessary to make the recovery of the gas economically attractive.

Another objective of gas processing is to lower the Btu content of the gas by extracting heavier components to meet a maximum allowable heating limit set by a gas sales contract. If the gas is too rich in heavier components, the gas will not work properly in burners that are designed for lower heating values. A common maximum limit is 1100 Btu per SCF. Thus, if the gas is rich in propane and heavier components it may have to be processed to lower the heating value, even in cases where it may not be economical to do so.

This chapter briefly describes the basic processes used to separate LPG and NGL liquids from the gas and to fractionate them into their various components. It is beyond the scope of this text to discuss detailed design of gas processing plants.

ABSORPTION/LEAN OIL

The oldest kind of gas plants are absorption/lean oil plants, where a kerosene type oil is circulated through the plant as shown in Figure 9-1. The "lean oil" is used to absorb light hydrocarbon components from the gas. The light components are separated from the rich oil and the lean oil is recycled.

Typically the inlet gas is cooled by a heat exchanger with the outlet gas and a cooler before entering the absorber. The absorber is a contact tower, similar in design to the glycol contact tower explained in Chapter 8. The lean absorber oil trickles down over trays or packing while the gas flows upward countercurrent to the absorber oil. The gas leaves the top of the absorber while the absorber oil, now rich in light hydrocarbons from the gas, leaves the bottom of the absorber. The cooler the inlet gas stream the higher the percentage of hydrocarbons which will be removed by the oil.

Rich oil flows to the rich oil de-ethanizer (or de-methanizer) to reject the methane and ethane (or the methane alone) as flash gas. In most lean

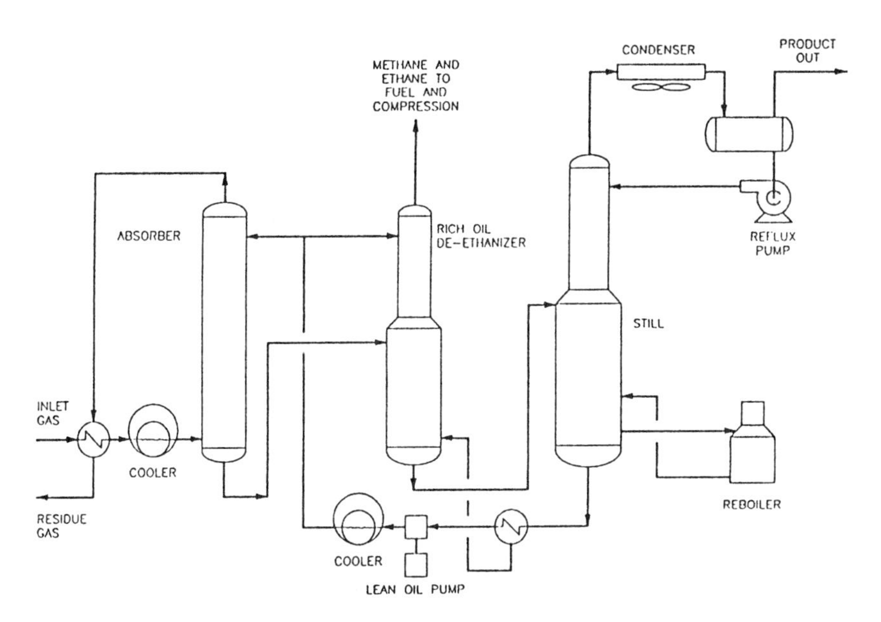

Figure 9-1. Simplified flow diagram of an absorption plant.

oil plants the ROD unit rejects both methane and ethane since very little ethane is recovered by the lean oil. If only methane were rejected in the ROD unit, then it may be necessary to install a de-ethanizer column downstream of the still to make a separate ethane product and keep ethane from contaminating (i.e., increasing the vapor pressure of) the other liquid products made by the plant.

The ROD is similar to a cold feed stabilizing tower for the rich oil. Heat is added at the bottom to drive off almost all the methane (and most likely ethane) from the bottoms product by exchanging heat with the hot lean oil coming from the still. A reflux is provided by a small stream of cold lean oil injected at the top of the ROD. Gas off the tower overhead is used as plant fuel and/or is compressed. The amount of intermediate components flashed with this gas can be controlled by adjusting the cold lean oil reflux rate.

Absorber oil then flows to a still where it is heated to a high enough temperature to drive the propanes, butanes, pentanes and other natural gas liquid components to the overhead. The still is similar to a crude oil stabilizer with reflux. The closer the bottom temperature approaches the boiling temperature of the lean oil the purer the lean oil which will be recirculated to the absorber. Temperature control on the condenser keeps lean oil from being lost with the overhead.

Thus the lean oil, in completing a cycle, goes through a recovery stage where it recovers light and intermediate components from the gas, a rejection stage where the light ends are eliminated from the rich oil and a separation stage where the natural gas liquids are separated from the rich oil.

These plants are not as popular as they once were and are rarely, if ever, constructed anymore. They are very difficult to operate, and it is difficult to predict their efficiency at removing liquids from the gas as the lean oil deteriorates with time. Typical liquid recovery levels are:

$C_3 \cong 80\%$
$C_4 \cong 90\%$
$C_{5+} \cong 98\%$

REFRIGERATION

In a refrigeration plant the inlet gas is cooled to a low enough temperature to condense the desired fraction of LPG and NGL. Either freon or propane is used as the refrigerant. Figure 9-2 shows a typical refrigeration plant.

The free water must be separated and the dew point of the gas lowered before cooling the feed to keep hydrates from forming. It is possible to

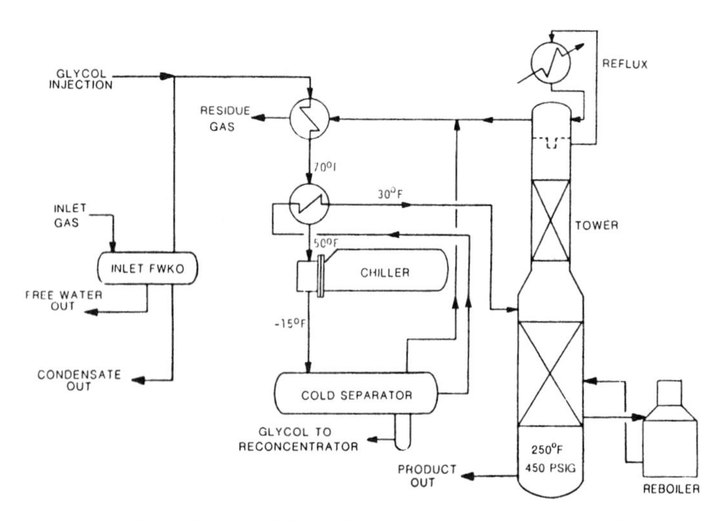

Figure 9-2. Simplified flow diagram of a refrigeration plant.

dehydrate the gas with TEG or mole sieves to the required dew point. It is more common to lower the hydrate temperature by injecting glycol in the gas after separation of free water. The glycol and water separate in the cold separator where they are routed to a regenerator, the water is boiled off and the glycol is circulated back to be injected into the inlet stream. Some glycol will be lost with time and will have to be made up. The most common glycol used for this service is ethylene glycol because of its low cost and the fact that at the low temperatures it is not lost to the gas phase.

The chillcr is usually a kettle type exchanger. Freon (which is cooled in a refrigeration cycle to –20°F) is able to cool the gas to approximately –15°F. Propane, which can be cooled to –40°F, is sometimes used if lower gas temperatures and greater recovery efficiences are desired.

The gas and liquid are separated in the cold separator, which is a three-phase separator. Water and glycol come off the bottom, hydrocarbon liquids are routed to the distillation tower and gas flows out the top. If it is desirable to recover ethane, this still is called a de-methanizer. If only propane and heavier components are to be recovered it is called a de-ethanizer. The gas is called "plant residue" and is the outlet gas from the plant.

The tower operates in the same manner as a condensate stabilizer with reflux. The inlet liquid stream is heated by exchange with the gas to approximately 30°F and is injected in the tower at about the point in the tower where the temperature is 30°F. By adjusting the pressure, number of trays, and the amount of reboiler duty, the composition of the bottoms liquid can be determined.

By decreasing the pressure and increasing the bottoms temperature more methane and ethane can be boiled off the bottoms liquid and the RVP of the liquid stream decreased to meet requirements for sales or further processing. Typical liquid recovery levels are:

$C_3 \cong 85\%$
$C_4 \cong 94\%$
$C_{5+} \cong 98\%$

These are higher than for a lean oil plant. It is possible to recover a small percentage of ethane in a refrigeration plant. This is limited by the ability to cool the inlet stream to no lower than –40°F with normal refrigerants.

Most refrigeration plants use freon as the refrigerant and limit the lowest temperature to –20°F. This is because the ANSI piping codes require special metallurgy considerations below –20°F to assure ductility.

Cryogenic Plants

Figure 9-3 shows a typical cryogenic plant where the gas is cooled to −100°F to −150°F by expansion through a turbine or Joule-Thompson (J-T) valve. In this example liquids are separated from the inlet gas at 100°F and 1,000 psig. It is then dehydrated to less than 1 ppm water vapor to assure that hydrates will not form at the low temperatures encountered in the plant. Typically, a mole sieve dehydrator is used.

The gas is routed through heat exchangers where it is cooled by the residue gas, and condensed liquids are recovered in a cold separator at approximately –90°F. These liquids are injected into the de-methanizer at a level where the temperature is approximately –90°F. The gas is then expanded (its pressure is decreased from inlet pressure to 225 psig) through an expansion valve or a turboexpander. The turboexpander uses the energy removed from the gas due to the pressure drop to drive a compressor, which helps recompress the gas to sales pressure. The cold gas (−150°F) then enters the de-methanizer column at a pressure and temperature condition where most of the ethanes-plus are in the liquid state.

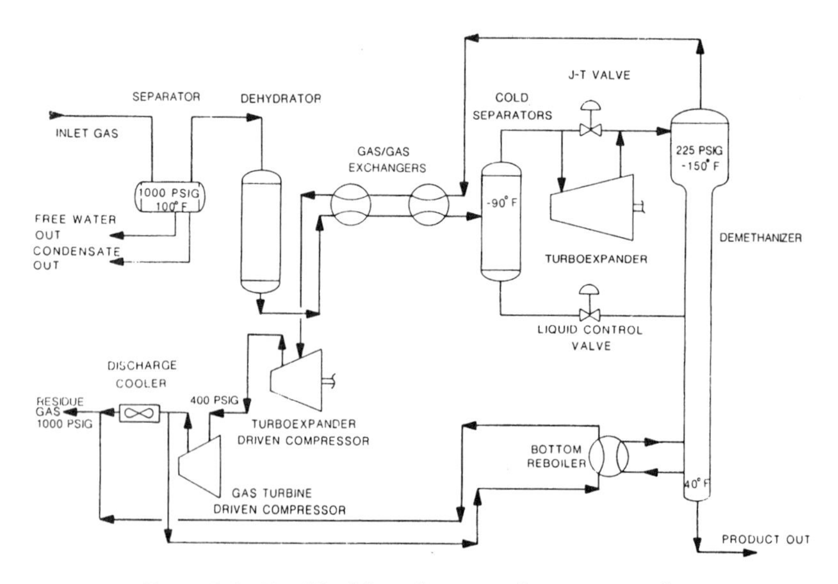

Figure 9-3. Simplified flow diagram of a cryogenic plant.

The de-methanizer is analogous to a cold feed condensate stabilizer. As the liquid falls and is heated, the methane is boiled off and the liquid becomes leaner and leaner in methane. Heat is added to the bottom of the tower using the hot discharge residue gas from the compressors to assure that the bottom liquids have an acceptable RVP or methane content.

The gas turbine driven compressor is required since there are energy losses in the system. The energy generated by expanding the gas from 600 psig to 225 psig in the turbo-expander cannot be 100% recovered and used to recompress the residue gas from 225 psig to 600 psig. In this particular plant it is only capable of recompressing the gas to 400 psig. Thus, even if the inlet gas and sales gas were at the same pressure, it would be necessary to provide some energy in the form of a compressor to recompress the gas.

Because of the lower temperatures that are possible, cryogenic plants have the highest liquid recovery levels of the plants discussed. Typical levels are:

$C_2 > 60\%$
$C_3 > 90\%$
$C_{4+} \cong 100\%$

CHOICE OF PROCESS

Because of the greater liquid recoveries, cryogenic plants are the most common designs currently being installed. They are simple to operate and easy to package, although somewhat more expensive than refrigeration plants. Refrigeration plants may be economical for rich gas streams where it is not desired to recover ethane. Lean oil plants are expensive and hard to operate. They are rarely designed as new plants anymore. Existing lean oil plants are sometimes salvaged, refurbished and moved to new locations.

Fractionation

The bottoms liquid from any gas plant may be sold as a mixed product. This is common for small, isolated plants where there is insufficient local demand. The mixed product is transported by truck, rail, barge or pipeline to a central location for further processing. Often it is more economical to separate the liquid into its various components and sell it as

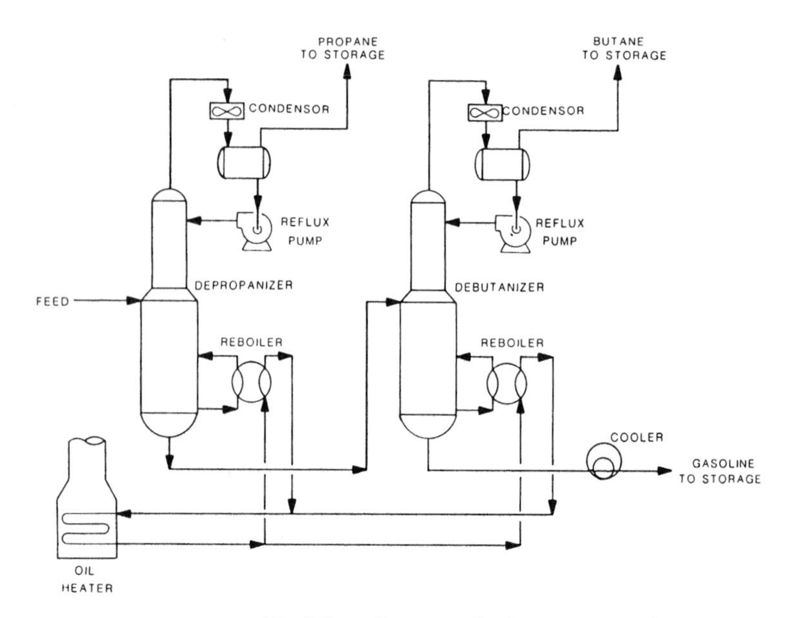

Figure 9-4. Simplified flow diagram of a fractionation plant.

ethane, propane, butane, and natural gasoline. The process of separating the liquids into these components is called fractionation.

Figure 9-4 shows a typical fractionation system for a refrigeration or lean oil plant. The liquid is cascaded through a series of distillation towers where successively heavier and heavier components (fractions) are separated as overhead gas. In this figure the liquid from the still of an absorption plant or the de-methanizer (or de-ethanizer) tower of an expansion or refrigeration plant is routed to a de-propanizer. If there is too high a fraction of butanes-plus in the propane, this can be reduced by adjusting the de-propanizer pressure upward or reflux condensing temperature downward. If the vapor pressure of the propane exceeds the required specification this means that the fraction of methane and ethane in the inlet stream is too high. This fraction can be adjusted downward by increasing the temperature or decreasing the operating pressure of the still or tower that feeds liquid to the de-propanizer.

The de-butanizer works in a similar manner. The upstream tower (de-propanizer) determines the maximum vapor pressure of the butane product. If the concentration of propane-minus is too large in the inlet stream, the vapor pressure of the butane overheads will be too high. Similarly, the concentration of pentanes-plus in the butane will depend upon the reflux condensing temperature and tower operating pressure. If the pentanes-plus exceed specifications, further reflux cooling or a higher operating pressure will be needed to condense pentanes-plus from the butane overheads.

The temperature at the base of the de-butanizer determines the vapor pressure of the gasoline product. If its vapor pressure is too high, the temperature must be increased or the tower pressure decreased to drive more butanes-minus out of the bottoms liquids.

If the feed to the fractionator contains recoverable ethane, such as is likely to be the case with a cryogenic plant, then a de-ethanizer tower would be installed upstream of the de-propanizer.

Design Considerations

The design of any of the distillation processes discussed requires choosing an operating pressure, bottoms temperature, reflux condenser temperature and number of trays. This is normally done using any one of several commercially available process simulation programs which can perform the iterative calculations discussed in Chapter 6.

Some typical parameters for design are shown in Table 9-4. The actual optimum to use for any given process will vary depending on actual feed properties, product specifications, etc.

In Table 9-4 the actual number of trays are included. This is because complete equilibrium between vapor and liquid is normally not reached on each tray. For calculation purposes the number of theoretical flashes may be quite a bit less than the number of trays. For smaller diameter

Table 9-3
Gas Caloric Heating Cost Basis Evaluation for Liquids Recovery

Assumed Value of Gas >>>			$2.00/MMBtu	$3.00/MMBtu
Gas Component	Net Heating Value Btu/SCF	SCF/Gallon	Equivalent Value $/Gallon	Equivalent Value $/Gallon
Ethane	1618	37.5	0.1213	0.1820
Propane	2316	36.4	0.1686	0.2529
Butane	3010	31.8	0.1915	0.2872
Pentane	3708	27.7	0.2054	0.3081

Table 9-4
Typical Fractionator-Absorber/Stripper Design Number of Trays

Tower	Pressure Range psig	Approximate Ranges Shown: Actual Trays Above Main Feed Number	Actual Trays Below Main Feed Number
Lean Oil Plant			
Absorber	200–1100	24–30	20–50
Rich Oil De-methanizer	450–600	20–30	20–50
Rich Oil De-ethanizer	175–300	24–30	20–50
Rich Oil Still	85–160	12–60	16–60
Refrigeration Plant			
De-methanizer	550–650	14–30	26–30
De-ethanizer	350–500	10–70	20–70
De-propanizer	200–300	17–70	18–70
De-butanizer	70–100	18–70	15–70

towers packing is used instead of trays. Manufacturers supply data for their packing material which indicates the amount of feet of packing required to provide the same mass transfer as a standard bubble cap tray.

Some recent advances in structured packing are being used by some operators in larger diameter towers where they would have normally used trays. The structured packing is said to allow both smaller diameter and less height of tower.

Once the operating conditions are established for a tower, its diameter and height can be chosen using data available from tray and packing manufacturers. The details of tower diameter selection, tray spacing, and internal design are beyond the scope of this text.

CHAPTER

10

*Compressors**

Compressors are used whenever it is necessary to flow gas from a lower pressure to a higher pressure system. Flash gas from low-pressure vessels used for multistage stabilization of liquids, oil treating, water treating, etc., often exists at too low a pressure to flow into the gas sales pipeline. Sometimes this gas is used as fuel and the remainder flared or vented. Often it is more economical or it is necessary for environmental reasons to compress the gas for sales. In a gas field, a compressor used in this service is normally called a "flash gas compressor." Flash gas compressors are normally characterized by low throughput rate and high differential pressure.

The differential pressure is expressed in terms of overall compressor ratio, R_T, which is defined as:

$$R_T = \frac{P_d}{P_s} \tag{10-1}$$

where R_T = overall compressor ratio
P_d = discharge pressure, psia
P_s = suction pressure, psia

Flash gas compressors typically have an overall compressor ratio in the range of 5 to 20.

*Reviewed for the 1999 edition by John H. Galey of Paragon Engineering Services, Inc.

In some marginal gas fields, and in many larger gas fields that experience a decline in flowing pressure with time, it may be economical to allow the wells to flow at surface pressures below that required for gas sales. In such cases a "booster compressor" may be installed. Booster compressors are typically characterized by low overall compressor ratio (on the order of 2 to 5) and relatively high throughput.

Booster compressors are also used on long pipelines to restore pressure drop lost to friction. The design of a long pipeline requires trade-off studies between the size and distance between booster compressor stations and the diameter and operating pressure of the line.

The use of large compressors is probably more prevalent in oil field facilities than in gas field facilities. Oil wells often require low surface pressure and the gas that flashes off the oil in the separator must be compressed in a flash gas compressor. Often a gas lift system is needed to help lift the oil to the surface. As described in Volume 1, a "gas lift compressor" must compress not only the formation gas that is produced with the oil, but also the gas-lift gas that is recirculated down the well. Gas lift compressors are characterized by both high overall compressor ratios and relatively high throughputs.

Often, other forms of artificial lift are used to produce oil wells such as downhole submersible pumps and rod pumps that require that most of the formation gas be separated downhole and flowed up the annulus between the tubing and the casing. When it is economical to recover this gas, or when the gas must be recovered for environmental reasons, a "casinghead gas compressor" will be installed. These are sometimes called "casing vapor recovery (CVR) units" or just "vapor recovery units (VRU)." Casinghead compressors are typically characterized by low suction pressure (0 to 25 psig). They often discharge at low pressure (50 to 300 psig) into the suction of a booster or flash gas compressor or into a low-pressure gas gathering system that gathers gas from several locations to a central compressor station.

Vapors from tanks and other atmospheric equipment may be recovered in a "vapor recovery compressor" (VRU). Vapor recovery compressors have very low suction pressure (0 to 8 ounces gauge) and typically have low flow rates. They normally discharge into the suction of a flash gas compressor.

This chapter presents an overview of the types of compressors, considerations for selecting a type of compressor, a procedure for estimating horsepower and number of stages, and some process considerations for both reciprocating and centrifugal compressors. Chapter 11 discusses reciprocating compressors in more detail, as this is the most common type used in oil and gas field compression.

TYPES OF COMPRESSORS

Volume 1 explains that pumps can be classified as either positive-displacement or kinetic. The same is true for compressors. In a positive displacement compressor the gas is transported from low pressure to high pressure in a device that reduces its volume and thus increases its pressure. The most common type of positive displacement compressors are reciprocating and rotary (screw or vane) just as was the case for pumps. Kinetic compressors impart a velocity head to the gas, which is then converted to a pressure head in accordance with Bernoulli's Law as the gas is slowed down to the velocity in the discharge line. Just as was the case with pumps, centrifugal compressors are the only form of kinetic compressor commonly used.

Reciprocating Compressors

A reciprocating compressor is a positive-displacement machine in which the compressing and displacing element is a piston moving linearly within a cylinder. Figure 10-1 shows the action of a reciprocating compressor.

In Position 1 the piston is moving away from the cylinder head and the suction valve is open, allowing the cylinder pressure to equal suction pressure and gas to enter the cylinder. The discharge valve is closed. At Position 2 the piston has traveled the full stroke within the cylinder and the cylinder is full of gas at suction pressure. The piston begins to move to the left, closing the suction valve. In moving from Position 2 to Position 3, the piston moves toward the cylinder head and the volume is reduced. This increases pressure until the cylinder pressure is equal to the discharge pressure and the discharge valve opens. The piston continues to move to the end of the stroke near the cylinder head, discharging gas. Pressure in the cylinder is equal to discharge pressure from Position 3 to Position 4. As the piston reverses its travel the gas remaining within the cylinder expands until it equals suction pressure and the piston is again in Position 1.

Reciprocating compressors are classified as either "high speed" or "slow speed." Typically, high-speed compressors run at a speed of 900 to 1200 rpm and slow-speed units at speeds of 200 to 600 rpm.

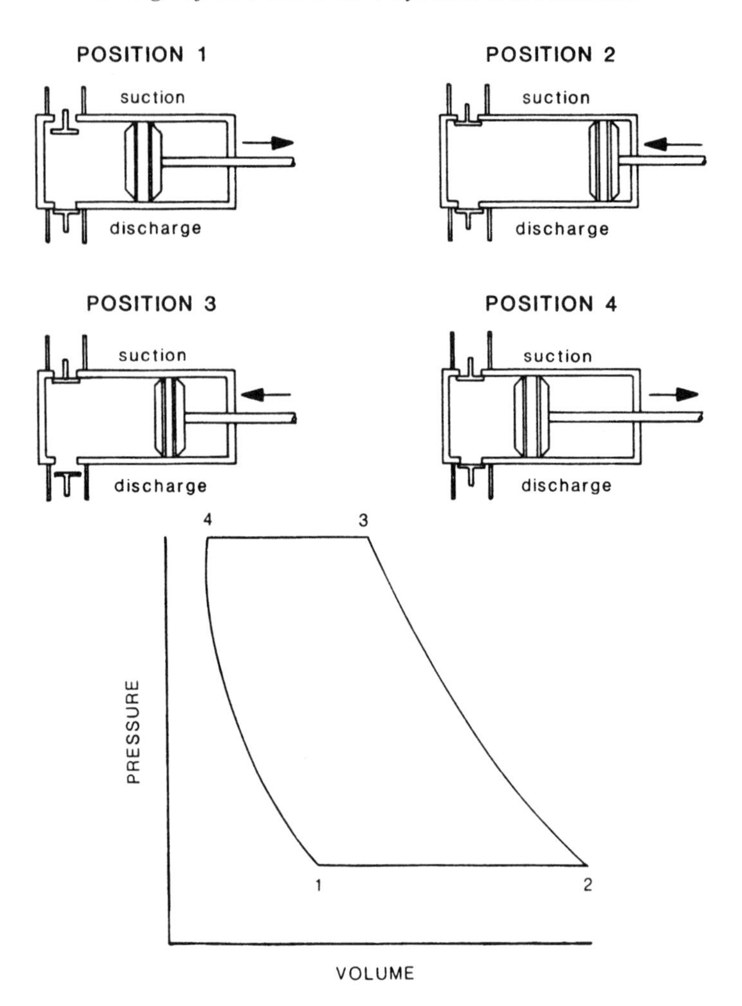

Figure 10-1. Reciprocating compressor action.

Figure 10-2 shows a high-speed compressor frame and cylinders. The upper compressor is called a two throw machine because it has two cylinders attached to the frame and running off the crank shaft. The lower compressor is a four-throw machine because it has four cylinders attached to the frame. The number of "throws" refers to the number of pistons.

As pointed out in Volume 1, Chapter 3, a compressor may have any number of stages. Each stage normally contains a suction scrubber to separate any liquids that carry over or condense in the gas line prior to the compressor cylinder (or case for centrifugal compressors). When gas

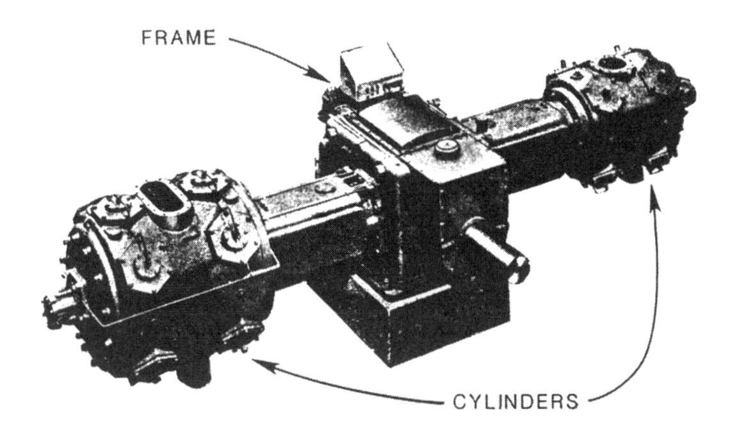

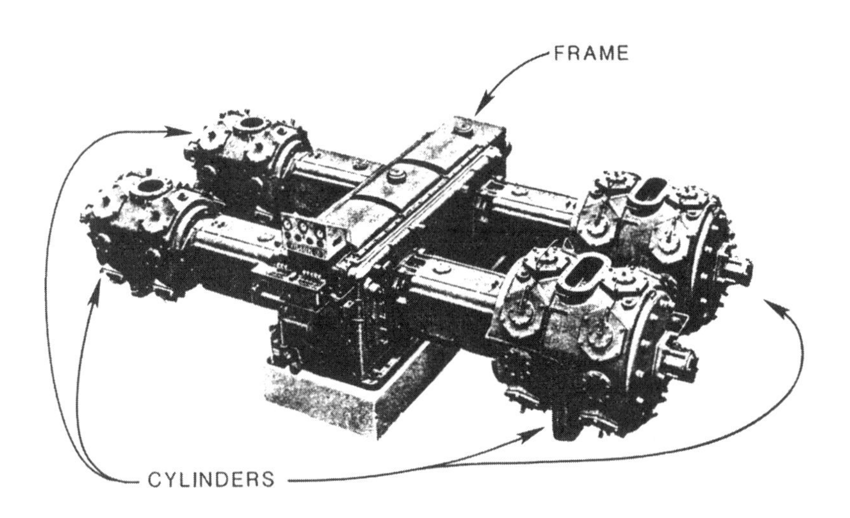

Figure 10-2. High-speed reciprocating compressor frames and cylinders. (*Courtesy of Dresser-Rand Company.*)

is compressed, its temperature increases. Therefore, after passing through the cylinder the gas is usually cooled before being routed to another suction scrubber for another stage of compression. A stage of compression thus consists of a scrubber, cylinder, and after-cooler. (The discharge from the final cylinder may not be routed to an after-cooler.)

The number of throws is not the same as the number of stages of compression. It is possible to have a two-stage, four-throw compressor. In this case there would be two sets of two cylinders working in parallel. Each set would have a common suction and discharge.

High-speed units are normally "separable." That is, the compressor frame and driver are separated by a coupling or gear box. This is opposed to an "integral" unit where power cylinders are mounted on the same frame as the compressor cylinders, and the power pistons are attached to the same drive shaft as the compressor cylinders.

High-speed units are typically engine or electric motor driven, although turbine drivers have also been used. Engines or turbines can be either natural gas or diesel fueled. By far the most common driver for a high-speed compressor is a natural gas driven engine.

Figure 10-3 shows a high-speed engine-driven compressor package. The unit typically comes complete on one skid with driver, compressor, suction scrubbers and discharge coolers for each stage of compression and all necessary piping and controls. On large units (>1,000 hp plus) the cooler may be shipped on a separate skid.

The major characteristics of high-speed reciprocating compressors are:

Size

- Numerous sizes from 50 hp to 3000 hp.
- 2, 4, or 6 compressor cylinders are common.

Advantages

- Can be skid mounted.
- Self-contained for easy installation and easily moved.
- Low cost compared to low-speed reciprocating units.
- Easily piped for multistage compression.
- Size suitable for field gathering offshore and onshore.
- Flexible capacity limits.
- Low initial cost.

Disadvantages

- High-speed engines are not as fuel efficient as integral engines (7,500 to 9,000 Btu/bhp-hr).

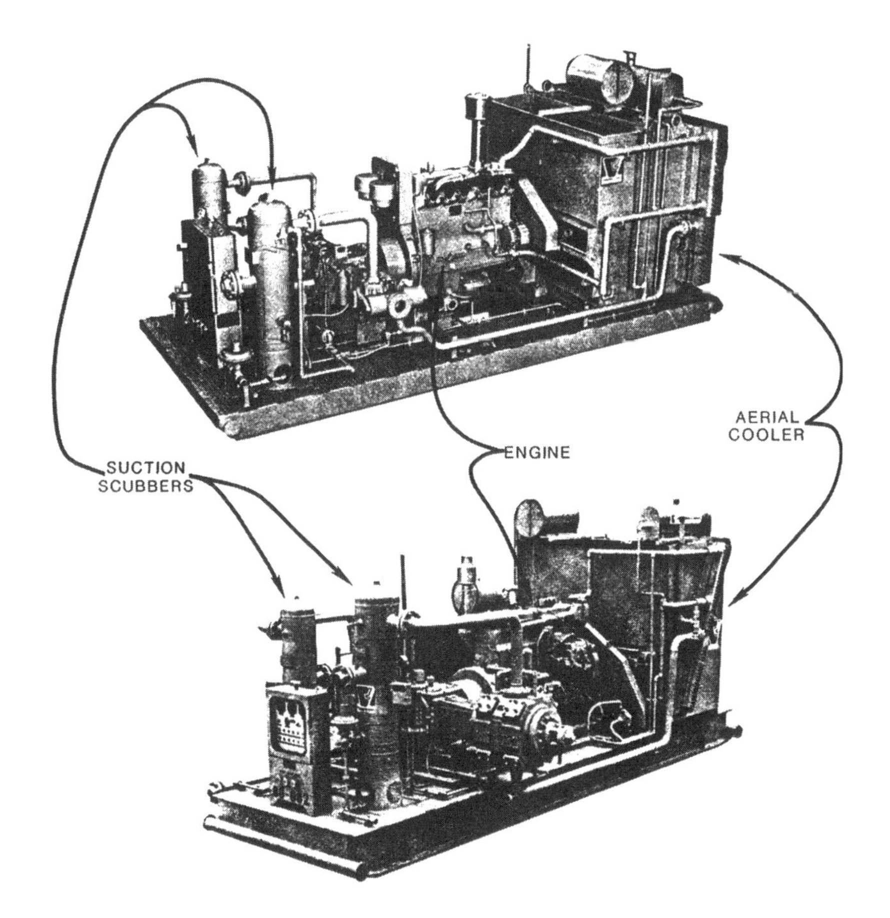

Figure 10-3. High-speed reciprocating compressor packages. (*Courtesy of Dresser-Rand Company.*)

- Medium range compressor efficiency (higher than centrifugal; lower than low-speed).
- Short life compared to low-speed.
- Higher maintenance cost than low-speed or centrifugal.

Low-speed units are typically integral in design as shown in Figure 10-4. "Integral" means that the power cylinders that turn the crank shaft are in the same case (same housing) as the cylinders that do the compressing of the gas. There is one crank shaft. Typically, integrals are con-

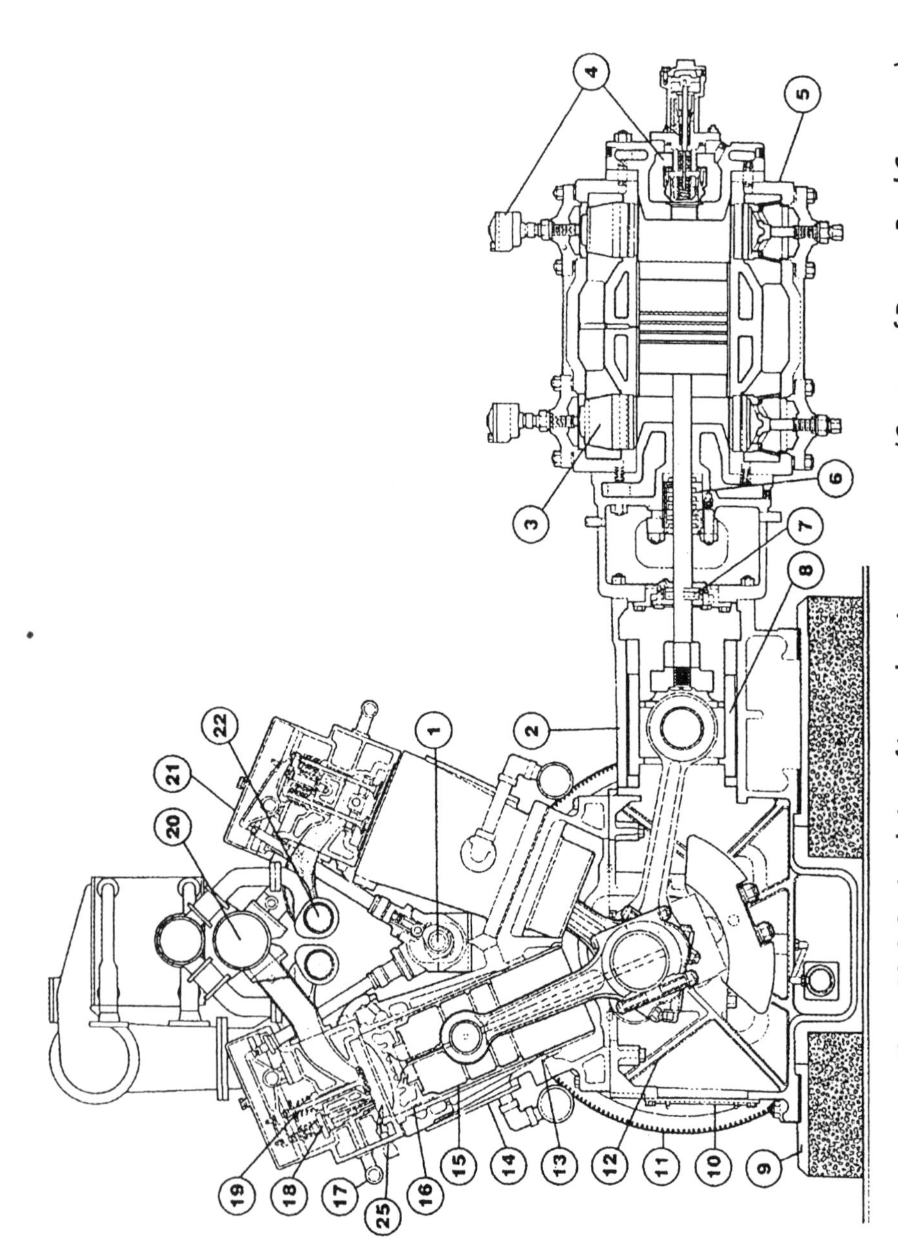

Figure 10-4. Sectional view of integral engine compressor. (*Courtesy of Dresser-Rand Company.*)

1. Fully-enclosed camshaft is built in sections for easy removal.
2. Crosshead guides are cast integrally with engine frame.
3. The best compressor valves for any service selected from the unmatched Dresser-Rand line, including famous gas-cushioned Channel Valves.
4. Clearance pockets and other types of capacity control devices are available to suit any application.
5. Compressor cylinders (dry or water-cooled) of cast iron, nodular iron, or forged steel are engineered to suit required pressures and capacities.
6. Full-floating packing adjusts itself in operation and assures best seal with minimum wear. Packing is pressure-lubricated and vented.
7. Oil wiper rings remove excess oil from piston rod and seal the frame.
8. Crossheads, running in bored guides, have shim-adjusted babbitted shoes at top and bottom and either full-floating or fixed crosshead pins. Suitable for addition of balance weights.
9. Simple, low-cost foundation, made possible by the smaller size, lighter weight, and smooth running blance of the KVSR.
10. Large frame openings give ample and unrestricted access to crankcase.
11. Flywheel-mounted ring gear for starting motors permits cranking with either air or gas between 150 psi and 225 psi supply pressure.
12. Rigid alloy-iron frame and top are well reinforced with cast-in ribs. Integrally cast bulkheads hold main bearings on both sides of every crankthrow. Keys, double-bolting, and tie rods are used to secure the frame and frame top together as a solid structure.
13. To assure oil-tight joints between power cylinders and frame top, the frame top is bored, and the precision-machined cylinder extension is provided with O-rings.
14. Cylinder water jackets are provided with removable cover plates for inspection.
15. Pistons are precision-ground for a perfect fit in a honed cylinder liner bore. Long-skirt, lightweight piston reduces wear.
16. For sustained low oil consumption, narrow, deep-groove piston rings conform easily to liner walls. Top compression ring is chrome-plated to condition liners during break-in.
17. Fuel gas headers, one for each bank of power cylinders, are controlled by common automatic valve and safety devices. Orifice plates equalize the distribution of gas to each of the cylinders. Individual adjustments are eliminated.
18. Simple fuel injection valves are operated from single camshaft.
19. Long-life special alloy valves have chromium-plated stems, hardened shrink-fit valve seats, and replaceable guides.
20. Common air inlet manifold conducts air from turbochargers to each power cylinder.
21. Large covers give easy access to valve gear, exclude dust and dirt.
22. Water-jacketed exhaust manifolds eliminate expansion strains and keep engine room temperature down.
23. Fitted with a reliable Altronic II CPU solid state low tension breakerless ignition system.
24. The engine can be fitted with either hydraulic or electronic governor systems to control engine speed.
25. In recent years, each power head is fitted with a bolt-in pre-combustion chamber, which allows the engine to burn a very lean mixture, resulting in very low exhaust emissions

Figure 10-5. Integral engine compressor. ***(Courtesy of Cooper Industries Energy Services Group.)***

sidered low-speed units. They tend to operate at 400–600 rpm, although some operate as low as 200 rpm.

Figure 10-5 shows a very large integral compressor. This would be typical of compressors in the 2,000 hp to 13,000 hp size. The size of this unit can be estimated by the height of the handrails above the compressor cylinder on the walkway that provides access to the power cylinders. This particular unit has sixteen power cylinders (eight on each side) and four compressor cylinders.

It should be obvious that one of these large integrals would require a very large and expensive foundation and would have to be field erected. Often, even the compressor cylinders must be shipped separate from the frame due to weight and size limitations. Large integrals are also much more expensive than either high-speeds or centrifugals.

For this reason, even though they are the most fuel efficient choice for large horsepower needs, large integrals are not often installed in oil and gas fields. They are more common in plants and pipeline booster service where their fuel efficiency, long life, and steady performance outweigh their much higher cost.

There are some low horsepower (140 to 360) integrals that are normally skid mounted as shown in Figure 10-6 and used extensively in small oil fields for flash gas or gas-lift compressor service. In these units the power cylinders and compressor cylinders are both mounted horizontally

Figure 10-6. Small-horsepower skid-mounted integrals. (*Courtesy of Cooper Industries.*)

and opposed to each other. There may be one or two compressor cylinders and one to four power cylinders. They operate at very slow speed. Their cost and weight are more than similar sized high-speed separable units, but they have lower maintenance cost, greater fuel efficiency, and longer life than the high speeds.

The major characteristics of low-speed reciprocating compressors are:

Size

- Some one and two power cylinder field gas compressors rated for 140 hp to 360 hp.
- Numerous sizes from 2,000 hp to 4,000 hp.
- Large sizes 2,000 hp increments to 12,000 hp.
- 2 to 10 compressor cylinders common.

Advantages

- High fuel efficiency (6–8,000 Btu/bhp-hr).
- High efficiency compression over a wide range of conditions.
- Long operating life.
- Low operation and maintenance cost when compared to high speeds.

Disadvantages

- Usually must be field erected except for very small sizes.
- Requires heavy foundation.
- High installation cost.
- Slow speed requires high degree of vibration and pulsation suppression.

Vane-Type Rotary Compressors

Rotary compressors are positive-displacement machines. Figure 10-7 shows a typical vane compressor. The operation is similar to that of a vane pump shown schematically in Figure 10-10 of Volume 1, 2nd Edition (Figure 10-9 in 1st Edition). A number of vanes, typically from 8 to 20, fit into slots in a rotating shaft. The vanes slide into and out of the slots as the shaft rotates and the volume contained between two adjacent vanes and the wall of the compressor cylinder decreases. Vanes can be cloth impregnated with a phenolic resin, bronze, or aluminum. The more vanes the compressor has, the smaller the pressure differential across the vanes. Thus, high-ratio vane compressors tend to have more vanes than low-ratio compressors.

A relatively large quantity of oil is injected into the flow stream to lubricate the vanes. This is normally captured by a discharge cooler and after-scrubber and recycled to the inlet.

Figure 10-7. Vane-type rotary compressor. (*Courtesy of Dresser-Rand Company.*)

Vane compressors tend to be limited to low pressure service, generally less than 100 to 200 psi discharge. They are used extensively as vapor recovery compressors and vacuum pumps. Single-stage vane compressors can develop 27 in. Hg vacuums, two-stage compressors can develop 29.9 in. Hg, and three-stage compressors can develop even higher vacuums.

The major characteristics of vane compressors are:

Size

- Common sizes up to 250 bhp, but mostly used for applications under 125 bhp.
- Available in sizes to 500 bhp.
- Discharge pressures to 400 psig.
- Single- or two-stage in tandem on same shaft.

Advantages

- Good in vacuum service.
- No pulsating flow.
- Less space.
- Inexpensive for low hp vapor recovery or vacuum service.

Disadvantages

- Must have clean air or gas.
- Takes 5 to 20% more horsepower than reciprocating.
- Uses ten times the oil of a reciprocating. Usually install after-cooler and separator to recycle oil.

Helical-Lobe (Screw) Rotary Compressors

Screw compressors are rotary positive displacement machines. Two helical rotors are rotated by a series of timing gears as shown in Figure 10-8 so that gas trapped in the space between them is transported from the suction to the discharge piping. In low-pressure air service, non-lubricated screw compressors can deliver a clean, oil-free air. In hydrocarbon service most screw compressors require that liquid be injected to help provide a seal. After-coolers and separators are required to separate the seal oil and recirculate it to suction.

Screw compressors can handle moderate amounts of liquid. They can also handle dirty gases because there is no metallic contact within the casing.

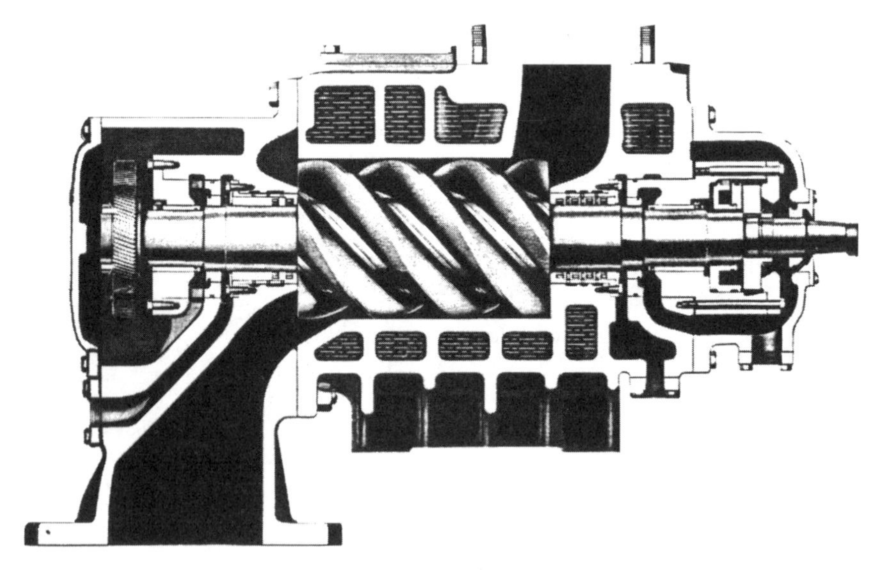

Figure 10-8. Screw-type rotary compressor. (*Courtesy of Dresser-Rand Company.*)

It tends to be limited to 250 psig discharge pressures and a maximum of 400 hp in hydrocarbon service, although machines up to 6,000 hp are available in other service. Screw compressors are not as good as vane compressors in developing a vacuum, although they are used in vacuum service.

Non-lubricated screw compressors have very close clearances and thus they are designed for limited ranges of discharge temperature, temperature rise, compression ratio, etc., all of which can cause changes in these clearances. Lubricated compressors have a somewhat broader tolerance to changes in operating conditions, but they are still more limited than reciprocating compressors.

The major characteristics of screw compressors are:

Size

- Up to 6,000 hp in air service, but more common below 800 hp.
- Up to 400 hp in hydrocarbon service.
- Discharge pressures to 250 psig.
- Single- or two-stage in tandem on same shaft.

Advantages

- Available as non-lubricated especially for air service.
- Can handle dirty gas.
- Can handle moderate amounts of liquids, but no slugs.
- No pulsating flow.
- At low discharge pressure (<50 psig) can be more efficient than reciprocating.

Disadvantages

- In hydrocarbon service needs seal oil with after-cooler and separator to recycle oil.
- At discharge pressure over 50 psig takes 10 to 20% more horsepower than reciprocating.
- Low tolerance to change in operating conditions of temperature, pressure, and ratio.

Centrifugal Compressors

Similar to multistage centrifugal pumps, centrifugal compressors, as shown in Figure 10-9, use a series of rotating impellers to impart velocity

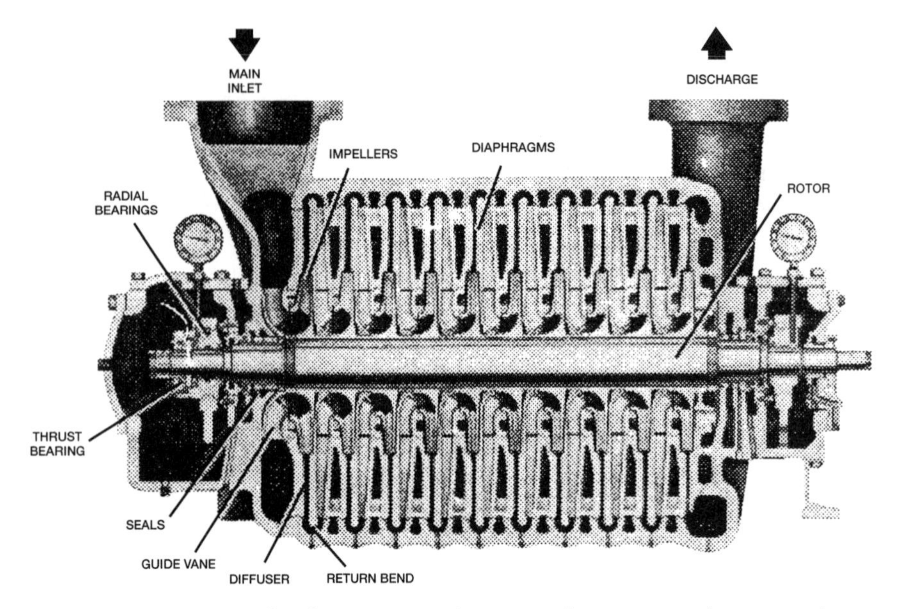

Figure 10-9. Centrifugal compressor. ***(Courtesy of Dresser-Rand Company.)***

head to the gas. This is then converted to pressure head as the gas is slowed in the compressor case. They are either turbine or electric motor driven and range in size from 1,000 hp to over 20,000 hp. Most larger compressors (greater than 4,000 hp) tend to be turbine-driven centrifugal compressors because there is such a first cost advantage in that size range over integrals. Centrifugal compressors have high ratios of horsepower per unit of space and weight, which makes them very popular for offshore applications.

As shown in Figure 10-10 they can be either horizontally split case or vertically split case (barrel). To develop the required gas velocities and head they must rotate at very high speeds (20,000 to 30,000 rpm), making the design of driver, gear, and compressor extremely important. Turbine drives are also high speed and a natural match for centrifugal compressors.

There is a disadvantage in centrifugal machines in that they are low efficiency. This means it requires more brake horsepower (bhp) to compress the same flow rate than would be required for a reciprocating compressor. If the compressor is driven with a turbine, there is even a greater disadvantage because the turbines are low in fuel efficiency. The net result is that turbine-driven centrifugal machines do not use fuel very

Figure 10-10. Horizontally split centrifugal compressor (top) and vertically split centrifugal compressor, barrel (bottom). (*Courtesy of Dresser-Rand Company.*)

efficiently. This fuel penalty can be overcome if process heat is needed. Waste heat can be recovered from the turbine exhaust, decreasing or eliminating the need to burn gas to create process heat.

As with electric motor and engine-driven high-speeds, turbine and electric motor-driven centrifugals can be easily packaged for use in oil and gas fields. They are very common in booster compressor service (high volume, low ratio) and for very high flow rate gas-lift service. Centrifugal compressors cannot be used for high ratio, low-volume applications.

The major characteristics of centrifugal compressors are:

Size

- Starts about 500 hp.
- 1,000 hp increments to 20,000 hp.

Advantages

- High horsepower per unit of space and weight.
- Turbine drive easily adapted to waste-heat recovery for high fuel efficiency.
- Easily automated for remote operations.
- Can be skid mounted, self-contained.
- Low initial cost.
- Low maintenance and operating cost.
- High availability factor.
- Large capacity available per unit.

Disadvantages

- Lower compressor efficiency.
- Limited flexibility for capacity.
- Turbine drives have higher fuel rate than reciprocating units.
- Large horsepower units mean that outage has large effect on process or pipeline capabilities.

SPECIFYING A COMPRESSOR

In specifying a compressor it is necessary to choose the basic type, the number of stages of compression, and the horsepower required. In order to do this the volume of gas, suction and discharge pressure, suction temperature, and gas specific gravity must be known.

The detailed calculation of horsepower and number of stages depends upon the choice of type of compressor, and the type of compressor depends in part upon horsepower and number of stages. A first approximation of the number of stages can be made by assuming a maximum compressor ratio per stage of 3.0 to 4.0 and choosing the number of stages such that:

$$R = \left(\frac{P_d}{P_s}\right)^{1/n} < 3.0 \text{ to } 4.0 \qquad (10\text{-}2)$$

where R = ratio per stage
n = number of stages
P_d = discharge pressure, psia
P_s = suction pressure, psia

A first approximation for horsepower can be made from Figure 10-11 or from the following equation:

$$BHP = 22\ R\ n\ F\ Q_g \qquad (10\text{-}3)$$

where BHP = approximate brake horsepower
R = ratio per stage
n = number of stages
F = an allowance for interstage pressure drop
= 1.00 for single-stage compression
1.08 for two-stage compression
1.10 for three-stage compression
Q_g = flow rate, MMscfd

Once the required horsepower and number of stages are estimated, a choice of compressor type can be made from the considerations included earlier. Some example selections are included in Table 10-1. The selections listed in this table are meant as common types that would normally be specified for the given conditions. It must be emphasized that these are *not* recommendations that should be accepted without consideration of the advantages and disadvantages listed earlier. In addition, local foundation conditions, type of drivers available, cost of fuel, availability of spare

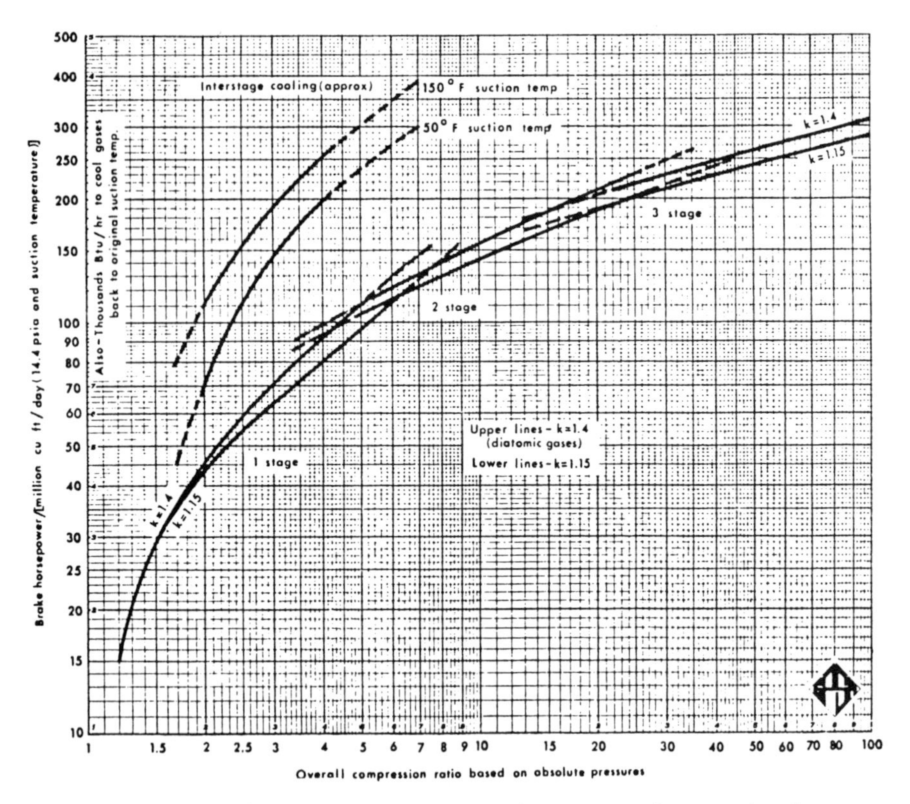

Figure 10-11. Curve for estimating compression horsepower. ***(Reprinted with permission from GPSA*** **Engineering Data Book,** ***10th Ed.)***

parts and personnel familiar with operating and maintenance, waste heat requirements, etc., could influence the selection for a specific installation.

Procedure for More Accurate Determination of Horsepower and Number of Stages

There are economic and operational reasons for considering an additional stage of compression. The addition of a stage of compression requires an additional scrubber, additional cylinder or case, and more complex piping and controls. In addition, there are some horsepower losses due to additional mechanical friction of the cylinder or rotating element and the increased pressure drop in the piping. This horsepower loss and additional equipment cost may be more than offset by the increased efficiency of compression.

Table 10-1
Example Compressor Type Selections

Service	Flow Rate MMscfd	R	n	Approx. bhp	Most Likely	Selection Alternate
Booster	100	2.0	1	4,400	Centrifugal	Integral (onshore only)
	10	2.0	1	440	High Speed	
Gas Lift	5	2.7	3	980	High Speed	
	20	2.7	3	3,920	Centrifugal	Integral (onshore only)
	100	2.7	3	19,602	Centrifugal	
Flash Gas	2	2.0	1	88	Screw	High Speed
	2	2.0	2	190	High Speed	Screw
	4	2.0	2	380	High Speed	
Vapor Recovery	0.1	4.0	1	9	Vane	Screw
	1.0	3.0	2	143	Screw	Vane
	2.0	3.0	2	286	High Speed	Screw

Figure 10-12 shows the pressure-volume curve for both single stage compression and two stage compression (neglecting interstage losses). By adding the second stage and cooling the gas from A to D before beginning the compression cycle in the second stage, the area under the curve is reduced by an amount equal to A-B-C-D. This represents the power saved by adding the second stage.

It is often even more important to add an additional stage in order to limit the discharge temperature of any one stage. It is clear from Figure 10-12 that because of the cooling that occured in the interstage (A to D) the gas at C is cooler than it would have been at point B.

The discharge temperature for any single stage of compression can be calculated from:

$$T_d = T_s \left(\frac{P_d}{P_s} \right)^{\frac{k-1}{k} \times \frac{1}{\eta}} \tag{10-4}$$

where T_d = stage discharge temperature, °R
T_s = stage suction temperature, °R
P_d = stage discharge pressure, psia
P_s = stage suction pressure, psia

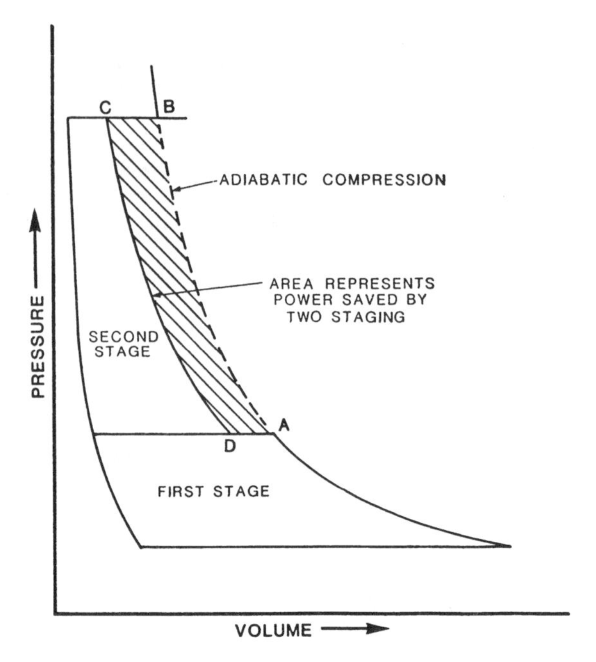

Figure 10-12. Horsepower reduction by multistaging (neglects interstage losses).

k = ratio of gas specific heats, C_p/C_v
η = polytropic efficiency
= 1.0 for reciprocating, 0.8 for centrifugal

It is desirable to limit discharge temperatures to below 250°F to 275°F to ensure adequate packing life for reciprocating compressors and to avoid lube oil degradation. At temperatures above 300°F eventual lube oil degradation is likely, and if oxygen is present ignition is even possible. Under no circumstances should the discharge temperature be allowed to exceed 350°F.

The discharge temperature can be lowered by cooling the suction gas and reducing the value of P_d/P_s, that is, by adding more stages of compression.

The brake horsepower per stage can be determined from:

$$\text{BHP} = 0.0857\,[Z_{av}]\left[\frac{(Q_g)(T_s)}{E}\right]\left[\frac{k\eta}{k-1}\right]\left[\left(\frac{P_d}{P_s}\right)^{\frac{k-1}{k\eta}} - 1\right] \qquad (10\text{-}5)$$

where BHP = brake horsepower per stage
Q_g = volume of gas, MMscfd
T_s = suction temperature, °R
Z_s = suction compressibility factor
Z_D = discharge compressibility factor
E = efficiency
high-speed reciprocating units — use 0.82
low-speed reciprocating units — use 0.85
centrifugal units — use 0.72
η = polytropic efficiency
k = ratio of gas specific heats, C_p/C_v
P_s = suction pressure of stage, psia
P_d = discharge pressure of stage, psia
$Z_{av} = (Z_s + Z_D)/2$

The total horsepower for the compressor is the sum of the horsepower required for each stage and an allowance for interstage pressure losses. It is assumed that there is a 3% loss of pressure in going through the cooler, scrubbers, piping, etc., between the actual discharge of the cylinder and the actual suction of the next cylinder. For example, if the discharge pressure of the first stage is 100 psia, the pressure loss is assumed to be 3 psia and the suction pressure of the next stage is 97 psia. That is, second stage suction pressure is not equal to the first stage discharge pressure.

The following procedure can now be used to calculate the number of stages of compression and the horsepower of the unit:

- First, calculate the overall compression ratio ($R_t = P_d/P_s$). If the compressor ratio is under 5, consider using one stage. If it is not, select an initial number of stages so that $R < 5$. For initial calculations it can be assumed that ratio per stage is equal for each stage.
- Next, calculate the discharge gas temperature for the first stage. If the discharge temperature is too high (more than 300°F), a large enough number of stages has not been selected or additional cooling of the suction gas is required. If the suction gas temperature to each stage cannot be decreased, increase the number of stages by one and recalculate the discharge temperature.
- Once the discharge temperature is acceptable, calculate the horsepower required, and calculate suction pressure, discharge temperature, and horsepower for each succeeding stage.

- If R > 3, recalculate, adding an additional stage to determine if this could result in a substantial savings on horsepower.

RECIPROCATING COMPRESSORS—PROCESS CONSIDERATIONS

Figure 10-13 is a generalized process flow diagram of a single stage reciprocating compressor. The following items should be considered in developing a process flow:

Recycle Valve

Most gas lift, flash gas, and vapor recovery compressors require a recycle valve because of the unsteady and sometimes unpredictable nature of the flow rate. Indeed there may be periods of time when there is no flow at all to the compressor.

At a constant speed, a constant volume of gas (at suction conditions of pressure and temperature) will be drawn into the cylinder. As the flow rate to the compressor decreases, the suction pressure decreases until the gas available expands to satisfy the actual volume required by the cylinder. When the suction pressure decreases, the ratio per stage increases and therefore the discharge temperature increases. In order to keep from having too high a discharge temperature, the recycle valve opens to help fill the compressor cylinder volume and maintain a minimum suction pressure.

Flare Valve

As flow rate to the compressor increases, the suction pressure rises until the volume of gas at actual conditions of temperature and pressure compressed by the cylinder equals the volume required by the cylinder. A flare valve is needed to keep the suction pressure from rising too high and overpressuring the suction cylinder, creating too high a rod load or increasing the horsepower requirements beyond the capability of the driver (see Chapter 11 for further discussion).

The flare valve also allows production to continue momentarily if a compressor shuts down automatically. Even in booster service it may be beneficial to allow an operator to assess the cause of the compressor shutdown before shutting in the wells. In flash gas or gas-lift service, it is almost always beneficial to continue to produce the liquids while the

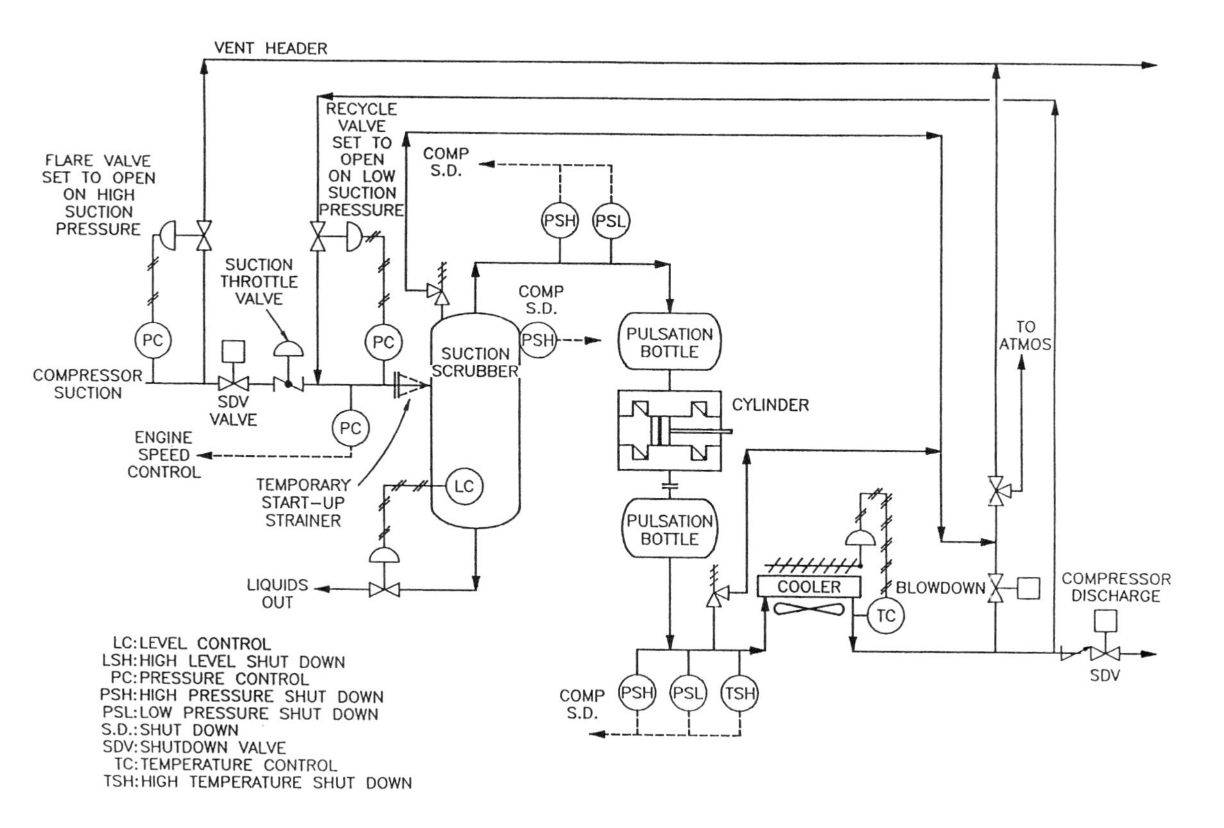

Figure 10-13. Example process flow diagram of reciprocating compressor.

cause of the compressor shutdown is investigated. The flare valve must always be installed upstream of the suction shutdown valve.

Suction Pressure Throttle Valve

A suction pressure throttling valve can also be installed to protect the compressor from too high a suction pressure. This is typically a butterfly valve that is placed in the suction piping. As flow rate to the compressor increases, the valve will close slightly and maintain a constant suction pressure. This will automatically limit the flow rate to exactly that rate where the actual volume of gas equals that required by the cylinder at the chosen suction pressure setting. It will not allow the suction pressure to increase and the compressor cylinder to thus handle more flow rate.

The pressure upstream of the suction valve will increase until sufficient back-pressure is established on the wells or equipment feeding the compressor to reduce the flow to a new rate in equilibrium with that being handled by the cylinder or until a flare valve or relief valve is actuated.

Suction throttle valves are common in gas-lift service to minimize the action of the flare valve. Flow from gas-lift wells decreases with increased back-pressure. If there were no suction valve, the flare valve may have to be set at a low pressure to protect the compressor. With a suction valve it may be possible to set the flare valve at a much higher pressure slightly below the working pressure of the low-pressure separator. The difference between the suction valve set pressure and the flare valve set pressure provides a surge volume for gas and helps even the flow to the compressor.

Speed Controller

A speed controller can help extend the operating range and efficiency of the compressor. As the flow rate increases, the compressor speed can be increased to handle the additional gas. Compressor speed will stabilize when the actual flow rate to be compressed equals the required flow rate for the cylinder at the preset suction pressure. As the flow rate decreases, the compressor slows until the preset suction pressure is maintained.

A speed controller does not eliminate the need for a recycle valve, flare valve, or suction throttling valve, but it will minimize their use. The recycle valve and suction throttling valve add arbitrary loads to the compressor and thus increase fuel usage. The flare valve leads to a direct waste of reservoir fluids and thus loss of income. For this reason, engine speed control is recommended for most medium to large size ($\geq$500 hp) reciprocating compressors where a constant flow rate cannot be ensured by the process.

Blowdown Valve

A blowdown valve relieves trapped pressure when the compressor is shut down due to a malfunction or for maintenance. The flowsheet shows an automatic blowdown valve.

Most operators require automatic blowdown valves so that if the compressor shuts down due to a malfunction, the trapped gas will not become a potential hazard. On some small onshore compressors some operators prefer manual valves to make it easier to restart the compressor. The compressor is only blown down for maintenance.

Often, the blowdown valve is routed to a closed flare system, which services other relief valves in the facility to ensure that all the gas is vented or flared at a safe location. In such instances, a separate manual blowdown valve piped directly to atmosphere, with nothing else tied in, is also needed. After the compressor is shut down and safely blown down through the flare system, the normal blowdown valve must be closed to block any gas that may enter the flare system from other relief valves. The manual blowdown valve to atmosphere protects the operators from small leaks into the compressor during maintenance operations.

Suction and Discharge Shut-down Valves, Discharge Check

These valves isolate the compressor. Most operators require both shutdown valves to be automatic. Some operators use manual valves on small onshore compressors. If the compressor is in a building it is preferable to locate the valves outside the building.

Relief Valve on Each Cylinder Discharge

Each cylinder discharge line should have a relief valve located upstream of the cooler. Like all reciprocating devices, the piston will continue to increase pressure if flow is blocked. The relief valve assures that nothing is overpressured. It must be located upstream of the coolers as ice can form in the coolers, blocking flow.

Pulsation Bottles

Each cylinder should have suction and discharge pulsation bottles to dampen the acoustical vibrations caused by the reciprocating flow.

Discharge Coolers

These cool the interstage gas. They may also be required to cool the discharge gas prior to gas treating or dehydration or to meet pipeline specifications. Typically, aerial coolers are used in these situations.

Suction Scrubbers

Suction scrubbers are required on the unit suction to catch any liquid carry-overs from the upstream equipment and any condensation caused by cooling in the lines leading to the compressor. They are also required on all the other stages to remove any condensation after cooling. On each suction scrubber a high level shut-down is required so that if any liquids do accumulate in the suction scrubber, the compressor will automatically shut down before liquids carry-over to the compressor cylinders.

CENTRIFUGAL COMPRESSORS—SURGE CONTROL AND STONEWALLING

Surge is the most important process design consideration for centrifugal compressors. The surge condition occurs when the compressor does not have enough flow to produce sufficient head. At this point, the gas in the discharge piping flows back into the compressor momentarily. This lowers the back-pressure of the system, establishing forward flow at a temporarily low head. The cycling from zero, or even backward flow to forward flow, is called "surge" and is very detrimental to the compressor bearings and seals. Most compressors can only sustain a very few cycles of surge before severe mechanical problems develop.

Surge may be caused by an increase in head requirement or a loss in throughput. Figure 10-14 shows the capacity curves for a typical compressor. The surge line for this particular compressor is shown. Any combination of speed, pressure, and flow rate to the right of the surge line is acceptable. Typically, a surge control line offsetting the theoretical surge limit given by the manufacturer is used to establish set points for a control system adjusting speed and recycle as shown in Figure 10-16.

A stonewall or choked flow condition occurs when sonic velocity is reached at the exit of a compressor wheel. When this point is reached, flow through the compressor cannot be increased even with further

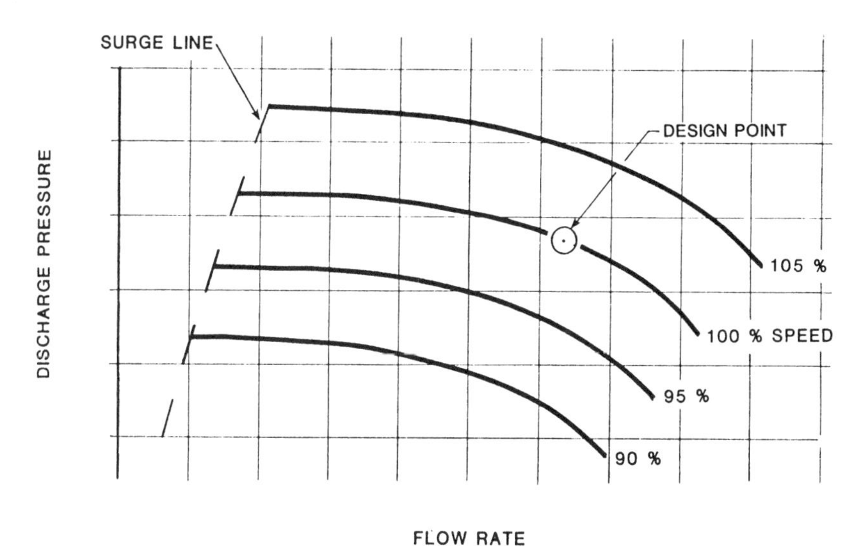

Figure 10-14. Typical centrifugal compressor curve showing surge.

increase in suction pressure. If this occurs, the suction pressure will rise. Operation in this region will cause excessive use of horsepower, occasionally to the point of overload, and frequent flaring.

If higher flow rates are desired, modifications to the impeller must be made, as shown in Figure 10-15.

CENTRIFUGAL COMPRESSORS PROCESS CONSIDERATIONS

Figure 10-16 is a generalized process flow diagram of a single-stage centrifugal compressor. The following items should be considered in developing a process flow:

Recycle (Surge Control) Valve

A recycle valve is needed for surge control as well as for the conditions listed above for reciprocating compressors. At constant speed the head-capacity relationship will vary in accordance with the performance curve. For a constant compressor speed:

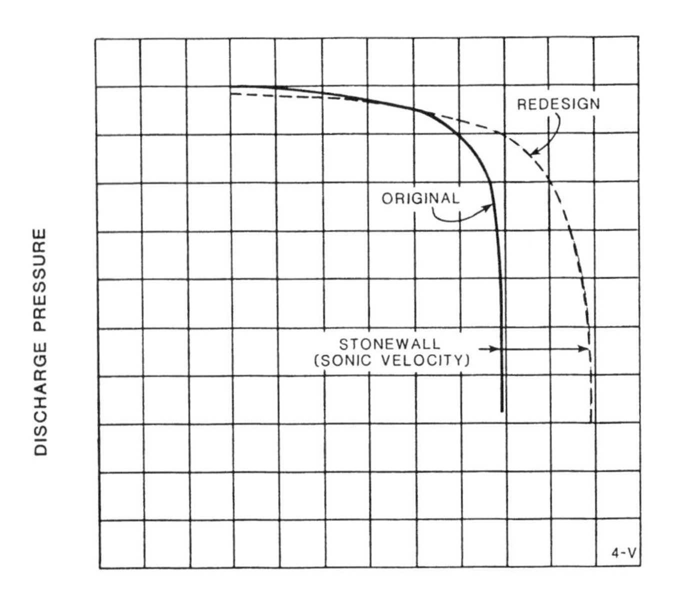

Figure 10-15. Graphic illustration of a "stonewall," or a choked flow condition.

- If the flow rate to the compressor decreases, the compressor approaches the surge point and a recycle valve is needed.
- If the suction pressure decreases, and discharge pressure remains constant, the compressor head must increase, approaching the surge point in the process.

Flare Valve

As suction pressure increases or discharge pressure decreases, the compressor head requirement will decrease and the flow rate will increase. A flare valve will avoid stonewalling or overranging driver horsepower.

Suction Pressure Throttle Valve

A throttling device can also be placed in the suction piping to protect against overpressure or to limit the horsepower demand to the maximum available from the driver.

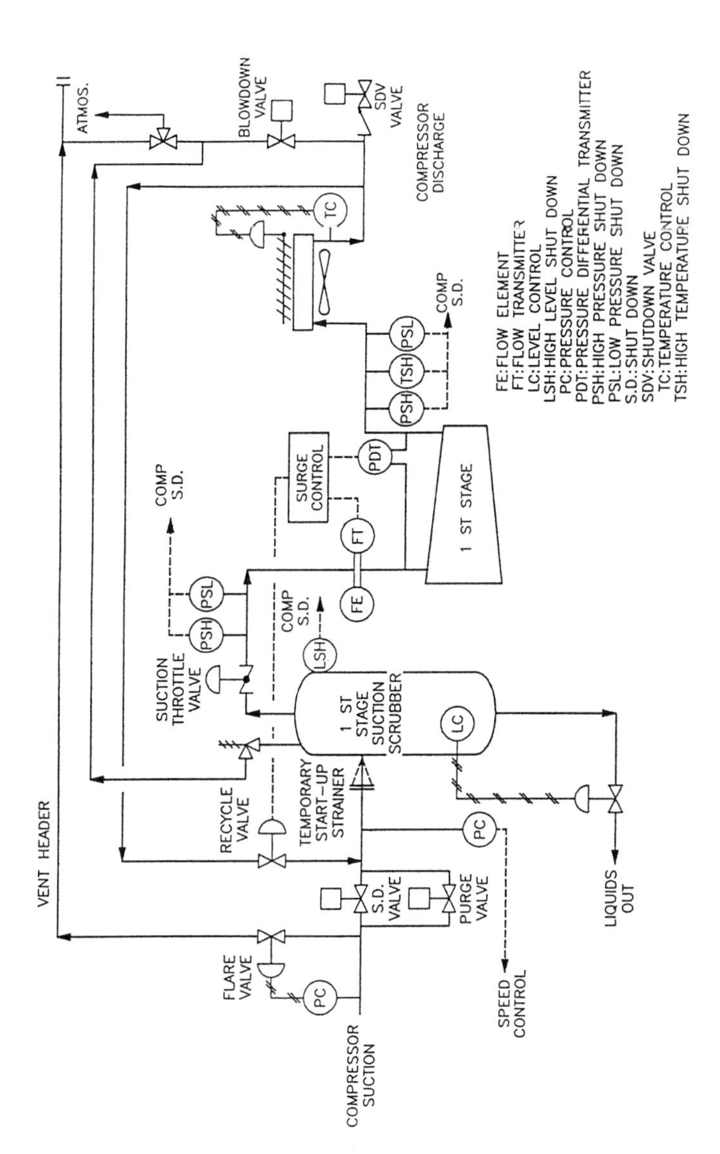

Figure 10-16. Example process flow diagram of centrifugal compressor.

Speed Controller

A speed controller is needed in conjunction with the surge control system. A new head-capacity curve is established for each speed, as shown in Figure 10-14.

Inlet Guide Vanes

The performance curve can also be shifted to match the process requirements by variable inlet guide vanes. Located at the compressor inlet, these vanes change the direction of the velocity entering the first-stage impeller. By changing the angle at which these vanes direct the flow at the impeller, the shape of the head capacity curve can be changed.

As more velocity change is added to the inlet gas, the performance curve steepens with very little efficiency loss. Extreme changes in process conditions cannot be accommodated. The high cost of inlet guide vanes limits use to very large compressors where small improvements in efficiency can bring large rewards.

Blowdown Valve

Blowdown valves must be installed in centrifugal compressors for the same reasons as in reciprocating compressors. They must be designed with more care than those on reciprocating units, since centrifugal compressors have oil film seals where the shaft goes through the case. These seals only work if the shaft is rotating. If the compressor shuts down, pressure must be relieved from the case before the shaft speed decreases to the point where the seal no longer will contain pressure. This requires careful attention to manufacturer furnished data as well as overall flare system design.

Suction and Discharge Shutdown Valves and Discharge Check Valves

These devices are required to isolate the compressor for the same reasons they are required for reciprocating compressors.

Discharge Check Valve (Each Stage)

In reciprocating compressors the compressor valves themselves act as check valves, preventing backflow from high-pressure stages to lower pressure stages. Multistage centrifugal compressors require check valves on each stage to isolate each surge loop, as well as to prevent backflow during unusual operating conditions.

Relief Valves

The compressor can operate at any point on the performance curve. For the maximum value of suction pressure, the pressure rise across the machine at the surge control point must be less than the system pressure rating. If not, a relief valve should be installed.

Suction Shut-down Bypass (Purge) Valve

The suction shut-down bypass valve is used to purge the piping system of air prior to compressor start-up. This valve is small to prevent high gas purge rates from spinning the impellers.

Discharge Coolers and Suction Scrubbers

These items are required for the reasons discussed under reciprocating compressors.

CHAPTER

11

Reciprocating Compressors*

The previous section discussed the various types of compressors, their selection, and process flow. This chapter presents greater detail concerning the major components, performance, operational and installation considerations, and standard specifications for reciprocating compressors.

For normal production facilities, reciprocating compressors far outnumber the other types, and it is necessary for the facility engineer to understand the details of reciprocating compressor design. In very large horsepower ranges or booster compressor situations, centrifugal compressors are common. These are not discussed in more detail in this book, because these large installations are normally the responsibility of rotating machinery experts.

COMPONENTS

Figure 11-1 is a cutaway that shows the various components of a reciprocating compressor. To understand how to specify and maintain a compressor properly, it is necessary to have a better understanding of the construction of the major components.

---

*Reviewed for the 1999 edition by Lonnie W. Shelton of Paragon Engineering Services, Inc.

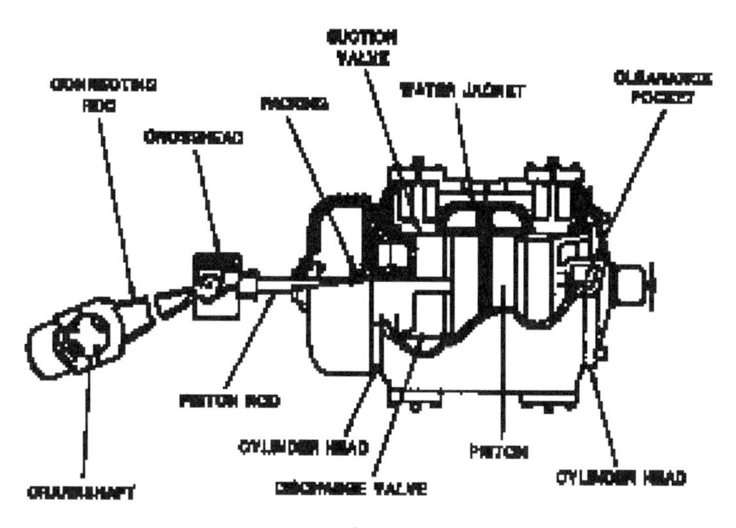

Figure 11-1. Cutaway view of typical reciprocating compressor.

Frame

The compressor frame, shown in Figure 11-2, is a heavy, rugged casting containing all the rotating parts and on which the cylinders and crossheads are mounted. All frames are rated by the compressor manu-

Figure 11-2. The compressor frame is the rugged casting that contains the rotating parts and on which the cylinders and crossheads are mounted. (*Courtesy of Dresser-Rand Company.*)

facturers for a maximum continuous horsepower, speed (rpm), and rod load. The rated horsepower is determined by the maximum horsepower that can be transmitted through the crankshaft to the compressor cylinders. The rod load is the force imposed on the piston rod by the pressure differential between the two ends of the piston.

Each frame is designed for a maximum number of cylinders. The frame itself does not indicate the number of stages or the duty of the compressor. An individual frame can be used for many different sizes of compressor cylinders and for a wide range of applications. Frames are typically classified as separable (balanced-opposed) or integral-type, as shown in Figure 11-3.

Separable (balanced-opposed) frames are characterized by an adjacent pair of crank throws 180° out of phase. The frame is separate from the

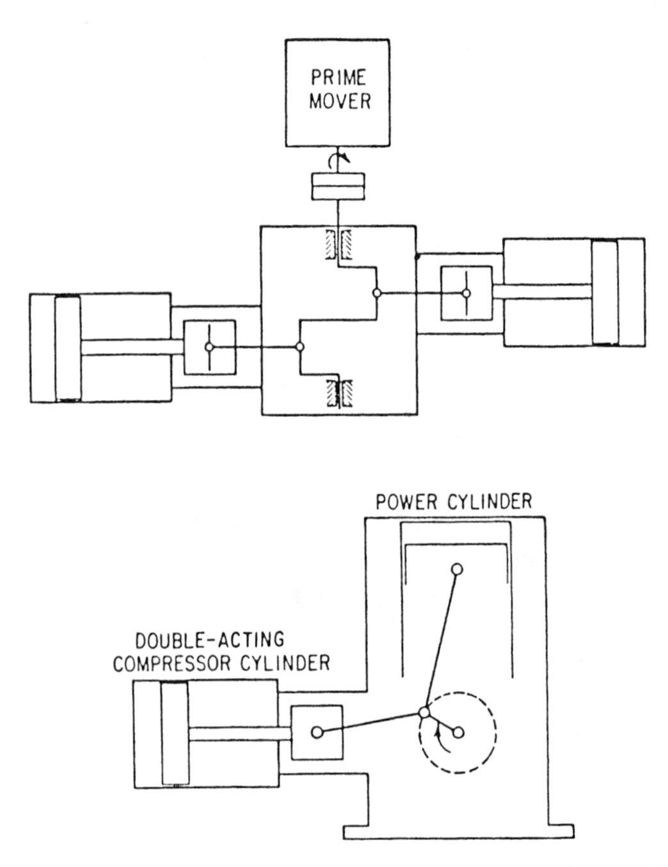

Figure 11-3. Separable (balanced-opposed) compressor (top) and integral-type gas engine compressor (bottom).

driver. Integral-type frames are characterized by having compressor cylinders and power cylinders mounted on the same frame and driven by the same crankshaft.

Cylinder

A cylinder is a pressure vessel that holds the gas during the compression cycle. There are two basic types:

1. Single-acting cylinders are those where compression occurs only once per crankshaft revolution.
2. Double-acting cylinders are those where compression occurs twice per cranksheet revolution.

Figure 11-4 is a cut-away drawing of a compressor with single-acting cylinders. True single-acting cylinders are typical of low horsepower air compressors. Single-acting process compressors are typically double-act-

Figure 11-4. Single acting cylinders. ***(Courtesy of Dresser-Rand.)***

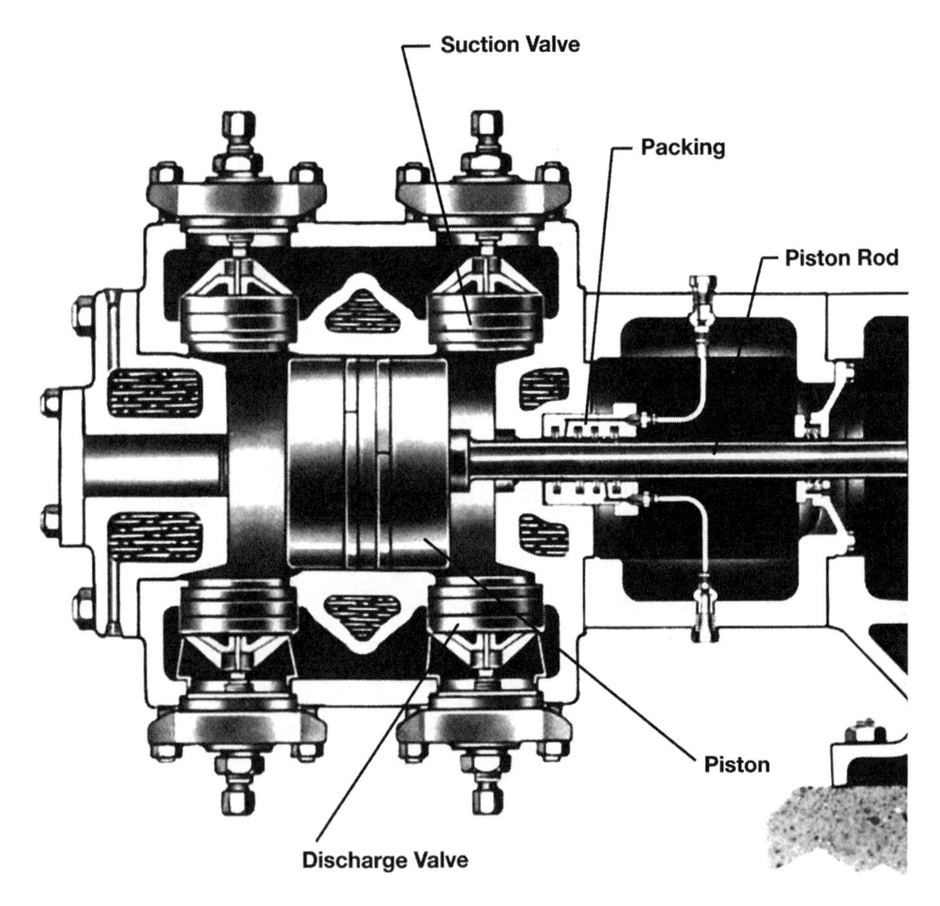

Figure 11-5. Typical double-acting compressor cylinder. (*Courtesy of Dresser-Rand Company.*)

ing cylinders with the outer end suction valves removed. Figure 11-5 is a cut-away of a double-acting cylinder.

Cylinders are made of different kinds of materials. Generally, cast iron is used for cylinder operating pressures up to 1,000 to 1,200 psig, nodular iron or cast steel for operating pressure in the 1,000 to 2,500-psig range, and forged steel for pressures greater than 2,500 psig.

Like all pressure vessels, the cylinder has a maximum allowable working pressure (MAWP). The maximum allowable working pressure of the cylinder determines the setting of the relief valve that is downstream of the cylinder. The MAWP of the cylinder should be a minimum of 10% or 50 psi greater than its operating pressure.

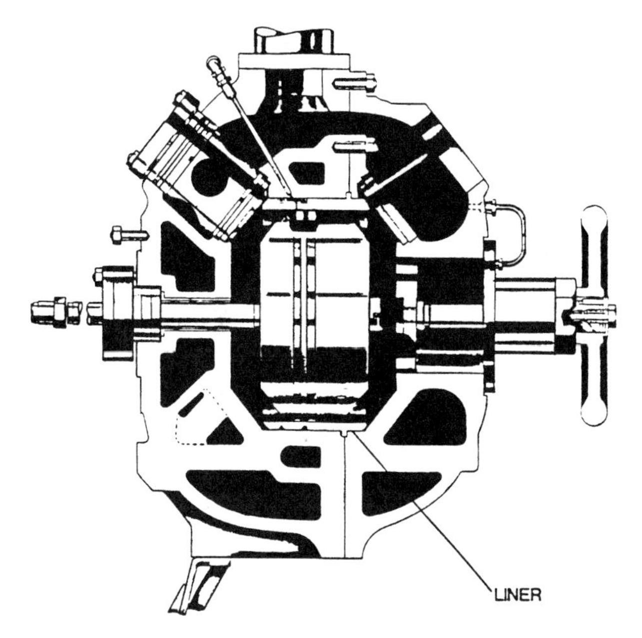

Figure 11-6. Cut-away view showing cylinder liner. (*Courtesy of Dresser-Rand Company.*)

A cylinder liner such as that shown in Figure 11-6 may be used to help prolong the life of the cylinder and improve operating flexibility. Any damage caused by the action of the piston or heat generated by compression will affect the cylinder liner, which may be removed and replaced. As the surface of the liner wears, it is much easier and quicker to repair it than to repair the cylinder itself. In addition, liners enable the diameter of the piston to be varied without changing the cylinder and thus provide flexibility to respond to different conditions of pressure and flow rate.

The disadvantages of liners are that they increase the clearance (discussed in more detail below) by increasing the distance between the piston and the valve, and they decrease the bore of the cylinder. Therefore, the cylinder will have less capacity and lower efficiency (at high ratios) than if there were no liner.

Special Compressor Cylinder Construction

Many variations and combinations of cylinder types and arrangements are available from the compressor manufacturers. The compressor manufacturer will generally make its selection based on the most economical combination it has available.

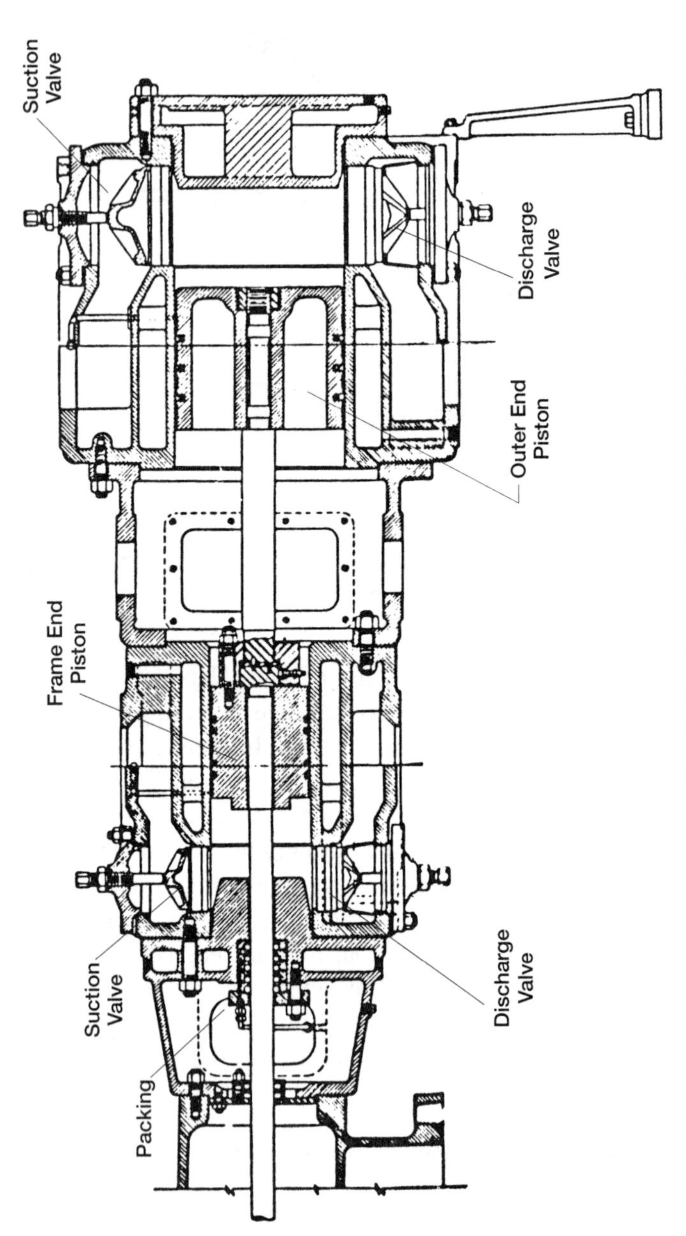

Figure 11-7. Steeple cylinder. (*Courtesy of Dresser-Rand Company.*)

Figure 11-7 is a cut-away of a steeple cylinder. This cylinder design is actually two single-acting cylinders coupled together with different-size pistons on the same piston rod. This arrangement allows two stages of compression on the same compressor throw and is usually used in low capacity, low rod load applications.

Another variation is the tandem cylinder. The tandem cylinder arrangement again allows two stages of compression on the same compressor throw but uses two double-acting cylinders separated by a second distance piece. This arrangement is usually used in low rod load applications where higher capacity is required.

Figure 11-8 is a cut-away of the latest innovation in compressor cylinder design. In this design, the two suction valves and the two discharge valves are installed inside the compressor cylinder bore. The suction valves are stationary and located at each end of the cylinder. The discharge valves are connected to the piston rod to form the piston; thus the name valve-in-piston design. This design offers the advantages of lower clearances (thus higher efficiencies), reduced sources of fugitive emissions, fewer replacement parts, simpler maintenance procedures, and reduced weight.

Distance Pieces

A distance piece provides the separation of the compressor cylinder from the compressor frame as shown in Figure 11-9. At the top of the figure is a standard distance piece.

The piston rod moves back and forth through packing that is contained within the distance piece. The packing keeps the compressed gas from leaking out of the cylinder through the piston rod opening. As the rod passes through the packing it is lubricated. As it goes back and forth, the rod is in contact with the frame lube oil and with the cylinder lube oil and gas. Thus, oil carry-over may occur on the rod from the cylinder to the crankcase. Impurities picked up by the oil from the gas being compressed could contaminate crankcase oil.

In a single-compartment distance piece, the frame end and the cylinder end contain packing. The space between the cylinder packing and the frame diaphragm and packing is sufficiently long to assure that no part of the rod enters both the cylinder and the frame. This minimizes contamination between the gas being compressed and the oil that is used to lubricate the crankcase. There are drains and vents off the distance piece and off the packing, so if there is a packing failure, the high-pressure gas has

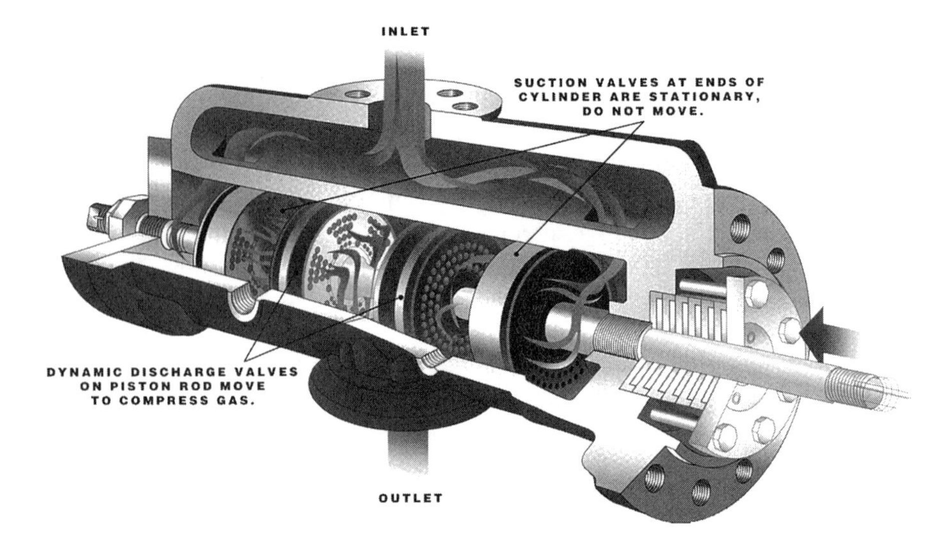

Figure 11-8. Valve-in-piston double-acting compressor cylinder. (*Courtesy of Dresser-Rand Company.*)

a place to vent and not build up pressure that could leak through the frame packing into the crankcase. An oil slinger as shown in Figure 11-9 may be added to further reduce the amount of cylinder lube oil migrating down the rod into the crankcase.

A two-compartment distance piece may be used for toxic gases, but it is not very common. In this configuration, no part of the rod enters both the crankcase and the compartment adjacent to the compressor cylinder. That is, even if there were one failure, the crankcase oil cannot be contaminated with the toxic gas.

Crosshead, Rods, and Crankshaft

The crosshead converts the rotating motion of the connecting rod to a linear, reciprocating motion, which drives the piston as shown in Figure 11-10. The crosshead is provided with top and bottom guide shoes, which ride on lubricated bearing surfaces atached to he compressor frame. In addition, balance weights may be attached to the crosshead to reduce unbalanced forces and moments. The connecting rod connects the crankshaft to the crosshead. The piston rod connects the crosshead to the piston.

The crankshaft rotates about the frame axis, driving the connecting rod, crosshead, piston rod, and piston.

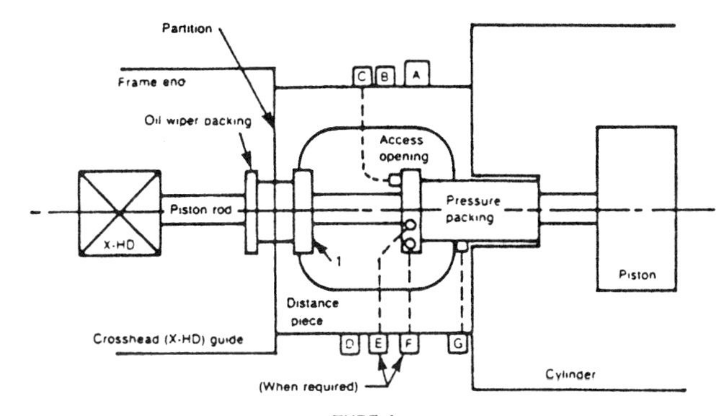

TYPE A
Short single-compartment distance piece
(may be integral with crosshead guide or distance piece)

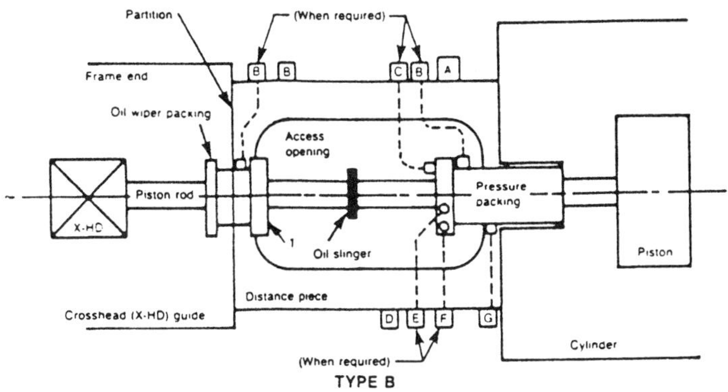

TYPE B
Long single-compartment distance piece
(sufficient length for oil slinger travel)

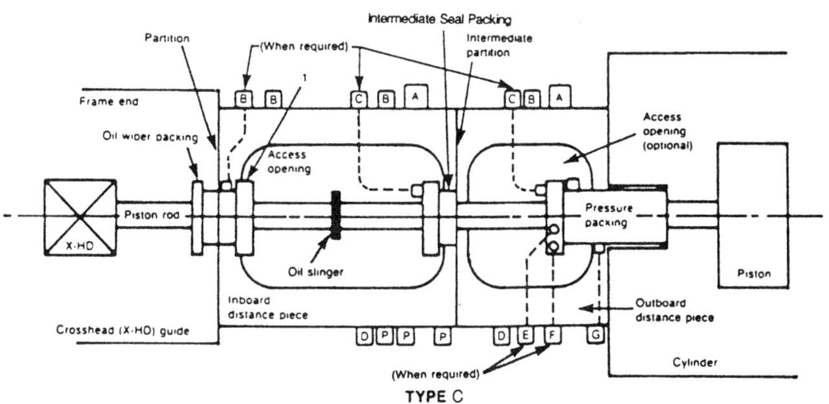

TYPE C
Short two-compartment or double distance piece arrangement
(inboard distance piece of sufficient length for oil slinger travel)

Figure 11-9. API type distance pieces. ***(Reprinted with permission from API, Std. 618, 3rd Ed., Feb. 1986.)***

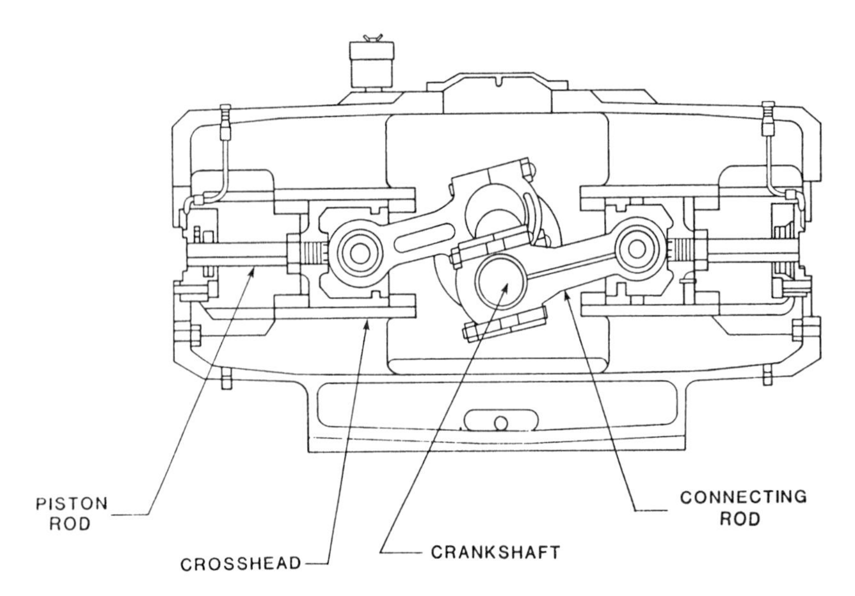

Figure 11-10. The crosshead converts the rotating motion of the connecting rod to a linear, reciprocating motion, which drives the piston. (*Courtesy of Dresser-Rand Company.*)

Piston

The piston is located at the end of the piston rod and acts as a movable barrier in the compressor cylinder. It is generally made from materials such as aluminum or cast iron and has a hollow center. Small-diameter high-pressure cylinders may be provided with a combined piston and rod machined from a single piece of bar stock.

To reduce friction and improve compression efficiency, the piston will be provided with segmented compression rings as shown in Figure 11-11. To prevent piston-to-bore contact, the piston may also be provided with removable wear bands that are in continuous contact with the cylinder wall. The compression rings and wear bands are replaced at regular intervals and typically made from soft materials such as brass, Micarta, Teflon, and the newer thermoplastics.

Bearings

Most field compressors use hydrodynamic type or "journal" bearings. As shown in Figure 11-12, oil enters into the bearing from supply holes strategically placed along the bearing circumference and builds up an oil

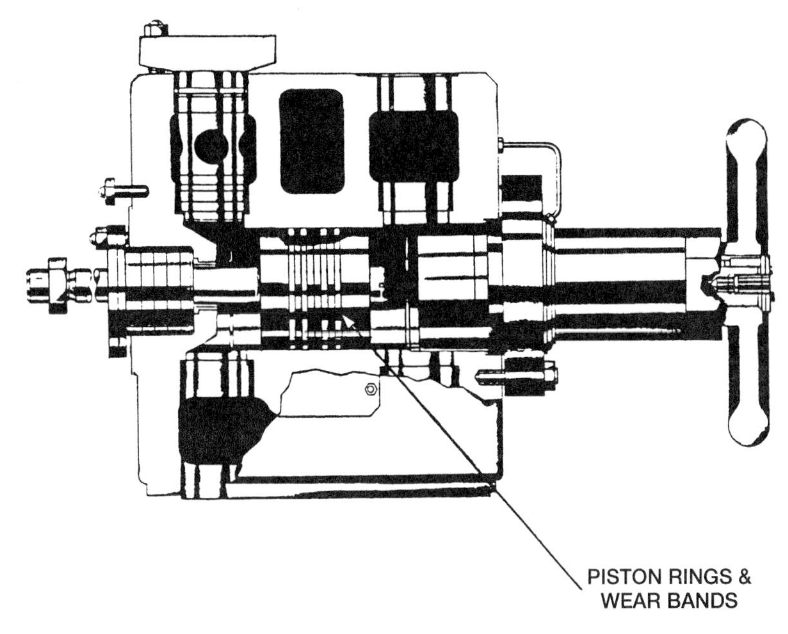

Figure 11-11. Piston rings and wear bands are made of material that is softer than the cylinder wall with which they are in constant contact, so they must be replaced regularly. (*Courtesy of Dresser-Rand Company.*)

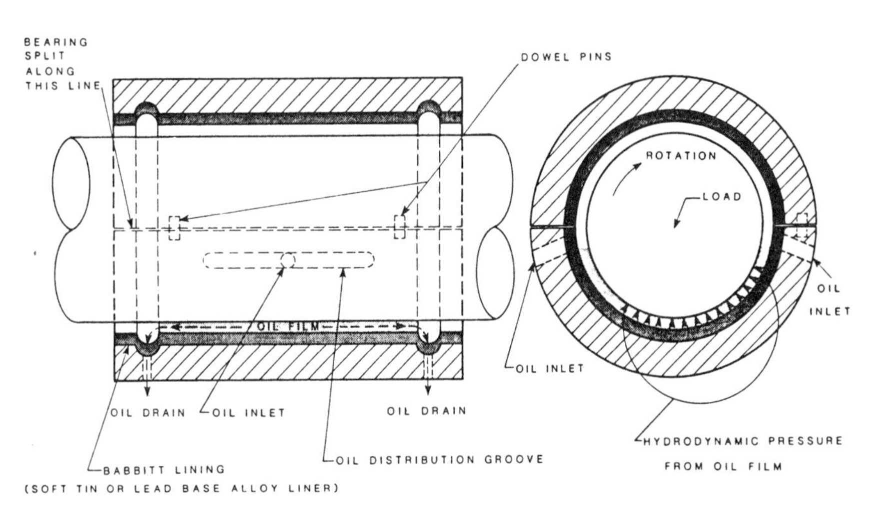

Figure 11-12. Journal bearings allow axial and circumferential oil flow along the bearing.

film between the stationary and rotating parts of the bearing. The oil flows axially and circumferentially along the bearing, then out the ends. As the oil flows through the bearing, the load compresses the oil film, generating the high pressure within the bearing that supports the load while allowing for rotation.

In a reciprocating compressor the bearing locations are:

- Main bearings—between crankshaft and frame
- Crank pin bearing—between crankshaft and connecting rod
- Wrist pin bearing—between connecting rod and crosshead
- Crosshead bearing (shoe)—underneath the crosshead

Packing

Packing provides the dynamic seal between the cylinder and the piston rod. It consists of a series of Teflon rings mounted in a packing case, which is bolted to the cylinder. The piston rod moves in a reciprocating motion through this case. Figure 11-13 shows a typical packing case. The packing case is constructed of a number of pairs of rings, as shown in Figure 11-14.

The gas pressure is higher on one side of each ring. This compresses the rings against the sealing area. Each pair of rings consists of one radial

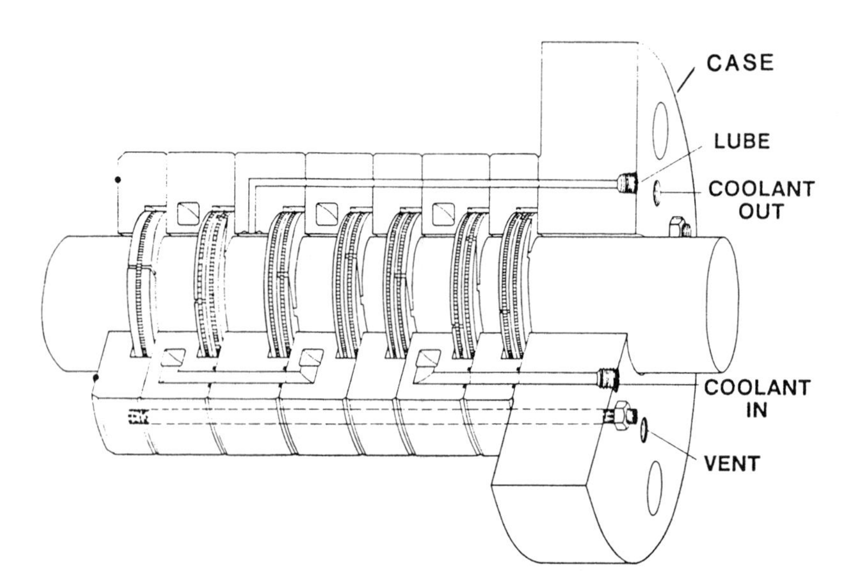

Figure 11-13. Cut-away view of packing case. (*Courtesy of C. Lee Cook, a Dover Resources Company.*)

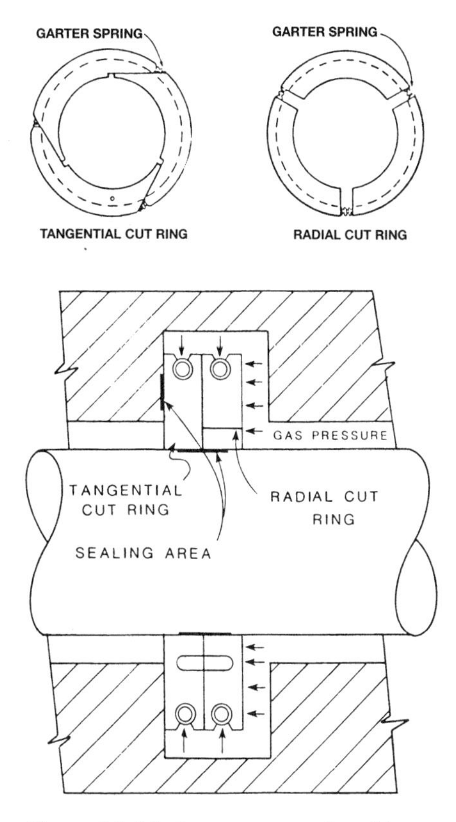

Figure 11-14. Compressor rod packing.

cut ring and one tangential cut ring. The radial cut ring is installed toward the cylinder (pressure) side. Gas flows around the front face of the radial cut ring and then around the outside of both rings. Since the ring outside diameter is greater than the ring inside diameter, a squeezing force is exerted on the rod. This seals the path between the rings and the rod. The radial cuts are positioned in the ring assembly so that they do not line up with the tangential cuts. Cylinder pressure will force the ring assembly against the packing case lip, thus preventing flow around the rings.

The amount of pressure differential one set of rings can withstand is limited. Therefore, several pairs must be installed to handle typical field gas compression applications. The basic design of the packing is left up to the manufacturer.

Lubrication is needed to reduce friction and provide cooling. Lubricating oil, which must be finely filtered to prevent grit from entering the case, is generally injected in the second ring assembly. The pressure differential moves the oil along the shaft.

A separate cooling system may be required for high-pressure service (5,000 psi) or where high compression ratios and long packing cases are installed.

Compressor Valves

The compressor valves control the flow of gas into and out of the compressor cylinder. All valves are similar in that the differential pressure across the seat must be greater than the balance spring force before gas may flow through the valve. The lift characteristics, seat area, and flow areas determine the advantages of each design. In older compressor stations, channel valves (Figure 11-15) were commonly used. Channel valves are now considered obsolete and used only in small air compressors In today's market, there are three common types of valves—poppet valves (Figure 11-16), ring valves (Figure 11-17), and plate valves (Figure 11-18).

Poppet valves are typically used for low compression ratio applications, such as pipeline booster compressors. As the pressure differential increases across each of the individual poppets, they lift and allow gas to pass through the flow openings in the stop plate.

Ring valves are typically used for slow speed, high pressure process compressors. Instead of individual poppets, these valves use concentric rings, which open and close the valve ports.

Plate valves are typically used for high speed separable compressors. Plate valves are similar to ring valves with the rings connected by ribs. Instead of individual elements opening and closing, all valve ports open and close at the same time.

Valve type and size should be specified by the compressor manufacturer. Normally, the manufacturer will quote a valve velocity, which can be calculated from:

$$V = 288D/A \tag{11-1}$$

where V = average gas velocity, ft/min
D = cylinder displacement, ft^3/min
A = product of the actual lift, the valve opening periphery, and the number of inlet or discharge valves per cylinder, $in.^2$

At lower velocities the valve has less pressure drop and thus has less maintenance associated with it. Velocities calculated from this equation can be used to compare valve designs.

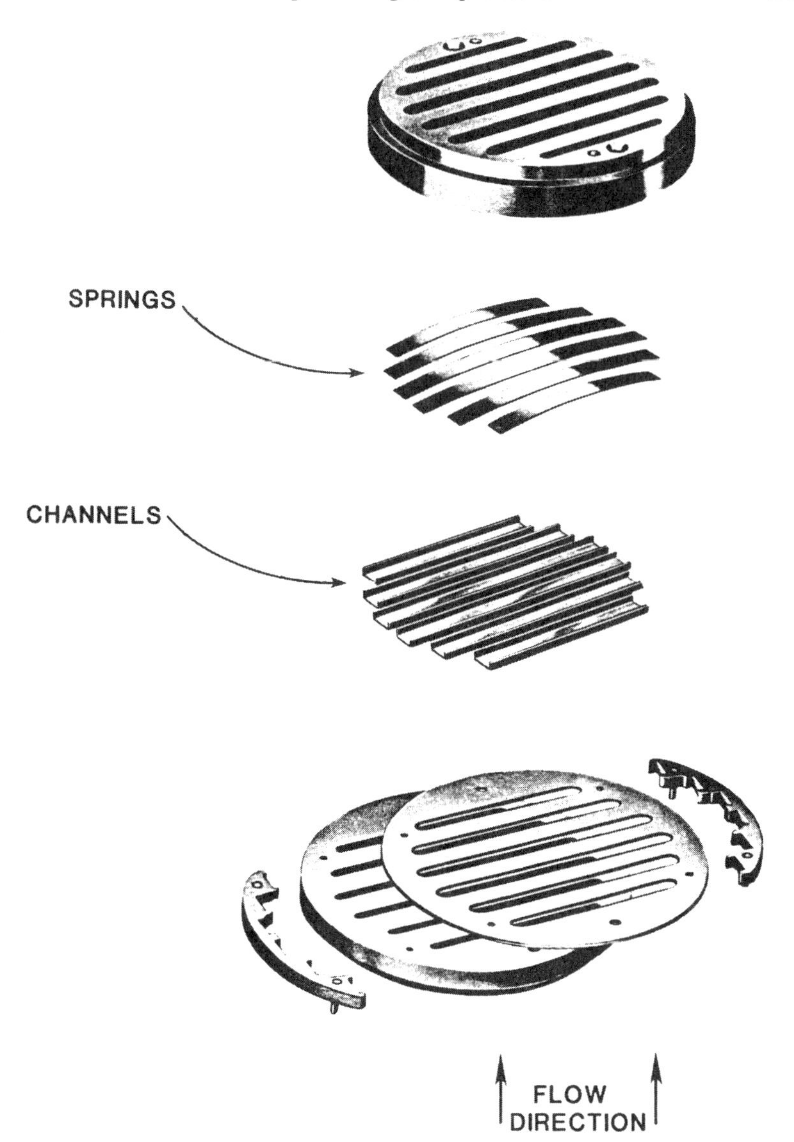

Figure 11-15. Channel valves. ***(Courtesy of Ingersoll-Rand Company.)***

In addition to valve velocity, the manufacturer can furnish the effective flow area of the valve. This area is determined by measuring the pressure drop across the valve with a known flow rate and then calculating an equivalent orifice area that provides the same pressure drop. Valves with larger effective flow areas have less pressure drop and better efficiencies. The effects of the seat area, the lift area, and the flow paths are automati-

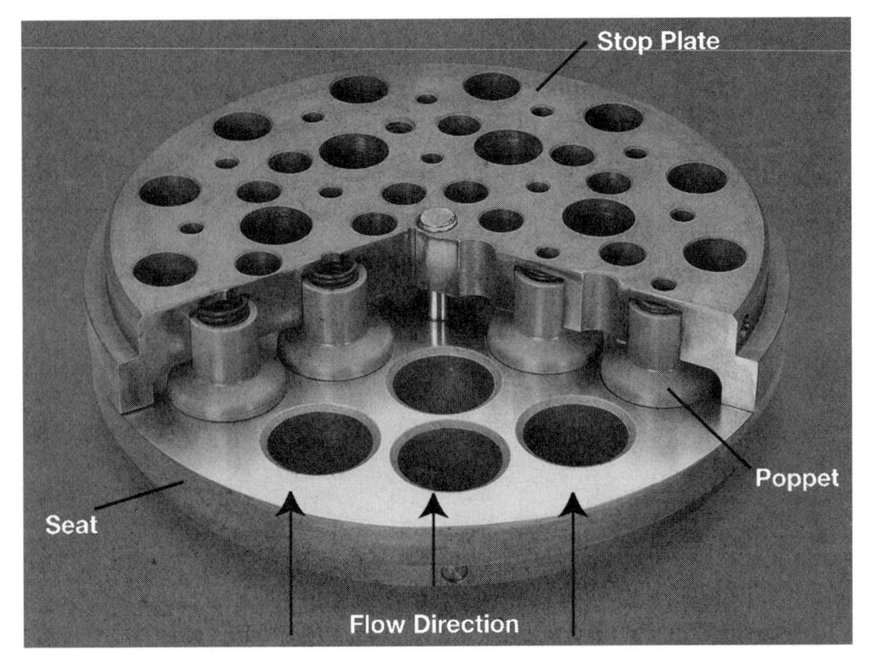

Figure 11-16. Cut-away view of poppet valve. (*Courtesy of Dresser-Rand Company.*)

cally included when the effective flow area is used to compare compressor valves. This in turn provides a better comparison of valve performance than just looking at valve velocity.

In addition to valve efficiency, the following should be considered in valve selection: ease of maintenance, durability, and spare parts required.

Capacity Control Devices

Reciprocating compressor capacity may easily be adjusted by changing compressor speed, changing compressor cylinder clearance, unloading compressor cylinder inlet valves, recycling gas from unit discharge to unit suction, or a combination of these methods. All these methods may be accomplished either manually by the operator or automatically by the control panel.

The use of speed control and/or a recycle valve is covered in Chapter 10. Our discussion in this chapter will concentrate on cylinder inlet valve unloaders and changing cylinder clearance.

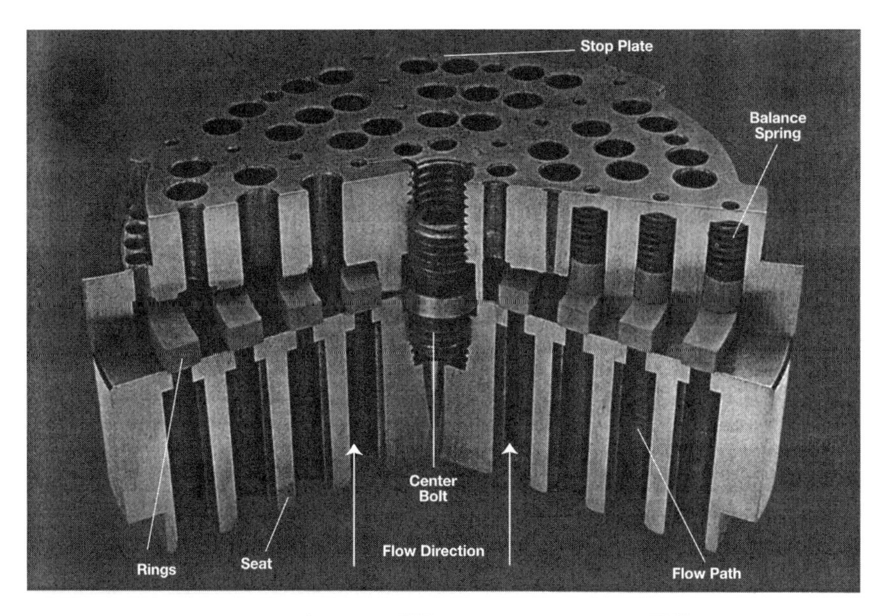

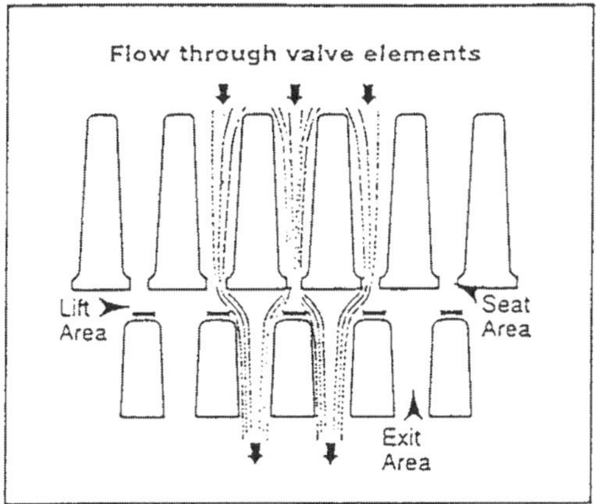

Figure 11-17. Cut-away view of ring valve. (*Courtesy of Dresser-Rand Company.*)

Valve Unloaders

Inlet valve unloaders are used to deactivate a cylinder end and reduce its capacity to zero. Two of the more common types of unloaders are depressor-type unloaders and plug-type unloaders. Depressor-type unloaders hold the inlet valve open during both the suction and discharge

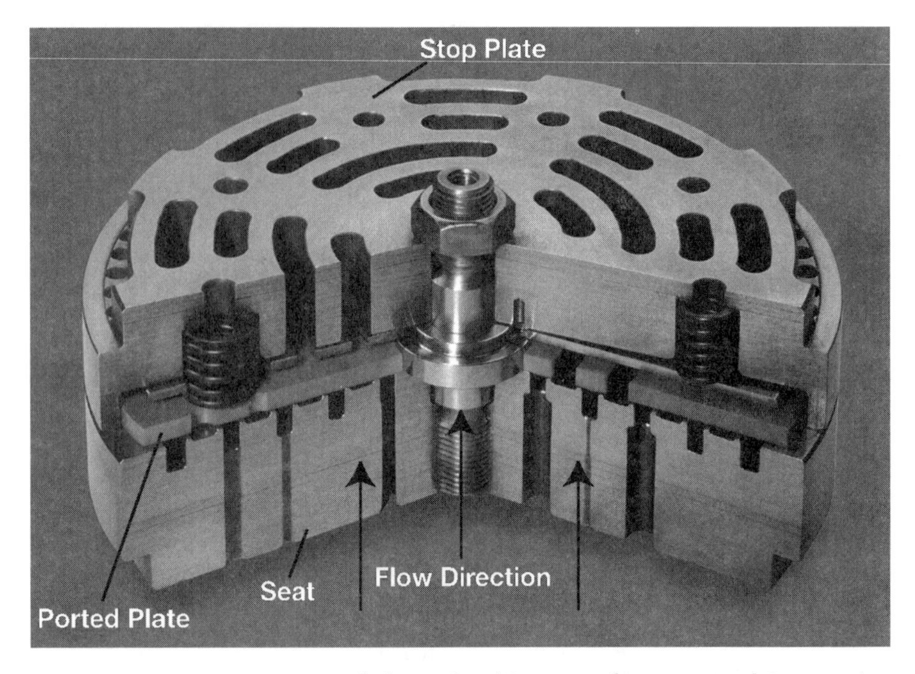

Figure 11-18. Cut-away view of plate valve. (*Courtesy of Dresser-Rand Company.*)

strokes so that all the gas is pushed back through the inlet valves on the discharge stroke. Plug-type unloaders open a port to bypass the inlet valve and connect the cylinder bore directly with the inlet gas passage.

Compressors may be set up with inlet valve unloaders to be used for both capacity control and reducing the compression load during starting and up-set conditions. For example, the capacity of a single stage compressor with two compressor cylinders may be reduced by 25% by unloading the outer end of one cylinder and by another 25% by unloading the outer end of the second cylinder. Unloading the frame ends of the cylinders at the same time that the outer ends are unloaded will reduce the flow to zero and is recommended only for start-up due to excessive heat build-up inside the cylinder after extended operation.

The compressor manufacturer must be consulted if the cylinder is to be run single acting with the frame end unloaded. Many times rod load reversal and proper lubrication may not be achieved while running single acting with the frame end unloaded.

When a cylinder end is deactivated, the pulsation levels in the piping system can increase significantly. If a cylinder may be operated with unloaders, it needs to be analoged in both its operating mode and its unloaded mode.

Cylinder Clearance

Clearance is the volume remaining in a cylinder end when the piston is at the end of its stroke. This is the sum of the volume between the head of the cylinder and the piston, and the volume under the valve seats. The total clearance is expressed in percent of the total piston displacement, normally between 4 and 30%.

As the piston starts its suction stroke, the gas that remains in the cylinder in the fixed and added clearance areas expands until the pressure in the cylinder is equal to the pressure in the line outside of the cylinder. The greater the clearance, the longer it takes for the suction valves to open and the less new gas enters the cylinder. Therefore, less gas will be compressed as cylinder clearance is increased.

End clearance is required to keep the piston from striking the compressor head or crank end. Some small clearance is also required under suc-

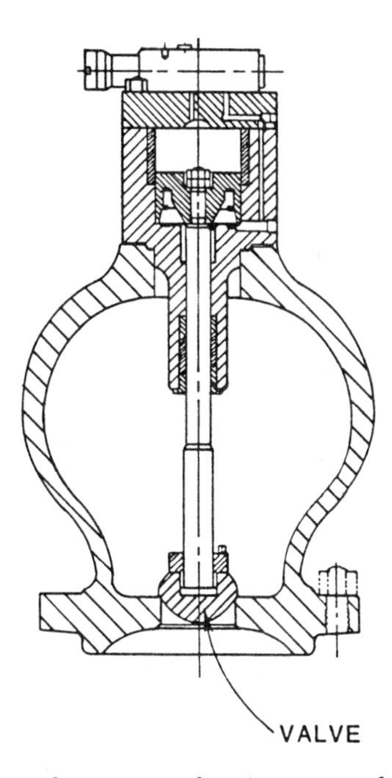

Figure 11-19. Fixed volume clearance pocket. (*Courtesy of Dresser-Rand Company.*)

tion and discharge valves so that the valves can be removed and reinstalled.

These clearances are called fixed clearances and can be adjusted by:

- Removing a small portion of the end of the compressor piston
- Shortening the projection of the cylinder heads into the cylinder
- Installing spacer rings between cylinder head and body or under the valves

Variable clearance that can be changed very readily can be built into the cylinder. Figure 11-19 is an example of a fixed volume clearance pocket mounted on the cylinder. This type is separated from the cylinder by a valve that can be opened and closed from the outside.

Fixed clearance can also be added to the outer end of the cylinder by adding a fabricated clearance bottle with the desired volume. To change the performance of the cylinder the clearance can be changed by shutting down the compressor, unbolting one bottle, and installing another bottle with a different volume. It is very easy in that respect to add clearance and subtract clearance from a cylinder if the cylinder is set up to receive clearance bottles.

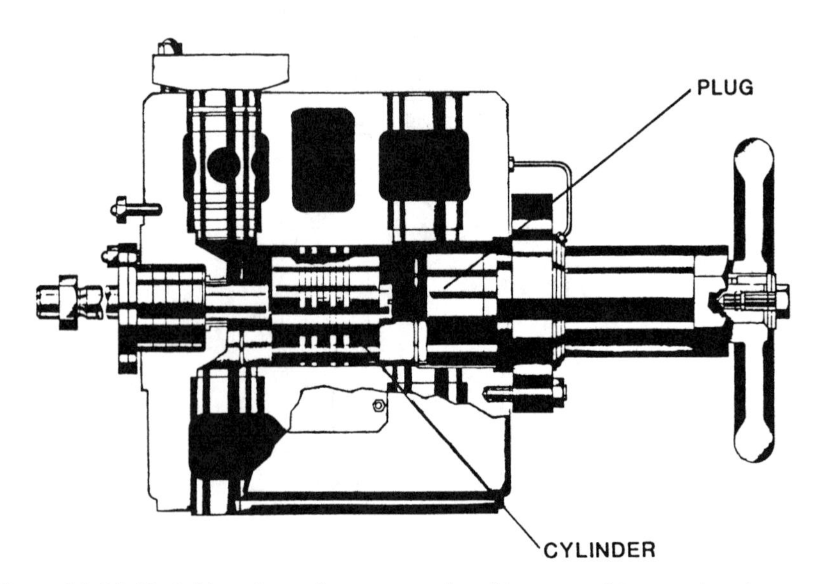

Figure 11-20. Variable volume clearance pocket. (*Courtesy of Dresser-Rand Company.*)

More flexibility can be obtained with a variable volume clearance pocket such as that shown in Figure 11-20. This is a plug built into the outer cylinder head. When moved, the clearance volume of the outer end of the cylinder changes.

Clearance is normally expressed as a percent or fraction of cylinder displacement. It is given by:

Single acting cylinder (head end clearance)

$$\% \, C = \frac{C_{HE}}{(d_c^2)(.7854)(s)} \times 100 \tag{11-2}$$

Double acting cylinder (average clearance)

$$\% \, C = \frac{(C_{HE}) + (C_{CE})}{[d_c^2 + (d_c^2 - d_r^2)](.7854)(s)} \times 100 \tag{11-3}$$

where $\% \, C$ = cylinder clearance, fraction
d_c = cylinder diameter, in.
d_r = rod diameter, in.
C_{HE} = head end clearance, in^3
C_{CE} = crank end clearance, in^3
s = stroke length, in.

Note: C_{HE} and C_{CE} can be obtained from the manufacturer.

CYLINDER SIZING

Typically, in specifying a unit, the suction and discharge pressures, capacity (MMscfd), inlet temperature, and gas properties are given. The actual sizing of the cylinders is left to the manufacturer from his specific combinations of standard cylinders, pistons, and liners. However, once a proposal is received from a manufacturer, sometimes it is beneficial to check the cylinder sizing and make sure that indeed the compressor will perform. Sometimes it is necessary to size a new cylinder for an existing compressor or to verify that an existing compressor will perform in a different service.

The capacity of the cylinder is a function of piston displacement and volumetric efficiency. This is in turn a function of cylinder clearance, compression ratio, and gas properties.

Piston Displacement

The actual volume of the cylinder that is swept by the piston per minute is piston displacement. It can be calculated from:

Single acting cylinder (head end displacement)

$$PD = \frac{(d_c^2)(s)(rpm)}{2{,}200} \tag{11-4}$$

Single acting cylinder (crank end displacement)

$$PD = \frac{(d_c^2 - d_r^2)(s)(rpm)}{2{,}200} \tag{11-5}$$

where PD = piston displacement, cfm
s = stroke length, in.
rpm = compressor speed, rpm
d_c = diameter of cylinder, in.
d_r = diameter of rod, in.

Double acting cylinder (sum of head end and crank end displacement)

$$PD = \frac{[2(d_c)^2 - (d_r)^2](s)(rpm)}{2{,}200} \tag{11-6}$$

Throughput can be changed directly by changing the piston displacement. This can be accomplished by changing the speed, by removing or deactivating suction valves, or by converting a double acting cylinder to single acting.

In addition, piston displacement and thus throughput can be changed by changing cylinder liners, installing sleeves, or boring the cylinder.

Volumetric Efficiency

The flow rate is not directly equal to the piston displacement. Volumetric efficiency is the ratio of actual volumetric flow at inlet temperature and pressure conditions to piston displacement. It is given by:

$$E_v = 96 - R - C\left[(R^{1/k})\frac{Z_s}{Z_d} - 1\right] \tag{11-7}$$

where E_v = stage volumetric efficiency, %
R = compression ratio (P_d/P_s) of the compressor stage (based on absolute pressure)
C = cylinder clearance, percent of piston displacement
Z_s = compressibility factor at suction, psia
Z_d = compressibility factor at discharge, psia
k = ratio of specific heats, C_p/C_v

It can be seen from Equation 11-7 that as R is increased, and as clearance is increased, volumetric efficiency is reduced. The relationship of volumetric efficiency and clearance is important, because it allows variable clearances (both fixed volume and adjustable volume pockets) to be used to control capacity and obtain the maximum use of available driver horsepower.

Cylinder Throughput Capacity

Using a known piston displacement and efficiency, the gas throughput can be calculated from:

$$q_a = E_v \times PD \tag{11-8}$$

where q_a = gas throughput at suction conditions of temperature and pressure, ft³/min
E_v = volumetric efficiency
PD = piston displacement, ft³/min

or from:

$$q_g = 35.4 \frac{q_a P_s}{T_s Z_s} \tag{11-9}$$

where q_g = gas throughput at standard conditions, scfm
P_s = suction pressure, psia
T_s = suction temperature, °R
Z_s = compressibility at suction conditions

or from:

$$Q_g = 0.051 \frac{q_a P_s}{T_s Z_s} \tag{11-10}$$

where Q_g = gas throughput, MMscfd

Compressor Flexibility

To enhance compressor flexibility it is desirable to design into the compressor the capability of operating at other than the original design conditions. If it is desired to provide flexibility to operate at lower suction pressures:

- Add the ability to increase speed or
- Add variable clearance initially, when the suction pressure is high; at lower suction pressure clearance is removed to increase actual throughput.

To provide flexibility for an eventually higher suction temperature:

- Add the ability to increase speed or
- Add variable clearance to reduce capacity initially at low temperature; at high temperature clearance is removed to increase actual throughput.

To provide flexibility for eventually higher discharge pressures:

- Ensure there are enough stages at the outset so that the ratio per stage can increase and still keep within the discharge temperature limits or
- Use variable clearance to adjust ratio per stage.

To provide flexibility for increased throughput:

- Add the ability to increase speed or
- Add variable clearance to reduce capacity initially when the throughput is low; clearance is removed to increase throughput.

ROD LOAD

The allowable rod load depends on rod diameter and material, and will be quoted by the manufacturer. The actual load can be calculated from the following equations if the geometry is known:

Single-acting cylinder, head end

$$RL_c = a_p\,(P_d - P_u) + a_r\,P_u \tag{11-11a}$$

$$RL_u = a_p\,(P_u - P_s) - a_r\,P_u \tag{11-11b}$$

Single-acting cylinder, crank end

$$RL_c = a_p\,(P_u - P_s) + a_r\,P_s \qquad (11\text{-}12a)$$

$$RL_t = a_p\,(P_d - P_u) - a_r\,P_d \qquad (11\text{-}12b)$$

Double-acting cylinder

$$RL_c = a_p\,(P_d - P_s) + a_r\,P_s \qquad (11\text{-}13)$$

$$RL_t = a_p\,(P_d - P_s) - a_r\,P_d \qquad (11\text{-}14)$$

where RL_c = rod load in compression, lb
RL_t = rod load in tension, lb
a_p = cross-sectional area of piston, in.^{2}
P_d = discharge pressure, psia
P_s = suction pressure, psia
P_u = pressure in unloaded area, psia
a_r = cross-sectional area of rod, in.^{2}

The calculations shown above provide the gas load imposed on the rod (and crosshead bushing) by the compressor cylinder piston. To provide a reasonable crosshead pin bushing life, the rod loading at the crosshead bushing must change from compression to tension during each revolution. This is commonly referred to as "rod reversal" and allows oil to lubricate and cool one side of the bushing while load is being applied to the other side of the bushing.

A single-acting, head end cylinder will not have load reversal if suction pressure is applied to the crank end. Similarly, if discharge pressure is applied to the head end of a single-acting, crank end cylinder, load reversal will not occur.

In addition to the gas load, the rod and crosshead pin bushing is subject to the inertia forces created by the acceleration and deceleration of the compressor reciprocating mass. The inertia load is a direct function of crank radius, the reciprocating weight, and speed squared. The total load imposed on the crosshead pin and bushing is the sum of the gas load and the inertia load and is referred to as the "combined rod load."

The combined rod load should be checked anytime the gas loads are approaching the maximum rating of the compressor frame or anytime rod reversal is marginal or questionable.

COOLING AND LUBRICATION SYSTEMS

Compressor Cylinder Cooling

Traditional compressor cylinder designs require cooling water jackets to promote uniform distribution of heat created by gas compression and friction. Some of the perceived advantages of water-cooled cylinders are reduced suction gas preheat, better cylinder lubrication, prolonged parts life, and reduced maintenance.

Operating experience during the last 30 years has proven that compressor cylinders designed without cooling water jackts (non-cooled) can successfully operate in most natural gas compession applications. Some of the perceived advantgaes of non-cooled cylinders are simplified cylinder designs that reduce cost and improve efficiency, reduced initial system costs due to reductions in the cooling water system, improved valve accessibility, and reduced weight.

Many manufacturers, users, and compressor applications still require that compressor cylinders be supplied with liquid-cooled cylinders. Figure 11-21 includes schematics of several types of liquid coolant systems.

In static systems, the cooling jackets are normally filled with a glycol and water mixture to provide for uniform heat distribution within the cylinder. This system may be used where the ΔT of the gas is less than 150°F and discharge gas temperature is less than 190°F.

Thermal siphons use the density differences between the hot and the cold coolants to establish flow. This system may be used where the ΔT of the gas is less than 150°F and discharge gas temperature is less than 210°F.

Forced coolant systems using a mixture of glycol and water are the most common for natural gas compressors. Normally, the compressor cylinder cooling system and compressor frame lube oil cooling system is combined. A single pump is used to circulate the coolant through the cylinders and the lube oil heat exchanger and then to an aerial cooler where the heat is dissipated.

When forced coolant systems are used, care must be taken to provide the coolant at the proper temperature. If the cylinder is too cool, liquids could condense from the suction gas stream. Thus, it is desirable to keep the coolant temperature 10°F higher than that of the suction gas. If the cylinder is too hot, gas throughput capacity is lost due to the gas heating and expanding. Therefore, it is desirable to limit the coolant temperature to less than 30°F above that of the suction gas.

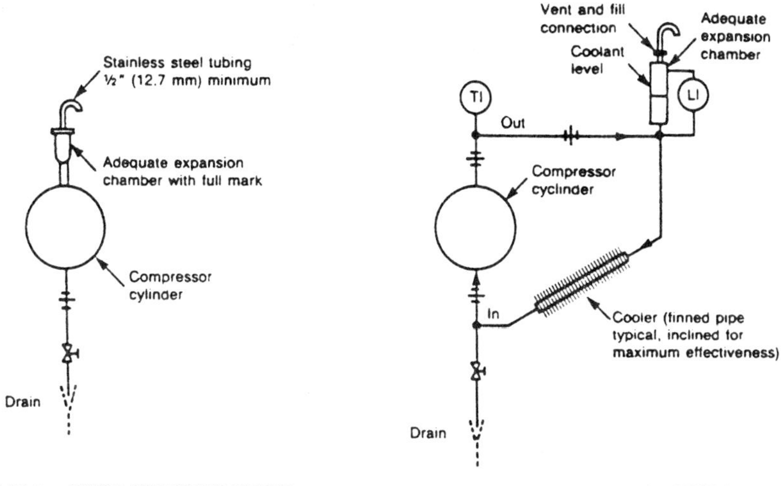

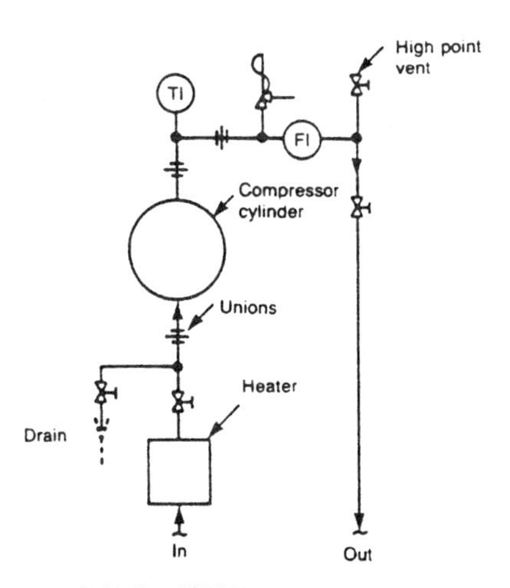

Figure 11-21. Cylinder cooling systems. ***(Reprinted with permission from API, Std. 618, 3rd Ed., Feb. 1986.)***

Frame Lubrication System

The frame lubrication system circulates oil to the frame bearings, connecting rod bearings, crosshead shoes, and can also supply oil to the packing and cylinder lubrication system. Splash lubrication systems are the least expensive and are used in small air compressors. Forced-feed systems are used for almost all oilfield gas compression applications.

Figure 11-22 shows a splash lubrication system where an oil ring rides loosely and freely on the rotating shaft, dipping into the oil sump as it rotates. The ring rotates because of its contact with the shaft, but at a slower speed. The oil adheres to the ring until it reaches the top of the journal when it flows onto the shaft.

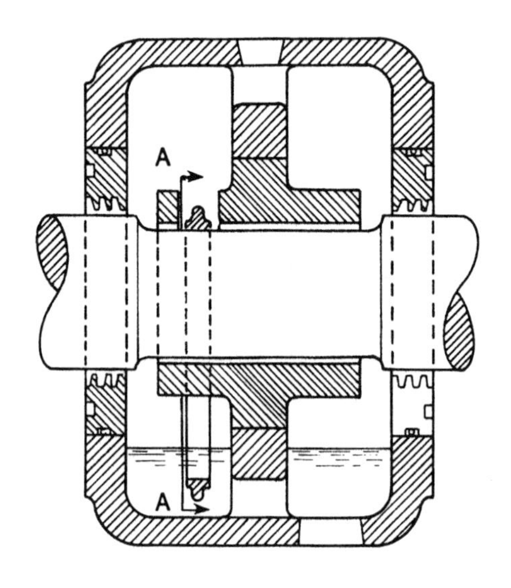

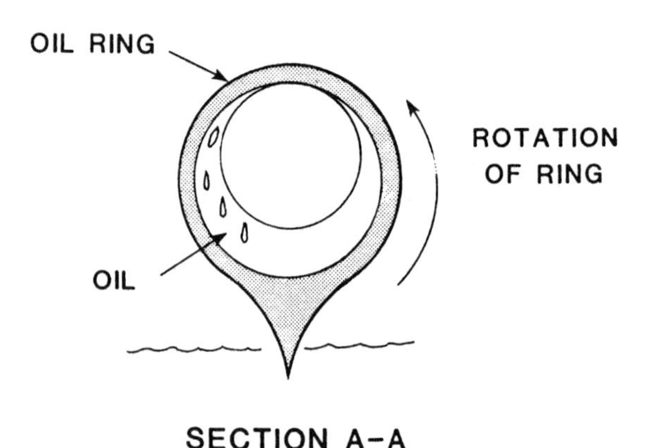

Figure 11-22. Splash lubrication system (oil slinger).

In a forced-feed lubrication system, a pump circulates lubricating oil through a cooler and filter to a distribution system that directs the oil to all the bearings and crosshead shoes. Figure 11-23 is a schematic of a typical system. The details of any one system will vary greatly. Major components and considerations of a forced feed lubrication system are as follows:

- Main oil pump
  - Driven from crankshaft.
  - Should be sized to deliver 110% of the maximum anticipated flow rate.
- Auxiliary pump
  - Backup for the main oil pump.
  - Electric motor driven.

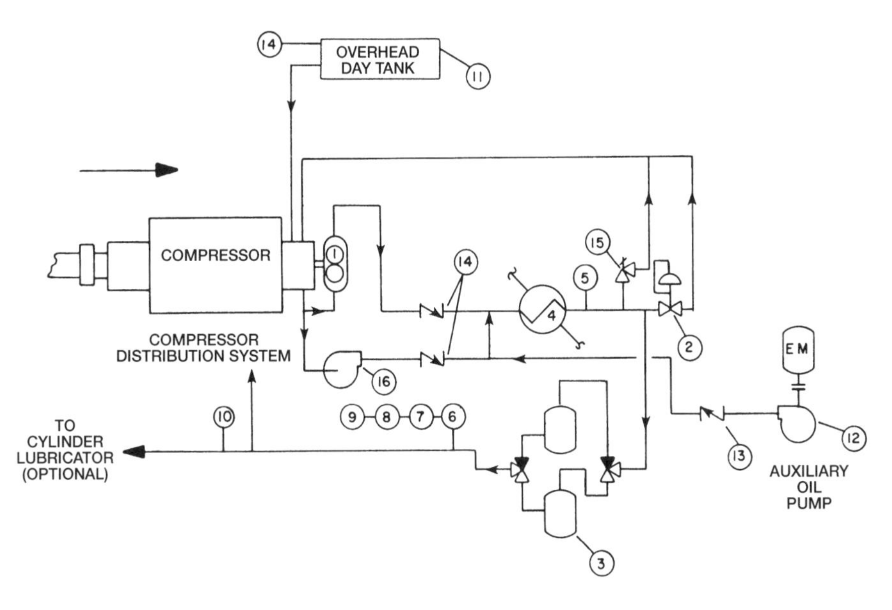

1. SHAFT DRIVEN MAIN OIL PUMP
2. PRESSURE-REGULATING VALVE
3. FULL-FLOW FILTER (dual)
4. OIL COOLER
5. TEMPERATURE GAGE
6. PRESSURE GAGE
7. LOW-PRESSURE ALARM & SHUTDOWN SWITCH
8. STARTUP SWITCH FOR MAIN PUMP
9. AUXILIARY PUMP PRESS. SWITCH (optional)
10. PIPING (supply & return)
11. OIL RESERVOIR (day tank)
12. AUXILIARY OIL PUMP
13. CHECK VALVE
14. OIL LEVEL INDICATOR
15. RELIEF VALVE
16. PRE-LUBE PUMP

NOTE: THIS ILLUSTRATION IS A TYPICAL SCHEMATIC AND DOES NOT CONSTITUTE ANY SPECIFIC DESIGN, NOR DOES IT INCLUDE ALL DETAILS (for example, vents and drains).

Figure 11-23. Forced-feed lubrication system.

  - Should start automatically when supply pressure falls below a certain level.
- Pre-lube pump
  - Manual or automatic.
  - Prevents running bearings dry at start.
- Oil cooler
  - Keeps oil temperature below 165°F.
  - Can use shell-and-tube exchanger with jacket cooling water or air-cooled exchanger.
  - Sized for 110% of the maximum anticipated duty.
- Oil filter
  - Dual, full flow, with isolation valves arranged so switching can occur without causing a low-pressure shutdown.
  - Size should be determined by vendor; in lieu of other information use API 618 requirements.
- Overhead day tank
  - Sized to handle one month of oil consumption.
  - Should be equipped with a level indicator.
- Piping
  - Stainless steel downstream of filters.
  - No galvanizing.
  - No socket welding or other pockets that can accumulate dirt downstream of filter.
  - Carbon steel lines should be pickled, passivated, and coated with rust inhibitor.
  - Lube oil system from pump discharge to the distribution system should be flushed with lube oil at 160°F–180°F. Oil should flow across a 200 mesh screen and flushing should cease when no more dirt or grit is found on the screen.

Packing/cylinder lubrication can be provided from a forced feed compressor lube oil system. For very cold installations, immersion heaters and special lube oils must be considered. If the lube oil temperature gets too cold, the oil becomes too viscous and does not flow and lubricate properly.

Cylinder/Packing Lubrication System

The flow required to lubricate the packing and cylinders is quite small, and the pressure necessary to inject the lubricant at these locations is quite high. Therefore, small plunger pump (force-feed lubricators) systems are used. The force-feed lubricators are usually driven by the compressor crankshaft.

The two basic types of cylinder lubrication systems are the pump-to-point system and the divider-block system. The pump-to-point system provides each lubrication point with its own lubricator pump. Thus, if the compressor cylinders and packing require six lubrication points, the lubricator box would be supplied with six cam driven pumps. The divider-block system uses one or more lubricator pumps to supply a divider block, which then distributes the flow to each of the lubrication points. The two systems are sometimes combined such that each stage of compression is provided with its own pump and a divider block to distribute the flow between the cylinders and packing of that particular stage.

Oil is supplied to this system from the frame lube oil system or from an overhead tank. This oil comes in contact with and thus contaminates the gas being compressed. Gas/oil compatibility should be checked.

PIPE SIZING CONSIDERATIONS

Because of the reciprocating action of the piston, care must be exercised to size the piping to minimize acoustical pulsations and mechanical vibrations. As a rule of thumb, suction and discharge lines should be sized for a maximum actual velocity of 30 ft/sec (1,800 ft/min) to 42 ft/sec (2,500 ft/min). Volume 1 contains the necessary formulas for determining pressure drop and velocity in gas piping.

Analog or digital simulators can be used to establish the pulsation performance of any compressor piping system in detail. API 618 Section 3.9.2 provides guidelines for piping pulsation and vibration control based on compressor discharge pressure and horsepower. In practice, many operators do not "analog" compressors of 1,000 horsepower or lower, but rather rely on extrapolations from proven designs. For larger horsepower sizes or where unusual conditions (e.g., unloading and loading cylinders) exist, an analog is recommended.

For smaller, high-speed compressors the piping sizing rules of thumb discussed above, in conjunction with pulsation bottles sized from Figure 11-24, should be sufficient for individual field compressors. These rules of thumb can also be used for preliminary sizing of piping and bottles in preparation for an analog study.

To minimize pipe vibrations it is necessary to design pipe runs so that the "acoustic length" of the pipe run does not create a standing wave that

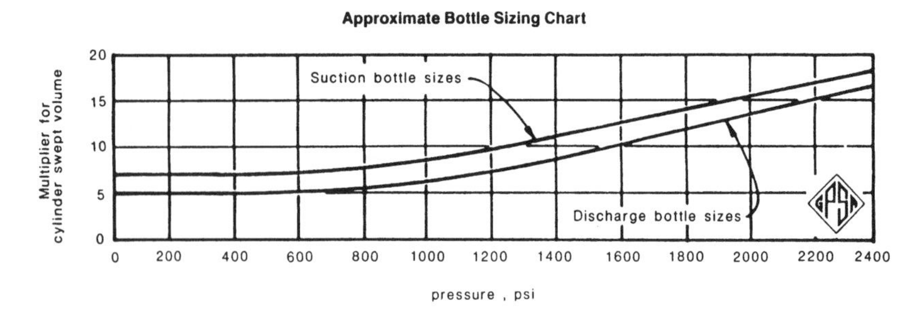

Figure 11-24. Pulsation bottle sizing chart (approximation). ***(Reprinted with permission from GPSA*** **Engineering Data Book,** ***10th Ed.)***

amplifies the pressure pulsations in the system. The acoustic length is the total overall length from end point to end point including all elbows, bends, and straight pipe runs. Typical pipe runs with respect to acoustic length are considered to be:

- Pipe length from suction pipeline to suction scrubber
- Pipe length from scrubber to suction pulsation dampeners
- Pipe length from discharge pulsation dampeners to cooler
- Pipe length from cooler to scrubber
- Pipe length from discharge scrubber to pipeline

The end of a pipe run can be classified as either "open" or "closed." Typically, closed ends are where the pipe size is dramatically reduced, as at orifice plates and at short length flow nozzles. A typical open end is where the pipe size is dramatically increased.

Where the pipe run contains similar ends (closed-closed or open-open), prohibited pipe lengths are:

$0.5\lambda, \lambda, 1.5\lambda, 2\lambda \ldots$

where λ = acoustic wave length, ft

Where the pipe run contains dissimilar ends (closed-open or open-closed), prohibited pipe lengths are:

$0.25\lambda, 0.75\lambda, 1.25\lambda, 1.75\lambda \ldots$

The wave length may be calculated from:

$$\lambda = 13{,}382 \frac{\left(\frac{kT}{MW}\right)^{1/2}}{R_c} \tag{11-16}$$

where λ = acoustic wavelength, ft
k = ratio of specific heats, dimensionless
T = gas temperature, °R
MW = molecular weight of gas
R_c = compressor speed, rpm

Mechanical vibration of pipe is handled in the same manner as for reciprocating pumps (Volume 1, Chapter 12). Normally, if the pipe support spacing is kept short, the pipe is securely tied down, the support spans are not uniform in length, and fluid pulsations have been adequately dampened, mechanical pipe vibrations will not be a problem. It is good practice to ensure that the natural frequency of all pipe spans is higher than the calculated pulsation frequency. The pulsation frequency is given by:

$$f_p = \left(\frac{R_c}{60}\right) n \tag{11-17}$$

where f_p = cylinder pulsation frequency, cps
n = 1 for single-acting cylinders and n = 2 for double-acting
R_c = speed of compressor, rpm

Refer to Volume 1, Chapters 8 and 9 for the calculations of natural frequency of pipe.

Foundation Design Considerations

Satisafactory compressor installations many times depend on how well the foundation or support structure was designed. An inadequate foundation design can result in equipment damage due to excessive vibration. The money saved by cutting corners on foundation design effort may be spent many times in costs associated with high maintenance and lost production.

Due to the basic design of the compressor, its rotating and reciprocating masses produce inertia forces and moments tha cannot be completely eliminated and must be absorbed by the foundation. The manufacturer has the ability to minimize the magnitude of these forces and moments by adding counterweights to the crossheads but cannot totally eliminate them.

In addition to the unbalanced forces and moments, the foundation must absorb the moments produced by the gas torque. This is the torque created by the gas pressure forces as the compressor goes through a revolution. The compressor manufacturer must provide the magnitude of the resulting forces and moments and the gas torques.

Typically foundation design engineers have only used the compressor unbalanced forces and moments in their design calculations. Recent experience has found that the moments created by the gas torque can have a significant impact on foundation design. Detailed information and good design practices for compressor support structures and foundations may be found in *Design of Structures and Foundations for Vibrating Machines* by Suresh Arya, Michael O'Neill, and George Pincus.

For complex offshore structures or where foundations may be critical, finite-element analysis computer programs with dynamic simulation capability can be used to evaluate foundation natural frequency and the forced vibration response.

Industry Standard Specifications

As previously discussed in this chapter and in Chapter 10, reciprocating compressors are generally classified as either low-speed (integral) compressors or high-speed (separable) compressors. API has provided a standard and specification for each type of compressor to help the user and the facility engineer provide reliable compressor installations.

API Standard 618 "Reciprocating Compressors for Petroleum, Chemical, and Gas Industry Services" covers moderate- to low-speed compressors in critical services. Integral compressors and low-speed, long stroke balanced-opposed compressors with speeds from 200 to 600 rpm generally fall into this type of construction. The use of this standard with high-speed packaged separable compressors generally results in pages of exceptions by the compressor packager.

API Specification 11P "Specification for Packaged Reciprocating Compressors for Oil and Gas Production Services" covers packaged high-speed separable compressors with speeds from 600 to 1,200 rpm. The majority of reciprocating compressors sold in today's market fall into this category.

The user and facilitiy engineer must determine the critical nature of each installation and determine the type of construction desired. He or she must consider such things as intended service, compressor location, the consequences of downtime, and frequency of up-set or abnormal conditions.

When specifying compressor packages to API 11P, it may be necessary to specify certain sections of API 618 to ensure satisfactory installations. An example of this would be the supply of multiple compressors to be located in pipeline booster stations. In this case, an analog or digital pulsation and vibration study per API 618 Section 3.9 would be advisable to improve reliability and to minimize system problems and potential damage caused by gas pulsations and interaction between the individual compressor packages.

Fugitive Emissions Control

One of the growing environmental concerns for both new and existing reciprocating compressor installations is fugitive emissions. Fugitive emissions are the leakage of volatile organic compounds (VOCs) into the atmosphere. The local environmental regulations should be checked at the beginning of the compression project to avoid delays and field modifications.

The major source of fugitive emissions from a gas compressor cylinder is the piston rod packing. Other sources of fugitive emissions are around the cylinder valve covers, unloader covers, unloader actuator packing, and clearance pocket gasket and actuator packing.

Fugitive emissions can be reduced by supplying improved O-ring seal designs along with piston rod packing cases and actuator stem seal designs that utilize an inert buffer gas purge. The purge gas and VOCs can then be collected and sent to either a flare or vapor recovery system. The compressor manufacturer must advise the maximum allowable backpressure on the compressor components. A typical compressor cylinder inert buffer gas arrangement is shown in Figure 11-25.

EXAMPLE PROBLEM

Given:

Late in the field life it is desirable to compress the 100 MMscfd for the example field downstream of the separator from 800 psig at 100°F to 1,000 psig. An engine-driven separable compressor is available from surplus. The engine is rated for 1,600 hp at 900 rpm. Horsepower is proportional to speed. The compressor frame has six 7-in. bore by 6.0-in. stroke double-acting cylinders with a minimum clearance of 17.92%, a rod load limit of 25,000 lb, and rod diameter of 1.75 in. Assume $k = 1.26$, $Z_s = 0.88$, and $Z_d = 0.85$.

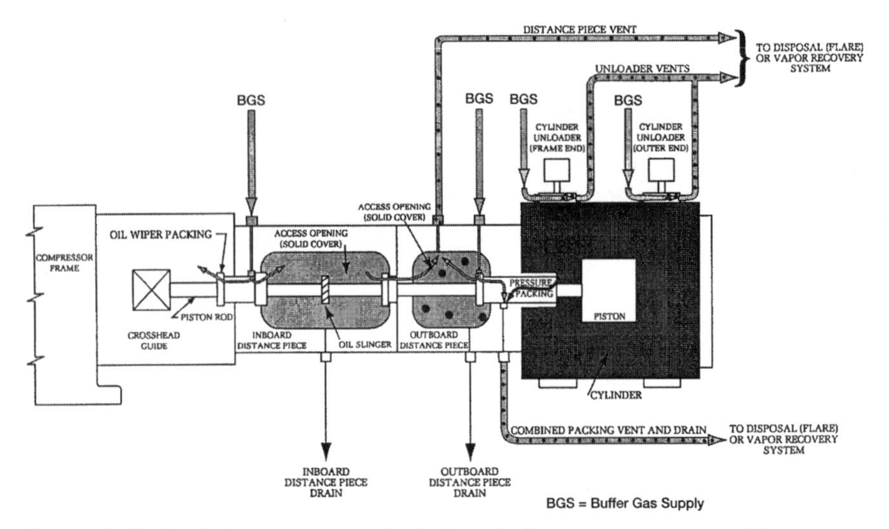

Figure 11-25. Typical compressor cylinder inert buffer gas arrangement. (*Courtesy of Dresser-Rand Company.*)

Compute discharge temperature, volumetric efficiency, required clearance, rod load, and required horsepower for the given conditions. Also calculate the lowest suction pressure at which this unit can compress 100 MMscfd.

Solution:

1. Calculate the gas discharge temperature

$$T_d = T_s \left(\frac{P_d}{P_s}\right)^{\frac{k-1}{k} \times \frac{1}{f}}$$

$$T_d = (560)\left(\frac{1{,}015}{815}\right)^{\frac{1.26-1}{1.26} \times \frac{1}{1}}$$

$T_d = 586°R$
$T_d = 126°F$

2. Calculate the volumetric efficiency

$$E_v = 96 - R - C\left[(R^{\frac{1}{k}})\frac{Z_s}{Z_d} - 1\right]$$

$$R = \frac{1{,}015}{815} = 1.245$$

$$E_v = 96 - 1.245 - 17.92\left[(1.245^{\frac{1}{1.26}})\left(\frac{0.88}{0.85}\right) - 1\right]$$

$$E_v = 90.6\%$$

3. Calculate the required clearance

$$PD = \frac{[2(d_c)^2 - (d_r)^2](s)(rpm)}{2{,}200}$$

$$PD = \frac{[2(7)^2 - (1.75)^2](6)(900)}{2{,}200}$$

$$PD = 233$$
$$PD_{tot} = (6)(233) = 1{,}398 \text{ cfm}$$
$$q_a = (E_v)(PD)$$
$$q_a = (.906)(1{,}398)$$
$$q_a = 1{,}267 \text{ cfm at suction conditions}$$

Convert to standard conditions:

$$Q_g = \frac{(0.051)(q_a\, P_s)}{T_s\, Z_s}$$

$$Q_g = \frac{(0.051)(1{,}267)(815)}{(560)(0.88)}$$

$$Q_g = 106.9 \text{ MMscfd}$$

At the present operating condition, the throughput is too high. One can decrease throughput by reducing speed, increasing clearance, which will reduce volumetric efficiency, using a thicker cylinder liner to reduce cylinder volume, or lowering suction pressure.

(a) Calculate required rpm to give desired throughput:

$$100 = \frac{(0.051)(q_a)(815)}{(560)(0.88)}$$

$$q_a = 1{,}186 \text{ cfm}$$

$$PD_{tot} = \frac{q_a}{0.906} = \frac{1{,}186}{0.906} = 1{,}309 \text{ cfm}$$

$$PD = \frac{PD_{tot}}{6} = \frac{1{,}309}{6} = 218 \text{ cfm}$$

$$\text{rpm} = \frac{(PD)(2{,}200)}{6[2(d_c)^2 - (d_r)^2]}$$

$$\text{rpm} = \frac{(218)(2{,}200)}{6[2(7)^2 - (1.75)^2]}$$

rpm = 842 or ≃ 850 rpm is the suitable speed.

(b) Calculate the clearance that would be needed to reduce the throughput from 106.9 MMscfd to 100 MMscfd:

Keep PD_{tot} constant, but lower the efficiency.

$$E_v = \frac{q_a}{PD} = \frac{1{,}186}{1{,}398} = 0.848$$

Now back calculate for the clearance that must be added to produce this volumetric efficiency.

$$84.8 = 96 - (1.245) - C\left[(1.245)^{1/1.26}\left(\frac{0.88}{0.85}\right) - 1\right]$$

$$C = 43.0\%$$

(c) Calculate the size liner required to reduce piston displacement: Assume E_v remains constant. This may have to be determined once a drawing of the specific cylinder and liner is available. However, it should not vary greatly. The PD required is:

$$PD = \frac{1{,}186}{0.906} = 1{,}309 \text{ cfm}$$

$$\frac{1{,}309}{6} = \frac{[2(d_c)^2 - (1.75)^2](6)(900)}{2{,}200}$$

$$d_c = 6.78 \text{ in.}$$

4. Calculate the rod load

$$RL_c = a_p (P_d - P_s) + a_r P_s$$

$$RL_c = \pi (7/2)^2 (1{,}015 - 815) + \pi (1.75/2)^2 (815) = 9{,}657 \text{ lb}$$

$$RL_t = a_p (P_d - P_s) - a_r P_d$$

$$RL_t = \pi (7/2)^2 (1{,}015 - 815) - \pi (1.75/2)^2 (1{,}015) = -5{,}256 \text{ lb}$$

The calculated rod load for both the compression and tension modes are within the 25,000-1b maximum rod load limit.

5. Calculate the required horsepower needed for the given conditions:

$$BHP = 0.0857 [Z_{av}] \left[\frac{Q_g T_s}{E}\right] \left[\frac{k}{k-1}\right] \left[\left(\frac{P_d}{P_s}\right)^{\frac{k-1}{k}} - 1\right]$$

$$Z_{av} = \left(\frac{0.88 + 0.85}{2}\right) = 0.865$$

$$E = \text{Adiabatic efficiency} \times \text{mechanical efficiency}$$
$$E = 0.87 \times 0.94 = 0.82$$

$$BHP = (0.0857)[0.865] \left[\frac{(100)(560)}{0.82}\right] \left[\frac{1.26}{1.26 - 1}\right]$$

$$\times \left[\left(\frac{1{,}015}{815}\right)^{\frac{1.26-1}{1.26}} - 1\right]$$

$$BHP = 1{,}137 \text{ hp}$$

This is less than is available from our engine.

6. Calculate the lowest suction pressure.

   If we use the minimum clearance,

   $$q_a = 1{,}267 \text{ cfm}$$

   $$Q_g = (0.051)\frac{q_a \, P_s}{T_s \, Z_s}$$

   Assume $Z_S = 0.88$ at the new lower P_S.

   $$P_s = \frac{(100)\,(560)\,(0.88)}{(0.051)\,(1{,}267)} = 762.6 \text{ psia}$$

   $$P_s = 747.9 \text{ psig}$$

   It would be possible to recalculate this by choosing a new value for Z_S and calculating a new E_V for this condition, but the results will not change materially. By inspection, neither horsepower, rod load, nor discharge temperature will limit this suction pressure.

CHAPTER

12

Mechanical Design of Pressure Vessels*

Previous chapters of this book, as well as Volume 1, discuss concepts for determining the diameter and length of various pressure vessels. Volume 1 examined the various codes and equations for choosing the wall thickness of piping. This chapter addresses the selection of design pressure rating and wall thickness of pressure vessels. It also presents a procedure for estimating vessel weight and includes some example design details.

The purpose of this chapter is to present an overview of simple concepts of mechanical design of pressure vessels that must be understood by a project engineer specifying and purchasing this equipment. Most pressure vessels in the U.S. and many in other parts of the world are designed and inspected according to the American Society of Mechanical Engineers' Boiler and Pressure Vessel Code (ASME Code). Because the ASME Code contains much more detail than can be covered in a single chapter of a general textbook such as this one, the project engineer should have access to a copy of the ASME Code and should become

---

*Reviewed for the 1999 edition by K. S. Chiou of Paragon Engineering Services, Inc.

familiar with its general contents. In particular, Section VIII of the code, "Pressure Vessels," is particularly important. Countries that do not use the ASME Code have similar documents and requirements. The procedures used in this chapter that refer specifically to the ASME Code are generally applicable in other countries, but should be checked against the applicable code.

In federal water of the U.S. and in a few states, all pressure vessels must be designed and inspected in accordance with the ASME Code. In many states, however, there is no such requirement. It is possible to purchase "non-code" vessels in these states at a small savings in cost. Non-code vessels are normally designed to code requirements (although there is no certainty that this is true), but they are not inspected by a qualified code inspector nor are they necessarily inspected to the quality standards dictated by the code. For this reason, the use of non-code vessels should be discouraged to assure vessel integrity.

DESIGN CONSIDERATIONS

Design Temperature

The maximum and minimum design temperatures for a vessel will determine the maximum allowable stress value permitted for the material to be used in the fabrication of the vessel. The maximum temperature used in the design should not be less than the mean metal temperature expected under the design operating conditions. The minimum temperature used in the design should be the lowest expected in service except when lower temepratures are permitted by the rules of the ASME Code. In determining the minimum temperature, such factors as the lowest operating temperature, operational upset, auto-refrigeration, ambient temperature, and any other source of cooling should all be considered. If necessary, the metal temperature should be determined by computation using accepted heat transfer procedures or by measurement from equipment in service under equivalent operating conditions.

Design Pressure

The design pressure for a vessel is called its "Maximum Allowable Working Pressure" (MAWP). In conversation this is sometimes referred to simply as the vessel's "working pressure." The MAWP determines the setting of the relief valve and must be higher than the normal pressure of the process contained in the vessel, which is called the "operating pressure" of the vessel. The operating pressure is fixed by process conditions. Table 12-1 recommends a minimum differential between operating pressure and MAWP so that the difference between the operating pressure and the relief valve set pressure provides a sufficient cushion. If the operating pressure is too close to the relief valve setting, small surges in operating pressure could cause the relief valve to activate prematurely.

Some vessels have high-pressure safety sensors (PSH) that shut in the inflow if a higher-than-normal pressure is detected. The use of safety sensors is discussed in more detail in Chapter 14. The differential between the operating pressure and the PSH sensor set pressure should be as indicated in Table 12-1, and the relief valve should be set at least 5% or 5 psi, whichever is greater, higher than the PSH sensor set pressure. Thus, the minimum recommended MAWP for a vessel operating at 75 psig with a PSH sensor would be 105 psig (75 + 25 + 5); the PSH sensor is set at 100 psig and the relief valve is set at 105 psig.

Often, especially for small vessels, it is advantageous to use a higher MAWP than is recommended in Table 12-1. It may be possible to increase the MAWP at little or no cost and thus have greater future flexibility if process changes (e.g., greater throughput) require an increase in operating pressure.

The MAWP of the vessel cannot exceed the MAWP of the nozzles, valves, and pipe connected to the vessel. As discussed in Volume 1, Chapter 9, pipe flanges, fittings and valves are manufactured in accordance with industry standard pressure rating classes. Table 12-2 is a summary of the more detailed Table 9-11 in Volume 1 (1st Edition: Table 9-9) and presents the MAWP of carbon steel fittings manufactured in

Table 12-1
Setting Maximum Allowable Working Pressures

Operating Pressure	Minimum Differential Between Operating and MAWP
Less than 50 psig	10 psi
51 psig to 250 psig	25 psi
251 psig to 500 psig	10% of maximum operating pressure
501 psig to 1000 psig	50 psi
1001 psig and higher	5% of maximum operating pressure

Vessels with high-pressure safety sensors have an additional 5% or 5 psi, whichever is greater, added to the minimum differential.

accordance with American National Standards Institute (ANSI) specification B16.5.

If the minimum MAWP calculated from Table 12-1 is close to one of the ANSI MAWP listed in Table 12-2, it is common to design the pressure vessel to the same MAWP as the ANSI class. For example, the 105-psig pressure vessel previously discussed will have nozzles, valves and fittings attached to it that are rated for 285 psig (ANSI Class 150). The increase in cost of additional vessel wall thickness to meet a MAWP of 285 psig may be small.

Often, a slightly higher MAWP than that calculated from Table 12-1 is possible at almost no additional cost. Once a preliminary MAWP is selected from Table 12-1, it is necessary to calculate a wall thickness for the shell and heads of the pressure vessel. The procedure for doing this is described in the following section. The actual wall thickness chosen for the shell and heads will be somewhat higher than that calculated, as the shells and heads will be formed from readily available plates. Thus, once the actual wall thickness is determined, a new MAWP can be specified for essentially no additional cost. (There will be a marginal increase in cost to test the vessel to the slightly higher pressure.)

This concept can be especially significant for a low-pressure vessel where a minimum wall thickness is desired. For example, assume the calculations for a 50-psig MAWP vessel indicate a wall thickness of 0.20 in., and it is decided to use ¼-in. plate. This same plate might be used if a MAWP of 83.3 psig were specified. Thus, by specifying the higher MAWP (83.3 psig), additional operating flexibility is available at essentially no increase in cost. Many operators specify the MAWP based on

Table 12-2
Summary ANSI Pressure Ratings Material Group 1.1

	MAWP, psig	
Class	**–20°F to 100°F**	**100°F to 200°F**
150	285	250
300	740	675
400	990	900
600	1480	1350
900	2220	2025
1500	3705	3375
2500	6170	5625

process conditions in their bids and ask the vessel manufacturers to state the maximum MAWP for which the vessel could be tested and approved.

Maximum Allowable Stress Values

The maximum allowable stress values to be used in the calculation of the vessel's wall thickness are given in the ASME Code for many different materials. These stress values are a function of temperature. Section VIII of the ASME Code, which governs the design and construction of all pressure vessels with operating pressures greater than 15 psig, is published in two divisions. Each sets its own maximum allowable stress values. Division 1, governing the design by Rules, is less stringent from the standpoint of certain design details and inspection procedures, and thus incorporates a higher safety factor of 4. For example, if a 60,000 psi tensile strength material is used, the maximum allowable stress value is 15,000 psi. On the other hand, Division 2 governs the design by Analysis and incorporates a lower safety factor of 3. Thus, the maximum allowable stress value for a 60,000 psi tensile strength material will become 20,000 psi.

Many companies require that all their pressure vessels be constructed in accordance with Division 2 because of the more exacting standards. Others find that they can purchase less expensive vessels by allowing manufacturers the choice of either Division 1 or Division 2. Normally, manufacturers will choose Division 1 for low-pressure vessels and Division 2 for high-pressure vessels.

The maximum allowable stress values at normal temperature range for the steel plates most commonly used in the fabrication of pressure vessels are given in Table 12-3. For stress values at higher temperatures and for other materials, the latest edition of the ASME Code should be referenced.

Determining Wall Thickness

The following formulas are used in the ASME Code Section VIII, Division 1 for determining wall thickness:

Wall thickness—cylindrical shells

$$t = \frac{Pr}{SE - 0.6P} \qquad (12\text{-}1)$$

Table 12-3
Maximum Allowable Stress Value for Common Steels

Metal Temperature		Not Lower Than Not Exceeding	ASME Section VIII Div. 1 −20°F 650°F	Div. 2 −20°F 100°F
Carbon	SA-516	Grade 55	13,800	18,300
Steel		Grade 60	15,000	20,000
Plates		Grade 65	16,300	21,700
and		Grade 70	17,500	23,300
Sheets	SA-285	Grade A	11,300	15,000
		Grade B	12,500	16,700
		Grade C	13,800	18,300
	SA-36		12,700	16,900
Low	SA-387	Grade 2, cl. 1	13,800	18,300
Alloy		Grade 12, cl.1	13,800	18,300
Steel		Grade 11, cl.1	15,000	20,000
Plates		Grade 22, cl.1	15,000	20,000
		Grade 21, cl.1	15,000	20,000
		Grade 5, cl.1	13,900	20,000
		Grade 2, cl.2	17,500	23,300
		Grade 12, cl.2	16,300	21,700
		Grade 11, cl.2	18,800	25,000
		Grade 22, cl.2	17,700	25,000
		Grade 21, cl.2	17,700	25,000
		Grade 5, cl.2	17,400	25,000
	SA-203	Grade A	16,300	21,700
		Grade B	17,500	23,300
		Grade D	16,300	21,700
		Grade E	17,500	23,300
High	SA-240	Grade 304	11,200	20,000
Alloy		Grade 304L	—	16,700
Steel		Grade 316	12,300	20,000
Plates		Grade 316L	10,200	16,700

Wall thickness—2:1 ellipsoidal heads

$$t = \frac{Pd}{2SE - 0.2P} \tag{12-2}$$

Wall thickness—hemispherical heads

$$t = \frac{Pr}{2SE - 0.2P} \tag{12-3}$$

Wall thickness—cones

$$t = \frac{Pd}{2\cos\alpha\,(SE - 0.6P)} \tag{12-4}$$

where S = maximum allowable stress value, psi
t = thickness, excluding corrosion allowance, in.
P = maximum allowable working pressure, psig
r = inside radius before corrosion allowance is added, in.
d = inside diameter before corrosion allowance is added, in.
E = joint efficiency, see Table 12-4 (Most vessels are fabricated in accordance with Type of Joint No. 1.)
α = ½ the apex of the cone

Figure 12-1 defines the various types of heads. Most production facility vessels use ellipsoidal heads because they are readily available, normally less expensive, and take up less room than hemispherical heads.

Cone-bottom vertical vessels are sometimes used where solids are anticipated to be a problem. Most cones have either a 90° apex ($\alpha = 45°$) or a 60° apex ($\alpha = 30°$). These are referred to respectively as a "45°" or "60°" cone because of the angle each makes with the horizontal. Equation 12-4 is for the thickness of a conical head that contains pressure. Some operators use internal cones within vertical vessels with standard ellipsoidal heads as shown in Figure 12-2. The ellipsoidal heads contain the pressure, and thus the internal cone can be made of very thin steel.

Table 12-4 lists joint efficiencies that should be used in Equations 12-1 to 12-4. This is Table UW-12 in the ASME Code.

Table 12-5 lists some of the common material types used to construct pressure vessels. Individual operating companies have their own standards, which differ from those listed in this table.

Corrosion Allowance

Typically, a corrosion allowance of 0.125 in. for non-corrosive service and 0.250 in. for corrosive service is added to the wall thickness calculated in Equations 12-1 to 12-4.

INSPECTION PROCEDURES

All ASME Code vessels are inspected by an approved Code inspector. The manufacturer will supply Code papers signed by the inspector. The

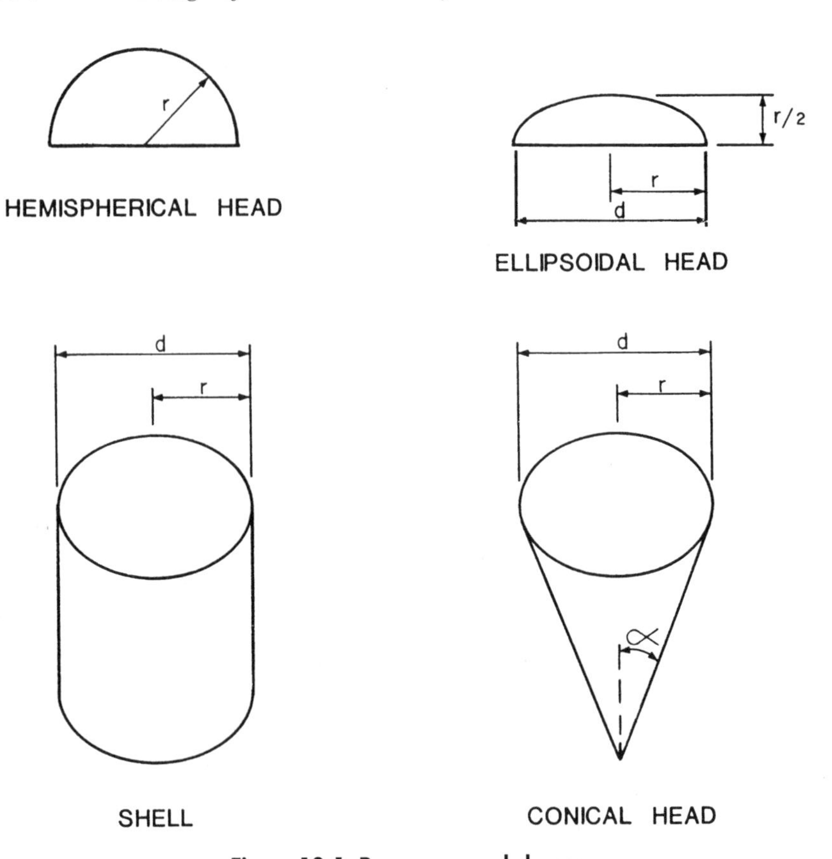

Figure 12-1. Pressure vessel shapes.

nameplate on the vessel will be stamped to signify it has met the requirements of the Code. One of these requirements is that the vessel was tested to 1.5 times MAWP. However, this is only *one* of the requirements. The mere fact that a vessel is tested to 1.5 times MAWP does not signify that it has met all the design and quality assurance safety aspects of the Code.

It must be pointed out that a Code stamp does not necessarily mean that the vessel is fabricated in accordance with critical nozzle dimensions or internal devices as required by the process. The Code inspector is only interested in those aspects that relate to the pressure-handling integrity of the vessel. The owner must do his own inspection to assure that nozzle locations are within tolerance, vessel internals are installed as designed, coatings are applied properly, etc.

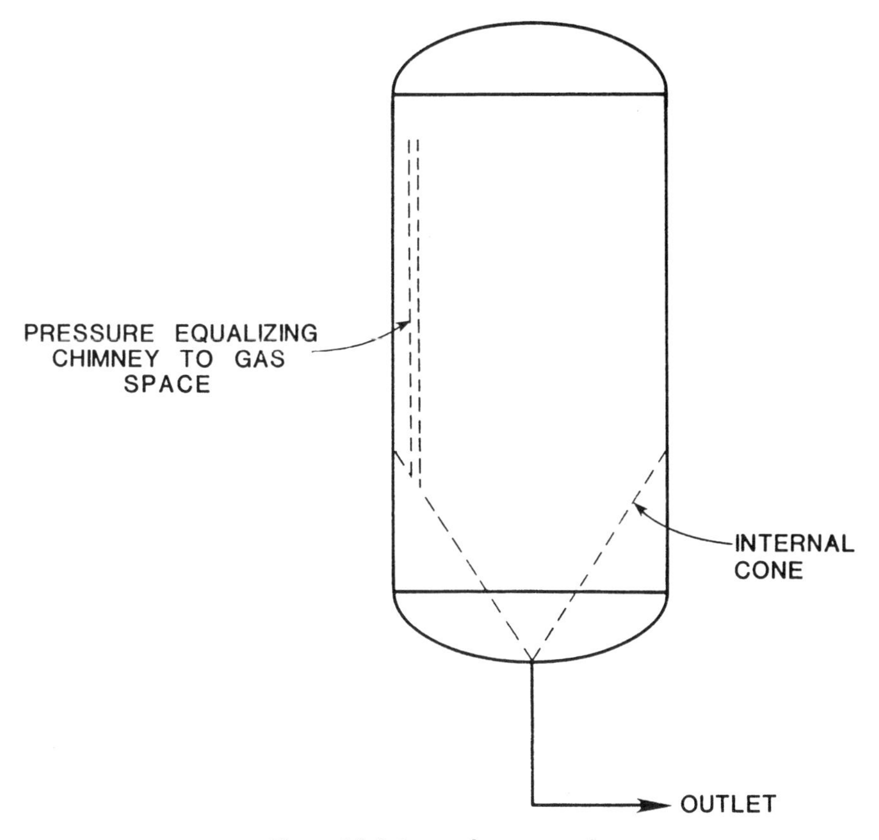

Figure 12-2. Internal cone vessel.

ESTIMATING VESSEL WEIGHTS

It is important to be able to estimate vessel weights, since most cost estimating procedures start with the weight of the vessel. The vessel weight, both empty and full with water, may be necessary to adequately design a foundation or to assure that the vessel can be lifted or erected once it gets to the construction site.

The weight of a vessel is made up of the weight of the shell, the weight of the heads, and the weight of internals, nozzles, pedestals, and skirts. The last two terms are defined in Figure 12-3.

(*text continued on page 339*)

Table 12-4
Maximum Allowable Joint Efficiencies for Arc and Gas Welded Joints

No.	Type of Joint Description	Limitations	(a) Fully Radiographed[1]	(b) Spot Examined	(c) Not Spot Examined[3]
1	Butt joints as attained by double-welding or by other means that will obtain the same quality of deposited weld metal on the inside and outside weld surfaces to agree with the requirements of UW-35. Welds using metal backing strips that remain in the place are excluded.	None	1.00	0.85	0.70
2	Single-welded butt joint with backing strip other than those included under (1).	(a) None except as in (b) below (b) Butt weld with one plate offset—for circumferential joints only, see UW-13(c) and Fig. UW-13.1(k)	0.90	0.80	0.65
3	Single-welded butt joint without use of backing strip	Circumferential joints only, not over ⅝-inch thick and not over 24-in. outside diameter.	—	—	0.60
4	Double full fillet lap joint.	Longitudinal joints not over ⅜-in. thick. Circumferential joints not over ⅝-in. thick.	—	—	0.55

(*table continued on next page*)

5	Single full fillet lap joints with plug welds conforming to UW-17.	(a)	Circumferential joints[4] for attachment of heads not over 24-in. outside diameter to shells not over ½ in. thick.	—	—	0.50
		(b)	Circumferential joints for the attachment to shells of jackets not over ⅝ in. in nominal thickness where the distance from the center of the plug weld to the edge of the plate is not less than 1½ times the diamter of the hole for the plug.			
6	Single full fillet lap joints without plug welds.	(a)	For the attachment of heads convex to pressure to shells not over ⅝-in. required thickness, only with use of fillet weld on inside of shell; or	—	—	0.45
		(b)	For attachment of heads having pressure on either side to shells not over 24-in. inside diameter and not over ¼-in. required thickness with fillet weld on outside of head flange only.			

[1]*See UW-12(a) and UW-51.*
[2]*See UW-12(b) and UW-52.*
[3]*The maximum allowable joint efficiencies shown in this column are the weld joint efficiencies multiplied by 0.80 (and rounded off to the nearest 0.05) to effect the basic reduction in allowable stress required by the Division for welded vessels that are not spot examined. See (UW-12(c)).*
[4]*Joints attaching hemispherical heads to shells are excluded.*

Table 12-5
Materials Typically Specified

	Low Pressure	Common Steel T > –20°F	NACE MR-01-75	Low Temp –50°F < T < –20°F	Low Temp T < –50°F	High CO_2 Service
Plate	SA-36 SA-285-C	SA-516-70	SA-516-70	SA-516-70	SA-240-304	SA-240-316L
Pipe	SA-53-B	SA-106-B	SA-106-B	SA-333-6	SA-312 TP-304	SA-312 TP-316L
Flanges and Fittings	SA-105	SA-105 SA-181-1	SA-105 SA-181-1	SA-350-LF1	SA-182 F-304	SA-182 F-316L
Stud Bolts	SA-193-B7	SA-193-B7	SA-193-B7M	SA-320-L7	SA-193-B8	SA-193-B8M
Nuts	SA-194-2H	SA-194-2H	SA-194-2M	SA-194-4	SA-194-8A	SA-194-8MA

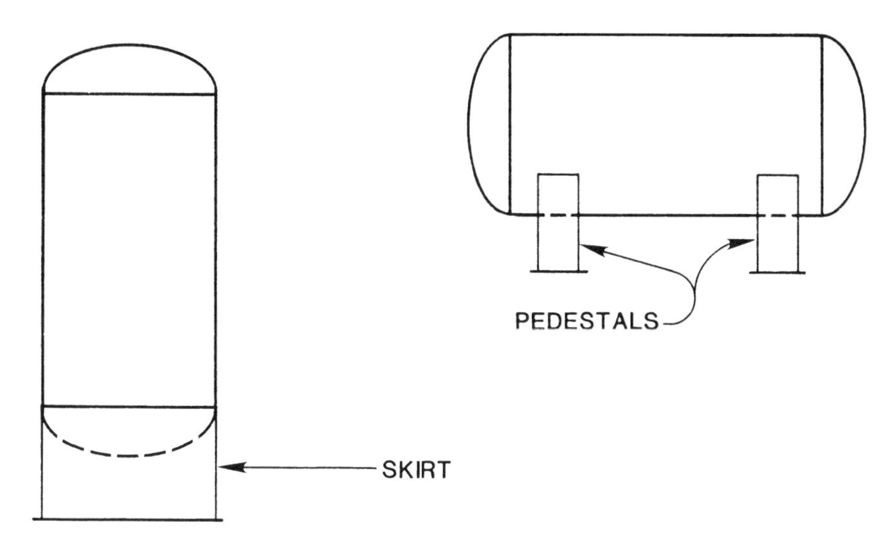

Figure 12-3. Vessel support devices.

(*text continued from page 335*)

The shell weight can be estimated from:

$$W = 11dtL \qquad (12\text{-}5)$$

where W = weight, lb
d = ID, in.
t = wall thickness, in.
L = shell length, ft

The weight of one 2:1 ellipsoidal head is approximately:

$$W \cong 0.34\,td^2 + 1.9\,td \qquad (12\text{-}6)$$

The weight of a cone is:

$$W = \frac{0.23td^2}{\sin\alpha} \qquad (12\text{-}7)$$

α = one half the cone apex angle

The weight of nozzles and internals can be estimated at 5 to 10% of the sum of the shell and head weights. The weight of a skirt can be estimated as the same weight per foot as the shell with a length given by Equation 12-8 for an ellipsoidal head and Equation 12-9 for a conical head.

$$L = \frac{0.25d}{12} + 2 \tag{12-8}$$

$$L = \frac{0.5\,d}{12 \tan \alpha} + 2 \tag{12-9}$$

where L = skirt length, ft

The weight of pedestals for a horizontal vessel can be estimated as 10% of the total weight of the vessel.

SPECIFICATION AND DESIGN OF PRESSURE VESSELS

Pressure Vessel Specifications

Most companies have a detailed general specification for the construction of pressure vessels, which defines the overall quality of fabrication required and addresses specific items such as:

- Code compliance
- Design conditions and materials
- Design details
  - Vessel design and tolerances
  - Vessel connections (nozzle schedules)
  - Vessel internals
  - Ladders, cages, platforms, and stairs
  - Vessel supports and lifting lugs
  - Insulation supports
  - Shop drawings
- Fabrication
  - General
  - Welding
  - Painting
  - Inspection and testing
  - Identification stamping

- Drawings, final reports, and data sheets
- Preparation for shipment

A copy of this specification is normally attached to a bid request form, which includes a pressure vessel specification sheet such as the one shown in Figure 12-4. This sheet contains schematic vessel drawings and pertinent specifications and thus defines the vessel in enough detail so the manufacturer can quote a price and so the operator can be sure that all quotes represent comparable quality. The vessel connections (nozzle schedules) are developed from mechanical flow diagrams. It is not necessary for the bidder to know the location of the nozzles to submit a quote or even to order material.

Shop Drawings

Before the vessel fabrication can proceed, the fabricator will develop complete drawings and have these drawings approved by the representative of the engineering firm and/or the operating company. These drawings are called shop drawings. They will show detailed vessel design and fabrication/welding, nozzle schedules and locations, details of vessel internals, and other accessories. Examples are shown in Figures 12-5 through 12-13. Some typical details are discussed below.

Nozzles

Nozzles should be sized according to pipe sizing criteria, such as those provided in API RP 14E. The outlet nozzle is generally the same size as the inlet nozzle. To prevent baffle destruction due to impingement, the entering fluid velocity is to be limited as:

$$V_{IN} \leq (3{,}500/\rho_f)^{1/2} \qquad (12\text{-}10)$$

where V_{IN} = maximum inlet nozzle fluid velocity, ft/sec
ρ_f = density of the entering fluid, lb/ft^3

If an interior centrifugal (cyclone) separator is used, the inlet nozzle size should be the same size as the pipe. If the internal design requires

(text continued on page 346)

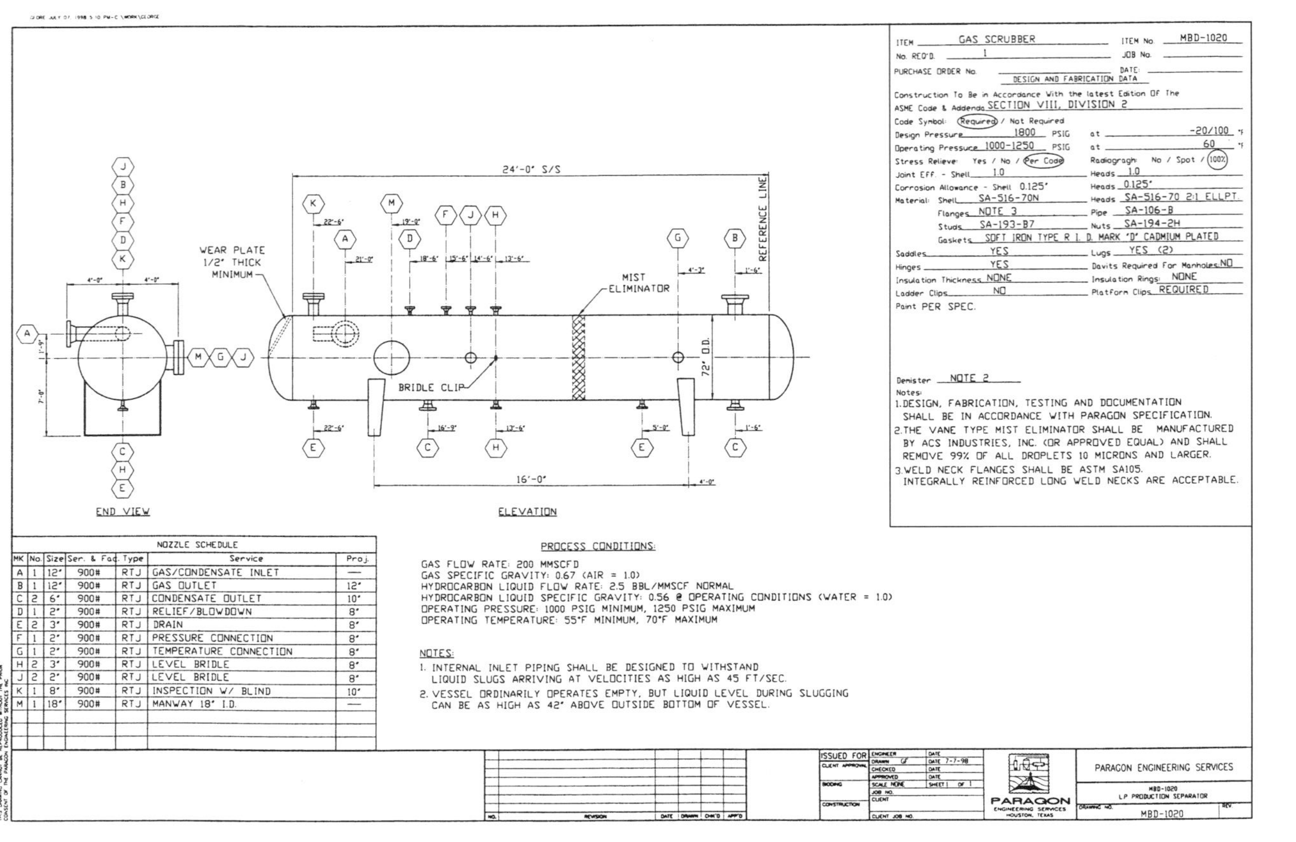

ITEM GAS SCRUBBER ITEM No. MBD-1020
No. REQ'D. 1 JOB No.
PURCHASE ORDER No. DATE:
DESIGN AND FABRICATION DATA
Construction To Be in Accordance With the latest Edition Of The ASME Code & Addenda SECTION VIII, DIVISION 2
Code Symbol: (Required) / Not Required
Design Pressure 1800 PSIG at -20/100 °F
Operating Pressure 1000-1250 PSIG at 60 °F
Stress Relieve: Yes / No / (Per Code) Radiograph: No / Spot / (100%)
Joint Eff. - Shell 1.0 Heads 1.0
Corrosion Allowance - Shell 0.125" Heads 0.125"
Material: Shell SA-516-70N Heads SA-516-70 2:1 ELLPT.
Flanges NOTE 3 Pipe SA-106-B
Studs SA-193-B7 Nuts SA-194-2H
Gaskets SOFT IRON TYPE R I. D. MARK "D" CADMIUM PLATED
Saddles YES Lugs YES (2)
Hinges YES Davits Required For Manholes NO
Insulation Thickness NONE Insulation Rings: NONE
Ladder Clips NO Platform Clips REQUIRED
Paint PER SPEC.

Demister NOTE 2
Notes:
1. DESIGN, FABRICATION, TESTING AND DOCUMENTATION SHALL BE IN ACCORDANCE WITH PARAGON SPECIFICATION.
2. THE VANE TYPE MIST ELIMINATOR SHALL BE MANUFACTURED BY ACS INDUSTRIES, INC. (OR APPROVED EQUAL) AND SHALL REMOVE 99% OF ALL DROPLETS 10 MICRONS AND LARGER.
3. WELD NECK FLANGES SHALL BE ASTM SA105. INTEGRALLY REINFORCED LONG WELD NECKS ARE ACCEPTABLE.

NOZZLE SCHEDULE

MK	No.	Size	Ser. & Fac.	Type	Service	Proj.
A	1	12"	900#	RTJ	GAS/CONDENSATE INLET	—
B	1	12"	900#	RTJ	GAS OUTLET	12"
C	2	6"	900#	RTJ	CONDENSATE OUTLET	10"
D	1	2"	900#	RTJ	RELIEF/BLOWDOWN	8"
E	2	3"	900#	RTJ	DRAIN	8"
F	1	2"	900#	RTJ	PRESSURE CONNECTION	8"
G	1	2"	900#	RTJ	TEMPERATURE CONNECTION	8"
H	2	3"	900#	RTJ	LEVEL BRIDLE	8"
J	2	2"	900#	RTJ	LEVEL BRIDLE	8"
K	1	8"	900#	RTJ	INSPECTION W/ BLIND	10"
M	1	18"	900#	RTJ	MANWAY 18" I.D.	—

PROCESS CONDITIONS:
GAS FLOW RATE: 200 MMSCFD
GAS SPECIFIC GRAVITY: 0.67 (AIR = 1.0)
HYDROCARBON LIQUID FLOW RATE: 2.5 BBL/MMSCF NORMAL
HYDROCARBON LIQUID SPECIFIC GRAVITY: 0.56 @ OPERATING CONDITIONS (WATER = 1.0)
OPERATING PRESSURE: 1000 PSIG MINIMUM, 1250 PSIG MAXIMUM
OPERATING TEMPERATURE: 55°F MINIMUM, 70°F MAXIMUM

NOTES:
1. INTERNAL INLET PIPING SHALL BE DESIGNED TO WITHSTAND LIQUID SLUGS ARRIVING AT VELOCITIES AS HIGH AS 45 FT/SEC.
2. VESSEL ORDINARILY OPERATES EMPTY, BUT LIQUID LEVEL DURING SLUGGING CAN BE AS HIGH AS 42" ABOVE OUTSIDE BOTTOM OF VESSEL.

PARAGON ENGINEERING SERVICES
MBD-1020 LP PRODUCTION SEPARATOR
MBD-1020
PARAGON ENGINEERING SERVICES HOUSTON, TEXAS
DATE 7-7-98 SCALE NONE SHEET 1 OF 1

Figure 12-4. Example of pressure vessel specification sheet.

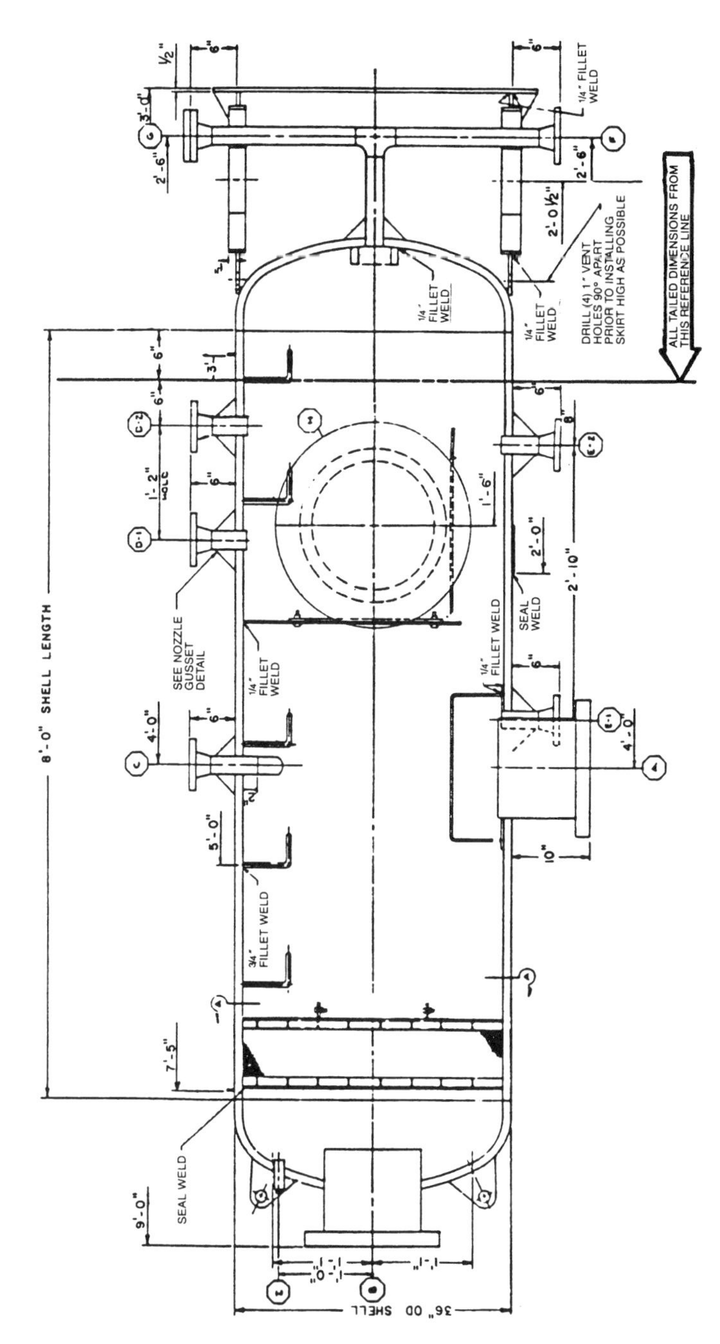

Figure 12-5. Example of pressure vessel shop drawing.

OUTSIDE PROJECTION, INCHES USING WELDING NECK FLANGE						
NOM. PIPE SIZE	PRESSURE RATING OF FLANGE LB					
	150	300	600	900	1500	2500
2	6	6	6	8	8	8
3	6	6	8	8	8	10
4	6	8	8	8	8	12
6	8	8	8	10	10	14
8	8	8	10	10	12	16
10	8	8	10	12	14	20
12	8	8	10	12	16	22
14	8	10	10	14	16	
16	8	10	10	14	16	
18	10	10	12	14	18	
20	10	10	12	14	18	
24	10	10	12	14	20	

(Diagram label: outside projection)

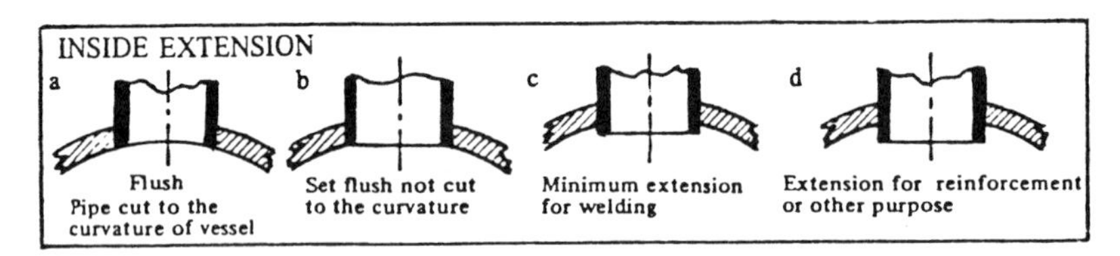

Figure 12-6. Nozzle projections. *(Reprinted with permission from* **Pressure Vessel Handbook,** *Publishing, Inc., Tulsa.)*

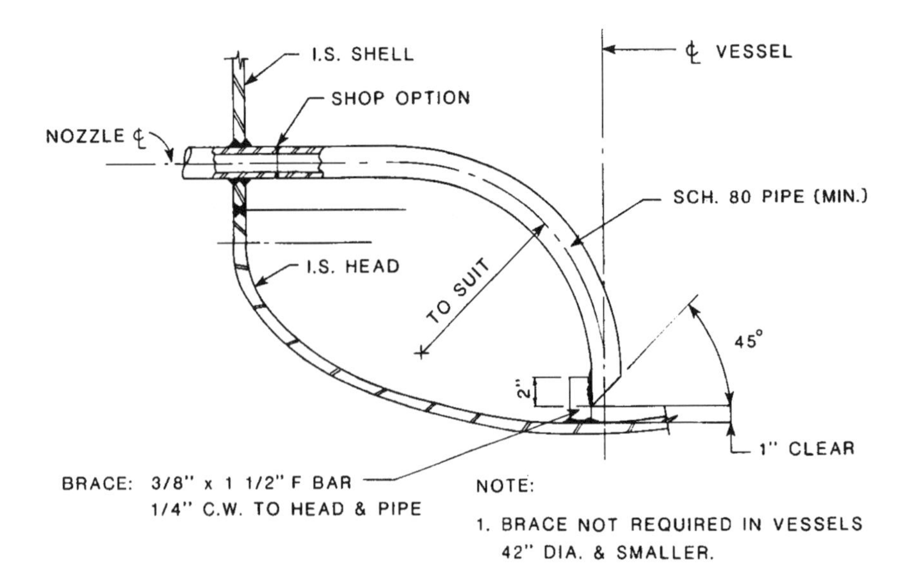

Figure 12-7. Siphon drain.

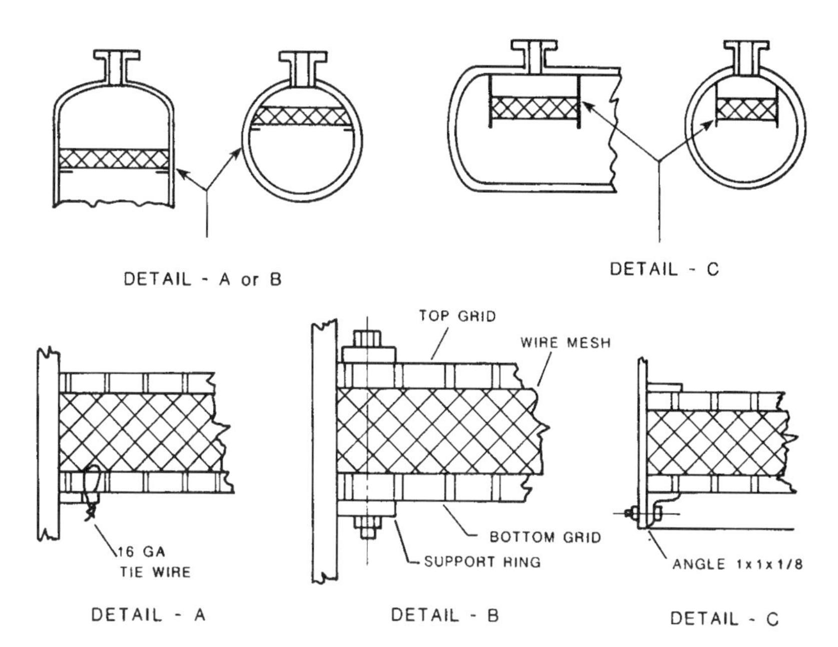

Figure 12-8. Example of supports for mist extractors. ***(Reprinted with permission from* Pressure Vessel Handbook, *Publishing, Inc., Tulsa.)***

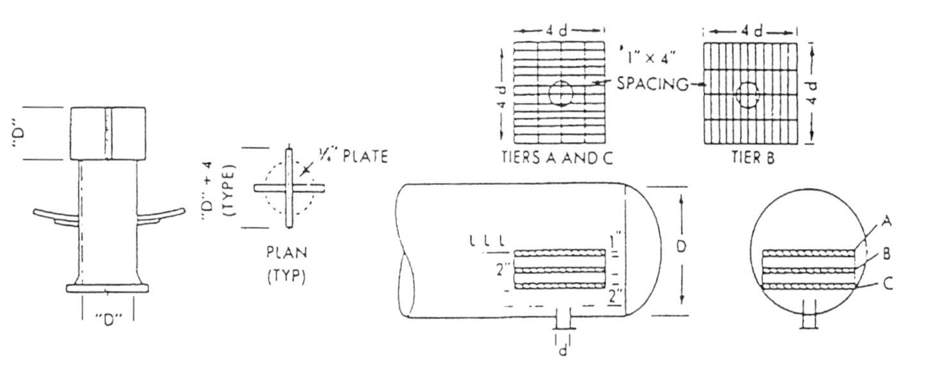

Figure 12-9. Examples of Vortex Breaker Details. ***(Source: Copyright © International Training & Development.)***

(*text continued from page 341*)

the smallest inlet and exit pressure losses possible, the nozzle size should be increased.

Vortex Breaker

As liquid flows out of the exit nozzle, it will swirl and create a vortex. Vortexing would carry the gas out with the liquid. Therefore, all liquid outlet nozzles should be equipped with a vortex breaker. Figure 12-9 shows several vortex breaker designs. Additional designs can be found in the *Pressure Vessel Handbook.* Most designs depend on baffles around or above the outlet to prevent swirling.

Manways

Manways are large openings that allow personnel access to the vessel internals for their maintenance and/or replacement. Vessels 36 in. and larger should have a minimum of one 18-in. manway. Vessels 30 in. and smaller should have two 4-in. flanged inspection openings. Manway cover davit should be provided for 12-in. and larger manways for safe and easy opening and closing of the cover. Figure 12-10 shows an example of a horizontal manway cover davit and sleeve details.

Vessel Supports

Small vertical vessels may be supported by angle support legs, as shown in Figure 12-11. Larger vertical vessels are generally supported by a skirt support, as shown in Figure 12-12. At least two (2) vent holes, 180° apart, should be provided at the uppermost location in the skirt to prevent the accumulation of gas, which may create explosive conditions. Horizontal vessels are generally supported by a pair of saddle type supports.

Ladder and Platform

Ladder and platform should be provided if operators are required to climb up to the top of the vessel regularly. An example is shown in Figure 12-13.

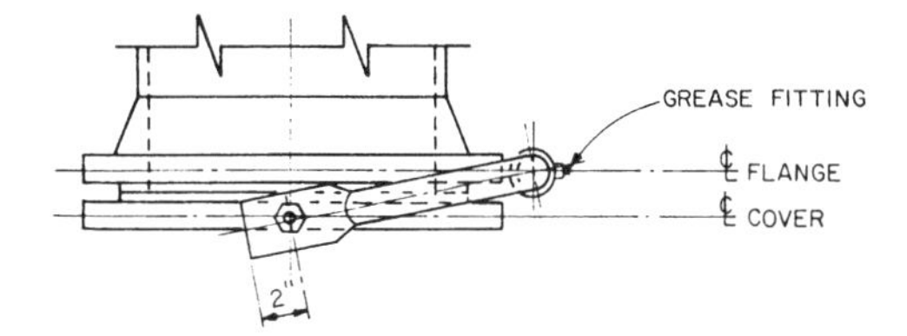

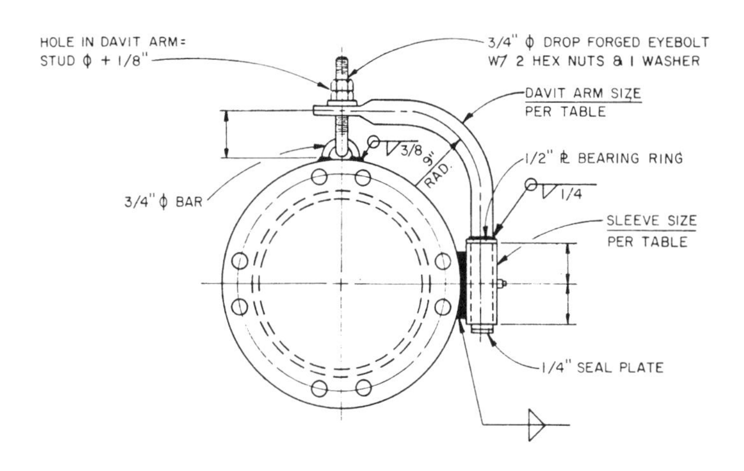

DAVIT SIZE	1 1/2 S/80	2 S/80	2 1/2 S/80	3 S/80
SLEEVE SIZE	2 S/80	2 1/2 S/40	3 S/40	3 1/2 S/40
MANWAY COVER SIZE & RATING	16 150#	24 150#	24 300#	20 600#
	18 150#	18 300#	16 600#	24 600#
	20 150#	20 300#	18 600#	16 900#
	16 300#			18 900#
				20 900#
	ON ANY COVER NOT EXCEEDING:			
	325# IN WEIGHT	525# IN WEIGHT	850# IN WEIGHT	1200# IN WEIGHT

Figure Figure 12-10. Example of horizontal manway cover davit and sleeve detail.

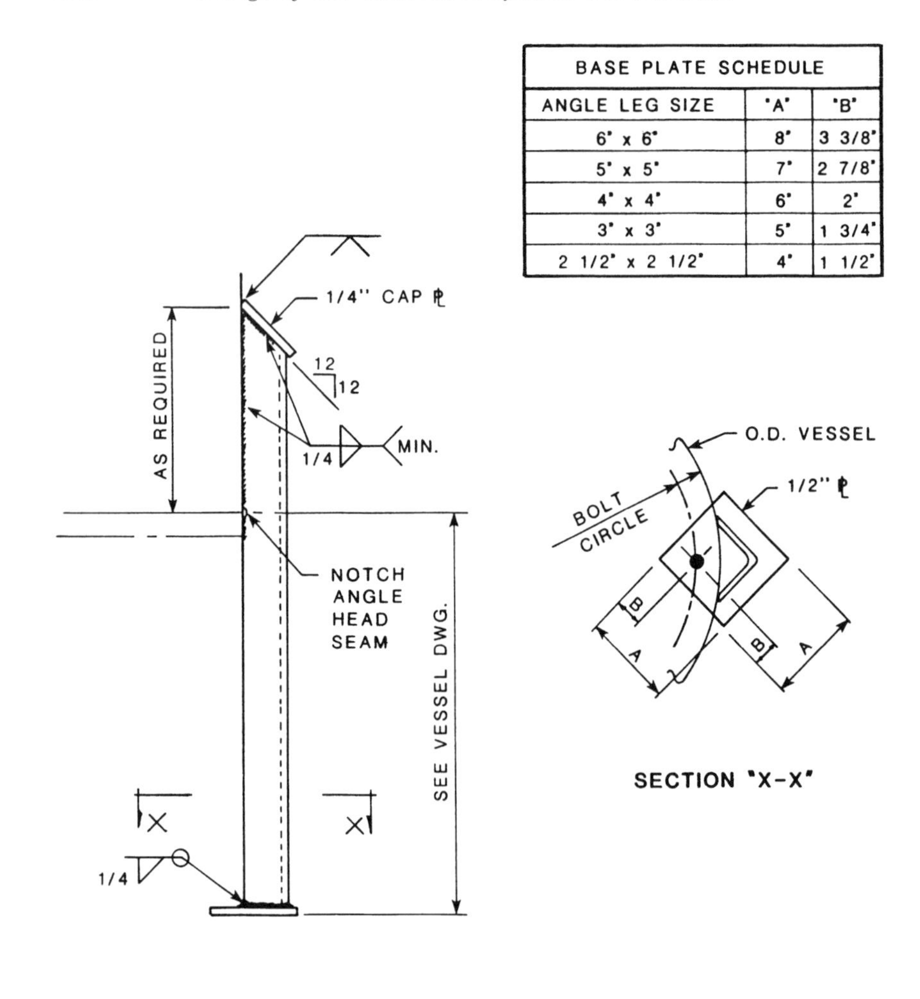

BASE PLATE SCHEDULE		
ANGLE LEG SIZE	"A"	"B"
6" x 6"	8"	3 3/8"
5" x 5"	7"	2 7/8"
4" x 4"	6"	2"
3" x 3"	5"	1 3/4"
2 1/2" x 2 1/2"	4"	1 1/2"

Figure 12-11. Angle support legs.

Pressure Relief Devices

All pressure vessels should be equipped with one or more pressure safety valves (PSVs) to prevent overpressure. This is a requirement of both the ASME Code and API RP 14C (refer to Chapter 14). The PSV should be located upstream of the mist extractor. If the PSV is located downstream of the mist extractor, an overpressure situation could occur when the mist extractor becomes plugged isolating the PSV from the high pressure, or the mist extractor could be damaged when the relief

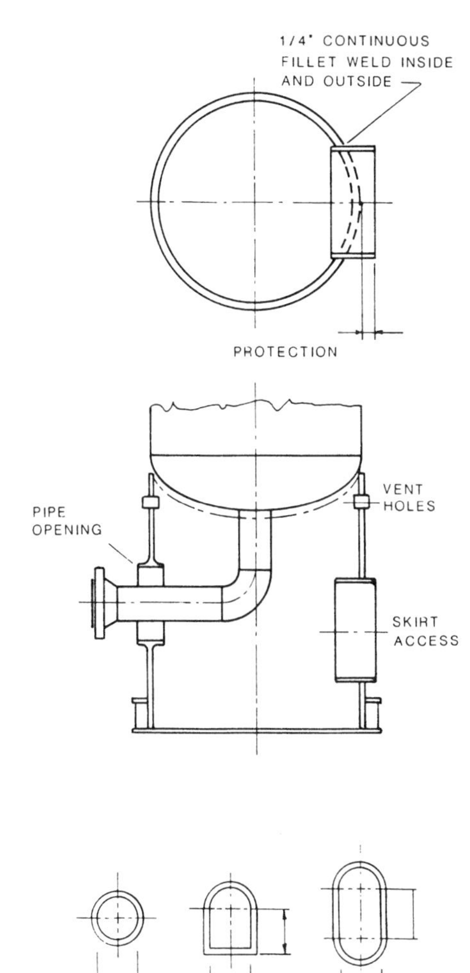

VENT HOLES

In service of hydrocarbons or other combustible liquids or gases the skirts shall be provided with minimum of two 2 inch vent holes located as high as possible 180 degrees apart. The vent holes shall clear head insulation. For sleeve may be used coupling or pipe.

ACCESS OPENINGS

The shape of access openings may be circular or any other shapes. Circular access openings are used most frequently with pipe or bent plate sleeves. The projection of sleeve equals to the thickness of fireproofing or minimum 2 inches. The projection of sleeves shall be increased when necessary for reinforcing the skirt under certain loading conditions.

Diameter (D) = 16–24 inches

PIPE OPENINGS

The shape of pipe openings are circular with a diameter of 1 inch larger than the diameter of flange. Sleeves should be provided as for access openings.

Figure 12-12. Skirt openings. (*Reprinted with permission from* Pressure Vessel Handbook, *Publishing, Inc., Tulsa.)*

valve opens. Rupture discs are sometimes used as a backup relief device for the PSV. The disc is designed to break when the internal pressure exceeds the set point. Unlike the PSV, which is self-closing, the rupture disc must be replaced if it has been activated.

Corrosion Protection

Pressure vessels handling salt water and fluids containing signficiant amounts of H_2S and CO_2 require corrosion protection. Common corro-

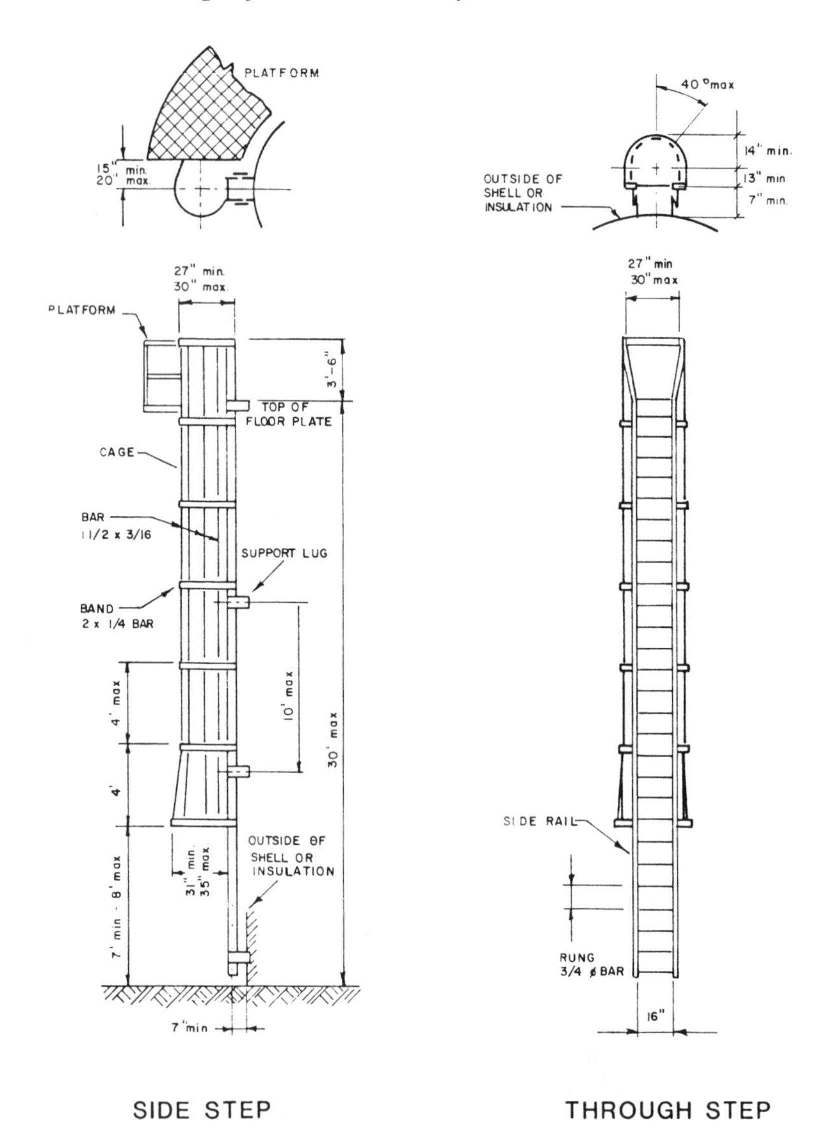

Figure 12-13. Ladders. ***(Reprinted with permission from*** **Pressure Vessel Handbook,** ***Publishing, Inc., Tulsa.)***

sion protection methods include internal coatings with synthetic polymeric materials and galvanic (sacrificial) anodes. All pressure vessels that handle corrosive fluids should be monitored periodically. Ultrasonic surveys can locate discontinuities in the metal structure, which will indicate corrosion damages.

EXAMPLE PROBLEM 12-1

Determine the weight for the following free-water knockout. It is butt weld fabricated with spot x-ray and to be built to Division 1. A conical head (bottom of the vessel) is desired for ease in sand removal. Compare this weight to that of a vessel without the conical section and that to a vessel with a ¼-in. plate internal cone.

Design pressure	= 125 psig
Maximum operating temperature	= 200°F
Corrosion allowance	= ¼ in.
Material	= SA516 Grade 70
Diameter	= 10 ft
Seam-to-seam length above the cone	= 12 ft
Cone apex angle	= 60°

Solution:

Case I—Cone bottom

(a) Shell:

$$t = \frac{Pr}{(SE - 0.6P)}$$

$S = 17{,}500$ psi (Table 12-3)
$E = 0.85$ (Table 12-4)

$$t = \frac{(125)(60)}{(17{,}500)(0.85) - (0.6)(125)} = 0.507 \text{ in.}$$

$$\text{Required thickness} = 0.507 + 0.250 = 0.757 \text{ in.}$$

$$\text{Use } \frac{13}{16}\text{-in. plate } (0.8125)$$

$$W = 11\,d\,t\,L$$

$$= (11)(12)(0.8125)(12) = 12{,}870 \text{ lb}$$

(b) Head:

$$t = \frac{(125)(120)}{(2)(17{,}500)(0.85) - (0.2)(125)} = 0.505 \text{ in.}$$

$$\text{Required thickness} = 0.505 + 0.250 = 0.755 \text{ in.}$$

$$\text{Use } \frac{13}{16}\text{-in. head } (0.8125)$$

$$W = (0.34)(0.8125)(120)^2 + (1.9)(0.8125)(120)$$
$$= 4{,}163 \text{ lb}$$

(c) Cone:

$$t = \frac{Pd}{2\cos\alpha\,(SE - 0.6P)}$$

$$t = \frac{(125)(120)}{(2\cos 30)(17{,}500 \times 0.85 - 0.6 \times 125)} = 0.585 \text{ in.}$$

$$\text{Required thickness} = 0.585 + 0.250 = 0.835 \text{ in.}$$

$$\text{Use } \frac{7}{8}\text{-in. plate } (0.875)$$

$$W = \frac{(0.23)(0.875)(120)^2}{\sin 30} = 5{,}796 \text{ lb}$$

(d) Skirt:

$$\text{Height} = \frac{5}{\tan 30} = 8.66 \text{ ft}$$

Allow 2 ft for access

Height = 11 ft. Assume it is ½-in. plate.

$$W = (11)(120)(0.5)(11) = 7{,}260$$

(e) Summary:

Shell	12,870
Head	4,163
Cone	5,796
Skirt	7,260
	30,089
Misc	5,000
	35,089 lb

Case II—Ellipsoidal head

(a) Skirt:

$$L = \frac{0.25\,d}{12} + 2$$

$$= \frac{(0.25)(120)}{12} + 2$$

$$= 4.50\text{ ft}$$

$$W = (11)(120)(0.5)(4.5) = 2{,}970\text{ lb}$$

(b) Summary:

Shell	12,870
Head	4,163
Head	4,163
Skirt	2,970
	24,166
Misc.	5,000
	29,166 lb

Case III—Internal cone

(a) Internal Cone:

$$W = \frac{(0.23)(0.25)(120)^2}{\sin 30} = 1{,}656\text{ lb}$$

(b) Shell:

$$\text{Height of cone} = \frac{(10/2)}{\tan 30} = 8.7\text{ ft}$$

$$\text{Length of shell} = 12 + 8.7 = 20.7\text{ ft}$$

$$\text{Weight of shell} = (11)(120)(0.8125)(20.7) = 22{,}200\text{ lb}$$

(c) Summary:

Shell	22,200
Head	4,163
Head	4,163
Skirt	2,970
Cone	1,656
	35,152
Misc.	5,000
	40,152 lb

CHAPTER

13

*Pressure Relief**

The most important safety devices in a production facility are the pressure relief valves, which ensure that pipes, valves, fittings, and pressure vessels can never be subjected to pressures higher than their design pressures. Relief valves must be designed to open rapidly and fully, and be adequately sized to handle the total flow of gas and liquids that could potentially cause an overpressure situation. They relieve the pressure by routing this stream to a safe location where it can be vented to atmosphere or burned.

As long as pressure, level, and temperature control devices are operating correctly, the safety system is not needed. If the control system malfunctions, then pressure, level, and temperature safety switches sense the problem so the inflow can be shut off. If the control system fails and the safety switches don't work, then relief valves are needed to protect against overpressure. Relief valves are essential because safety switches *do* fail or can be bypassed for operational reasons. Also, even when safety switches operate correctly, shutdown valves take time to operate, and there may be pressure stored in upstream vessels that can overpressure downstream equipment while the system is shutting down. Relief valves are an essential element in the facility safety system.

---

*Reviewed for the 1999 edition by Mary E. Thro of Paragon Engineering Services, Inc.

RELIEF REQUIREMENTS

The ASME code requires every pressure vessel that can be blocked in to have a relief valve to alleviate pressure build up due to thermal expansion of trapped gases or liquids. In addition, the American Petroleum Institute Recommended Practice (API RP) 14C, "Analysis, Design, Installation and Testing of Basic Surface Safety Systems on Offshore Production Platforms," recommends that relief valves be installed at various locations in the production system; and API RP 520, "Design and Installation of Pressure Relieving Systems in Refineries," recommends various conditions for sizing relief valves.

In production facility design, the most common relieving conditions are (1) blocked discharge, (2) gas blowby, (3) regulator failure, (4) fire, (5) thermal, and (6) heat exchanger tube rupture. Relief valve design flow rates are commonly determined as follows.

1. *Blocked Discharge.* It is assumed that *all* outlets from the vessel are shut in and the total inlet flow stream (gas and liquids) must flow out through the relief valve. This condition could occur, for example, if the equipment has been shut in and isolated and the operator opens the inlet before opening the outlet valves.
2. *Gas Blowby.* A gas blowby condition is the most critical and sometimes overlooked condition in production facility design. It assumes that there is a failure of an upstream control valve feeding the pressure vessel and that the relief valve must handle the *maximum* gas flow rate into the system during this upset condition. For example, if the liquid control valve on a high pressure separator were to fail open, all the liquid would dump to the downstream lower-pressure vessel. Then the gas from the high pressure separator would start to flow to the downstream vessel. The lower pressure vessel's relief valve must be sized to handle the total gas flow rate that will fit through the liquid dump valve in a full open position. We normally assume (conservatively) that the upstream vessel pressure is the PSV set point (or less conservatively at its operating pressure) and the downstream vessel is at its PSV set point. The resulting gas flow rate may be larger than the design gas flow rate to the high-pressure separator inlet. The rate can be reduced by a choke or other restrictions in the line. In that case, the rate would be the maximum rate that would fit through the choke at the appropriate vessel pressures. Most accidents involving overpressuring of low pressure separators

are a result of relief valves not being adequately sized to handle the gas blowby condition.

Note: Liquid dump valves are normally fail closed to prevent gas blowby. That means that in case of loss of instrument gas or air pressure, the spring will drive the valve to the closed position. However, the valve can mechanically fail because the level controller malfunctions or the seat cuts out due to solids erosion, which would cause it to fail in an open configuration.

3. *Regulator Failure.* It is assumed that a pressure control valve or regulator fails in the full open position. Regulator failure could occur where a regulator is used to feed gas from a high pressure line to a fuel gas scrubber. Normally, the regulator only opens enough to keep the pressure in the scrubber constant. If the valve fails open, the users can't take the excess gas, so the pressure in the scrubber goes up until the relief valve opens. The feed to the regulator is assumed to be (conservatively) at the upstream relief valve set point, and the downstream vessel will pressurize to its relief valve set point.
4. *Fire.* The relief valve must be sized to handle the gases evolving from liquids if the equipment is exposed to an external fire. A procedure for calculating this is presented in API RP 520. This condition may be critical for large, low-pressure vessels and tanks but does not normally govern for other pressure vessels.
5. *Thermal.* Thermal relief is needed in a vessel or piping run that is liquid-packed and can be isolated, for example pig launchers and meter provers. Liquid is subject to thermal expansion if it is heated. It is also incompressible. The thermal expansion due to heating by the sun from a nighttime temperature of 80°F to a sun-heated temperature of 120°F can be enough to rupture piping or a vessel. The required capacity of thermal relief valves is very small.
6. *Tube Rupture.* It is common for a heat exhanger to have a high-pressure fluid in the tubes and a lower-pressure rated shell. If there is a break in one of the tubes, the higher pressure fluid will leak to the shell, resulting in overpressure. It is conservative to assume a tube is completely split with choked flow from both sides of the break.

A vessel may be subject to more than one condition under different failure scenarios. For example, a low pressure separator may be subject to blocked discharge, gas blowby from the high pressure separator, and fire. Only one of these failures is assumed to happen at any time. The relief valve size needs to be calculated for each pertinent relieving rate

Table 13-1
Maximum Rate Relieving Scenarios

Vessel	Relieving Scenarios
Production Separators	Blocked Discharge
Test Separators	Blocked Discharge
Low Pressure Separators	Blocked Discharge or Blowby
Glycol Contact Tower	Fire
Oil Treater	Gas Blowby or Fire
Utility or Fuel Gas Scrubber	Regulator Failure
Heat Exchanger	Tube Rupture
Compressor Scrubber	Fire
Compressor Discharge	Blocked Discharge

and the largest size used. The usual controlling cases for common vessels and piping are shown in Table 13-1.

A vessel can only be overpressured if the upstream vessel has a higher pressure than the vessel in question. A compressor scrubber with a MAWP of 285 that gets flow from a 285 MAWP separator does not need to have a relief valve sized for blocked discharge. The upstream relief valve will keep the upstream separator pressure from going higher than 285, so there is no way it can overpressure the downstream scrubber. The scrubber PSV only needs to be sized for fire.

The same concept applies to a glycol contact tower, except for one small trick. Contact towers often handle glycol at temperatures above 100°F, so the MAWP is reduced for the higher temperature (e.g. 1,440 psig at 130°F rather than 1,480 psig at 100°F). In order to keep from sizing the contact tower relief valve for the full wellstream gas flow, the separator relief valve is usually set at the glycol tower MAWP (1,440 psig) even though the separator MAWP may actually be higher (1,480 psig).

Good engineering judgment should be used to determine the relief rate when the separator MAWP is higher than the well SITP. Unexpected things can happen with a well. Production reservoirs at different pressures within the well bore can communicate in unexpected ways (for example, as the result of a poor cement job). Where flow is coming from a well, it is a good idea to provide an extra margin of safety. If the vessel MAWP is not significantly higher than the well SITP, the relief valve should be sized for blocked discharge of the full production rate.

Figure 13-1 shows the various relationships between MAWP and the relief valve set pressure. The primary relief valve should be set to open at no more than 100% of MAWP and to relieve the worst case flow rates,

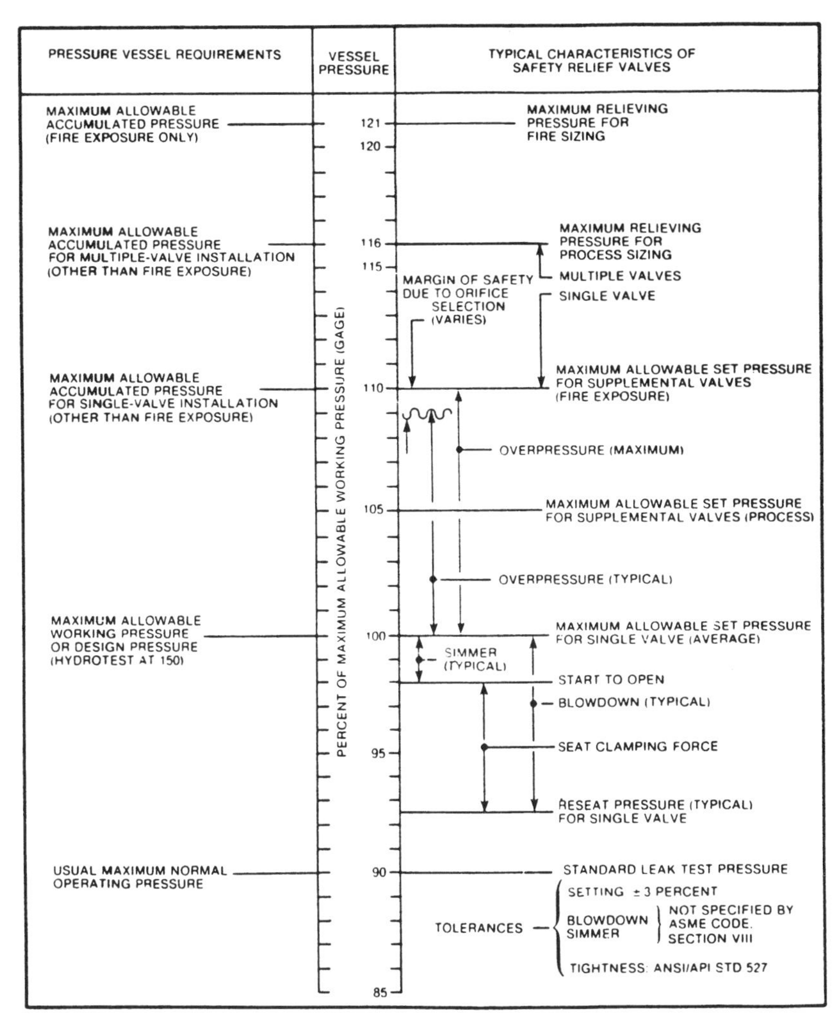

NOTES:
1. The operating pressure may be any lower pressure required.
2. The set pressure and all other values related to it may be moved downward if the operating pressure permits.
3. This figure conforms with the requirements of the *ASME Boiler and Pressure Vessel Code*, Section VIII, "Pressure Vessels," Division 1.
4. The pressure conditions shown are for safety relief valves installed on a pressure vessel (vapor phase).

Figure 13-1. Various relationships between MAWPs and relief valve set pressure. (*Reprinted with permission from API RP 521.*)

not counting fire (i.e., blocked discharge or gas blowby), at a pressure of 1.10 MAWP. If two relief valves are used to handle the worst case flow rates, the first must be set no higher than 100% MAWP and the second at 1.05 MAWP. They must relieve the worst case flow rates, not counting fire, at 1.16 MAWP. The maximum pressure for relieving fire relief rates is 1.21 MAWP. Thus, under relief conditions, the pressure in the vessel may actually exceed MAWP. This buildup of pressure in the vessel above the MAWP as the relief valve opens is called "overpressure." This is taken into account by the various safety factors in the ASME Code and is one of the reasons the vessel is originally tested to 1.5 MAWP.

The relief valve must be installed so that gases are routed to a safe location. In small facilities and remote locations this is accomplished with a simple "tail pipe," which points the discharge vertically upward and creates a jet in excess of 500 feet per second. The jet action dilutes the discharge gases to below the lower flammable limit in approximately 120 pipe diameters. Liquids may fall back on the equipment.

In large facilities and offshore platforms where the escaping gases and liquids could present a source of pollution or ignition, it is common to route the relief valve discharges into a common "header" that discharges at a remote safe location. Often a vent scrubber is installed in this header to separate the bulk of the liquids and to minimize the possibility of liquid discharges to atmosphere.

TYPE OF DEVICES

Valves that activate automatically to relieve pressure are called "safety valves," "relief valves," or "safety relief valves." Safety valves are spring loaded and characterized by a rapid full opening or "pop" action. They are used primarily for steam or air service. Sometimes they are referred to as "pop valves." Relief valves are spring loaded and open more slowly. They reach full opening at 25% over set pressure and are used primarily for liquid services. Safety relief valves can be either spring loaded or pilot operated and are designed to provide full opening with little overpressure. Most automatically-actuating relief devices used in production facilities are actually safety relief valves; however, they are commonly referred to as relief valves or safety valves. In this book the term "relief valve" is used in the generic sense of any automatically-actuating pressure relieving device.

There are three types of relief valves: conventional, balanced-bellows, and spring loaded.

Conventional Relief Valves

Figure 13-2 shows a cross section of a conventional relief valve and Figure 13-3 is a schematic that shows the valve's operation. Convention-

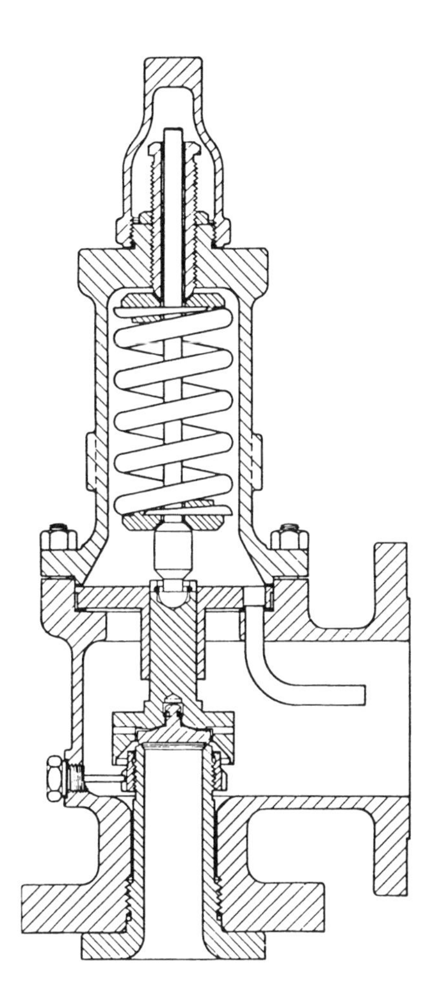

Figure 13-2. Conventional safety-relief valve. ***(Courtesy of API.)***

al relief valves can be used anywhere that back-pressure in the relief header is low. They are common onshore where relief valves are fitted with individual tail pipes. On offshore platforms, they are used mainly as small threaded valves for fire and thermal relief and for liquid relief around pumps. In a conventional relief valve, a spring holds a disk closed against the vessel pressure. A bonnet covers the spring and is vented to the valve outlet. The outlet pressure P_2 acts on both sides of the disk, balancing the pressure across the disk except for the portion of the disk open to the vessel pressure P_1. The net opening force is equal to P_1 times the area over which P_1 acts. The closing force is the spring force F_S plus P_2

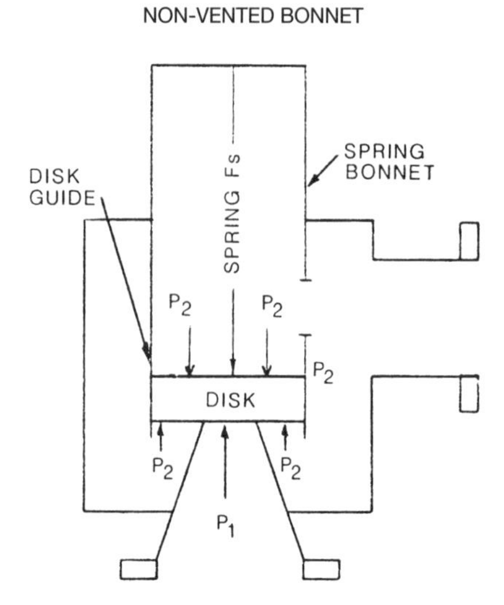

Figure 13-3. Operation of conventional safety-relief valve. ***(Reprinted with permission from API RP 520.)***

times the same area where P_1 acts. When the open area times the difference in pressures, P_1 minus P_2, equals the spring force, the valve begins to open. Increasing the pressure on the back of the disk, P_2 or the back-pressure, will hold the valve closed. "Back-pressure" is the pressure that builds up in the relief piping and at the outlet of the relief valve. It consists of constant back-pressure in the system, back-pressure due to other relief valves relieving, and self-imposed back-pressure due to the valve itself relieving. If P_2 increases because the valve is installed in a header system with other valves, then the amount of pressure in the vessel (the set point) required to overcome the spring force increases.

Conventional relief valves should only be used where the discharge is routed independently to atmosphere, or if installed in a header system, the back-pressure build-up when the device is relieving must be kept below 10% of the set pressure so the set point is not significantly affected. The set point increases directly with back-pressure.

Conventional relief valves may be equipped with lifting levers or screwed caps. The lifting lever permits mechanical operations of the valve for testing or clean-out of foreign material from under the seat. Screwed caps prevent leakage outside of the valve, but also prevent overriding the spring if foreign material or ice become lodged under the disc.

Balanced-Bellows Relief Valves

Balanced relief valves are spring-loaded valves that contain a bellows arrangement to keep back-pressure from affecting the set point. Figure 13-4 shows a cross section of a balanced relief valve, and Figure 13-5 is a schematic that shows how the valve operates. The bonnet is vented to atmosphere and a bellows is installed so that the back-pressure acts both downward and upward on the same area of the disc. Thus, the forces created by the back-pressure always cancel and do not affect the set point.

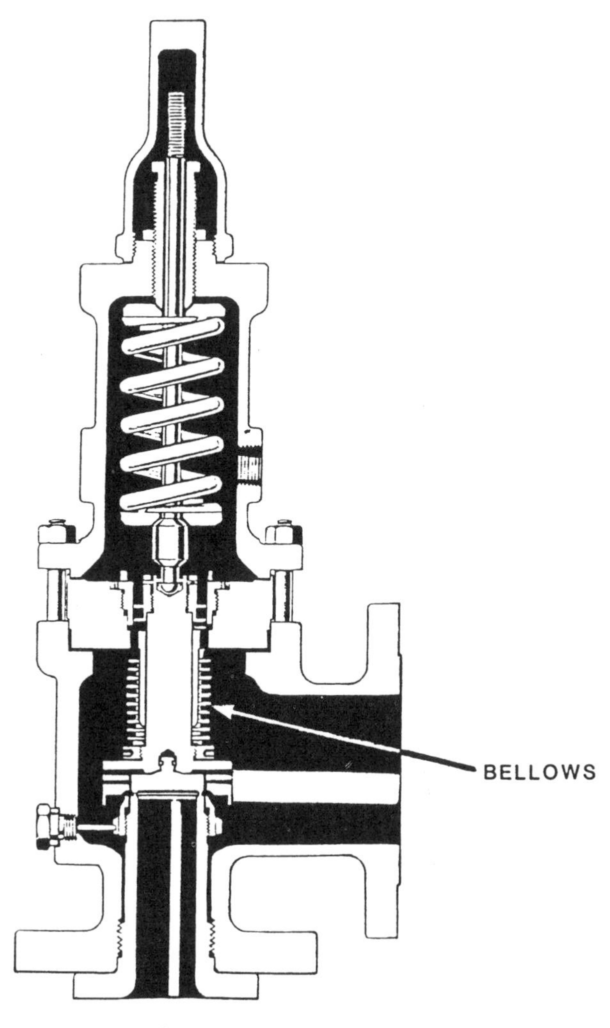

Figure 13-4. Balanced-bellows relief valve. (*Courtesy of API.*)

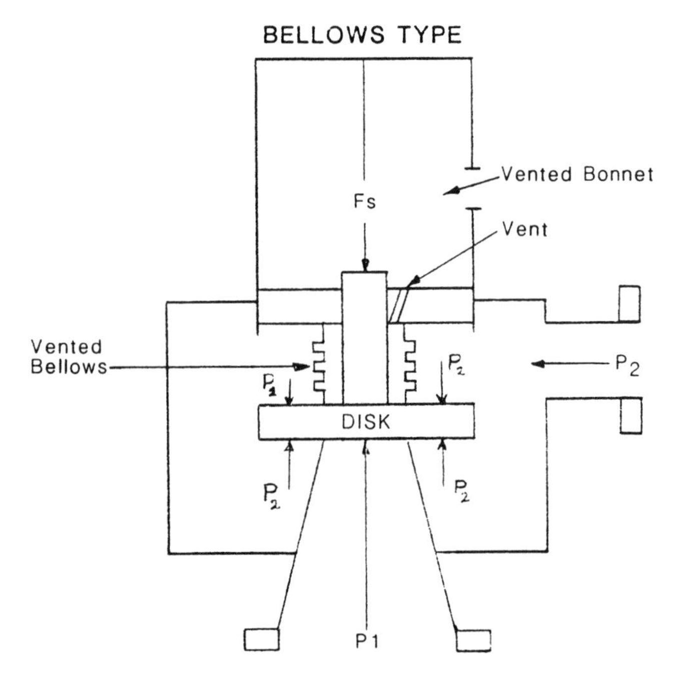

Figure 13-5. Operation of balanced-bellows safety-relief valve. ***(Reprinted with permission from API 520.)***

Balanced bellows type valves are normally used where the relief valves are piped to a closed flare system and the back-pressure exceeds 10% of the set pressure, where conventional valves can't be used because back-pressure is too high. They are also used in flow lines, multiphase lines, or for paraffinic or asphaltic crude, where pilot-operated valves can't be used due to possible plugging of the pilot line. An advantage of this type of relief valve is, for corrosive or dirty service, the bellows protects the spring from process fluid. A disadvantage is that the bellows can fatigue, which will allow process fluid to escape through the bonnet. For H_2S service, the bonnet vent must be piped to a safe area.

Pilot-Operated Relief Valves

Pilot-operated relief valves use the pressure in the vessel rather than a spring to seal the valve and a pilot to activate the mechanism. Figure 13-6 is a schematic of a typical pilot-operated valve. A piece of tubing communicates pressure between the relief valve inlet and pilot. When this pressure is below the set pressure of the pilot, the pilot valve is in the position

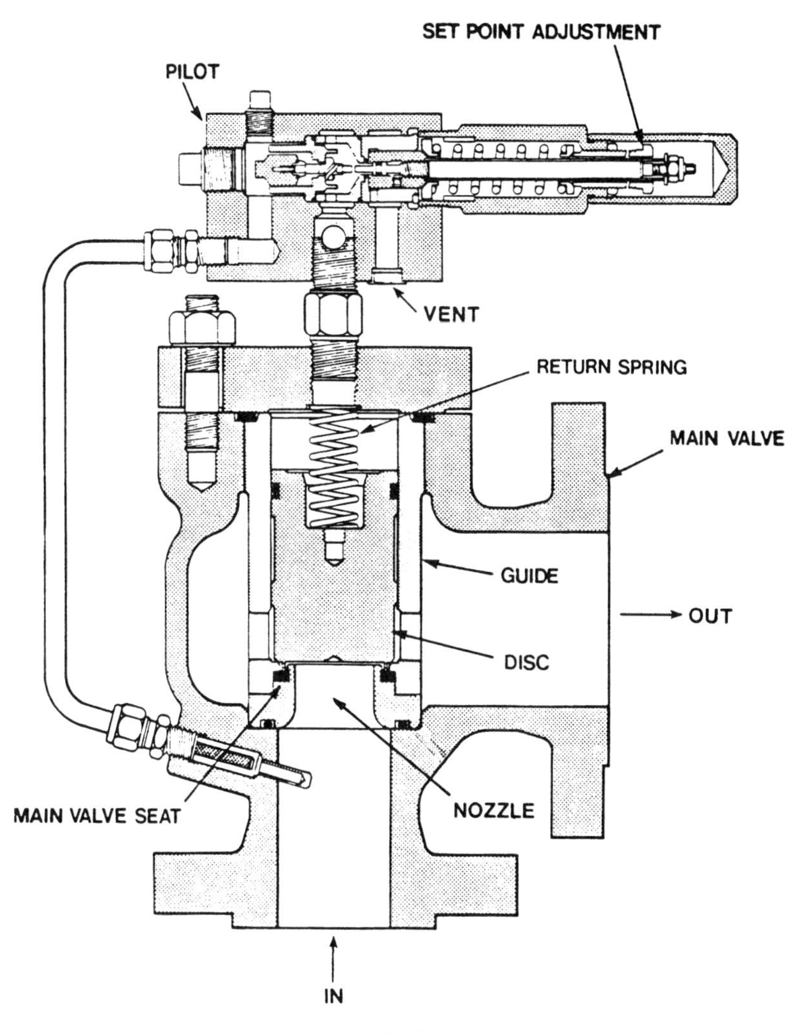

Figure 13-6. Typical pilot-operated valve.

shown, and there is pressure communication between the inlet pressure and the top of the disc. Since the disc has approximately 25% greater area on the top than in the throat of the nozzle, there is a net closing force on the disc equal to the difference in magnitude of the areas times the vessel pressure. The closer the vessel pressure gets to the set point, the greater the closing force, and thus "simmer," which can occur in spring loaded valves near the set point, is eliminated. When the set point is reached, the pilot shifts to the right, blocking the pressure from the vessel, venting pressure from above the disc, and allowing the disc to rise.

Pilot-operated valves have the advantage of allowing operations near the set point with no leakage, and the set position is not affected by back-pressure. However, they will not function if the pilot fails. If the sensing line fills with hydrates or solids, the valve will open at 25% over the pressure trapped above the disc (usually the normal operating pressure of the vessel). For this reason they should be used with care in dirty gas service and liquid service. They are used extensively offshore where all the platform relief valves are tied into a single header because up to 50% back-pressure will not affect the valve capacity.

A disadvantage of pilot operated valves is that, if there is no pressure in the vessel, back-pressure could cause the disc to lift. This could occur if a vessel was shut-in and depressured for maintenance, the relief valve was installed in a header, and another valve in the header was opened, building back-pressure. Figure 13-7 shows an arrangement of two check valves in the sensing system to assure that the higher of the vessel pressure or the header pressure is always present above the disc. This is called "backflow protection." A backflow preventer should be specified if the vessel could be subject to vacuum, such as a compressor suction scrubber, or where the back-pressure in a relief header can exceed the relief valve set pressure when other valves are relieving.

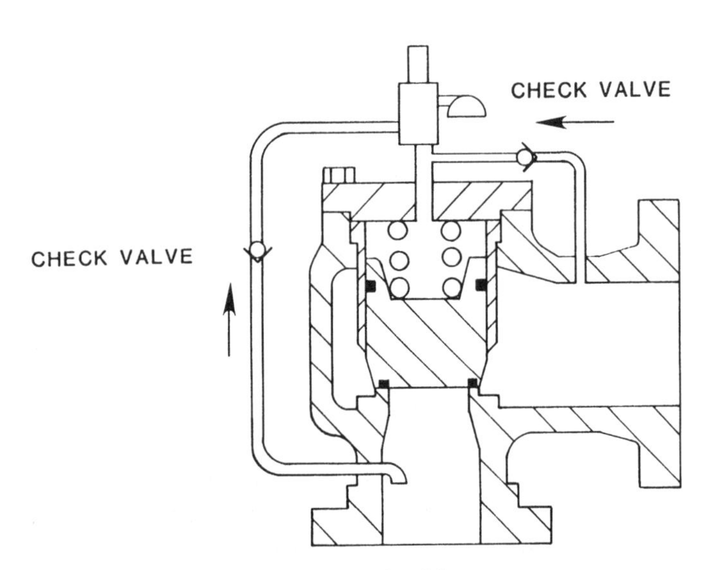

Figure 13-7. Check valve backflow preventer.

Rupture Discs

These are thin diaphragms held between flanges and calibrated to burst at a specified static inlet pressure. Unlike relief valves, rupture discs cannot reseal when the pressure declines. Once the disc ruptures, any flow into the vessel will exit through the disc, and the disc must be replaced before the pressure vessel can be placed back in service. Rupture discs are manufactured in a variety of materials and with various coatings for corrosion resistance.

The most common disc materials are aluminum, monel, inconel, and stainless steel, but other materials or coatings, such as carbon, gold and plastic, are available.

Rupture discs may be used alone, but they are normally used as a backup to a relief valve set to relieve at approximately 115% MAWP. This ensures that the disc ruptures only if the relief valve fails or in the unlikely event that the pressure rises above 110% MAWP and the relief valve does not have enough capacity.

Rupture discs are also used below relief valves to protect them from corrosion due to vessel fluids. The rupture disc bursts first and the relief valve immediately opens. The relief valve reseals, limiting flow when the pressure declines. When this configuration is used, it is necessary to monitor the pressure in the space between the rupture disk and the relief valve, either with a pressure indicator or a high pressure switch. Otherwise, if a pinhole leak develops in the rupture disk, the pressure would equalize on both sides, and the rupture disk would not rupture at its set pressure because it works on differential pressure.

VALVE SIZING

Most relief valve manufacturers have software available for sizing relief valves. These programs are relatively easy to use. Understanding how relief valves are sized and inputting the correct information into the program are essential for calculating correct sizes.

Critical Flow

The flow of a compressible fluid through an orifice is limited by critical flow. Critical flow is also referred to as choked flow, sonic flow, or Mach 1. It can occur at a restriction in a line such as a relief valve orifice or a choke, where piping goes from a small branch into a larger header, where pipe size increases, or at the vent tip. The maximum flow occurs at sonic velocity, which exists as long as the pressure drop through the orifice is greater than the critical pressure drop given by:

$$\frac{P_{cf}}{P_1} = \left(\frac{2}{k+1}\right)^{\frac{k}{k-1}} \tag{13-1}$$

where k = specific heat ratio, C_p/C_v
P_{cf} = critical flow outlet pressure, psia
P_1 = inlet relieving pressure, psia

For gases with specific heat ratios of approximately 1.4, the critical pressure ratio is approximately 0.5. For hydrocarbon service, this means that if the back-pressure on the relief valve is greater than 50% of the set pressure, then the capacity of the valve will be reduced. In other words, if the pressure in the relief piping at the valve outlet is greater than half the set pressure, then a larger relief valve will be required to handle the same amount of fluid.

As long as the pressure ratio exceeds the critical-pressure ratio, the throughput will vary with the inlet pressure and be independent of outlet pressure. For example, a relief valve set at 100 psi will have the same gas flow through it as long as the back-pressure is less than approximately 50 psi.

Effects of Back-Pressure

Back-pressure can affect either the set pressure or the capacity of a relief valve. The *set pressure* is the pressure at which the relief valve begins to open. *Capacity* is the maximum flow rate that the relief valve will relieve. The set pressure for a conventional relief valve increases directly with back-pressure. Conventional valves can be compensated for *constant* back-pressure by lowering the set pressure. For self-imposed back-pressure—back-pressure due to the valve itself relieving—there is no way to compensate. In production facility design, the back-pressure is usually not constant. It is due to the relief valve or other relief valves relieving into the header. Conventional relief valves should be limited to 10% back-pressure due to the effect of back-pressure on the set point.

The set points for pilot-operated and balanced-bellows relief valves are unaffected by back-pressure, so they are able to tolerate higher back-pressure than conventional valves. For pilot-operated and balanced-bellows relief valves, the capacity is reduced as the back-pressure goes above a certain limit.

For balanced-bellows relief valves, above about 35% back-pressure, the back-pressure affects the stiffness of the bellows and decreases the relief valve's capacity. Relief valves can be designed for higher back-pressure by increasing the size so that when the capacity is reduced the resulting size is adequate. The manufacturer's suggested correction for back-pressure should be used when available. API RP 520 offers a generic back-pressure correction factor for balanced-bellows relief valves shown in Figure 13-8. The back-pressure correction factor is calculated using gage pressure. Balanced-bellows valves may be limited by the manufacturer to a back-pressure lower than 35% due to the design strength of the bellows.

All relief valves are affected by reaching critical flow, which corresponds to a back-pressure of about 50% of the set pressure. Pilot-operated relief valves can handle up to 50% back-pressure without any significant effect on valve capacity. Back-pressure correction factors can be obtained from the relief valve manufacturers for back-pressures above 50%. API RP 520 gives a generic method for sizing a pilot-operated relief valve for sub-critical flow.

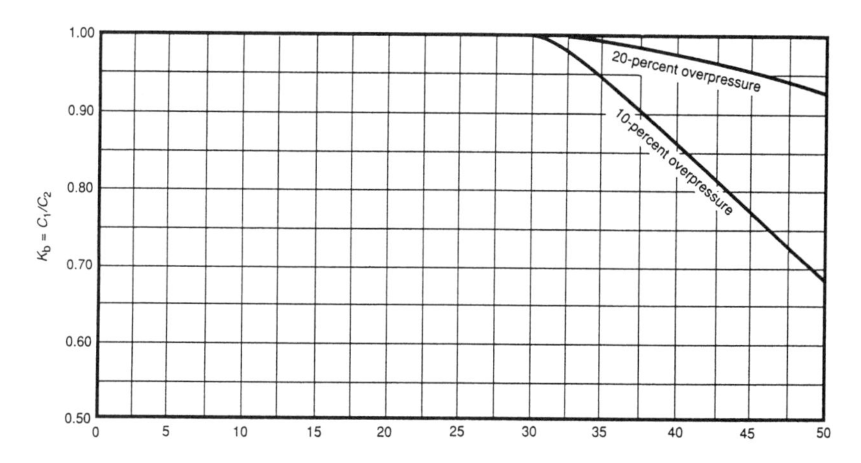

Percent of gauge back-pressure = $P_B/P_S \times 100$
C_1= capacity with back-pressure.
C_2= rated capacity with zero back-pressure.
P_B= back-pressure, in pounds per square inch gauge.
P_S= set pressure, in pounds per square inch gauge.

Note: The curves above represent a compromise of the values recommended by a number of relief valve manufacturers and may be used when the make of the valve or the actual critical flow pressure point for the vapor or gas is unknown. When the make is known, the manufacturer should be consulted for the correction factor. These curves are for set pressures of 50 pounds per square inch gauge and above. They are limited to back-pressure below critical flow pressure for a given set pressure. For subcritical flow back-pressures below 50 pounds per square inch gauge, the manufacturer must be consulted for values of K_b.

Figure 13-8. Back-pressure sizing factor, K_b, for balanced-bellows pressure relief valves—vapors and gases.

In summary, the back-pressure for relief valves should be limited to the following values unless the valve is compensated. We do not recommend using a relief valve with higher back-pressure than shown below without consulting a person knowledgeable in relief valve sizing and relief system design.

Type of relief valve	Maximum back-pressure As % of set pressure	Limiting factor	Units
Conventional	10%	Set pressure	psia
Balanced-bellows	35%	Capacity	psig
Pilot-operated	50%	Capacity	psia

The relief piping design pressure is an additional limit to back-pressure. Relief piping is usually designed as ANSI 150 piping with a MAWP of 285 psig. Relief valves with ANSI 600 inlets usually have outlet flanges rated ANSI 150. A pilot-operated relief valve set at 1,480 psig could have a back-pressure of 740 psig without affecting the valve's capacity, but that would overpressure the relief piping so the allowable back-pressure is limited to 285 psig. For this reason, ANSI 900 and above relief valves often have ANSI 300 outlet flanges to allow for higher back-pressure in the relief piping.

Flow Rate for Gas

The flow rate for gas through a given orifice area or the area required for a given flow rate is obtained by:

$$Q_M = \frac{6.32\, aK_d\, P_1\, C\, K_b}{(Z\, T\, W)^{1/2}} \tag{13-2}$$

$$a = \frac{Q_M\, (Z\, T\, M\, W)^{1/2}}{6.32\, C\, K_d\, P_1\, K_b} \tag{13-3}$$

where Q_M = maximum flow, scfm
a = actual orifice area, in.^{2}
K_d = valve coefficient of discharge (from valve manufacturer)
Farris and Consolidated spring-operated,
$K = .975$
AGCO type 23 and 33 pilot-operated,
$K = 0.92$

P_1 = flowing pressure, psia (set pressure + overpressure + 14.7); Overpressure is normally 10% of set pressure.
MW = molecular weight of gas
Z = compressibility factor
C = gas constant based on ratio of specific heats, C_p/C_v, or k (See Table 13-2)
T = flowing temperature, °R
K_b = back-pressure correction factor

Table 13-2
Gas Constant, C, as Function of Ratio of Specific Heat, C_p/C_v, or k

k	C	k	C	k	C	k	C
1.01	317[a]	1.31	348	1.61	373	1.91	395
1.02	318	1.32	349	1.62	374	1.92	395
1.03	319	1.33	350	1.63	375	1.93	396
1.04	320	1.34	351	1.64	376	1.94	397
1.05	321	1.35	352	1.65	376	1.95	397
1.06	322	1.36	353	1.66	377	1.96	398
1.07	323	1.37	353	1.67	378	1.97	398
1.08	325	1.38	354	1.68	379	1.98	399
1.09	326	1.39	355	1.69	379	1.99	400
1.10	327	1.40	356	1.70	380	2.00	400
1.11	328	1.41	357	1.71	381	—	—
1.12	329	1.42	358	1.72	382	—	—
1.13	330	1.43	359	1.73	382	—	—
1.14	331	1.44	360	1.74	383	—	—
1.15	332	1.45	360	1.75	384	—	—
1.16	333	1.46	361	1.76	384	—	—
1.17	334	1.47	362	1.77	385	—	—
1.18	335	1.48	363	1.78	386	—	—
1.19	336	1.49	364	1.79	386	—	—
1.20	337	1.50	365	1.80	387	—	—
1.21	338	1.51	365	1.81	388	—	—
1.22	339	1.52	366	1.82	389	—	—
1.23	340	1.53	367	1.83	389	—	—
1.24	341	1.54	368	1.84	390	—	—
1.25	342	1.55	369	1.85	391	—	—
1.26	343	1.56	369	1.86	391	—	—
1.27	344	1.57	370	1.87	392	—	—
1.28	345	1.58	371	1.88	393	—	—
1.29	346	1.59	372	1.89	393	—	—
1.30	347	1.60	373	1.90	394	—	—

[a]*Interpolated value, since* c *becomes indeterminate as* k *approaches 1.00.*

Reprinted with permission from API 520.

Flow Rate for Liquids

The corresponding equations for liquid flow are the following:

Conventional Valve, Balanced-Bellows Valve, or Pilot-Operated Valve

The corresponding equations for liquid flow are the following:

$$L = \frac{38\, a\, K_d\, K_v\, K_w}{\left(\frac{SG}{\Delta P}\right)^{1/2}} \tag{13-4}$$

$$a = \frac{L\left(\frac{SG}{\Delta P}\right)^{1/2}}{38\, K_d\, K_v\, K_w} \tag{13-5}$$

where L = maximum liquid flow, gpm
a = actual orifice area, in.^{2}
SG = liquid specific gravity (water = 1)
ΔP = set pressure plus allowable overpressure minus back pressure, psig
K_d = valve coefficient of discharge (from manufacturer or use 0.65)
K_v = viscosity correction factor (See Figure 13-9)

where K_w = correction factor for back pressure (from manufacturer or curve, Figure 13-10) $K_w = 1$ for conventional and pilot-operated valves

To calculate Reynolds number for Figure 13-9

$$Re = \frac{2{,}800\, L\, (SG)}{\mu (a)^{1/2}} \tag{13-6}$$

$$Re = \frac{12{,}700\, L}{U(a)^{1/2}} \tag{13-7}$$

where Re = Reynolds number
μ = liquid viscosity, cp
U = liquid viscosity, Saybolt Universal Seconds

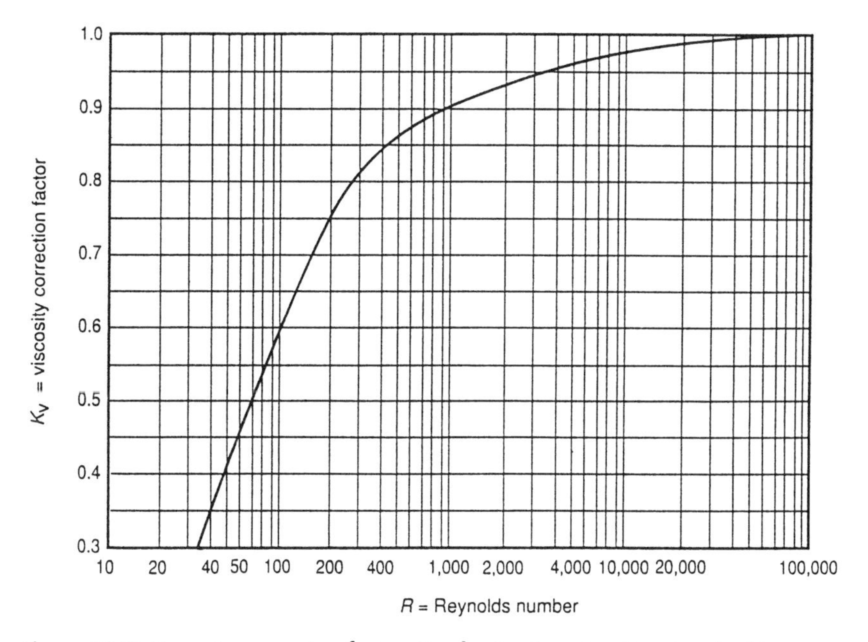

Figure 13-9. Capacity correction factor, K_v, due to viscosity. ***(Reprinted with permission from API 520.)***

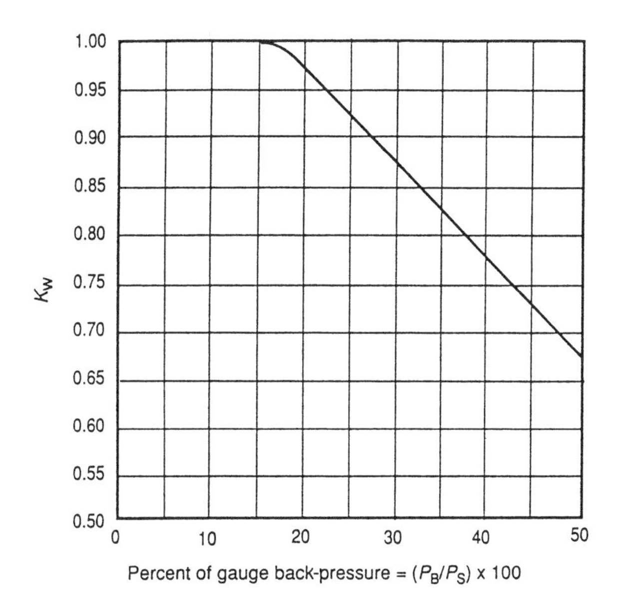

K_W = correction factor due to back-pressure.
P_B = back-pressure, in pounds per square inch gauge.
P_S = set pressure, in pounds per square inch gauge.

Note: The curve above represents values recommended by various manufacturers. This curve may be used when the manufacturer is not known. Otherwise, the manufacturer should be consulted for the applicable correction factor.

Figure 13-10. Back-pressure correction factors for 25% overpressure on balanced-bellows pressure relief valves in liquid service. ***(Reprinted with permission from API 520.)***

Note that a preliminary orifice size must be determined in order to calculate Reynolds number. If the viscosity correction is significant, it may be necessary to iterate to get a final size.

Two-Phase Flow

There are no precise formulas for calculating orifice area for two-phase flow. The common convention is to calculate the area required for the gas flow as if there were no liquid present and the area required for the liquid flow as if there were no gas present. The two areas are then added to approximate the area required for two-phase flow.

Standard Sizes

Relief valves are most often sold using the standard orifice sizes shown in Table 13-3. Once a required area is calculated from the applicable formulas, the next larger standard orifice size is specified.

INSTALLATION

Each relief valve should be equipped with inlet piping no smaller than the valve inlet flange size, and inlet piping should be as short as practi-

Table 13-3
Standard Orifice Areas and Designations

Orifice	Area in.^{2}
D	0.110
E	0.196
F	0.307
G	0.503
H	0.785
J	1.287
K	1.838
L	2.853
M	3.60
N	4.34
P	6.38
Q	11.05
R	16.0
T	26.0

cal. Inlet piping should be designed so that the pressure drop from the source to the relief valve inlet flange will not exceed 3% of the valve set pressure. Pressure drop larger than 3% may cause the relief valve to "chatter" (or rapidly open and close), which can damage the relief valve. The 3% pressure drop should include the losses due to the inlet from the vessel to the piping.

Relief valves vented to the atmosphere should have "tail pipes" equal to or larger in diameter than the relief valve outlet that extend vertically a minimum of one foot above building eaves, or eight feet above adjacent platforms on operating areas. The tail pipes should be provided with a drain located such that the exhaust through the drain hole does not impinge on vessels, piping, other equipment or personnel.

Discharges from relief valves are sometimes tied together in a common header so that the relief fluids can be routed to a safe location for venting or flaring. Piping should be installed in such a manner that liquid in the piping will drain into the header. Relief valves should be located above the header to prevent liquid from collecting at the outlet of the relief valve. In the event of a release, this liquid can be propelled at high speed, causing damage to the relief piping. If it is not possible to locate relief valves and branches above the main header, then branches should enter the header from the top to prevent liquid from draining out of the header into the branch. Branches below the main header, PSV outlets below the header, and unavoidable low spots in the piping should be equipped with a drain valve piped to a safe location.

Relief valves should be tested on a periodic basis even if not required by regulations. Pilot operating valves can be tested by sending a test signal to the pilot through a test connection in the pilot sensing line. Spring loaded relief valves must (1) be removed from service for testing, (2) be tested by subjecting the equipment being protected to set pressure, or (3) have an upstream block valve, which can isolate the relief valve from the equipment being protected, and a test connection between the block and relief valves installed. There is no industry consensus on which of these three test methods provides the highest level of safety. Therefore, some relief valves are installed with upstream block valves and some without.

If relief valves discharge to a common header, it is sometimes convenient to install downstream block valves so that the relief valve can be removed for repairs without shutdown of all equipment tied into the common header. Where either upstream or downstream block valves are used they should be full bore gate or ball valves with a device that enables them to be locked open and sealed. These are often referred to as "car-seal-open" valves. A lock out/tag out procedure should be in place to ensure that the block valves are not inadvertantly left closed.

Various arrangements employing three-way valves and multiple relief valves are sometimes used to provide the benefits of being able to isolate the relief valve for testing and maintenance without the disadvantage of decreasing safety through inadvertent closing of a block valve. Three-way valve arrangements are much more costly than block valves.

In so far as possible, relief valves should be located so as to be accessible from platforms. Relief valve connections to equipment and all relief piping should be designed to withstand the high impact forces that occur when the valve opens. Discharge piping supports should be arranged to minimize bending moments at the connection to the equipment being protected.

Vent Scrubber

A vent scrubber is a two-phase separator designed to remove the liquid from the relieving fluids before the gas is flared or vented. The liquid is returned to the process. Design of vent scrubbers is covered in *Volume 1, Surface Production Operations,* in the chapter about two-phase separators. A vent scrubber is sized as a standard two-phase separator with a liquid droplet size in the gas of 300–600 microns. The operating pressure depends on the pressure drop in the vent piping between the scrubber and the outlet to atmosphere. A reasonable range for operating pressure on a vent scrubber is 25 to 75 psig. For design pressure, if air migrates into the scrubber, forming an explosive mixture in the scrubber, and the mixture ignites due to lightning or static electricity, then the vessel will withstand the resulting detonation if the design pressure is 150 psig or higher. The liquid retention time can be short, but the scrubber needs to have liquid capacity to retain all the liquid produced while the shutdown valve is closing. The MMS has a requirement of 45 sec maximum for platform shutdown valves to close, so 1 min is a reasonable retention time.

Vent or Flare Tip

A pipe that releases gas to disperse into the atmosphere is called a vent. If the gas is burned at the tip, it is called a flare. In its simplest form, a vent or flare tip is a pipe. Sometimes the pipe diameter is reduced for the last 5 ft or so to increase exit velocity for better mixing with the air. The operating pressure of the vent scrubber can be adjusted by reducing the tip diameter to increase pressure drop across the tip. Fluidic seals also give increased pressure drop and can be used to reduce the infusion of air into the vent piping. There are a variety of vendor-supplied flare tips available with various back-pressures, velocities, and capacities.

Relief Header Design

The relief header is a system of piping connecting the outlets of all the relief valves into a common pipe or header that goes to the relief scrubber and then out the vent as shown in Figure 13-11. There are some general rules of thumb useful for sizing relief piping. The goal is to make the piping big enough so it doesn't restrict the flow of the relieving fluids and impose excessively high back-pressure on the relief valves flowing into it. However, designing very large diameter piping costs money and uses up valuable space on the platform.

The pressure in the relief piping is usually equal to atmospheric pressure as long as no relief valve is relieving. There is a common miscon-

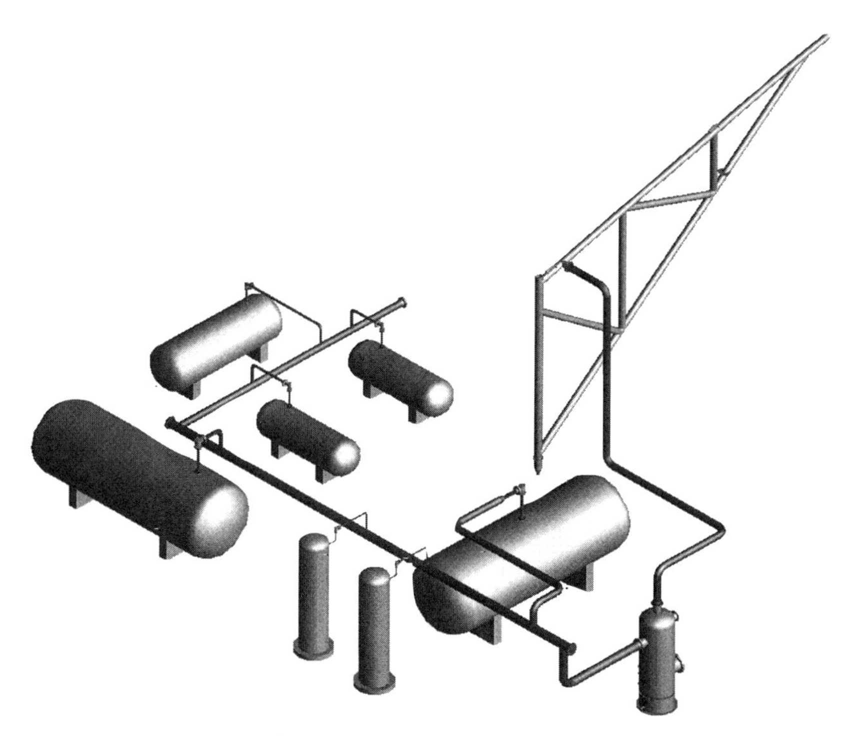

Figure 13-11. Schematic of a relief system showing pressure vessels, relief valves, relief header, vent scrubber, and vent boom. Process piping has been omitted for clarity.

ception that, since the relief piping is open to the atmosphere, the pressure remains atmospheric. However, when a relief valve goes off, the pressure in the relief piping increases due to the friction of the gas moving through the piping. There is pressure drop in relief piping, just like any other piping. The pressure at the vent tip is atmospheric, so the pressure at the relief valve outlet must be higher for the fluid to flow out. Generally, relief piping is designed for much higher velocity than process piping because the service is intermittent; therefore erosional velocity can be ignored, and higher noise levels are acceptable. The result of the higher velocity is much higher-pressure drops than in process piping.

The relief header is designed using a reasonable combination of simultaneous relieving cases. It is not reasonable to assume that all relief valves are relieving at design rates at the same time. It is not necessary to assume multiple failures happening at the same time, unless a single event can cause several valves to relieve at the same time. An example of this would be that a sales pipeline shutdown could cause the compressor to blow down plus the production separator relieving.

The back-pressure at the outlet of a relief valve depends on the relieving rate and the length and size of the outlet piping. It can't be determined from the vessel pressure and PSV orifice size because the flow is choked and is independent of the outlet pressure as long as this pressure is less than 50% of the set pressure. The two known quantities for relief piping back-pressure calculations are the flow rate, which is controlled by the PSV size, and the pressure at the outlet of the vent tip—atmospheric pressure. Unlike most other pressure calculations, relief systems require starting at the outlet and working backward to determine what pressure is required to push the gas out the tip. Once steady state is reached, no gas is accumulating in the header. All the gas that exits the relief valve must exit the vent tip. In order for the relief valve to operate at design flow rates, it should be in critical flow, so back-pressure must be less than 50% of the set pressure. We normally assume the relief valve is in critical flow and size the outlet piping to ensure the back-pressure is low enough. If the back-pressure is too high, then relief valves have reduced relieving capacity. If the back-pressure between the vent scrubber and the vent outlet is too low, then the low pressure, high volume gas will require an excessively large vent scrubber.

The relief piping should be segregated into pressure relief and atmospheric relief systems. If there is a wide range of set pressures on relief valves, the pressure relief piping is sometimes divided into high pressure and low pressure systems.

When the relieving scenarios are defined, assume line sizes, and calculate pressure drop from the vent tip back to each relief valve to assure that the back-pressure is less than or equal to allowable for each scenario. The velocities in the relief piping should be limited to 500 ft/sec, on the high pressure system and 200 ft/sec on the low pressure system. Avoid sonic flow in thc relief header because small calculation errors can lead to large pressure drop errors. Velocity at the vent or flare outlet should be between 500 ft/sec and MACH 1 to ensure good dispersion. Sonic velocity is acceptable at the vent tip and may be chosen to impose back-pressure on the vent scrubber.

As gas goes from the relief valve outlet to the main header and out the tip, the pressure in the piping decreases. As the pressure decreases, the gas expands, or its volume increases. As the gas volume increases, the velocity must increase to retain the same mass flow. Higher velocities result in higher pressure drops. In order to prevent large back-pressures and avoid sonic velocities in the vent header, the following rules of thumb are useful for sizing relief piping. Generally, the relief piping size should go up one to two pipe sizes from an ANSI 600 × 150 relief valve and two to three sizes for an ANSI 900 × 300 relief valve. If a relief line is run at the PSV outlet pipe size for more than 10 ft, the back-pressure will generally be too high. A common error is to install a compressor skid on an existing platform and run relief piping from the skid relief flange without increasing the piping diameter, no matter how far away the relief header or flare scrubber is. This will usually result in excessive back-pressure on the compressor relief valves. Back-pressure should be checked. The relief line usually needs to increase one to two pipe sizes at the skid limit.

Finally review the relief system design after the platform design is complete. Design changes are inevitable as a facility evolves from the initial concept to the final design. Equipment arrangement, vessel sizes and MAWP, and control valve sizes all may change as the facility design is worked out. Have a knowledgeable person check the final relief system design to verify that PSV and header sizes are adequate and no significant failure scenarios have been missed.

EXAMPLE PROBLEM 13-1

Given: Q_g = maximum flow = 10MMscfd
MW = molecular weight of gas = 23.2
Z = compressibility factor = 0.9334
k = ratio of specific heats = 1.245
T = flowing temperature = 100°F
P = set pressure = 285 psig
= back-pressure = 125 psig
= overpressure (10% set pressure) = 28.5 psi
Q_l = liquid rate = 360 bpd
SG = specific gravity of liquid (water = 1) = 0.63
K_d = valve coefficient of discharge for conventional and bellows = 0.975
= for pilot = 0.92
= for liquid service = 0.65
K_v = viscosity correction factor = 0.95

Calculate orifice size for:

1. Conventional safety relief valve.
2. Pilot-operated valve.
3. Balanced-bellows safety relief valve.

Solution

Conventional Safety Relief Valve

The percent absolute back-pressure for conventional and pilot-operated safety relief valves is:

$$\%\text{ absolute back-pressure} = \frac{\text{back-pressure (psia)}}{\text{set pressure} + \text{overpressure (psia)}} \times 100$$

$$= \frac{125 + 14.7}{285 + 28.5 + 14.7} \times 100$$

$$= 43\%$$

For a conventional relief valve, this back-pressure is too high because of the effect on set pressure. A conventional relief valve should not be used for this service unless the back-pressure can be reduced to less than 10%.

Pilot-operated Valve

1. Calculate orifice size for gas.

$$a = \frac{Q_m \, (ZT\,MW)^{1/2}}{6.32\, CK_d P_1 K_b}$$

where Q_m = maximum flow, scfm
K_d = valve coefficient of discharge = 0.92 for pilot-operated
P_1 = flowing pressure, psia
MW = molecular weight of gas = 23.2
Z = compressibility factor = 0.9334
C = gas constant based on ratio of specific heats C_P/C_v
T = flowing temperature, °R
K_b = back-pressure correction factor

$$Q_m = (10)(10^6)\frac{ft^3}{day} \times \frac{day}{24\ hr} \times \frac{hr}{60\ min} = 6944\,\frac{ft^3}{min}$$

$$P_1 = (285 + 28.5 + 14.7) = 328 \text{ psia}$$

$$C = \frac{341.22 + 342.19}{2} = 341.71 \text{ (From Table 13-2)}$$

$$T = 100 + 460 = 560°R$$

Determine K_b:

Percent absolute back-pressure is 43%, which is less than the 50% limit for pilot-operated valves.

$K_b = 1.0$ below 50% back-pressure

$$a = \frac{6{,}944\,[(0.9334)\,(560)\,(23.2)]^{1/2}}{6.32\,(341.71)\,(0.92)\,(328)\,(1.0)}$$

$$a = 1.173 \text{ in.}^2$$

2. Calculate orifice size for liquid.

$$a = \frac{L\sqrt{\frac{SG}{\Delta P}}}{38\ K_D\ K_V}$$

where L = maximum liquid flow, gpm
a = orifice area, in.^{2}
SG = liquid specific gravity
ΔP = set pressure + allowable overpressure – back-pressure, psig
K_v = viscosity correction factor
K_d = valve coefficient of discharge = 0.65 for liquid

$$L = 360\ \frac{\text{barrels}}{\text{day}} \times \frac{42\ \text{gal}}{\text{barrel}} \times \frac{\text{day}}{24\ \text{hr}} \times \frac{\text{hr}}{60\ \text{min}} = 10.5\ \text{gpm}$$

$SG = 0.63$
$\Delta P = 285 + 28.5 - 125 = 188.5$
$K_V = 0.95$
$K_d = 0.65$

$$a = \frac{(10.5)\sqrt{\frac{0.63}{188.5}}}{(38)(0.65)(0.73)(0.95)}$$

$$= 0.026$$

3. Calculate total orifice size.

$$a_{total} = a_L + a_G$$
$$= 0.026 + 1.173$$
$$= 1.199 \quad \text{Use size "J" orifice.}$$

Balanced-Bellows Safety Relief Valve

1. Calculate orifice size for gas.

$$a = \frac{Q_m\ (ZT\ MW)^{1/2}}{6.32\ CK_dP_1K_b}$$

where $Q_m = 6{,}944\ ft^3/min$
$K_d = 0.975$ for balanced bellows
$P_1 = 328$ psia
$MW = 23.2$
Z = compressibility factor = 0.9334
$C = 341.71$
$T = 560°R$

Determine K_b:

$$\text{Percent gage back-pressure} = \frac{\text{back-pressure, psig}}{\text{set pressure, psig}} \times 100$$

$$= \frac{125\,(100)}{285} = 43.9\%$$

$$K_b = 0.78 \text{ from Figure 13-8 with 10\% overpressure}$$

$$a = \frac{6{,}944\,[(0.9334)\,(560)\,(23.2)]^{1/2}}{6.32\,(341.71)\,(0.975)\,(328)\,(0.78)}$$

$$= 1.419\ \text{in.}^2$$

2. Calculate orifice size for liquid.

$$a = \frac{L\sqrt{\dfrac{SG}{\Delta P}}}{38\,K_D\,K_w\,K_V}$$

where K_w = correction factor for back-pressure
= 0.73 for 43.9% gauge back-pressure from Figure 13-10

$$a = \frac{(10.5)\sqrt{\dfrac{0.63}{188.5}}}{(38)\,(0.65)\,(0.73)\,(0.95)}$$

$$= 0.035\ \text{in.}^2$$

3. Calculate total orifice size.

$a_{total} = 0.035 + 1.419$
$= 1.454\ \text{in.}^2$ Use size "K" orifice.

EXAMPLE PROBLEM 13-2

Given:
Q_g = maximum flow = 50 MMscfd
MW = molecular weight of gas = 17.4
Z = compressibility factor = 0.9561
k = ratio of specific heats = 1.27
T = flowing temperature = 300°F
P = set pressure = 1,970 psig
= back-pressure = 500 psig
= overpressure (10% set pressure) = 197 psig
Q_l = liquid rate = 0 bpd
K_d = valve coefficient of discharge
= for pilot-operated = 0.92

1. Calculate orifice size for pilot-operated relief valve.
2. What are the inlet and outlet flange ratings?

Solution

1. Calculate orifice size for gas.

$$a = \frac{Q_m\,(ZT\,MW)^{1/2}}{6.32\,CK_dP_1K_b}$$

where Q_m = maximum flow, scfm
K_d = valve coefficient of discharge = 0.92
P_1 = flowing pressure, psia
MW = molecular weight of gas = 17.4
Z = compressibility factor = 0.9561
C = gas constant based on ratio of specific heats C_P/C_v
T = flowing temperature, °R
K_b = back-pressure correction factor

$$Q_m = \quad \frac{ft}{day} \times \frac{day}{hr} \times \frac{hr}{\quad} = \quad \frac{ft}{\quad}$$

P_1 = (1,970 + 197 + 14.7) = 2,182 psia
C = 344.13 (from Table 13-2)
T = 300 + 460 = 760°R

Determine percent back-pressure:

$$Percent\ absolute\ back\ \ pressure = \frac{back}{set\ pressure + overpressure\ \ psia} \times$$

$$= \left(\frac{\quad + \quad}{\quad + \quad + \quad}\right) \times$$

$$=$$

$K_b = 1.0$ since percent back-pressure is less than 50%

$$a = \frac{34,722\,[(0.956)\,(760)\,(17.4)]^{1/2}}{6.32\,(344.13)\,(0.92)\,(2182)\,(1.0)}$$

$$= 0.8942\ \text{in.}^2 \quad \text{Use Size "J" orifice.}$$

2. The inlet flange rating is ANSI Class 900 because the relief valve inlet of 1,970 psig falls within the pressure rating of Class 900. The outlet flange rating is ANSI Class 300 because with a back-pressure of 500 psig the pressure rating of ANSI Class 150 (maximum pressure of 285 psig at 100°F) is inadequate.

CHAPTER

14

*Safety Systems**

This chapter discusses overall safety analysis techniques for evaluating production facilities, describes the concepts used to determine where safety shutdown sensors are required, and provides background and insight into the concept of a Safety and Environmental Management Program.

To develop a safe design, it is necessary to first design and specify all equipment and systems in accordance with applicable codes and standards. Once the system is designed, a process safety shutdown system is specified to assure that potential hazards that can be detected by measuring process upsets are detected, and that appropriate safety actions (normally an automatic shutdown) are initiated. A hazards analysis is then normally undertaken to identify and mitigate potential hazards that could lead to fire, explosion, pollution, or injury to personnel and that cannot be detected as process upsets. Finally, a system of safety management is implemented to assure the system is operated and maintained in a safe manner by personnel who have received adequate training.

Safety analysis concepts are discussed in this chapter by first describing a generalized hazard tree for a production facility. From this analysis, decisions can be made regarding devices that could be installed to monitor process upset conditions and to keep them from creating hazards.

---

*Reviewed for the 1999 edition by Benjamin T. Banken of Paragon Engineering Services, Inc.

This analysis forms the basis of a widely used industry consensus standard, American Petroleum Institute, Recommended Practice 14C, *Analysis, Design, Installation, and Testing of Basic Surface Systems for Offshore Production Platforms* (RP14C), which contains a procedure for determining required process safety devices and shutdowns. The procedures described here can be used to develop checklists for devices not covered by RP14C or to modify the consensus checklists presented in RP14C in areas of the world where RP14C is not mandated.

While RP14C provides guidance on the need for process safety devices, it is desirable to perform a complete hazards analysis of the facility to identify hazards that are not necessarily detected or contained by process safety devices and that could lead to loss of containment of hydrocarbons or otherwise lead to fire, explosion, pollution, or injury to personnel. The industry consensus standard, American Petroleum Institute Recommended Practice 14J, *Design and Hazards Analysis for Offshore Facilities* (RP14J), provides guidance as to the use of various hazards analysis techniques.

The final portion of this chapter describes the management of safety using Safety and Environmental Management Programs (SEMP) as defined in API RP75, *Recommended Practices for Development of a Safety and Environmental Management Program for the Outer Continental Shelf (OCS) Operations and Facilities,* and using a Safety Case approach as is commonly done in the North Sea.

HAZARD TREE

The purpose of a hazard tree is to identify potential hazards, define the conditions necessary for each hazard, and identify the source for each condition. Thus, a chain of events can be established that forms a necessary series of required steps that results in the identified hazard. This is called a "hazard tree." If any of the events leading to the hazard can be eliminated with absolute certainty, the hazard itself can be avoided.

A hazard tree is constructed by first identifying potential hazards. Starting with the hazard itself, it is possible to determine the conditions necessary for this hazard to exist. For these conditions to exist, a source that creates that condition must exist and so forth. Using this reasoning, a hierarchy of events can be drawn, which becomes the hazard tree. In a hazard analysis an attempt is made, starting at the lowest level in the tree, to see if it is possible to break the chain leading to the hazard by eliminating one of the conditions. Since no condition can be eliminated with absolute certainty, an attempt is made to minimize the occurrence of each of the steps in each chain leading to the hazard so that the overall probability of the hazard's occurrence is within acceptable limits.

This process is perhaps best illustrated by a simple example. Figure 14-1 shows a hazard tree developed for the "hazard" of injury while walking down a corridor in an office. The conditions leading to injury are identified as collision with others, tripping, hit by falling object, and total building failure. The sources leading to each condition are listed under the respective condition. Some of the sources can be further resolved into activities that could result in the source. For example, if no soil boring was taken this could lead to "inadequate design," which would lead to "building failure," which could lead to "injury."

It is obvious that it is impossible to be absolutely certain that the hazard tree can be broken. It is, however, possible to set standards for ceiling design, lighting, door construction, etc., that will result in acceptable frequencies of collision, tripping, etc., given the severity of the expected injury from the condition. That is, we could conclude that the probability of building failure should be lower than the probability of tripping because of the severity of injury that may be associated with building failure.

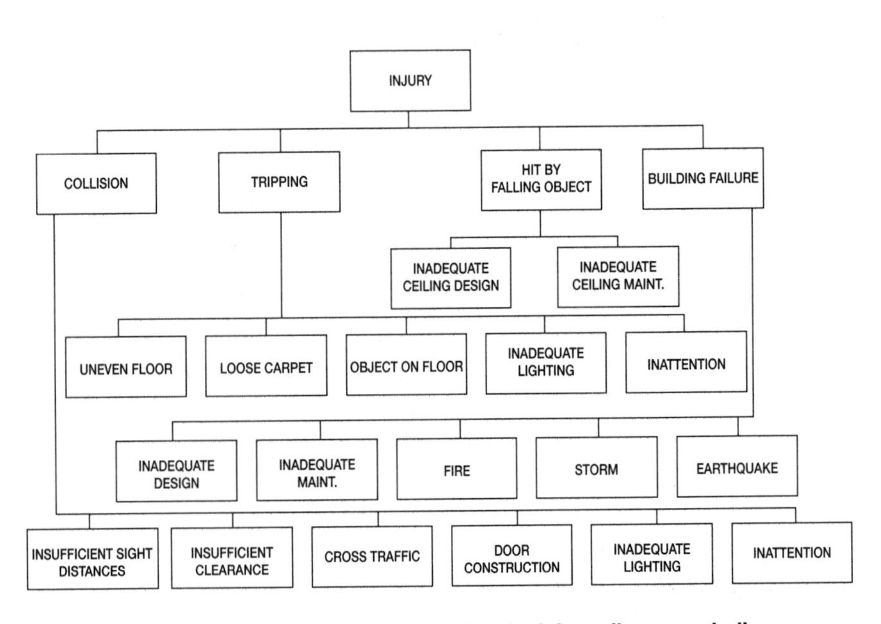

Figure 14-1. Hazard tree for injury suffered while walking in a hallway.

It should be obvious from this discussion that the technique of creating a hazard tree is somewhat subjective. Different evaluators will likely classify conditions and sources differently and may carry the analysis to further levels of sources. However, the conclusions reached concerning building design, maintenance, layout of traffic patterns, lighting, etc., should be the same. The purpose of developing the hazard tree is to focus attention and help the evaluator identify all aspects that must be considered in reviewing overall levels of safety.

It is possible to construct a hazard tree for a generalized production facility, just as it is possible to construct a hazard tree for a generalized hallway. That is, Figure 14-1 is valid for a hallway in Paragon Engineering Services' offices in Houston, in Buckingham Palace in London, or in a residence in Jakarta. Similarly, a generalized hazard tree constructed for a production facility could be equally valid for an onshore facility or an offshore facility, no matter what the specific geographic location.

Figure 14-2 is a hazard tree for a generalized production facility. The hazards are identified as "oil pollution," "fire/explosion," and "injury." Beginning with injury, we can see that the hazards of fire/explosion and oil pollution become conditions for injury since they can lead to injury as well as being hazards in their own right. The tree was constructed by beginning with the lowest level hazard, oil pollution. Oil pollution occurs as a result of an oil spill but only if there is inadequate containment. That is, if there is adequate containment, there cannot be oil pollution. Onshore, dikes are constructed around tank farms for this reason. Offshore, however, and in large onshore facilities it is not always possible to build containment large enough for every contingency. The requirement for drip pans and sumps stems from the need to reduce the probability of oil pollution that could result from small oil spills.

One source of an oil spill could be the filling of a vessel that has an outlet to atmosphere until it overflows. Whenever inflow exceeds outflow, the tank can eventually overflow. Another source is a rupture or sudden inability of a piece of equipment to contain pressure. Events leading to rupture are listed in Figure 14-2. Note that some of these events can be anticipated by sensing changes in process conditions that lead to the rupture. Other events cannot be anticipated from process conditions.

Other sources for oil spills are listed. For example, if a valve is opened and the operator inadvertently forgets to close it, oil may spill out of the system. If there is not a big enough dike around the system, oil pollution will result. It is also possible for oil to spill out the vent/flare system. All pressure vessels are connected to a relief valve, and the relief valve dis-

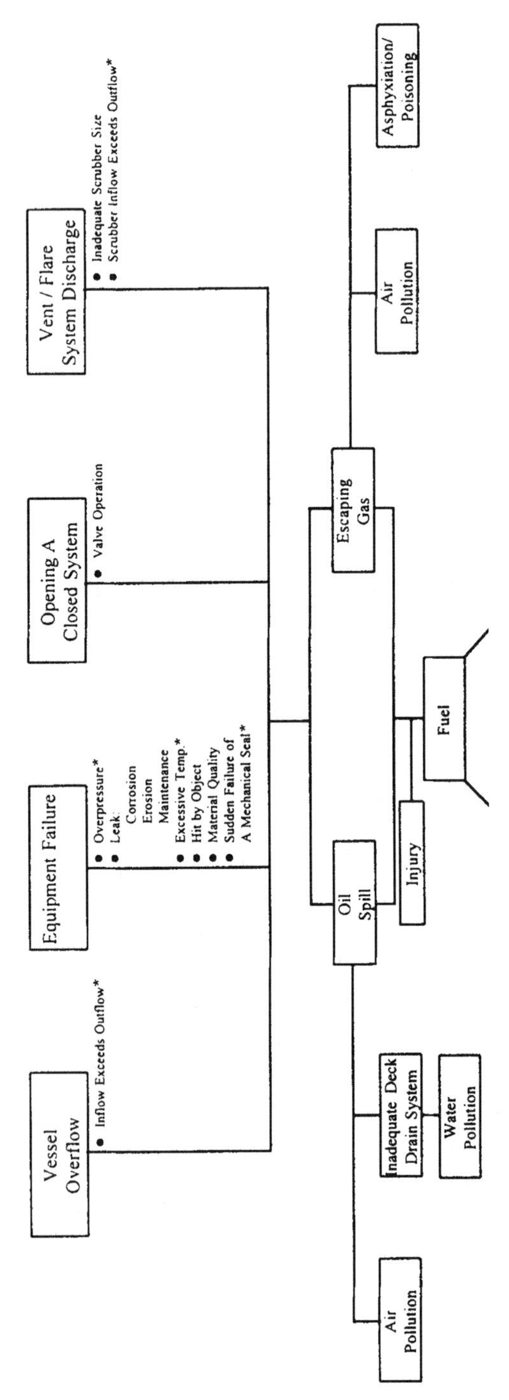

Figure 14-2. Hazard tree for production facility. ***(Source: API RP14.)***

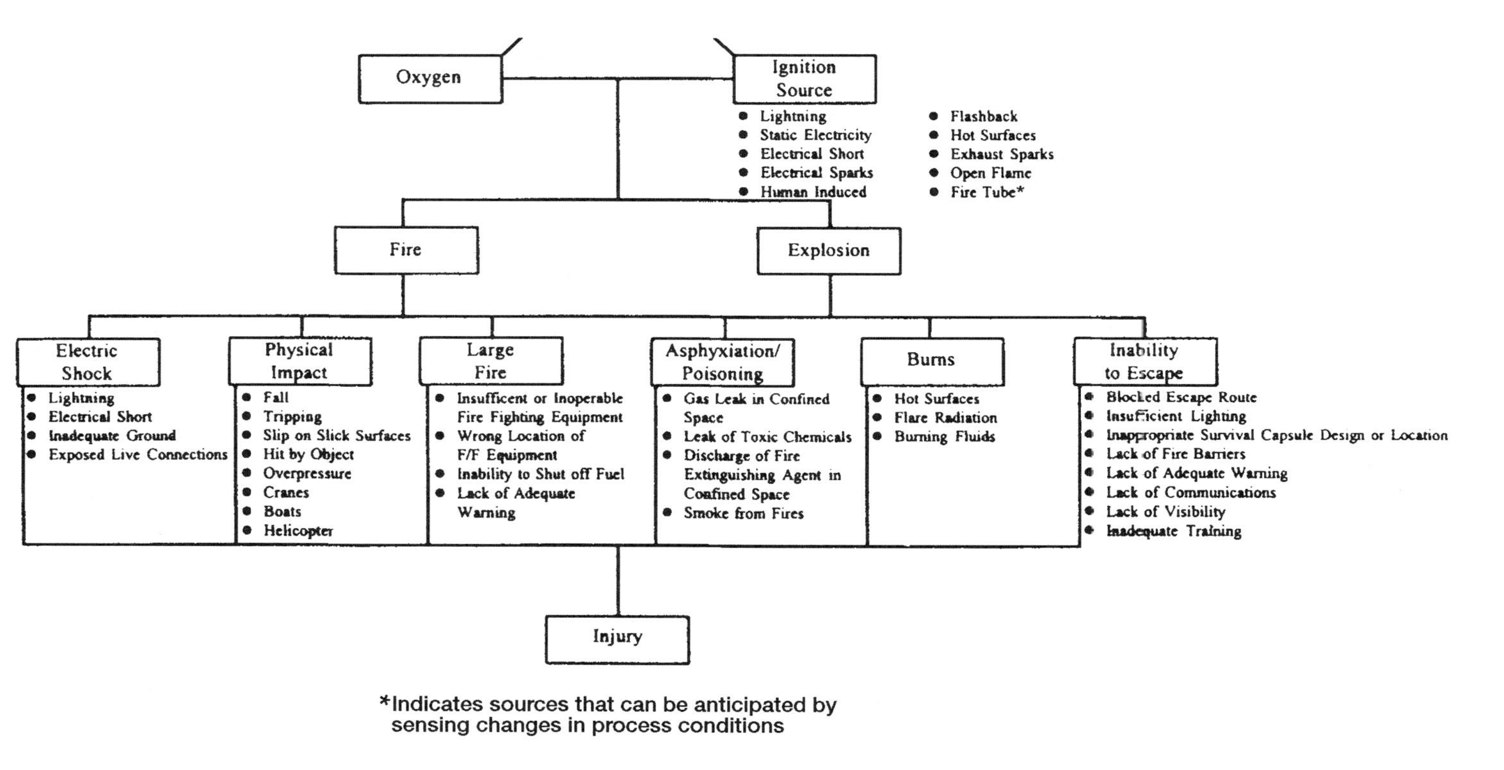

Figure 14-2. Continued

charges out a vent or flare system. If the relief scrubber is not adequately sized, or if it does not have a big enough dump rate, oil will go out the vent system.

Fire and explosion are much more serious events than pollution. For one thing, fire and explosion can create catastrophes that will lead to pollution anyway, but for another thing, they can injure people. We clearly want to have more levels of safety (that is, a lower probability of occurrence) in the chain leading to fire or explosion than is necessary in the chain leading to pollution. That is, whatever the acceptable risk for oil pollution, a lower risk is required for fire or explosion.

For fire or explosion to occur, fuel, an ignition source, oxygen, and time to mix them all together are needed. If any of these elements can be eliminated with 100% assurance, the chain leading to fire or explosion will be broken. For example, if oxygen can be kept out of the facility, then there can be no fire or explosion. Eliminating oxygen can be done inside the equipment by designing a gas blanket and ensuring positive pressure. For practical purposes it cannot be done outside the equipment, as a human interface with the equipment is desired.

Fuel cannot be completely eliminated, though the inventory of combustible fuels can be kept to a minimum. Oil and gas will be present in any production facility, and either an oil spill or escaping gas can provide the fuel needed. Escaping gas can result from rupture, opening a closed system, or gas that is normally vented. The amount of fuel present can be minimized by preventing oil spills and gas leaks.

Ignition sources are numerous, but it is possible to minimize them. Lightning and static electricity are common ignition sources in production facility, especially tank vents. It is not possible to anticipate the ignition by sensing changes in process conditions, but gas blankets, pressure vacuum valves, and flame arrestors can be installed to ensure that flame will not flash back into the tank and create an explosion. Electrical shorts and sparks are also sources of ignition. These are kept isolated from any fuel by a whole series of rules and regulations for the design of electrical systems. In the United States, the National Electrical Code and the API Recommended Practices for Electrical Systems (Chapter 17) are used to minimize the danger of these ignition sources. Human-induced ignition sources include welding and cutting operations, smoking, and hammering (which causes static electricity). Flash back is also a source of ignition. In some vessels a flame exists inside a fire tube. If a fuel source develops around the air intake for the fire tube, the flame can propagate outside the fire tube and out into the open. The flame would then become a source of ignition for any more fuel present and could lead to a fire or explosion. This is why flame arrestors are required on natural draft fire tubes.

Hot surfaces are another common source of ignition. Engine exhaust, turbine exhaust, and engine manifold on engine-driven compressors may be sufficiently hot to ignite oil or gas. A hot engine manifold can become a source of ignition for an oil leak. An engine exhaust can become a source of ignition for a gas escape.

Exhaust sparks from engines and burners can be a source of ignition. Any open flame on the facility can also be a source of ignition.

Fire tubes, especially in heater treaters, where they can be immersed in crude oil, can become a source of ignition if the tube develops a leak, allowing crude oil to come in direct contact with the flame. Fire tubes can also be a source of ignition if the burner controls fail and the tube overheats or if the pilot is out and the burner turns on when there is a combustible mixture in the tubes.

Because these ignition sources cannot be anticipated by sensing changes in process conditions and since oxygen is always present, a hazards analysis must concentrate on reducing the risk of oil spill and gas leak when any of these ignition sources is present. Or the hazards analysis must concentrate on reducing the probability that the ignition source will exist at the same location as an oil spill or gas leak.

Injury is always possible by fire, explosion, or the other conditions listed in Figure 14-2. A fire can lead directly to injury, but normally there needs to be several contributory events before the fire becomes large enough to lead to injury. For example, if a fire develops and there is sufficient warning, there should be sufficient time to escape before injury results. If the fuel is shut off and there is enough fire-fighting equipment to fight the fire before it becomes large, the probability of injury is small.

When an explosion occurs, however, it can directly cause injury. A substantial cloud of gas can accumulate before the combustible limit reaches an ignition source. The force of the explosion as the cloud ignites can be substantial.

There are other ways to injure people, such as physical impact due to falling, tripping, slipping on a slick surface, or being hit by an object or by direct physical impact from a rupture. Asphyxiation can occur, especially when dealing with toxic chemicals.

Electric shock and burns can also lead to injury. Burns can occur by touching hot surfaces. They can also occur from radiation.

The probability of injury from any of these conditions is increased by an inability to escape. All the conditions tend to be more likely to lead to injury the longer people are exposed to the situation. Therefore, escape routes, lighting, appropriate selection of survival capsules or boats, fire barriers, etc., all lead to a reduction in injury.

DEVELOPING A SAFE PROCESS

In going through this hazard tree it can be seen that many of the sources and conditions leading to the three major hazards have nothing to do with the way in which the process is designed. Many sources cannot be anticipated by sensing a condition in the process. For example, it is not possible to put a sensor on a separator that keeps someone who is approaching the separator to perform maintenance from falling. Another way of stating this is that many of the sources and conditions identified on the hazard tree require design considerations that do not appear on mechanical flow diagrams. The need for proper design of walkways, escape paths, electrical systems, fire-fighting systems, insulation on piping, etc., is evident on the hazard tree. In terms of developing a process safety system, only those items that are starred in the hazard tree can be detected and therefore defended against.

This point must be emphasized because it follows that a production facility that is designed with a process shut-in system as described in API RP14C is not necessarily "safe." It has an appropriate level of devices and redundancy to reduce the sources and conditions that can be anticipated by sensing changes in process conditions. However, much more is required from the design of the facility if the overall probability of any one chain leading to a hazard is to be acceptable. That is, API RP14C is merely a document that has to do with safety analysis of the process components in the production facility. It does not address all the other concerns that are necessary for a "safe" design.

The starred items in the hazard tree are changes in process conditions that could develop into sources and lead to hazards. These items are identified in Table 14-1 in the order of their severity.

Overpressure can lead directly to all three hazards. It can lead directly and immediately to injury, to fire or explosion if there is an ignition source, and to pollution if there is not enough containment. Therefore, we must have a very high level of assurance that overpressure is going to have a very low frequency of occurrence.

Fire tubes can lead to fire or explosion if there is a leak of crude oil into the tubes or failure of the burner controls. An explosion could be sudden and lead directly to injury. Therefore, a high level of safety is required.

Table 14-1
Sources Associated with Process System Changes

Source	Hazard	Contributing Source of Condition
Overpressure	Injury	None
	Fire/Explosion	Ignition Source
	Pollution	Inadequate Containage
Leak	Fire/Explosion	Ignition Source
	Oil Pollution	Inadequate Containage
Fire Tubes	Fire/Explosion	Fuel
Inflow Exceeds Outflow	Oil Pollution	Inadequate Containage
Excessive Temperature	Fire/Explosion	Ignition Source
	Oil Pollution	Inadequate Containage

Excessive temperature can lead to premature failure of an item of equipment at pressures below its design maximum working pressure. Such a failure can create a leak, potentially leading to fire or explosion if gas is leaked or to oil pollution if oil is leaked. This type of failure should be gradual, with warning as it develops, and thus does not require as high a degree of protection as those previously mentioned.

Leaks cannot lead directly to personal injury. They can lead to fire or explosion if there is an ignition source and to oil pollution if there is inadequate containment. Both the immediacy of the hazard developing and the magnitude of the hazard will be smaller with leaks than with overpressure. Thus, although it is necessary to protect against leaks, this protection will not require the same level of safety that is required to protect against overpressure.

Inflow exceeding outflow can lead to oil pollution if there is inadequate containment. It can lead to fire or explosion and thus to injury by way of creating an oil spill. This type of accident is more time-dependent and lower in magnitude of damage, and thus an even lower level of safety will be acceptable.

The hazard tree also helps identify protection devices to include in equipment design that may minimize the possibility that a source will develop into a condition. Examples would be flame arrestors and stack arrestors on fire tubes to prevent flash back and exhaust sparks, gas detectors to sense the presence of a fuel in a confined space, and fire detectors and manual shutdown stations to provide adequate warning and to keep a small fire from developing into a large fire.

PRIMARY DEFENSE

Before proceeding to a discussion of the safety devices required for the process, it is important to point out that the primary defense against hazards in a process system design is the use of proper material of sufficient strength and thickness to withstand normal operating pressures. This is done by designing the equipment and piping in accordance with accepted industry design codes. If this is not done, no sensors will be sufficient to protect from overpressure, leak, etc. For example, a pressure vessel is specified for 1,480 psi maximum working pressure, and its relief valve will be set at 1,480 psi. If it is not properly designed and inspected, it may rupture before reaching 1,480 psi pressure. The primary defense to keep this from happening is to use the proper codes and design procedures and to ensure that the manufacture of the equipment and its fabrication into systems are adequately inspected. In the United States, pressure vessels are constructed in accordance with the ASME Boiler and Pressure Vessel Code discussed in Chapter 12, and piping systems are constructed in accordance with one of the ANSI Piping Codes discussed in Volume 1.

It is also important to assure that corrosion, erosion, or other damage has not affected the system to the point that it can no longer safely contain the design pressure. Maintaining mechanical integrity once the system has been placed in service is discussed later in this chapter.

FAILURE MODE EFFECT ANALYSIS—FMEA

One of the procedures used to determine which sensors are needed to sense process conditions and protect the process is called a Failure Mode Effect Analysis—FMEA. Every device in the process is checked for its various modes of failure. A search is then made to assure that there is a redundancy that keeps an identified source or condition from developing for each potential failure mode. The degree of required redundancy depends on the severity of the source as previously described. Table 14-2 lists failure modes for various devices commonly used in production facilities.

In applying FMEA, a mechanical flow diagram must first be developed. As an example, consider the check valve on a liquid dump line. It can fail

Table 14-2
Failure Modes of Various Devices

Sensors		**Signal/Indicator**	
FTS	Fail to See	FTI	Fail to Indicate
OP	Operate Prematurely		
Check Valves		**Switch**	
FTC	Fail to Close (Check)	FS	Fail to Switch
Lin	Leak Internally	FC	Fail Close
Lex	Leak Externally	FO	Fail Open
Orifice Plates (Flow Restrictor)		**Engine**	
FTR	Fail to Restrict	FTD	Fail to Deliver
BL	Block	FXP	Deliver Excess Power
Pumps		**Transformer**	
FTP	Fail to Pump	FTF	Fail to Function
POP	Pump to Overpressurization		
LEX	Leak Externally		
Controllers		**General**	
FTCL	Fail to Control Level	OF	Overflow
FTCT	Fail to Control Temperature	NP	Not Processed
FTCF	Fail to Control Flow	NS	No Signal
OP	Operate Prematurely	FP	Fail to Power
FTCLL	Fail to Control Low Level	MOR	Manual Override
FTCHL	Fail to Control High Level	NA	Not Applicable
FTRP	Fail to Reduce Pressure		
FTCP	Fail to Control Pressure	**Rupture Disc**	
FTAA	Fail to Activate Alarms	RP	Rupture Prematurely
		FTO	Fail to Open
Valves		LEX	Leak Externally
FO	Fail Open		
FC	Fail Close	**Meter**	
FTO	Fail to Open	FTOP	Fail to Operate Properly
FTC	Fail to Close		
Lin	Leak Internally	LEX	Leak Externally
Lex	Leak Externally	BL	Block
		Timer	
		FTAP	Fail to Activate Pump
		FTSP	Fail to Stop Pump

one of three ways—it can fail to close, it can leak internally, or it can leak externally. The FMEA will investigate the effects that could occur if this particular check valve fails to close. Assuming this happens, some redundancy that keeps a source from developing must be located in the system. Next, the process would be evaluated for the second failure mode, that is, what occurs if the check valve leaks internally. Next, the process would be evaluated for the third failure mode of this check valve. Check valves are easy. A controller has nine failure modes, and a valve has six.

In order to perform a complete, formal FMEA of a production facility, each failure mode of each device must be evaluated. A percentage failure rate and cost of failure for each mode for each device must be calculated. If the risk discounted cost of failure is calculated to be acceptable, then there are the proper numbers of redundancies. If that cost is not acceptable, then other redundancies must be added until an acceptable cost is attained.

It is obvious that such an approach would be lengthy and would require many pages of documentation that would be difficult to check. It is also obvious that such an approach is still subjective in that the evaluator must make decisions as to the consequences of each failure, the expected failure rate, and the acceptable level of risk for the supposed failure.

This approach has been performed on several offshore production facilities with inconsistent results. That is, items that were identified by one set of evaluators as required for protection in one design were not required by another set of evaluators in a completely similar design. In addition, potential failure of some safety devices on one facility caused evaluators to require additional safety devices as back-up, while the same group in evaluating a similar installation that did not have the initial safety devices at all did not identify the absence of the primary safety device as a hazard or require back-up safety devices.

It should be clear that a complete FMEA approach is not practical for the evaluation of production facility safety systems. This is because (1) the cost of failure is not as great as for nuclear power plants or rockets, for which this technology has proven useful; (2) production facility design projects cannot support the engineering cost and lead time associated with such analysis; (3) regulatory bodies are not staffed to be able to critically analyze the output of an FMEA for errors in subjective judgment; and most importantly, (4) there are similarities to the design of all production facilities that have allowed industry to develop a modified FMEA approach that can satisfy all these objections.

MODIFIED FMEA APPROACH

The modified FMEA approach evaluates each piece of equipment (not each device) as an independent unit, assuming worst case conditions of input and output. Separators, flowlines, heaters, compressors, etc., function in the same manner no matter the specific design of the facility. That is, they have level, pressure and temperature controls and valves. These are subject to failure modes that impact the piece of equipment in the same manner. Thus, if an FMEA analysis is performed on the item of equipment standing alone, the FMEA will be valid for that component in any process configuration.

Furthermore, once every process component has been analyzed separately for worst case, stand-alone conditions, there is no additional safety risk created by joining the components into a system. That is, if every process component is fully protected based on its FMEA analysis, a system made up of several of these components will also be fully protected.

It is even possible that the system configuration is such that protection furnished by devices on one process component can protect others. That is, devices that may be required to provide adequate protection for a component standing alone may be redundant once all components are assembled in a system. This procedure is outlined below:

1. For each piece of equipment (process component), develop an FMEA by assuming in turn that each process upset that could become a potential source occurs. That is, assume a control failure, leak, or other event leading to a process upset.
2. Provide a sensor that detects the upset and shuts-in the process before an identified source of condition develops. For example, if the pressure controller fails and the pressure increases, provide a high-pressure sensor to shut-in the process. If there is a leak and the pressure decreases, provide a low-pressure sensor to shut-in the process.
3. Apply FMEA techniques to provide an independent back-up to the sensor as a second level of defense before an identified hazard is created. The degree of reliability of the back-up device will be dependent upon the severity of the problem. For example, since overpressure is a condition that can lead to severe hazards, the back-up device should be extremely reliable. Typically, a high pressure sensor would be backed up by a relief valve. In this case a relief valve is actually more reliable than the high pressure sensor, but it has other detriments associated with it. Oil leakage, on the other hand, is not as severe. In this case, a drip pan to protect against oil pollution may be adequate back-up.
4. Assume that two levels of protection are adequate. Experience in applying FMEA analysis to production equipment indicates that in many cases only one level of protection would be required, given the degree of reliability of shutdown systems and the consequences

of failure. However, it is more costly in engineering time to document that only one level is required for a specific installation than it is to install and maintain two levels. Therefore, two levels are always specified.

5. Assemble the components into the process system and apply FMEA techniques to determine if protection devices on some components provide redundant protection to other components. For example, if there are two separators in series, and they are both designed for the same pressure, the devices protecting one from overpressure will also protect the other. Therefore, there may be no need for two sets of high pressure sensors.

The application of this procedure is best seen by performing an FMEA on a simple two-phase separator. Table 14-3 lists those process upsets that can be sensed before an undesirable event leading to a source of condition occurs. For overpressure, primary protection is provided by a high pressure sensor that shuts in the inlet (PSH). If this device fails, secondary protection is provided by a relief valve (PSV).

A large leak of gas is detected by a low-pressure sensor (PSL) that shuts in the inlet, and a check valve (FSV) keeps gas from downstream components from flowing backward to the leak. Similarly, a large oil leak is detected by a low-level sensor (LSL) and a check valve. Back-up protection is provided by a sump tank and its high-level sensor (LSH) for an oil leak. That is, before an oil spill becomes pollution there must be a

Table 14-3
FMEA of a Separator

Undesirable Event	Primary	Secondary
Overpressure	PSH	PSV
Large Gas Leak	PSL and FSV	ASH, Minimize Ignition Sources
Large Oil Leak	LSL and FSV	Sump tank (LSH)
Small Gas Leak	ASH, Minimize Ignition Source	Fire Detection
Small Oil Leak	Sump Tank (LSH)	Manual Observation
Inflow Exceeds Outflow	LSH	Vent Scrubber (LSH)
High Temperature	TSH	Leak Detection Devices

failure of a second sensor. Back-up protection for a gas leak then becomes the fire detection and protection equipment if the small leak were to cause a fire. There is no automatic back-up to the sump tank LSH for a small oil leak. Manual intervention, before containment is exceeded and oil pollution results, becomes the back-up.

The primary protection for high temperature, which could lower the maximum allowable working pressure below the PSV setting, is a high-temperature sensor (TSH), which shuts in the inlet or the source of heat. Back-up protection is provided by leak detection devices.

Inflow exceeding outflow is sensed by a high-level sensor (LSH). Back-up protection is furnished by the PSH (to keep the relief valve from operating) or an LSH in a downstream vent scrubber if the vessel gas outlet goes to atmosphere. That is, a vent scrubber must be installed downstream of any vessel that discharges directly to atmosphere.

Once the FMEA is completed, the specific system is analyzed to determine if all the devices are indeed needed. For example, if it is not possible for the process to overpressure the vessel, these devices are not required. If it is impossible to heat the vessel to a high enough level to effect its maximum working pressure, the TSH can be eliminated.

API RECOMMENDED PRACTICE 14C

The modified FMEA approach has been used by the API to develop RP14C. In this document ten different process components have been analyzed and a Safety Analysis Table (SAT) has been developed for each component. A sample SAT for a pressure vessel is shown in Table 14-4. The fact that Tables 14-3 and 14-4 are not identical is due to both the subjective natures of a Hazard Analysis and FMEA, and to the fact that RP14C is a consensus standard. However, although the rationale differs somewhat, the devices required are identical. (The "gas make-up system" in Table 14-4 is not really required by RP14C, as we shall see.)

The RP 14C also provides standard reasons allowing the elimination of certain devices when the process component is considered as part of an overall system. Figure 14-3 shows the Safety Analysis Checklist (SAC) for a pressure vessel. Each safety device is identified by the SAT (with the exception of "gas make-up system") is listed. It must either be installed or it can be eliminated if one of the reasons listed is valid.

(*text continued on page 405*)

Table 14-4
Safety Analysis Table (SAT)
Pressure Vessels

Undesirable Event	Cause	Detectable Condition At Component
Overpressure	Blocked or restricted outlet Inflow exceeds outflow Gas blowby (upstream component) Pressure control system failure Thermal expansion Excess heat input	High pressure
Underpressure (vacuum)	Withdrawals exceed inflow Thermal contraction Open outlet Pressure control system failure	Low pressure
Liquid Overflow	Inflow exceeds outflow Liquid slug flow Blocked or restricted liquid outlet Level control system failure	High liquid level
Gas Blowby	Liquid withdrawals exceed inflow Open liquid outlet Level control system failure	Low liquid level
Leak	Deterioration Erosion Corrosion Impact damage Vibration	Low pressure Low liquid level
Excess temperature	Temperature control system failure High inlet temperature	High temperature

Source: API RP 14C, 6th Edition, March 1998.

SAFETY ANALYSIS CHECKLIST (SAC)—PRESSURE VESSELS

a. High Pressure Sensor (PSH)
   1. PSH installed.
   2. Input is from a pump or compressor that cannot develop pressure greater than the maximum allowable working pressure of the vessel.
   3. Input source is not a wellhead flow line(s), production header, or pipeline, and each input source is protected by a PSH that protects the vessel.
   4. Adequately sized piping without block or regulating valves connects gas outlet to downstream equipment protected by a PSH that also protects the upstream vessel.
   5. Vessel is final scrubber in a flare, relief, or vent system and is designed to withstand maximum built-up back-pressure.
   6. Vessel operates at atmospheric pressure and has an adequate vent system.

b. Low Pressure Sensor (PSL)
   1. PSL installed.
   2. Minimum operating pressure is atmospheric pressure when in service.
   3. Each input source is protected by a PSL, and there are no pressure control devices or restrictions between the PSL(s) and the vessel.
   4. Vessel is scrubber or small trap, is not a process component, and adequate protection is provided by downstream PSL or design function (e.g., vessel is gas scrubber for pneumatic safety system or final scrubber for flare, relief, or vent system).
   5. Adequately sized piping without block or regulating valves connects gas outlet to downstream equipment protected by a PSL that also protects the upstream vessel.

c. Pressure Safety Valve (PSV)
   1. PSV installed.
   2. Each input source is protected by a PSV set no higher than the maximum allowable working pressure of the vessel, and a PSV is installed on the vessel for fire exposure and thermal expansion.
   3. Each input source is protected by a PSV set no higher than the vessel's maximum allowable working pressure, and at least one of these PSVs cannot be isolated from the vessel.

Figure 14-3. Safety analysis checklist for pressure vessels. *(Source: API RP 14C, 6th Edition, March 1998.)*

4. PSVs on downstream equipment can satisfy relief equipment of the vessel and cannot be isolated from the vessel.
5. Vessel is final scrubber in a flare, relief, or vent system, is designed to withstand maximum built-up back-pressure, and has no internal or external obstructions, such as mist extractors, back-pressure valves, or flame arrestors.
6. Vessel is final scrubber in a flare, relief, or vent system, is designed to withstand maximum built-up back-pressure, and is equipped with a rupture disk or safety head (PSE) to bypass any internal or external obstructions, such as mist extractors, back-pressure valves, or flame arrestors.

d. High Level Sensor (LSH)
1. LSH installed.
2. Equipment downstream of gas outlet is not a flare or vent system and can safely handle maximum liquid carry-over.
3. Vessel function does not require handling separated fluid phases.
4. Vessel is a small trap from which liquids are manually drained.

e. Low Level Sensor (LSL)
1. LSL installed to protect each liquid outlet.
2. Liquid level is not automatically maintained in the vessel, and the vessel does not have an immersed heating element subject to excess temperature.
3. Equipment downstream of liquid outlet(s) can safely handle maximum gas rates that can be discharged through the liquid outlet(s), and vessel does not have an immersed heating element subject to excess temperature. Restrictions in the discharge line(s) may be used to limit the gas flow rate.

f. Check Valve (FSV)
1. FSV installed on each outlet.
2. The maximum volume of hydrocarbons that could backflow from downstream equipment is significant.
3. A control device in the line will effectively minimize backflow.

g. High Temperature Sensor (TSH)

High temperature sensors are applicable only to vessels having a heat source.
1. TSH installed.
2. (Deleted in Second Edition.)
3. Heat source is incapable of causing excess temperature.

Figure 14-3. Continued.

(*text continued from page 401*)

The SAC list provides a handy shorthand for communicating which devices are required and the reasons why some may not be used. For example, for any pressure vessel there is either a PSH required, or a rationale numbered, A.4.a.2, A.4.a.3, A.4.a.4, A.4.a.5 or A.4.a.6 must be listed. It becomes a simple matter to audit the design by checking that each device is either present or an appropriate rationale listed.

The SAT and SAC for each process component are updated periodically by API and the most recent edition should be used in any design. Please note that for fired and exhaust heated components it may be necessary to include the devices required for a process tank or vessel as well as those required for the heating components.

For components not covered by RP 14C, SAT and SAC tables can be developed using the modified FMEA analysis procedure.

MANUAL EMERGENCY SHUTDOWN

The safety system should include features to minimize damage by stopping the release of flammable substances, de-energizing ignition sources, and shutting down appropriate equipment processes. This is accomplished by locating emergency shutdown (ESD) stations at strategic locations to enable personnel to shut down the production facility. These ESD stations should be well marked and located conveniently (50–100 feet) from protected equipment, with back-up stations located some greater distance (250–500 feet) away. A good choice for location is along all exit routes. At least two widely separated locations should be selected.

The ESD can either shut down the entire facility, or it can be designed for two levels of shutdown. The first level shuts down equipment such as compressors, lean oil pumps, and direct fired heaters, and either shuts in the process or diverts flow around the process by closing inlet/outlet block valves and opening bypass valves. The second level shuts down the remaining utilities and support facilities, including generators and electrical feeds.

ANNUNCIATION SYSTEMS

These systems give early warning of impending trouble to allow personnel to take corrective action prior to a shut-in, and provide information about the initial cause of a shut-in. They are a vital party of any large shutdown system design. On smaller systems, process alarms may be minimal as there may not be sufficient time for personnel to react to the alarm before an automatic shutdown is initiated.

Annunciator panels should be in a central location with alarm annunciators and shutdown annunciators grouped separately. The first alarm and the first shut-down normally sound a horn and are annunciated. This is called "first-out indication." Subsequent shutdown or alarm signals received by the panel are either not annunciated or are annunciated in a different manner so that the operator can determine the initiating cause of the process upset.

Alarm signals may come from the output signal used to control an operational valve. Shutdown signals should come from a completely separate instrument not dependent upon a normally used output signal for operation.

FUNCTION MATRIX AND FUNCTION CHARTS

One method used to summarize the required devices and show the function performed by each device is with a function matrix. Figure 14-4 is a completed function matrix chart for the simple process flow diagram shown in Figure 14-5. The function matrix is from RP 14C and is called a SAFE chart. Each component is listed in the left hand column with an identification number and description. Under "Device I.D.," each of the devices listed in the SAC is listed. If the device is not present, the appropriate SAC reference number is listed. If the SAC rationale requires that another device be present on another component, that device is listed under "Alternate Device," if applicable.

Listed across the top of the matrix are the various shutdown valves in the facility. A mark in each box indicates the function performed by each device to assure that it protects the process component. By comparing the functions performed by each device to the mechanical flowsheet, it is possible for an auditor to quickly ensure that the process component is indeed isolated.

A function matrix can also form the basis for the design of the logic necessary to carry out the functions that are to be performed when a sig-

(text continued on page 410)

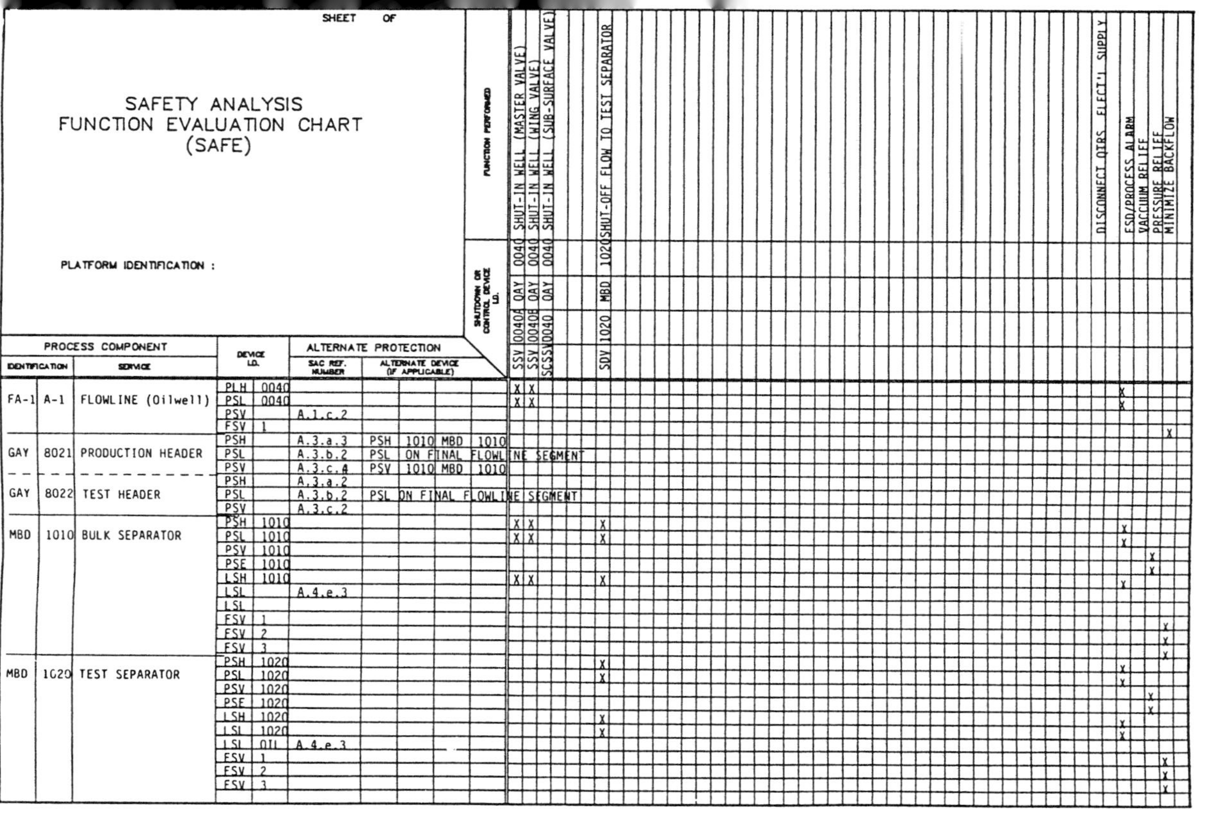

SAFETY ANALYSIS
FUNCTION EVALUATION CHART
(SAFE)

SHEET OF

PLATFORM IDENTIFICATION :

Process Component: Identification		Process Component: Service	Device I.D.		Alternate Protection: SAC Ref. Number	Alternate Protection: Alternate Device (if applicable)	SSV 0040A OAY 0040 Shut-in well (master valve)	SSV 0040B OAY 0040 Shut-in well (wing valve)	SCSSV 0040 OAY 0040 Shut-in well (sub-surface valve)	SDV 1020 MBD 1020 Shut-off flow to test separator	Disconnect QTRS elect'l supply	ESD/process alarm	Vacuum relief	Pressure relief	Minimize backflow
FA-1	A-1	FLOWLINE (Oilwell)	PLH	0040			X	X				X			
			PSL	0040			X	X				X			
			PSV		A.1.c.2										
			FSV	1											X
GAY	8021	PRODUCTION HEADER	PSH		A.3.a.3	PSH 1010 MBD 1010									
			PSL		A.3.b.2	PSL ON FINAL FLOWLINE SEGMENT									
			PSV		A.3.c.4	PSV 1010 MBD 1010									
GAY	8022	TEST HEADER	PSH		A.3.a.2										
			PSL		A.3.b.2	PSL ON FINAL FLOWLINE SEGMENT									
			PSV		A.3.c.2										
MBD	1010	BULK SEPARATOR	PSH	1010			X	X		X		X			
			PSL	1010			X	X		X		X			
			PSV	1010										X	
			PSE	1010										X	
			LSH	1010			X	X		X		X			
			LSL		A.4.e.3										
			LSL												
			FSV	1											X
			FSV	2											X
			FSV	3											X
MBD	1020	TEST SEPARATOR	PSH	1020						X		X			
			PSL	1020						X		X			
			PSV	1020										X	
			PSE	1020										X	
			LSH	1020						X		X			
			LSL	1020						X		X			
			LSL	OIL	A.4.e.3										
			FSV	1											X
			FSV	2											X
			FSV	3											X

Figure 14-4. Safety analysis function evaluation (SAFE) chart for process flow in Figure 14.5.

(figure continued on next page)

SHEET OF

SAFETY ANALYSIS
FUNCTION EVALUATION CHART
(SAFE)

PLATFORM IDENTIFICATION :

PROCESS COMPONENT IDENTIFICATION		PROCESS COMPONENT SERVICE	DEVICE I.D.		ALTERNATE PROTECTION SAC REF. NUMBER	ALTERNATE PROTECTION ALTERNATE DEVICE (IF APPLICABLE)	SSV 0040A OAY 0040 SHUT-IN WELL (MASTER VALVE)	SSV 0040B OAY 0040 SHUT-IN WELL (WING VALVE)	SCSSV 0040 OAY 0040 SHUT-IN WELL (SUB-SURFACE VALVE)	SDV 1020 MBD 1020 SHUT-OFF FLOW TO TEST SEPARATOR		DISCONNECT QTRS. ELECT'L SUPPLY	ESD/PROCESS ALARM	VACUUM RELIEF	PRESSURE RELIEF	MINIMIZE BACKFLOW
KAH	8010	DEPARTING PIPELINE	PSH	8010			X	X		X	X					
			PSL	8010			X	X		X	X					
			PSV		A.9.c.3	PSV 1010 and 1020										
			FSV	1											X	
			FSV	2											X	
ASH	1000	QUARTERS GAS DETECTION										X	X			
ESD		ESD STATIONS (MANUAL VALVES)					X	X	X	X			X			
TSE		FUSIBLE LOOP					X	X	X	X			X			

Figure 14-4. Continued.

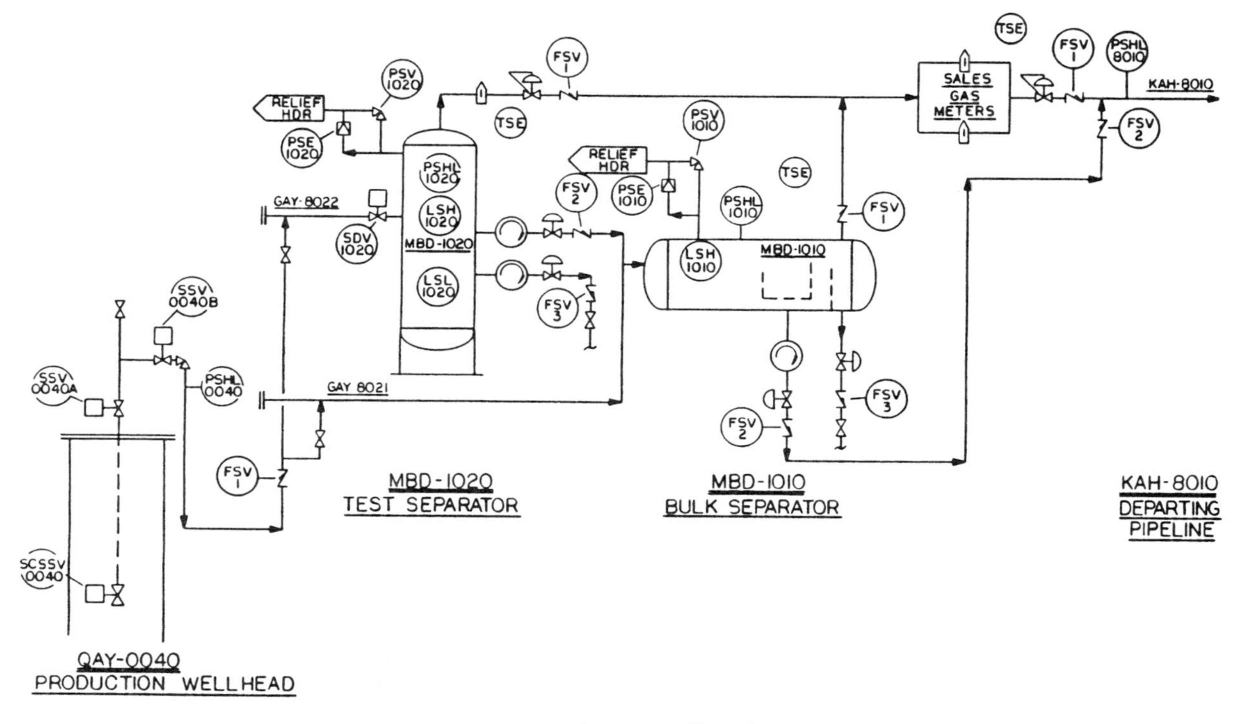

Figure 14-5. Simple process flow diagram.

(*text continued from page 406*)

nal is received from each device. More frequently, the function matrix is used to develop a "function chart" such as that shown in Figure 14-6, and the function chart is used for designing the logic. It is possible to develop a function chart directly from the facility flow diagram. However, some designers and regulatory agencies feel that it is better to develop a function matrix first to ensure that all devices required from the FMEA are considered and to clearly show the end devices causing the shutdown or alarm to occur.

In a function chart each sensing device is listed on the left side and a path is then drawn showing the route of the signal from the sensing device to the device that performs the shutdown or alarm function.

SYMBOLS

Table 14-5 shows symbols used in RP 14C to represent the various sensors and shutdown devices. Although these symbols are used extensively in U.S. production facilities, they are not used in other industries. They are widely used overseas and are understood by all who are involved in production facility design. In other countries and other industries, the ISA symbol system is more common.

Table 14-6 shows the system used in RP 14C for identifying equipment items. The RP 14C system enables a relief valve on a specific separator to be identified as:

PSV, MBD-1000

If there are two, they would be designated:

PSV, MBD-1000A and PSV, MBD-1000B

Many operators use a simpler system, using "V" for pressure vessel, "T" for tank, "P" for pump, "C" for compressor, and "E" for heat exchanger, in which case the relief valve would be designated:

PSV, V1000 or

PSV, V1000A and PSV, V1000B

(*text continued on page 418*)

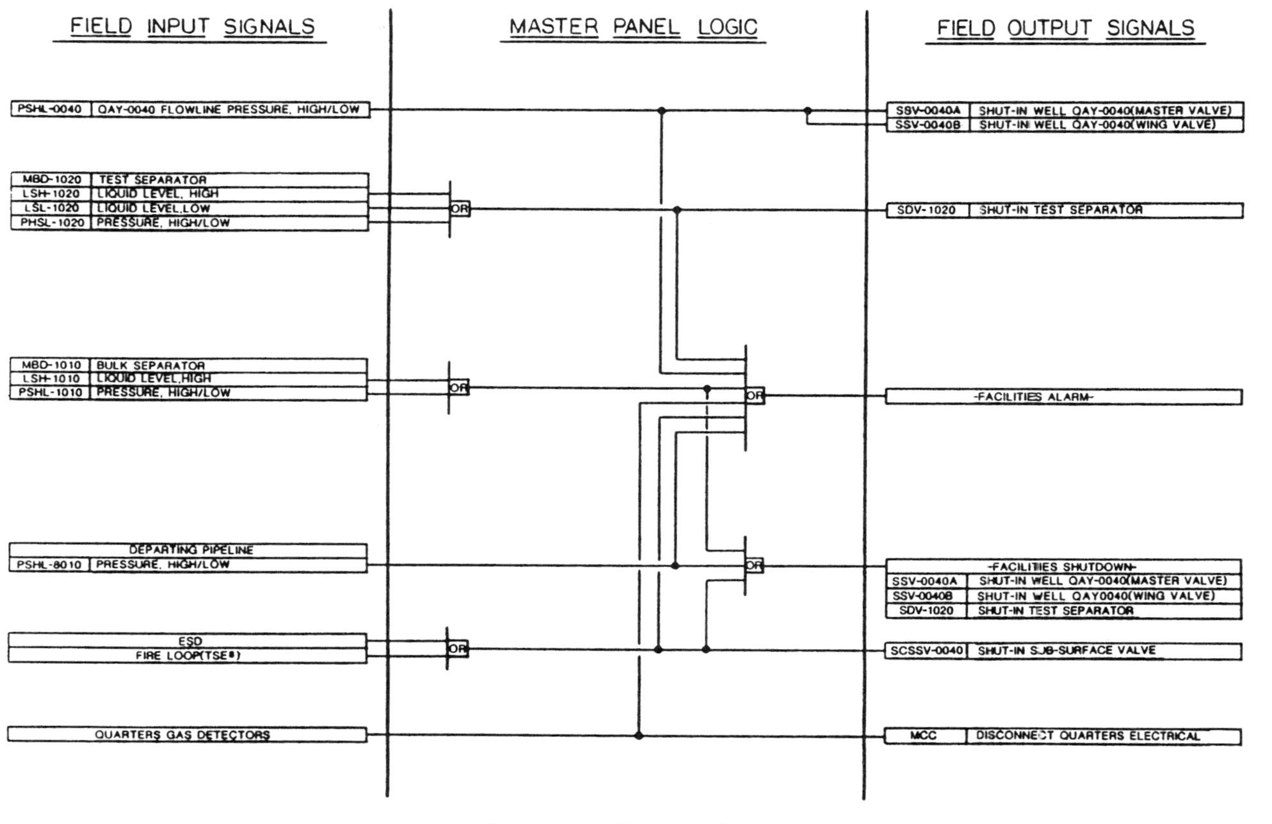

Figure 14-6. Function chart.

Table 14-5
Safety Device Symbols

Sensing and Self-Acting Devices				
	Safety Device Designation		Symbol	
Variable	Common	Instrument Society of America (I.S.A.)	Single Device	Combination Device
Backflow	Check Valve	Flow Safety Valve	FSV	
Burner Flame	Burner Flame Detector	Burner Safety Low	BSL	
Combustible Gas Concentration	Combustible Gas Detector	Analyzer Safety High	ASH	
Flow	High Flow Sensor	Flow Safety High	FSH	FSHL
	Low Flow Sensor	Flow Safety Low	FSL	
Level	High Level Sensor	Level Safety High	LSH	LSHL
	Low Level Sensor	Level Safety Low	LSL	

(table continued)

Pressure	High Pressure Sensor	Pressure Safety High	FSH	PSHL
	Low Pressure Sensor	Pressure Safety Low	PSL	
	Pressure Relief or Safety Valve	Pressure Safety Valve	PSV PSV	
	Rupture Disc or Safety Head	Pressure Safety Element	PSE	
Pressure or Vacuum	Pressure-Vacuum Relief Valve	Pressure Safety Valve	PSV	
	Pressure-Vacuum Relief Manhole Cover	Pressure Safety Valve	PSV	
	Vent	None		
Vacuum	Vacuum Relief Valve	Pressure Safety Valve	PSV	
	Rupture Disc or Safety Head	Pressure Safety Element	PSE	
Temperature	Fusible Material	Temperature Safety Element	TSE	
	High Temperature Sensor	Temperature Safety High	TSH	TSHL
	Low Temperature Sensor	Temperature Safety Low	TSL	
Flame	Flame or Stack Arrestor	None		

(table continued on next page)

Table 14-5 (Continued)
Safety Device Symbols

Sensing and Self-Acting Devices				
	Safety Device Designation		Symbol	
Variable	Common	Instrument Society of America (I.S.A.)	Single Device	Combination Device
Fire	Flame Detector (Ultraviolet/ Infrared)		USH	
	Heat Dector (Thermal)	Temperature Safety High	TSH	
	Smoke Detector (Ionization)		YSH	
	Fusible Material	Temperature Safety Element	TSE	
Combustible Gas Concentration	Combustible Gas Detector	Analyzer Safety High	YSH	

(table continued)

Actuated Valves

Service		Common Symbols		
Wellhead Surface Safety Valve or Underwater Safety Valve	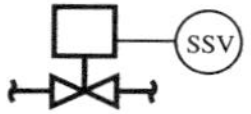 SSV	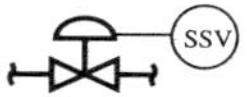 SSV	Note: Show "USV" for Underwater Safety Valves	
Blow Down Valve	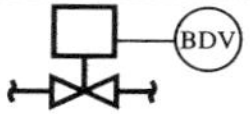 BDV	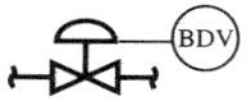 BDV	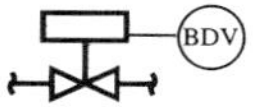 BDV	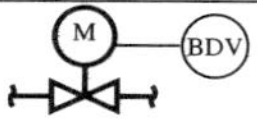 M BDV
All other Shut Down Valves	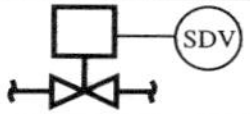 SDV	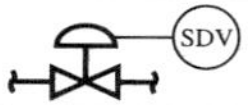 SDV	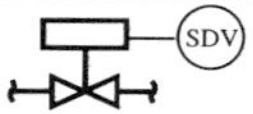 SDV	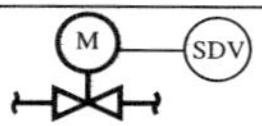 M SDV

Table 14-6
Component Identification

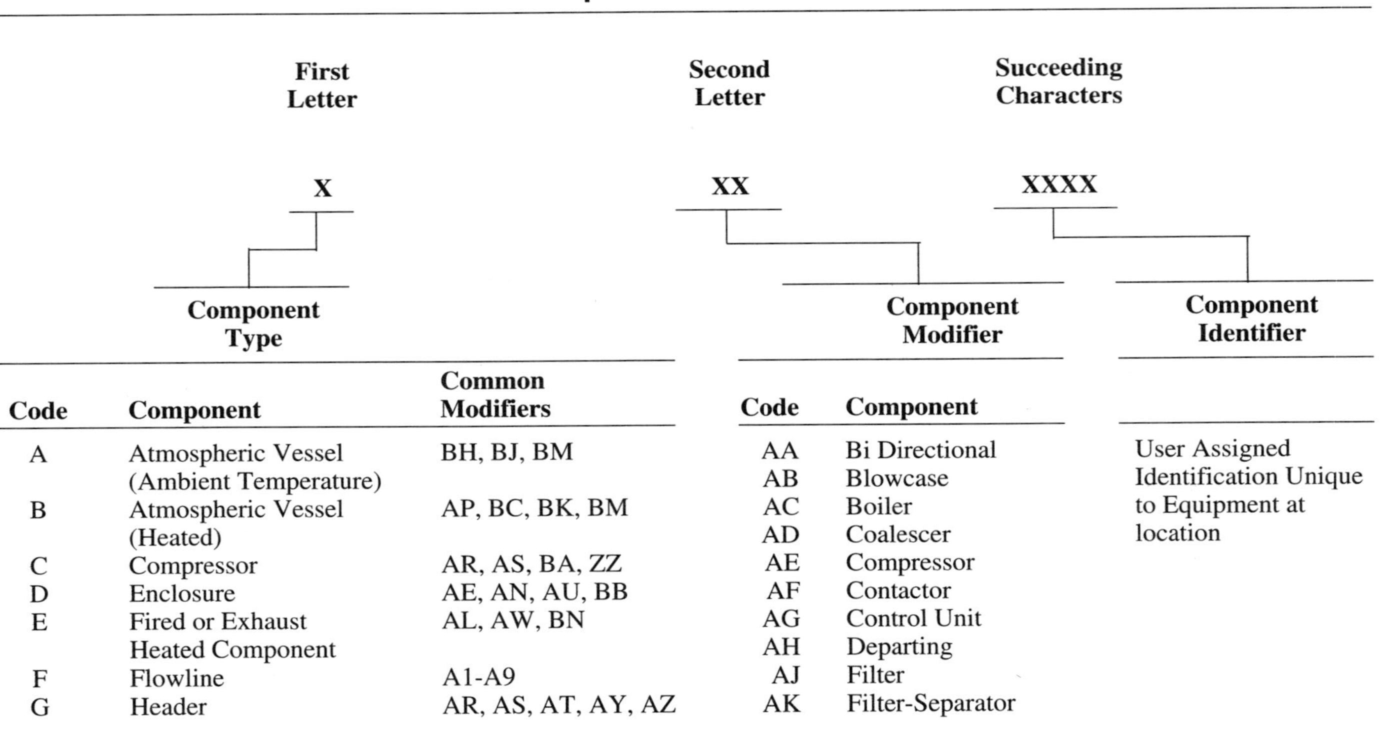

First Letter	Second Letter	Succeeding Characters
X	XX	XXXX
Component Type	Component Modifier	Component Identifier

Component Type

Code	Component	Common Modifiers
A	Atmospheric Vessel (Ambient Temperature)	BH, BJ, BM
B	Atmospheric Vessel (Heated)	AP, BC, BK, BM
C	Compressor	AR, AS, BA, ZZ
D	Enclosure	AE, AN, AU, BB
E	Fired or Exhaust Heated Component	AL, AW, BN
F	Flowline	A1-A9
G	Header	AR, AS, AT, AY, AZ

Component Modifier

Code	Component
AA	Bi Directional
AB	Blowcase
AC	Boiler
AD	Coalescer
AE	Compressor
AF	Contactor
AG	Control Unit
AH	Departing
AJ	Filter
AK	Filter-Separator

Component Identifier

User Assigned Identification Unique to Equipment at location

(*table continued*)

H	Heat Exchanger	BG	AL	Forced Draft
J	Injection Line	AR, AS, AT	AM	Freewater Knockout
K	Pipeline	AA, AH, AQ	AN	Generator
L	Platform	AG	AP	Heater
M	Pressure Vessel (Ambient Temperature)	AB, AD, AF, AJ, AK, AM, AV, BD, BF, BH, BJ, BL, BM	AQ	Incoming
			AR	Injection, Gas
			AS	Injection, Gas Lift
			AT	Injection, Water
N	Pressure Vessel (Heated)	AC, AF, AM, AP BC, BD, BG, BJ, BK	AU	Meter
			AV	Metering Vessel
			AW	Natural Draft
P	Pump	AX, BA, BE	AX	Pipeline
Q	Wellhead	AR, AT, AY, AZ	AY	Production, Hydrocarbon
Z	Other		AZ	Production, Water
			A1–A9	Flowline Segment
			BA	Process, Other
			BB	Pump
			BC	Reboiler
			BD	Separator
			BE	Service
			BF	Scrubber
			BG	Shell and Tube
			BH	Sump
			BJ	Tank
			BK	Treater
			BL	Volume Bottle
			BM	Water Treating
			BN	Exhaust Heated
			ZZ	Other

(text continued from page 410)

HAZARDS ANALYSIS

A hazards analysis is a systematic procedure for identifying potential hazards which could exist in a facility, evaluating the probability and consequence associated with the hazard, and either reducing the probability or mitigating the consequence so that the overall risk associated with the hazard is "acceptable." The different hazards analysis techniques can be applied at various stages during the course of the project to assess and mitigate potential hazards during design, construction and operations of the facility.

Types of Hazards Analysis

Hazards analysis techniques fall in two broad categories. Some techniques focus on hazards control by assuring that the design is in compliance with a pre-existing standard practice. These techniques result from prior hazards analysis, industry standards and recommended practices, results of incident and accident evaluations or similar facilities. Other techniques are predictive in that they can be applied to new situations where such pre-existing standard practices do not exist.

The most common hazards control technique is a "checklist." The checklist is prepared by experienced personnel who are familiar with the design, construction and operation of similar facilities. Checklists are relatively easy to use and provide a guide to the evaluator of items to be considered in evaluating hazards. API RP 14J has examples of two checklists which can be used to evaluate facilities of different complexity. Because production facilities are very similar and have been the subject of many hazard analyses, a checklist analysis to assure compliance with standard practice is recommended for most production facilities. The actual procedure by which the checklist is considered and the manner in which the evaluation is documented to assure compliance varies from case-to-case.

The most common predictive technique which is used to analyze facilities which contain new equipment or processes, or where there is an unusually high risk to personnel or the environment is the Hazard and Operability technique or "HAZOP." A HAZOP study requires a team of five to ten multi-discipline personnel consisting of representatives from engineering, operations, and health, safety, and environmental staff. The facility is broken down into "nodes" (usually a major piece of equipment and its associated piping, valves and instrumentation), and an experienced team leader guides the team through an analysis of each node using a predetermined list of "guide words" and "process parameters." For example, the guide word "LOW" and process parameter "PRESSURE" results in questions being asked as to potential causes for lower than design pressure at the node. If the condition is possible, effects are analyzed and, if necessary, methods of mitigation are added until the risk is deemed acceptable. Although this method is time consuming, it proves to be a thorough method of analysis and is effective for a new process which has never been analyzed before or for a known process which incorporates new equipment. However, a checklist should be used in conjunction with a HAZOP to assure that compliance with standard practice is not inadvertently overlooked by the HAZOP team.

Problems Commonly Encountered

There are several problem areas which seem to appear often in the results of hazards analyses. The most common are:

1. Relief Valve Sizes
   Relief valves are often seen to be undersized for the required relieving rate, due either to poor initial design or changes in the process conditions which occurred during design. The most common system problem is that the relief valve was adequately sized for blocked discharge but not sized for the flowrate that could occur as a result of a failure in the open position of an upstream control valve (i.e., gas blowby). See Chapter 13.
2. Open and Closed Drains
   Another common problem area is having open and closed drain systems tied together. Liquid which drains from pressure vessels "flash" at atmospheric pressures giving off gas. If this liquid flows in the same piping as open drains, the gas will seek the closest exit to atmosphere it can find, causing a potential fire hazard at any open drain in the system.
   Many accidents have occurred where gas has migrated through the drain system to an unclassified area where welding, or other hot work, was being performed. See Chapter 15.

3. Piping Specification Breaks
   Piping pressure ratings should be designed so that no matter which valve is closed, the piping is rated for any possible pressure it could be subjected to, or is protected by a relief valve.
   When a spec break is taken from a higher to a lower MAWP, there must be a relief valve on the lower pressure side to protect the piping from overpressure. The relief valve can be either on the piping or, more commonly, on a downstream vessel. Spec break problems most commonly occur where a block valve exists on a vessel inlet, or where a bypass is installed from a high pressure system, around the pressure vessel which has a relief valve, to a lower pressure system.
4. Electrical Area Classification
   Another common mistake often uncovered is electrical equipment which is not consistent with the design area classification. See Chapter 17.

SAFETY MANAGEMENT SYSTEMS

A hazards analysis by itself cannot assure that an adequate level for safety is provided for a facility unless the hazard analysis is included as part of a comprehensive safety management system. In the United States every facility handling highly hazardous chemicals, including some onshore production facilities and most gas plants, must have a Process Safety Management (PSM) Plan in place. Offshore operators have developed a voluntary safety management system presented in API RP 75, "Recommended Practices for Development of a Safety and Environmental Management Program for Outer Continental Shelf (OCS) Operations and Facilities" (SEMP), which describes the elements which should be included in a safety management plan.

The requirements of both PSM and SEMP are, from a practical standpoint, identical and thus, SEMP can easily be applied to onshore facilities as well as offshore facilities. The basic concepts of SEMP are as follows:

Safety and Environmental Information

Safety and environmental information is needed to provide a basis for implementation of further program components such as operating procedures and hazards analysis. Specific guidelines as to what information is needed are contained in API RP 14J.

Hazards Analysis

This subject is addressed in the previous section of this chapter. Specific guidelines for performing hazards analysis are contained in API RP 14J.

Management of Change

Management of Change is a program that helps to minimize accidents caused by changes of equipment or process conditions due to construction, demolition, or modification. Procedures should be set up to identify the various hazards associated with change. All changes, although sometimes minor, can result in accident and/or injury if proper steps are not implemented to make operators aware of the differences. Changes in facilities as well as changes in personnel should be managed to maintain the safety of all personnel and the environment.

Operating Procedures

The management program should include written facility operating procedures. These procedures should provide ample instruction for sound operation and be consistent with the safety and environmental information. Procedures should be reviewed and updated periodically to reflect current process operating practices. Procedures provide the means for education of new employees about the process and provide education to all employees on new equipment and practices.

Safe Work Practices

A disproportional amount of accidents occur during construction and major maintenance activities. Safe work practices are written with this in mind and, as a minimum, should cover the following:

- Opening of equipment or piping,
- Lockout and tagout of electrical and mechanical energy sources,
- Hot work and other work involving ignition sources,
- Confined space entry, and
- Crane operations.

Training

Training for new employees and contractors, and periodic training of existing employees is necessary to educate personnel to be able to perform their work safely and to be aware of environmental considerations. Training should address the operating procedures, the safe work practices, and the emergency response and control measures.

Assurance of Quality and Mechanical Integrity of Critical Equipment

Procedures for assurance of quality in the design, fabrication, installation, maintenance, testing and inspection for critical equipment are required. Safety requires that critical safety devices must operate as intended and process system components must be maintained to be able to contain design pressures.

Pre-startup Review

A pre-startup safety and environmental review should be performed on all modified or newly constructed facilities.

Emergency Response and Control

An Emergency Action Plan should be established, assigning an emergency control center and appropriate personnel for emergency response. Drills should be carried out to assure all personnel are familiar with these plans.

Investigation of Incidents

An investigation is required if an incident involving serious safety or environmental consequences or the potential for these consequences occurs. The purpose of such investigation is to learn from mistakes made and provide corrective action. Investigations should be performed by knowledgeable personnel and should produce recommendations for safer working conditions.

Audit of Safety and Environmental Management Program Elements

Periodically, the SEMP elements should be audited to evaluate the effectiveness of the program. Auditing should be conducted by qualified personnel through interviews and inspections. If audits consistently find no deficiencies in the program, then management should conclude that the audit is not in itself being done properly, as there are always improvements that can be made in a safety management system.

SAFETY CASE AND INDIVIDUAL RISK RATE

The overall system for safety described above can be called the "API System." It is based on a series of API Standards and Recommended Practices which can be summarized as a four step system with each succeeding step encompassing the preceding steps:

1. Design and maintain a system for process upset detection and shutdown—RP 14C.
2. Design and select hardware with known reliability and mechanical integrity to contain pressure and mitigate failure consequences—all other API RP 14 series standards.
3. Follow system design concepts, documentation needs and hazards analysis requirements—RP 14J.
4. Develop a management of safety system—RP 75.

This system has proven to provide adequate levels of safety in the Gulf of Mexico and other similar areas where it is possible to abandon the location during a catastrophic event. In the North Sea where harsh environmental conditions exist, a different approach to safety has evolved which is based on developing a Safety Case and calculating an Individual Risk Rate (IRR) to show that the risk to any individual working in the facility is As Low As Reasonably Practicable (ALARP).

A Safety Case is a narrative that literally makes the case that an adequate level of safety has been reached for an installation. It requires looking at all potential hazards which could lead to a loss of the installation, a loss of life, or a major pollution event. A risk analysis is performed on each hazard evaluating the probability of the event occurring and describing the magnitude of the consequences. A discussion is then given of the measure undertaken to lower the probability of occurrence or to mitigate the consequences and a "case" is made that the risk for the installation meets the ALARP safety criteria.

In the North Sea this is often done with detailed quantified risk assessments and the calculation of an overall IRR or risk of total loss of structure. Mitigation measures are incorporated until it can be shown that risk levels meet a minimum criteria *and* the cost of further mitigation has such high cost to benefit ratios that further mitigation is no longer "practicable."

These analyses tend to be rather long and complex and can negatively impact both project cycle time and cost. Indeed, as a check to assure that basic known safety concepts are not inadvertently overlooked in the pile of documentation which is necessary for a safety case, the safety case approach should include within it all the elements of the "API System." Even if a safety case is performed, it is still necessary to assure compliance with good practices and that all elements of a proper safety management system are included. Thus a common sense approach in the absence of government regulation would be to use the API System for most installations and, in those instances where there is a large concentration of personnel or where abandoning the location may be impossible due to weather or remoteness, to use a qualitative safety case to think through fire fighting and escape options.

CHAPTER

15

Valves, Fittings, and Piping Details*

The various items of equipment in the production facility are connected by valves, fittings, and piping to enable and control flow from one piece of equipment to another. Chapter 9 of Volume 1 discusses factors governing the choice of line size and wall thickness. This chapter describes the various types of valves and fittings commonly used in production facilities and presents some common piping details and specifications.

The specific piping details used in a project are normally contained in a company pipe, valve, and fittings specification, which addresses the following subjects:

1. Governing industry codes.
2. Material requirements for pipe, flanges, fittings, bolts, nuts, and gaskets.
3. Material and construction for each valve used in the piping.
4. Pipe schedule and end connection for each service and pressure rating.

---

*Reviewed for the 1999 edition by Jorge Zafra of Paragon Engineering Services, Inc.

5. Welding certification and inspection requirements.
6. Design details (if not already included in drawings). For example, branch connections, pipe support spacing and details, clearances, and accessibility.

In order to cover all the lines in the facility, pipe and valve tables are normally included. Each pipe pressure class is assigned a designation. Sometimes it is necessary to assign two classes for a single designation. For example, in Table 15-1 "A," "L," and "AA" are all ANSI 150 class, but they contain different fluids.

A separate table such as the example in Table 15-2 is prepared for each line designation. Each valve is assigned a designation on the flowsheets and explained in this table. The pipe, valves, and fittings table can specify acceptable valves by manufacturer and model number, by a generic description, or by a combination of the two as shown in the example. It should be pointed out that Tables 15-1 and 15-2 are examples from American Petroleum Institute Recommended Practice (API RP) 14E and are illustrative only. There are almost as many different formats for pipe, valve, and fittings tables as there are companies, and these examples are in no manner typical or recommended. Often, for simplicity, valve types are not described in the pipe, valve, and fittings specifications but on separate sheets for each valve designation, as discussed below under Valve Selection and Designation.

VALVE TYPES

The following descriptions are meant to briefly describe the various generic valve types. Within each type there are numerous different design details that separate one valve manufacturer's valves from the next. All of these specific models have good points and bad points; all can be used correctly and incorrectly. It is beyond the scope of this book to critique each valve manufacturer's design. However, the various valve salesmen will be more than pleased to contrast the benefits of their valve's features with that of the competition. The reader is cautioned that higher-cost valves do not necessarily mean better valves, and that expensive valves can result in a significant waste of money when a less expensive but adequate valve will perform satisfactorily.

Ball Valves

This is a quarter-turn on-off valve (Figure 15-1). A bore through the ball allows flow when it is lined up with the pipe and blocks flow when it

Table 15-1
Example Index of Pipe, Valves, and Fittings

Table	Service	Pressure Rating Classification
A	Non-corrosive Hydrocarbons and Glycol	150 lb ANSI
B	Non-corrosive Hydrocarbons and Glycol	300 lb ANSI
C	Non-corrosive Hydrocarbons and Glycol	400 lb ANSI
D	Non-corrosive Hydrocarbons and Glycol	600 lb ANSI
E	Non-corrosive Hydrocarbons and Glycol	900 lb ANSI
F	Non-corrosive Hydrocarbons and Glycol	1500 lb ANSI
G	Non-corrosive Hydrocarbons and Glycol	2500 lb ANSI
H	Non-corrosive Hydrocarbons	API 2000 psi
I	Non-corrosive Hydrocarbons	API 3000 psi
J	Non-corrosive Hydrocarbons	API 5000 psi
K	Non-corrosive Hydrocarbons	API 10000 psi
L	Air	150 lb ANSI
M	Water	125 lb Cast Iron
N	Steam and Steam Condensate	300 lb ANSI
O	Drains and Sewers	Atmospheric
P (Spare)		
Q (Spare)		
R (Spare)		
SV	Valves for Corrosive Service	General
AA	Corrosive Hydrocarbons	150 lb ANSI
BB	Corrosive Hydrocarbons	300 lb ANSI
CC (Not Prepared)	Corrosive Hydrocarbons	400 lb ANSI
DD	Corrosive Hydrocarbons	600 lb ANSI
EE	Corrosive Hydrocarbons	900 lb ANSI
FF	Corrosive Hydrocarbons	1500 lb ANSI
GG	Corrosive Hydrocarbons	2500 lb ANSI

is perpendicular to the pipe. A "regular port" ball valve has a bore diameter less than the pipe inside diameter, while a "full opening" ball valve has a ball diameter equal to the pipe inside diameter.

Ball valves are limited in temperature by the elastomer material used in their seats. Many designs have a secondary metal-to-metal seat to provide a seal in case of fire. Ball valves are not suitable as a throttling valve, but can be used on start-up and shut-down in the partially open position to bleed pressure into or out of a system.

These valves are the most common general-purpose on/off valves in production facilities.

(*text continued from page 430*)

Table 15-2
Example Specifications of Pipe, Valves, and Fittings

150-lb ANSI
Non-corrosive service[1]
Temperature range: –20 to 650°F
Maximum pressure: Depends on flange rating[2] at service temperature

Size Ranges	General Specifications	Platform Service
Pipe	Grade depends on service	ASTM A106, Grade B, Seamless[3]
¾-in. and smaller nipples	threaded and coupled	Schedule 160 or XXH
1½-in. and smaller pipe	threaded and coupled	Schedule 80 min
2-in.–3-in. pipe	beveled end	Schedule 80 min
4-in. and larger pipe	beveled end	See Table 2-4
Valves (Do not use for temperatures above maximum indicated.)		
Ball		
½-in. and smaller	1500 lb CWP ANSI 316 SS screwed, regular port, wrench operated, Teflon seat	Manufacturer's Figure No. _________ (300°F)
¾-in.–1½-in.	1500 lb CWP, CS, screwed, regular port, wrench operated, Teflon seat	Manufacturer's Figure No. _________ or Figure No. _________ (450°F)
2-in.–8-in.	150 lb ANSI CS RF flanged, regular port, lever or hand wheel operated, trunnion mounted	Etc.
10-in. and larger	150 lb ANSI CS RF flanged, regular port, gear operated, trunnion mounted	Etc.
Gate		
½-in. and smaller	2000 lb CWP, screwed, bolted bonnet, AISI 316 SS	Etc.
¾-in.–1½-in.	2000 lb CWP, screwed, bolted bonnet, forged steel	Etc.
2-in–12-in.	150 lb ANSI CS RF flanged, standard trim, hand wheel or lever operated	Etc.

Table 15-2 (Continued)
Example Specifications of Pipe, Valves, and Fittings

Globe		
1½-in. and smaller (Hydrocarbons)	2000 lb CWP CS screwed	Etc.
1½-in. and smaller (Glycol)	2000 lb CWP CS socketweld	Etc.
2-in. and larger	150 lb ANSI CS RF flanged, handwheel operated	Etc.
Check		
1½-in. and smaller	600 lb ANSI FS screwed, bolted bonnet[4], standard trim	Etc.
2-in. and larger	150 lb ANSI CS RF flanged, bolted bonnet[4], swing check, standard trim	Etc.
Reciprocating Compressor Discharge	300 lb ANSI CS RF flanged, piston, check, bolted bonnet[4]	Etc.
Lubricated Plug		
1½-in.–6-in.	150 lb ANSI CS RF flanged, bolted bonnet	Etc.
Non-lubricated Plug		
1½-in.–6-in.	150 lb ANSI CS RF flanged. bolted bpmmet	Etc.
Compressor Laterals		
	Use ball valves	
Needle		
¼-in.–½-in.	6000 lb CWP, bar stock screwed, AISI 316 SS	Etc.
Fittings		
Ells and Tees		
¾-in. and smaller	6000 lb FS screwed	ASTM A105
1-in.–1½-in.	3000 lb FS screwed	ASTM A105
2-in. and larger	Butt weld, seamless, wall to match pipe	ASTM A234, Grade WPB
Unions		
¾-in. and smaller	6000 lb FS screwed, ground joint, steel to steel seat	ASTM A105
1-in.–1½-in.	3000 lb FS screwed, ground joint, steel to steel seat	ASTM A105
2-in. and larger	Use flanges	

(*table continued on next page*)

Table 15-2 (Continued)
Example Specifications of Pipe, Valves, and Fittings

Couplings		
1-in. and smaller	6000 lb FS screwed	ASTM A105
1½-in.	3000 lb FS screwed	ASTM A105
Plugs		
1½-in. and smaller	Solid bar stock, forged steel	ASTM A105
2-in. and larger	X-Strong seamless, weld cap	ASTM A234, Grade WPB
Screwed Reducers		
¾-in. and smaller	Sch. 160 seamless	ASTM A105
1-in–1½-in.	Sch. 80 seamless	ASTM A105
Flanges		
1½-in. and smaller	150 lb ANSI FS RF screwed	ASTM A105
2-in. and larger	150 lb ANSI FS RF weld neck, bored to pipe schedule	ASTM A105
Bolting		
Studs	Class 2 fit, threaded over length	ASTM A193, Grade B7[4]
Nuts	Class 2 fit, heavy hexagon, semi-finish	ASTM A194, Grade 2H[4]
Gaskets	Spiral wound asbestos	Spiral Wound Mfg. Type ________ or Mfg. No. ________ w/AISI 304 SS windings
Thread Lubricant	Conform to API Bulletin 5A2	Mfg. No. ________

Notes:
[1]For glycol service, all valves and fittings shall be flanged or socketweld.
[2]API 5L, Grade B, Seamless may be substituted if ASTM A106, Grade B, Seamless is not available.
[3]Studs and nuts shall be hot-dip galvanized in accordance with ASTM A153.
*[4]Fittings and flanges that do not require normalizing in accordance with ASTM A105, due to size or pressure rating, shall be normalized when used for service temperatures from –20°F to 60°F. Fittings and flanges shall be marked HT, N, * or with some other appropriate marking to designate normalizing.*

(text continued from page 427)

Plug Valves

Plug valves are similar to ball valves except a cylinder with a more-or-less rectangular opening is used instead of a ball (Figure 15-2). Plug valves cannot be made full opening and are limited in temperature rating by the rating of their elastomer seats.

Figure 15-1. Cutaway of ball valve. (*Courtesy of Cameron Iron Works, Inc.*)

Figure 15-2. Cutaway of plug valve. (*Courtesy of Xomox Corp.*)

Plug valves usually require lubrication on a regular basis to seal. They tend to have a more unpredictable torque than ball valves and are thus hard to automate.

As with ball valves, plug valves are on/off valves and should only be used infrequently in throttling service. They tend to be less expensive than ball valves, but are not as popular because of their need for lubrication.

Gate Valves

A gate valve seals off flow when a slab is either raised (normal acting) or lowered (reversed acting) so that the hole through the slab no longer lines up with the pipe (Figure 15-3). Gates are harder to operate manually than balls or plugs as they take many turns of the handwheel to open or close the valve. However, this action is easier to automate with a power piston than the quarter-turn action required for balls and plugs.

An unprotected rising stem that is corroded or painted can make the valve difficult to operate. Stem protectors and keeping the stem packed in grease help alleviate this problem.

Gate valves are on/off valves and should never be used for throttling. Gate valves are usually less expensive than balls or plugs in high-pressure service or for large pipe sizes. They make excellent high-pressure flowline and pipe shut-in valves, but are not extensively used in normal facility piping 12 in. and smaller in diameter.

Butterfly Valves

These valves are relatively inexpensive, quarter-turn valves. The seal is made by a rotating disc that remains in the flowstream subject to erosion while in the open position (Figure 15-4). Except in low-pressure service, they should not be used to provide a leak-tight seal, but they can be used to throttle where a tight shut-off is not required.

Butterfly valves are particularly useful in low-pressure produced water service and as gas throttling valves on the inlets of compressors.

Globe Valves

The most common valve construction for throttling service, and thus for automatic control, is a globe valve (Figure 15-5). The movement of the stem up or down creates an opening between the disc and seat that allows fluid to pass through the valve. The greater the stem movement, the larger the annulus opening for fluid flow. Because only a small stem movement is

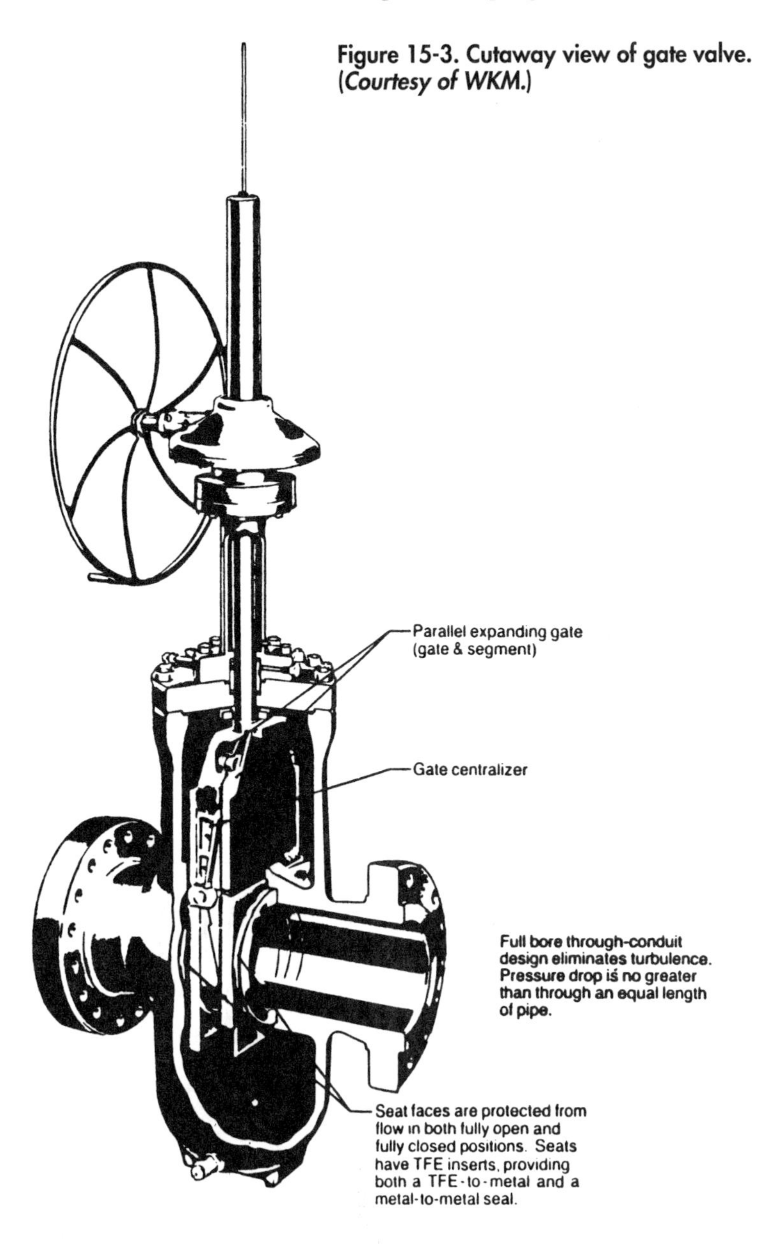

Figure 15-3. Cutaway view of gate valve. (*Courtesy of WKM.*)

required from the fully closed to fully open positions, globe valves are easy to automate with a diaphragm operator as shown in Figure 15-6.

Globe valves have metal-to-metal seats, which are easy to replace. Because of the erosive action of the fluid when the valve is throttling, they should not be used for on/off service. A tight seal may not be possible.

Figure 15-4. Butterfly valve. (*Courtesy of Keystone Valve USA, Inc.*)

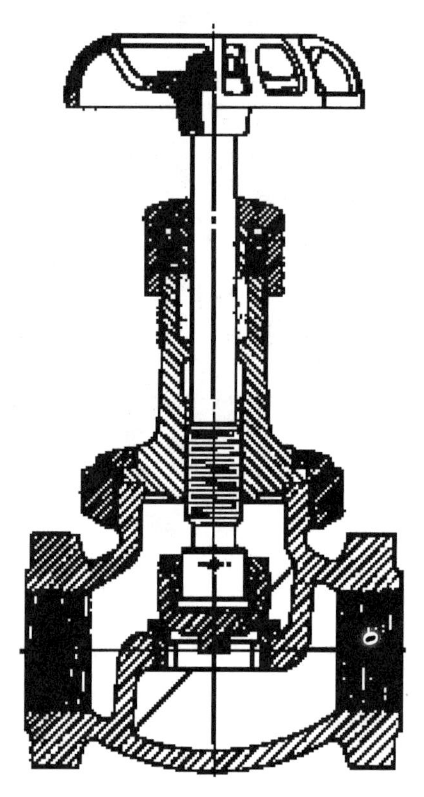

Figure 15-5. Cutaway view of globe valve. (*Courtesy of Jenkins Bros.*)

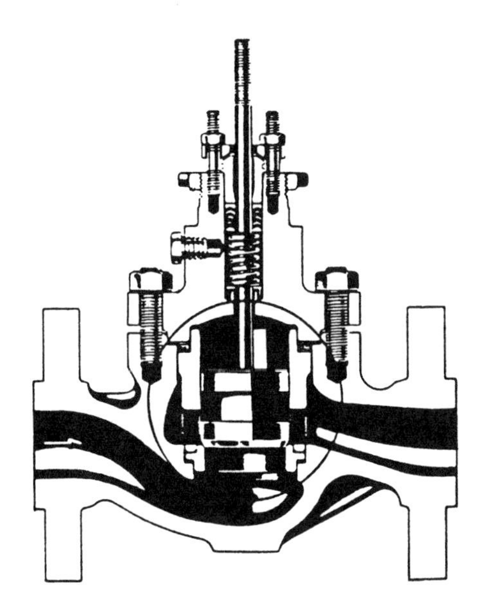
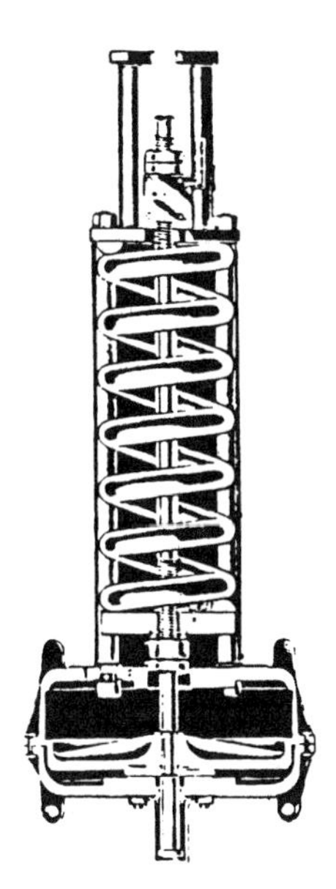

Figure 15-6. Typical single-port body control valve (left) and pneumatic actuator (right). (*Courtesy of Fisher Controls International, Inc.*)

There are many configurations of plugs and seats ("trim") that create different control responses. The design specification of control valves is beyond the scope of this book.

Diaphragm (Bladder) Valves

These special throttling valves use an elastomer diaphragm to restrict or stop flow (Figure 15-7). They are suitable for slurry service and make an excellent valve for sand drains. Unfortunately, they do not provide a reliable, positive shut-off and should be installed in series with a ball or other on/off valve if positive shut-off is required.

The elastomer selection limits the temperature rating of the valve.

Needle Valves

A type of miniature globe valve, needle valves are used in instrument systems for throttling of small volumes. They have metal to metal seats, but due to the small size, can be used for positive shut-off (Figure 15-8). Needle valves have small passageways that may plug easily and limit their use to very small flow rates.

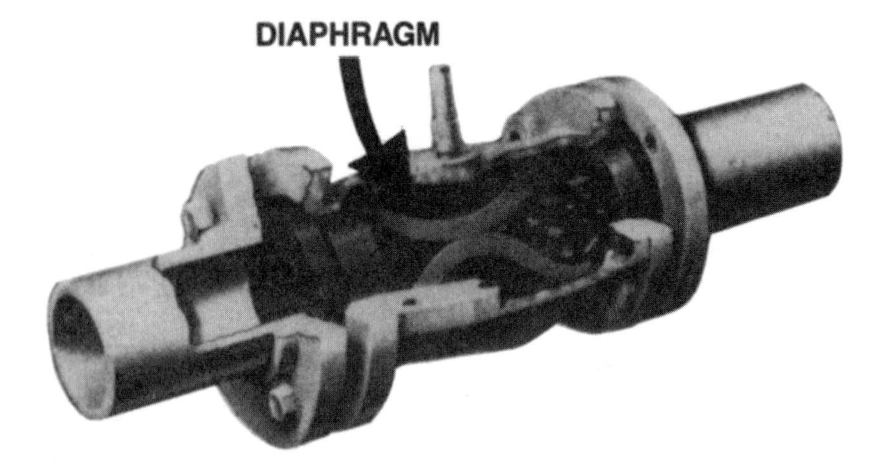

Figure 15-7. Diaphragm valve. (*Courtesy of Flexible Valve Corp.*)

Check Valves

Used to *restrict* reversal of flow, check valves should not be considered as positive shut-off valves when flow is reversed, since the seating element is always in the flow stream and subject to erosion (Figures 15-9 to 15-13). A section of a line should not be considered isolated if the only barrier to flow is a check valve. On the other hand, because they do restrict backflow to very low levels, check valves installed in appropriate locations can protect equipment and minimize damage in case of a leak in the upstream line. Some of the advantages and disadvantages of the various check valve configurations are as follows:

- Swing
  1. Suitable for non-pulsating flow.
  2. Not good for vertical upward flow.
  3. Available in wafer design for mounting between flanges.
- Split Disk
  1. Mounted between flanges.
  2. Springs subject to failure.
- Lift Plug and Piston
  1. Good for pulsating flow.
  2. Can be used in vertical upward flow.
  3. Easier to cut out in sandy service than full-opening swing.
  4. Subject to fouling with paraffin and debris.

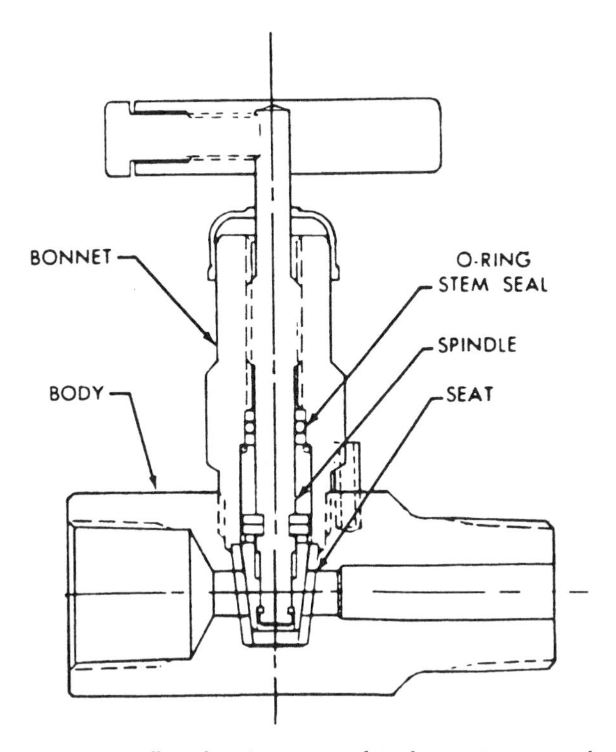

Figure 15-8. Needle valve. (*Courtesy of Anderson Greenwood and Co.*)

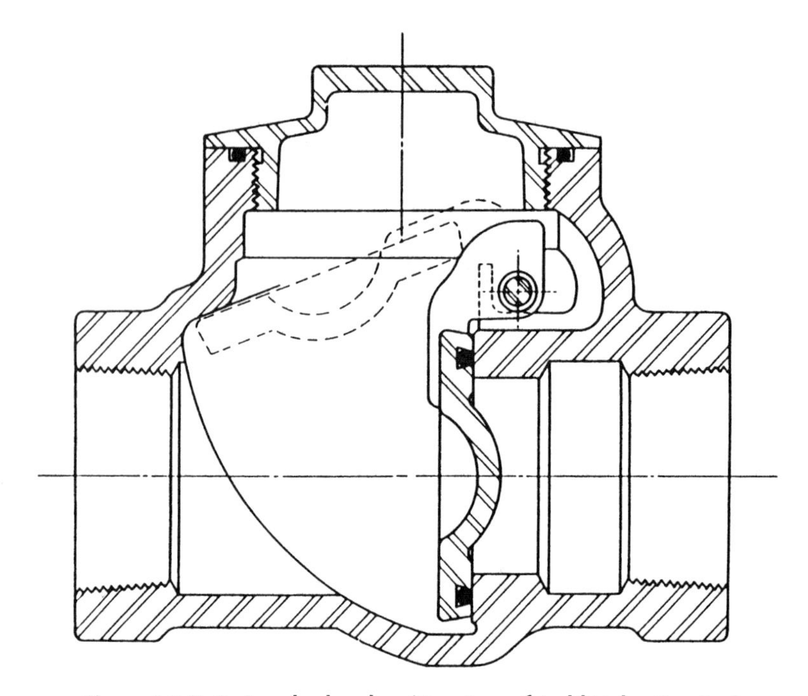

Figure 15-9. Swing check valve. (*Courtesy of Judd Valve Co., Inc.*)

- Ball
  1. Does not have a tendency to slam shut on flow reversal.
  2. Usually for sizes 1-in. and smaller.
  3. Can be used in vertical lines.

Valve Selection and Designation

Table 15-3 summarizes and compares the different valve types discussed in this chapter and highlights important properties that impact valve selection.

It is beneficial to designate valve types in schematic drawings of the facilities. The designation should indicate the type of valve (ball, gate, etc.) the type of end connection (flange, socketweld, threaded, etc.), the pressure rating class (ANSI 150, ANSI 600, API 2000, etc.) and the materials of construction. Table 15-4 shows a sample designation system. Using this system, the designation VBF-15-1 would indicate an ANSI 150 flanged ball valve. The specific attributes would then come from a pipe, valve, and fitting specification, such as Table 15-2, or from a separate valve specification for VBF-15-1, as shown in Table 15-5.

Table 15-3
Comparison of Valve Properties

Valve	Bubble Tight	Throttle	Where Used	Pig	Pressure Drop	Size
Ball	Yes	No On/Off	Isolation ubiquitous	Yes (Full)	Low	¼″–36″
Plug	Yes	No On/Off	Isolation Rare	No	Low	Rare Cheaper than ball
Gate	Yes	Some	Control, wellhead isolation, double block & bleed	Yes	Low	2″–Up Larger sizes cheaper than ball
Butterfly	Yes for low ΔP ANSI 150	Yes Gas Low ΔP	Isolation/Control	No	Low	2″–Up Larger sizes cheaper than globe
Globe	Not All	Yes	Control bypass, vent	No	High	2″–Up
Needle	Yes	Yes	Inst/Control	No Roddable		¼″–1½″
Check	No	No	To restrict reversal of flow Isolation	Swing check Valves only some cases	Low	½″–36″
Choke	Yes Adjustable choke only	Yes Adjustable choke only	Control	No	High	2″–9″ Bigger diameters special order

Courtesy of Paragon Engineering Services, Inc.

Table 15-4
Sample Valve Designation System

Each valve designation has four (4), and possibly five (5), parts.

(1) This part of each valve designation is always V, which stands for "valve."
(2) The second letter identifies valve type:

B = Ball
C = Check
D = Diaphragm
G = Gate
N = Needle
O = Globe
P = Plug
Y = Butterfly

(3) The third letter identifies end connections:

T = Threaded
S = Socketweld
F = Flanged
B = Buttweld

(4) The fourth part of each valve designation is a 2-, 3-, or 4- digit number indicating the highest ANSI or API class for which the valve can be used:

15 = ANSI 150
30 = ANSI 300
60 = ANSI 600
90 = ANSI 900
150 = ANSI 1500
250 = ANSI 2500
200 = 2000# API
300 = 3000# API
500 = 5000# API

(5) The fifth part of a valve designation, when used, is a modifier that distinguishes between two or more valves that have the same type and pressure rating but that are considered separately for some other reason.

Courtesy of Paragon Engineering Services, Inc.

CHOKES

Chokes are used to control flow where there is a large pressure drop. They can either be adjustable, where the opening size can be varied manually as shown Figure 15-14 and 15-15 or have a fixed size orifice. Due to the erosive nature of the fluid flow through a choke, they are constructed so beans, discs, and seats can be easily replaced.

Table 15-5
Sample Valve Table

Valve Designation:	VBF-15-1	
Service:	Hydrocarbons, Non-corrosive Glycol	
Type:	Ball Valve	
Rating:	ANSI 150	
	Design Temperature	Design Pressure
	–20° to 100°F	285 psig
	to 200°F	260 psig
	to 300°F	230 psig
Pressure Rating:	ANSI 150	
Body Material:	Carbon Steel	
Trim Material:	Hard Plated Carbon Steel Ball	
End Connection:	RF Flanged	
Valve Operator:	Lever through 8″, Gear Operated 10″ and larger	
Body Construction:	2″–4″: Floating Ball, Regular Port	
	6″ and larger: Trunnion Mounted Ball, Regular Port	
Trim Construction:	Renewable Seats, Removable Stem, Fire Safe	

Valve Comparison List

Manufacturer	Manufacturer's Fig. No.	Nominal Sizes
WKM	310-B100-CS-02-CS-HL	½″–4″
WKM	370CR-ANSI150RF21-AAF-21	6″–14″
Demco	121136X	2″–12″

PIPING DESIGN CONSIDERATIONS

Process Pressures

Maximum allowable working pressure (*MAWP*): Highest pressure to which the system can be subjected during operation. Thus, pressure is established by a relief device set pressure and must be less than or equal to the material strength limitations of equipment. This pressure establishes piping class for fittings and pipe wall thickness requirements, both of which are discussed in Volume 1.

Normal operating pressure: Anticipated process operating pressure used to determine pipe diameter requirements and pressure drop limitations for various operating conditions.

(*text continued on page 445*)

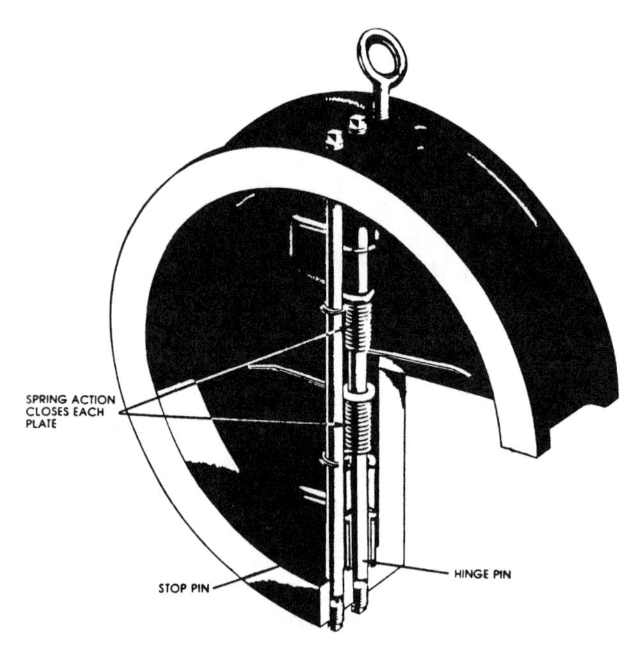

Figure 15-10. Wafer check valve. (*Courtesy of TRW Mission Drilling Products Division.*)

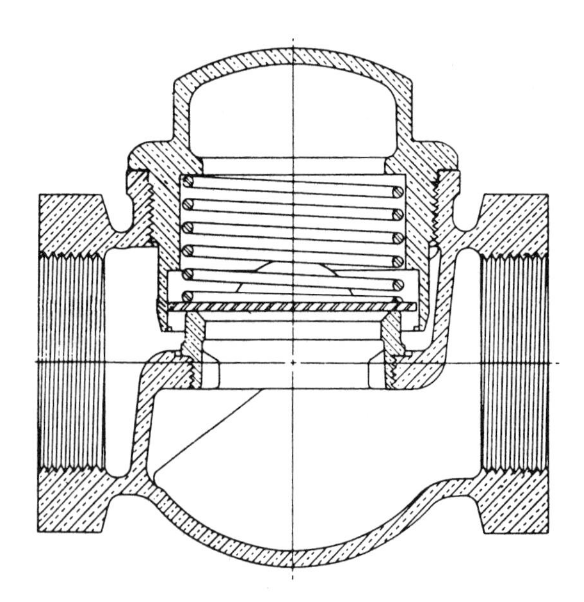

Figure 15-11. Lift check valve. (*Courtesy of Jenkins Bros.*)

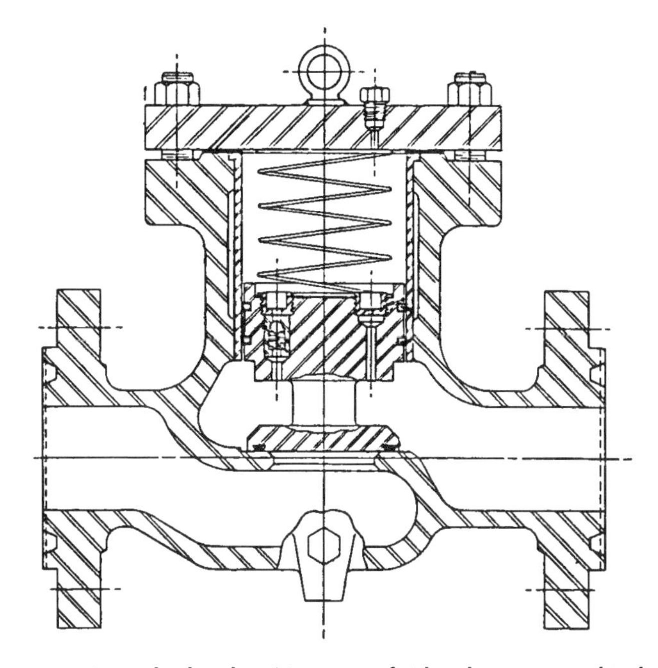

Figure 15-12. Piston check valve. (*Courtesy of Wheatley Pump and Valves, Inc.*)

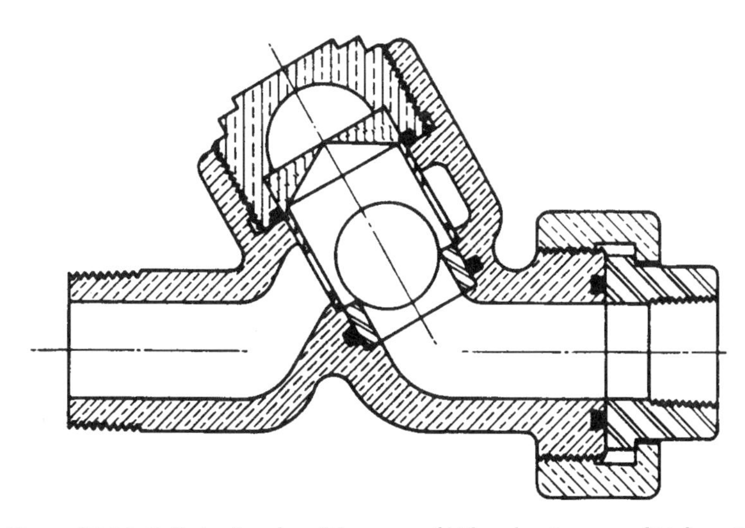

Figure 15-13. Ball check valve. (*Courtesy of Wheatley Pump and Valves, Inc.*)

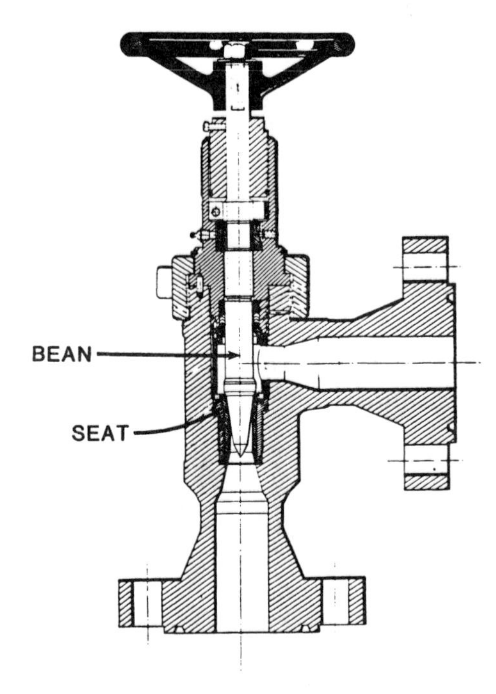

Figure 15-14. Plug and seat choke. *(Courtesy of Willis Control Division, Cameron Iron Works, Houston.)*

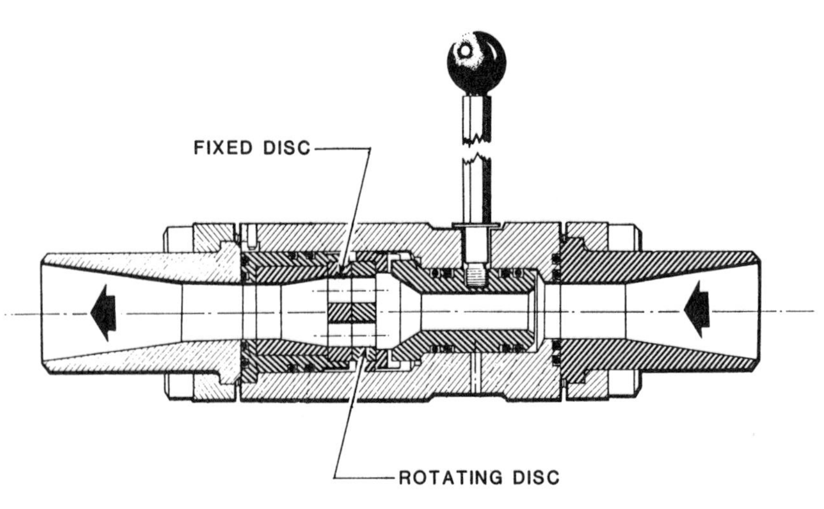

Figure 15-15. Rotating disc choke. *(Courtesy of Willis Control Division, Cameron Iron Works, Houston.)*

(text continued from page 441)

Future operation pressures: Sizing of lines must consider operating pressures expected as the reservoir depletes. Diameter requirement calculations should be made using both initial and future conditions to determine the governing case. Often in gas and two-phase lines the greatest flow velocity occurs late in life when flowing pressures are low even though flow rates may be lower than initial conditions.

Process Temperatures

Design temperature: Highest or lowest (depending upon which is controlling) temperature to which a line can be subjected during operation.

Normal operating temperature: Anticipated process operating temperature used to determine pipe diameter for various operating conditions.

Process Liquid Flow Rates

Liquid lines in production facilities are generally in either continuous or slugging service. Continuous duty lines should be sized to handle the average daily flow rate of the facility. An additional capacity is often added for surges. Lines in slugging service should be sized to accommodate actual flowing conditions. Design flow rates should be the maximum capacity that a line will accommodate within the design limits of velocity and pressure drop, both initially and in the future.

Process Gas Flow Rates

The sizing procedure for gas piping must take both high-pressure and low-pressure flow conditions into consideration if the operating pressure of the line changes over time.

Two-Phase Flow Rates

Whenever two-phase flow is encountered in facility piping it is usually in flowlines and interfield transfer lines. Some designers size liquid lines downstream of control valves as two-phase lines. The amount of gas involved in these lines is low and thus the lines are often sized as single-phase liquid lines. Oversizing two-phase lines can lead to increased slugging and thus as small a diameter as possible should be used; consistent with pressure drop available and velocity constraints discussed in Volume 1.

Viscosity: High viscosity crudes may flow in the laminar flow regime which causes high pressure drops. This is especially true of emulsions of water in high-viscosity crudes where the effective velocity of the mixture could be as much as ten times that of the base crude (see Volume 1).

Solids: Some wells produce large amounts of sand and other solids entrained in the fluid. Where solids are contained in the stream, sufficient velocity should be provided to assure they do not build up in the bottom of the pipe, causing higher than anticipated pressure drops or potential areas for corrosion. However, if the velocity is too high, erosion may occur. (See Volume 1.)

Fluid Compositions

The composition of a production fluid is usually not well defined. In most cases, only a specific gravity is known. Compositions are important to the prediction of physical properties of the fluid as it undergoes phase changes. Estimations can be made based only upon specific gravity, however, for good reliability, molecular compositions should be used when available.

Gases such as H_2S and CO_2 (acid gases) in the production streams are sometimes encountered. These gases are not only corrosive to piping, but many are harmful and possibly fatal upon contact with humans. Special care should be exercised in designing piping containing acid gases. Velocities above 30 to 50 ft/s should be avoided in piping containing acid gases to avoid affecting the ability of corrosion inhibitors to protect the metal. Special metallurgy may be needed to combat H_2S corrosion. (See Chapter 8.)

Handling Changing Operating Conditions

Each production facility has three categories of equipment whose design depends upon operating conditions:

1. Vessels and other mechanical equipment are the most difficult to change or alter after installation.
2. Piping is the next most difficult.
3. Instrumentation is the least difficult.

Often the facility is designed with equipment and piping that can handle the complete range of operating conditions, and with control valves selected so that their internals ("trim") can be substituted as operating conditions change. Sometimes the piping must be designed to allow addition of future pieces of equipment. This is especially true for compressors and water treating equipment that may not be needed initially.

The key to arriving at the most flexible system design lies in forecasting future operating conditions. Many engineers are not aware of the implications of future conditions and their effect upon initial design and long-term operation. Often some information is available on potential future scenarios, but the facility design engineer elects to design for a specific "most likely" forecast. This is unfortunate, as the designer should at least consider the sensitivity of the design and economic consequences to the whole range of possible forecasts.

Selecting Pipe Sizes

Basic steps in piping design are:

1. Establish operating conditions, i.e., flow rates, temperatures, pressures and compositions of fluid over the life of the system. This may involve several cases.
2. Using velocity as the limiting criterion, calculate allowable pipe internal diameter ranges using the criteria of Chapter 9, Volume 1.
3. If more than one standard pipe size is indicated, calculate the wall thickness for each standard pipe size based on required maximum allowable working pressure and select a standard wall thickness for each size.
4. Calculate maximum and minimum capacities for each size using velocity limits as criteria.
5. Estimate the pressure drop for each size and compare to the available pressure drop.
6. Arrange the information from the previous steps and determine which pipe size is best suited to all operating conditions.
7. As piping drawings are developed, re-evaluate those lines where estimated pressure drop was a criterion in size selection, taking into account the actual piping configuration and effects of control and piping components.
8. Proceed with design of pipe supports and stress analysis, if required.

It is also a good practice to verify design conditions and piping calculations just prior to release of the drawings for construction. System requirements sometimes change significantly during the course of a project. In most facility piping situations experienced designers can select size quickly without a formal tabulation of the steps just described. In certain cases, especially where pressure drop is an important consideration, a formal tabulation may be required.

GENERAL PIPING DESIGN DETAILS

Steel Pipe Materials

Most production facility piping is fabricated from ASTM A-106 Grade B or API 5L Grade B pipe, which is acceptable for sweet service and temperatures above –20°F. Between –20 °F and –50 °F, ANSI B31.3, "Chemical Plant and Petroleum Refinery Piping," allows this material to be used if the pressure is less than 25% of maximum allowable design and the combined longitudinal stress due to pressure, dead weight, and displacement strain is less than 6,000 psi. Below –50°F it is required that the pipe be heat treated and Charpy impact tested. Volume 1, Chapter 9 discusses the various common piping codes and methods for calculating maximum allowable pressure for various steels. Some common low-temperature steels include:

Steel	Minimum Temp. without Special Testing
A-333 Grade 1	– 50°F
A-334 Grade 1	– 50°F
A-312 TP 304L	–425°F
A-312 TP 316L	–325°F

For sour service, National Association of Corrosion Engineers (NACE) MR-01-75 requires that steel material have a Rockwell C hardness of less than 22 and contain less than 1% nickel to prevent sulfide stress cracking.

Figure 7-1 shows regions of H_2S concentration and total pressure where the provisions of NACE MR-01-75 govern. A-53 Grade B, A-106 Grade B, A-333 Grade 1, and API 5L Grades B and X-42 through X-65 are acceptable for use in the sulfide-stress cracking region.

Minimum Pipe Wall Thickness

From the standpoint of mechanical strength, impact resistance, and corrosion resistance, some operators prefer to establish a minimum wall thickness of approximately 0.20 in. Thus, they establish the following minimum pipe schedules (standard wall thickness), even though pressure containment calculations would indicate that smaller thicknesses are allowed:

¾ in. and smaller — Sch 160
2, 2½, and 3 in. — Sch 80
4 and 6 in. — Sch 40

ANSI B 31.3 requires threaded pipe that is 1½ in. and smaller be at least Sch 80 and that 2 in. and larger be at least Sch 40.

Pipe End Connections

Pipe, valve, and fittings tables must specify which size of each class of pipe is threaded, flanged, or socket welded. ANSI B31.3 provides no specific guidance except that it suggests that threads be avoided where corrosion, severe erosion, or cyclic loading is anticipated.

API RP 14E recommends:

- Pipe 1½ in. or less should be socket welded for:
  Hydrocarbon service above 600 ANSI
  Hydrocarbon service above 200°F
  Hydrocarbon service subject to vibration
  Glycol service
- Pipe 2 inches and larger should be flanged for:
  Hydrocarbon service
  Glycol service
- Utility piping 2 inches and smaller may be threaded.

A common practice onshore is to use threaded connections on 2-in. pipe or smaller, no matter what the service. It is also common to see threaded connections on 4-in. pipe and smaller in low pressure oil service.

Figure 15-16 shows three types of flange faces. Raised-face (RF) and flat-faced (FF) flanges use a donut-shaped flat gasket to create the pressure seal. Ring-joint flanges (RTJ) use a ring that fits into the circular notches in the face of the flange to effect the pressure seal. RTJ flanges create a more positive seal and are used for all API class flanges and for higher pressure ANSI classes. However, they are difficult to maintain, as they require the mating flanges to be spread to remove the ring. Raised-face flanges tend to form a tighter seal than flat-faced flanges and are used in steel piping. Flat-face flanges are used in cast-iron piping and in bolting to cast-iron and ductile-iron pumps, compressors, strainers, etc.

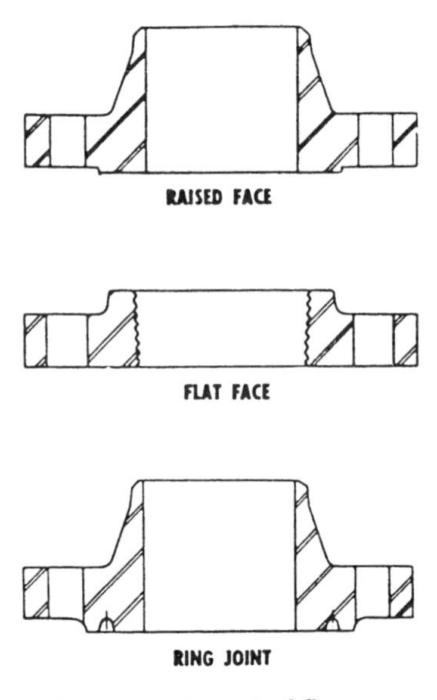

Figure 15-16. Typical flanges.

Bolting a raised-face flange to a flat-faced, cast-iron flange can create bending moments in the less ductile cast-iron flange, which could cause it to crack.

The ANSI specifications allow the use of both RF and RTJ flanges. API RP 14E recommends RTJ flanges for ANSI Class 900 and higher and recommends RTJ flanges be used in 600 ANSI service subject to vibration. Onshore it is common to use RF flanges for ANSI classes through 2500.

The hesitancy to use RF flanges at higher pressures may stem from an era when plain ¹⁄₁₆-in, asbestos gaskets were the only type available. Modern spiral-wound polytetrafluoroethylene (PTFE) filled with internal ring gaskets with 316 stainless-steel windings may create as positive a seal with RF flanges as is obtainable from RTJ flanges.

RTJ gaskets are normally cadmium-plated, soft iron or low carbon steel. Soft iron is used for ANSI 600 and 900 classes, and 304 or 316 stainless steel for higher classes.

Branch Connections

Where a branch connection is connected to a main run of pipe, it is necessary to specify the type of fitting required. ANSI B31.3 provides a procedure for calculating the amount of reinforcement needed to adequately support the branch connection. In accordance with this code, no reinforcement is needed where:

- A tee is used.
- A coupling is used, the branch size is 2 in. or less, *and* the branch size is less than ¼ diameter of the run.
- An integrally reinforced branch connection fitting that has been pressure tested (weld-o-let type) is used.

API RP 14E recommends that no reinforcement be used and presents a typical branch connection schedule (Table 15-6) to provide more mechanical strength than is required by ANSI B31.3. Most onshore operators use integrally reinforced branch-connection fittings or tees interchangeably.

Fiberglass Reinforced Pipe

The use of fiberglass reinforced pipe (FRP) and tanks has been on the increase in production facilities. Onshore applications include low-pressure flowlines, high-pressure water injection lines, oil treating systems, fire water systems, and produced water treating systems. Offshore applications include fire water and utility systems. The primary advantages are ease of field installation and non-corrosiveness. The American Petroleum Institute has developed specifications for fiberglass tanks (API Spec 12P) and fiberglass piping (API Spec 15LR).

Insulation

Insulation is normally required for personnel protection for pipe operating at higher than approximately 150°F or 200°F. Pipe operating at greater than approximately 400°F should be located and insulated to keep it from becoming an ignition source for spilled liquid hydrocarbons. Pipe operating at temperatures above approximately 900°F should be protected from coming into contact with combustible gases.

As described in Chapter 17, any surface in excess of 726°F in an electrically classified area should be insulated or isolated from gas sources. A normal rule of thumb and a requirement of some codes is to provide insulation or isolation barriers for surfaces hotter than 400°F that are located within electrically classified areas. Surfaces in electrically unclassified areas are only insulated or isolated if necessary for personnel protection.

Table 15-6
Branch Connection Schedule—Welded Piping

1	2	3	4	5	6	7	8	9	10	11	12	13	14	15	16
Nominal Branch Size (in.)	Nominal Run Size (in.)														
	½	¾	1	1½	2	2½	3	4	6	8	10	12	14	16	18
½	SWT	SWT	SWT	SWT	6SC	6SC	6SC	6SC	6SC	6SC	6SC	6SC	6SC	6SC	6SC
¾		SWT	SWT	SWT	SOL	6SC	6SC	6SC	6SC	6SC	6SC	6SC	6SC	6SC	6SC
1			SWT	SWT	SOL	SOL	SOL	6SC	6SC	6SC	6SC	6SC	6SC	6SC	6SC
1½				SWT	TR	SOL	SOL	SOL	6SC	6SC	6SC	6SC	6SC	6SC	6SC
2					T	RT	RT	RT	WOL	WOL	WOL	WOL	WOL	WOL	WOL
2½						T	RT	RT	WOL	WOL	WOL	WOL	WOL	WOL	WOL
3							T	RT	RT	WOL	WOL	WOL	WOL	WOL	WOL
4								T	RT	RT	WOL	WOL	WOL	WOL	WOL
6									T	RT	RT	RT	WOL	WOL	WOL
8										T	RT	RT	RT	RT	WOL
10											T	RT	RT	RT	RT
12												T	RT	RT	RT
14													T	RT	RT
16														T	RT
18															T

T—Straight Tee (Butt Weld)
RT—Reducing Tee (Butt Weld)
TR—Straight Tee and Reducer or Reducing Tee
WOL—Welded nozzle or equivalent (Schedule of Branch Pipe)
SOL—Socketweld couplings or equivalent—6000 lb Forged Steel
SWT—Socketweld Tee
6SC—6000 lb Forged Steel Socketweld Coupling (¾ inch and smaller threadbolts or screwed couplings may be used for sample, gage, test connection and instrumentation purposes)

Pipe Insulation Considerations

Materials

- Some commonly used insulating materials are calcium silicate, mineral slagwool, glass fiber, cellular glass, and polyurethane.
- Insulating material, such as magnesia, that if wet could deteriorate or cause corrosion of the insulated surface, should not be used.
- Certain heating fluids are not compatible with some insulating materials, and auto-ignition may occur. Caution should be exercised in selecting materials.

Vapor Barriers

- A vapor barrier should be applied to the outer surface of the insulation on cold piping.
- Insulation should be protected by sheet-metal jacketing from weather, oil spillage, mechanical wear, or other damage.
- If aluminum sheet metal is used for this purpose, insulation should be protected by a vapor barrier.

Sour Service

- To prevent H_2S from concentrating around the bolts, flanges should not be insulated in H_2S service.

Table 15-7 shows recommended insulation thicknesses from API RP 14E. Two of the most common types of acceptable insulation systems are:

1. Metal jacket—This type is primarily used on piping, heat exchangers, and other cylindrical shapes.
2. Blanket—This type is primarily used on irregular objects that are difficult to insulate due to irregular surface configurations—such as an expansion joint.

Examples of insulation and isolation installations are shown in Figures 15-17 through 15-24.

(text continued on page 461)

Table 15-7A
Typical Hot Insulation Thickness (in.)

1	2	3	4	5	6	7	8	9
Maximum Temperature (°F)	Nominal Pipe Size, inches							
	1½ & Smaller	2	3	4	6	8	10	12 & Larger
250	1	1	1	1½	1½	1½	1½	1½
500	1	1½	1½	1½	2	2	2	2
600	1½	1½	2	2	2	2½	2½	2½
750	2	2	2	2	2½	3	3	3

Table 15-7B
Typical Cold Insulation Thickness (in.)

1	2	3	4	5	6	7	8	9	10	11	12	13	14	15	16	17	18	19	20
	Nominal Pipe Size, in.																		
Minimum Temperature (°F)	½	¾	1	1½	2	2½	3	4	6	8	10	12	14	16	18	20	24	30	Flat Surf.
40	1	1	1	1	1	1	1	1	1	1½	1½	1½	1½	1½	1½	1½	1½	1½	1½
30	1	1	1	1½	1½	1½	1½	1½	1½	1½	1½	1½	1½	1½	1½	1½	1½	1½	1½
20	1½	1½	1½	1½	1½	1½	1½	1½	2	2	2	2	2	2	2	2	2	2	2
10	1½	1½	1½	1½	2	2	2	2	2	2	2½	2½	2½	2½	2½	2½	2½	2½	2½
0	1½	2	2	2	2	2	2	2½	2½	2½	2½	2½	2½	2½	2½	2½	3	3	3
−10	2	2	2	2	2	2½	2½	2½	2½	3	3	3	3	3	3	3	3	3	3½
−20	2	2	2	2½	2½	2½	2½	2½	3	3	3	3	3	3½	3½	3½	3½	3½	4

Table 15-7C
Typical Insulation For Personnel Protection
(Applicable Hot Surface Temperature Range (°F))

1	2	3	4	5	6	7	8
Nominal Pipe Size (in.)	Nominal Insulation Thickness (in.)						
	1	1½	2	2½	3	3½	4
½	160–730	731–1040	1041–1200	–	–		–
¾	160–640	641–940	941–1200	–	–	–	–
1	160–710	711–960	961–1200	–	–	–	–
1½	160–660	661–880	881–1200	–	–	–	–
2	160–640	641–870	871–1090	1091–1200	–	–	–
2½	160–620	621–960	961–1160	1161–1200	–	–	–
3	160–600	601–810	811–1000	1001–1200	–	–	–
4	160–600	601–790	791–970	971–1125	1126–1200	–	–
6	160–550	551–740	741–930	931–1090	1091–1200	–	–
8	–	160–740	741–900	901–1090	1091–1200	–	–
10	–	160–750	751–900	901–1060	1061–1200	–	–
12	–	160–740	741–900	901–1030	1031–1170	1171–1200	–
14	–	160–700	701–850	851–1000	1001–1130	1131–1200	–
16	–	160–690	691–840	841–980	981–1120	1121–1200	–
18	–	160–690	691–830	831–970	971–1100	1101–1200	
20	–	160–690	691–830	831–970	971–1100	1101–1200	–
24	–	160–680	681–820	821–960	961–1090	1091–1200	–
30	–	160–680	681–810	811–950	951–1080	1081–1200	–
Flat Surface*	160–520	521–660	661–790	791–900	901–1010	1011–1120	1121–1200

Application range also applies to piping and equipment over 30 inches in diameter.

Figure 15-17. Exhaust system on top of generator package. Insulation or barriers needed because location can be used as a work or storage area; otherwise insulation may not be necessary.

Figure 15-18. No insulation on the crane exhaust is necessary because it is isolated from personnel performing normal operations and is not in a classified area.

Figure 15-19. Insulation on this fire water pump is not necessary because it is not a hydrocarbon handling vessel and is not located in a classified area or work area.

Figure 15-20. Insulation of the generator package is necessary because the exhaust system is located in a work area.

Figure 15-21. Isolated compressor. Insulation is necessary because the compressor itself is a potential source of gas and requires the area to be classified.

Figure 15-22. Insulation is necessary because the compressor is a potential source of gas and requires the area to be classified.

Figure 15-23. Fire water pump insulation is not necessary because the exhaust is not in a work area and the fire water pump is not in a classified area (more than 10 ft from production equipment, oil storage, etc.)

Figure 15-24. Insulation is not necessary on the portion of the exhaust system extending outside the compressor building because it is not in a classified area and is not a work area. The inside portion needs insulation because it is in a classified area.

(*text continued from page 453*)

Insulation for personnel safety is required only when accidental contact of the hot surfaces could be made by personnel within normal work or walk areas. Isolation may be in the form of guards or barriers and, in special cases, warning signs.

Hot surfaces associated with natural gas compressors and pumps handling volatile flammable fluids should be insulated since the equipment itself is a source of hydrocarbon liquids or gases. Generators, electric motors, and engine-driven equipment such as fire water pumps, wireline units, welding machines, hydraulic equipment, and the like do not themselves cause the area to become classified from an electrical standpoint. However, they may be in a classified area due to other equipment and thus require insulation or barriers. Turbo-chargers, exhaust manifolds, compressor heads, expansion bottles and the like (including associated piping), which cannot be insulated without causing mechanical failure, are not normally insulated. In these cases, warning signs, barriers, gas detectors, or other methods for the protection of personnel and minimizing exposure to hydrocarbon liquids and gases are acceptable.

MISCELLANEOUS PIPING DESIGN DETAILS

Target Tees

Where 90° turns in piping are required, standard long radius ells (ell centerline radius equals 1.5 times pipe nominal diameter) are usually used. In sandy service, the sand has a tendency to erode the metal on the outside of the bend. Target tees, such as shown in Figure 15-25, are often specified for such service. The sand builds up against the bull plug and provides a cushion of sand that is constantly being eroded and subject to deposition by the sand in the flow stream.

Chokes

The flow of fluid leaving a choke is in the form of a high-velocity jet. For this reason it is desirable to have a straight run of pipe of at least ten pipe diameters downstream of any choke prior to a change in direction, so that the jet does not impinge on the side of the pipe.

Often on high-pressure wells two chokes are installed in the flowline—one a positive choke and the other an adjustable choke. The adjustable choke is used to control the flow rate. If it were to cut out, the positive choke then acts to restrict the flow out of the well and keep the well from damaging itself. Where there are two chokes, it is good piping practice to separate the chokes by 10 pipe diameters to keep the jet of flow formed by the first choke from cutting out the second choke. In practice this separation is not often done because of the expense of separating two chokes by a spool of pipe rated for well shut-in pressure. It is much less expensive to bolt the flanges of the two chokes together. No data has been collected to prove whether the separation of chokes is justified from maintenance and safety considerations.

Whenever a choke is installed, it is good piping practice to install block valves within a reasonable distance upstream and downstream so that the choke bean or disc can be changed without having to bleed down a long length of pipeline. A vent valve for bleeding pressure off the segment of the line containing the choke is also needed. This is particularly true in instances where a positive choke is installed at the wellhead and an adjustable choke is installed hundreds of feet away in a line heater. If block valves are not installed downstream of the positive choke and upstream of the adjustable choke, it would be necessary to bleed the entire flowline to atmosphere to perform maintenance on either choke.

Flange Protectors

The full faces of flanges never really touch due to the gaskets or rings that cause the seal. The space between the two flange faces is a very

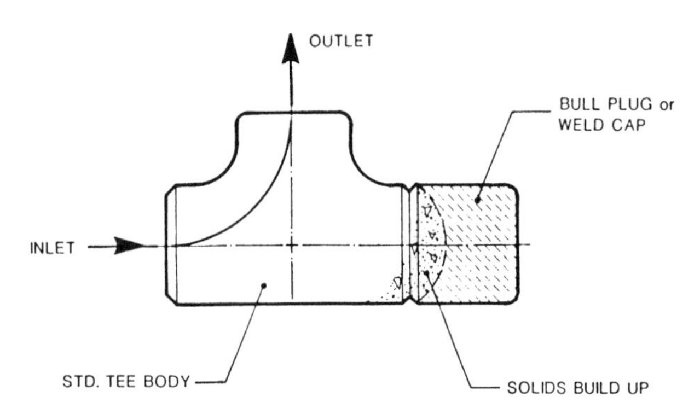

Figure 15-25. Target tee.

good spot for corrosion to develop, as shown in Figure 15–26. Flange protectors made of closed-cell soft rubber are sometimes used to exclude liquids from penetrating this area. Stainless-steel bands and grease fittings are also used.

Closed-cell flange protectors are much less expensive than stainless bands. However, if not installed properly they can actually accelerate corrosion if a path is created through the material to allow moisture to enter. Flange protectors should not be used in H_2S service. They may trap small leaks of sour gas and keep them from being dispersed in the atmosphere.

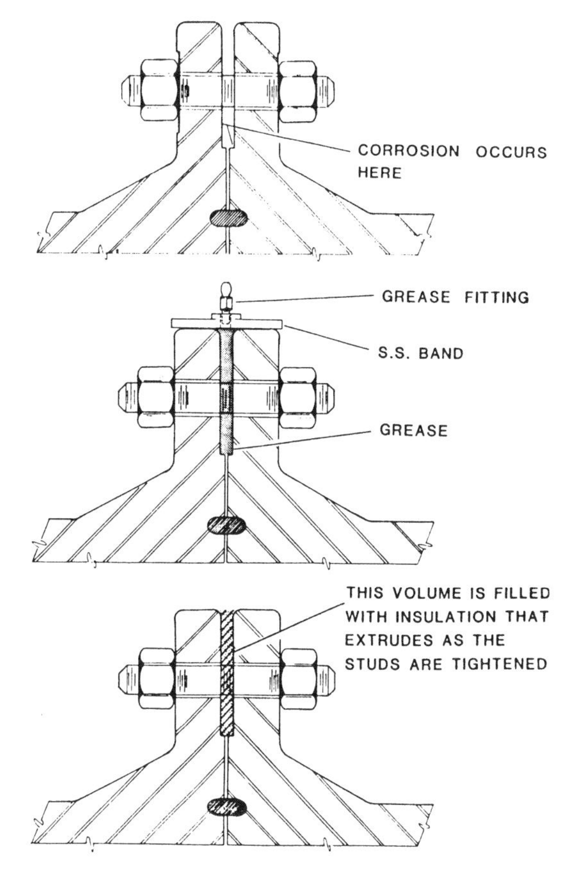

Figure 15-26. Flange protector types.

Vessel Drains

If vessel drain valves are used often, there is a tendency for these valves to cut out. As the valve is opened and shut, there is an instantaneous flow of a solid slurry across the valve that creates an erosive action. Figure 15-27 shows a tandem valve arrangement to minimize this potential problem. To drain the vessel, the throttling valve is shut and one or more drain valves are opened. These valves open with no flow going through them. Then the throttling valve is opened. To stop draining, the throttling valve is closed, flow goes to zero, and the drain valves are shut. The throttling valve will eventually cut out, but it can be easily repaired without having to drain the vessel.

Vessel drain systems can be very dangerous and deserve careful attention. There is a tendency to connect high-pressure vessels with low-pressure vessels through the drain system. If a drain is inadvertently left open, pressure can communicate through the drain system from the high-pressure vessel to the low-pressure vessel. If this is the case, the low pressure vessel relief valve must be sized for this potential gas blowby condition.

The liquid drained from a vessel may flash a considerable quantity of natural gas when it flows into an atmospheric drain header. The gas will find a way out of the piping system and will seek the closest exit to atmosphere that it can find. Thus, a sump collecting vessel drains must be vented to a safe location.

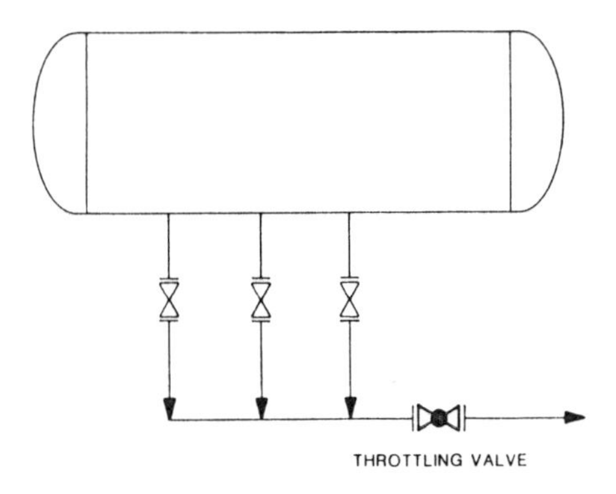

Figure 15-27. Drain valves for a separator.

Open Drains

Open, gravity drains should not be combined with pressure vessel drain systems. The gas flashing from vessel liquids may exit an open drain system at any point and create a hazard.

On open drain piping leaving buildings, a liquid seal should be installed as further protection to assure that gases flashing from liquids from other locations in the drain system will not exit the system in the building.

The elevation of gravity drain systems must be carefully checked to assure that liquids will flow to the collection point without exiting the piping at an intermediate low point.

Piping Vent and Drain Valves

At high points in piping, vent valves are required to remove air for hydrotesting and for purging the system. At low points, drain valves are required to drain liquids out of the system to perform maintenance. Normally, vent and drain valves are ½-in. or ¾-in. ball valves.

Control Stations

Whenever it is necessary to control the process level, pressure, temperature, etc., a control station is installed. A control station may be as simple as a single control valve or it may contain several control valves, block valves, bypass valves, check valves, and drain or vent valves.

Where there is a control valve, block valves are often provided so the control valve can be maintained without having to drain or bleed the pressure from the vessel. Typically, the safety-systems analysis would also call for a check valve at this point to prevent backflow. Drain or vent valves are often installed to drain liquid or bleed pressure out of the system so that the control valve can be maintained. In smaller installations drain and vent valves may not be provided and the line is depressured by backing off slightly on flange bolts (always leaving the bolts engaged until all pressure is released) or slowly unscrewing a coupling. This is not a good practice although it is often used for small-diameter, low-pressure installations.

Bypass valves are sometimes installed to allow the control valve to be repaired without shutting in production. On large, important streams the bypass could be another control valve station. Manual bypass valves are more common. The bypass valve could be a globe valve if it is anticipated that flow will be throttled through the valve manually during the bypass operation, or it could be an on/off valve if the flow is to be cycled. Because globe valves do not provide positive shutoff, often globe-bypass valves have a ball or other on/off valve piped in series with the globe valve.

The piping around any facility, other than the straight pipe connecting the equipment, is made up primarily of a series of control stations. Flow from one vessel goes through a control station and into a piece of pipe that goes to another vessel. In addition to considering the use of block valves, check valves, etc., all control stations should be designed so that the control valve can be removed and any bypass valve is located above or on a level with the main control valve. If the bypass is below the control valve, it provides a dead space for water accumulation and corrosion.

CHAPTER

16

Prime Movers*

Both reciprocating engines and turbines are used as prime movers in production facilities to directly drive pumps, compressors, generators, cranes, etc. Reciprocating engines for oil field applications range in horsepower from 100 to 3,500, while gas turbines range from 1,500 to in excess of 75,000.

Prime movers are typically fueled by natural gas or diesel. Dual fuel turbine units exist that can run on natural gas and can automatically switch to diesel. So-called "dual fuel" reciprocating engines run on a mixture of diesel and natural gas. When natural gas is not available, they can automatically switch to 100% diesel. Most prime movers associated with producing facilities are typically natural gas fueled due to the ready availability of fuel. Diesel fueled machines are typically used to provide stand-by power or power for intermittent or emergency users such as cranes, stand-by generators, firewater pumps, etc.

Due to the extremely wide variety of engines and turbines available, this discussion is limited to those normally used in production facilities. The purpose of this chapter is to provide facility engineers with an understanding of basic engine operating principles and practices as necessary for selection and application. The reader is referred to any of the many texts available on engine and turbine design for more in-depth discussion of design details.

---

*Reviewed for the 1999 edition by Santiago Pacheco of Paragon Engineering Services, Inc.

RECIPROCATING ENGINES

Reciprocating engines are available in two basic types—two-stroke or four-stroke cycle. Regardless of the engine type, the following four functions must be performed in the power cylinder of a reciprocating engine:

1. Intake—Air and fuel are admitted to the cylinder.
2. Compression—The fuel and air mixture is compressed and ignited.
3. Power—Combustion of the fuel results in the release of energy. This energy release results in increase in temperature and pressure in the cylinder. The expansion of this mixture against the piston converts a portion of the energy released to mechanical energy.
4. Exhaust—The combustion products are voided from the cylinder and the cycle is complete.

In this manner the chemical energy of the fuel is released. Some of the energy is lost in heating the cylinder and exhaust gases. The remainder is converted to mechanical energy as the expanding gases move the piston on the power stroke. Some of the mechanical energy is used to overcome internal friction or to sustain the process by providing air for combustion, circulating cooling water to remove heat from the cylinder, and circulating lube oil to minimize friction. The remainder of the energy is available to provide external work. The amount of external work that can be developed by the engine is termed its "brake horsepower" or bhp. The amount of work required to sustain the engine is termed its "friction horsepower" or fhp. The work developed by the power cylinders is termed the "indicated horsepower" or ihp. The indicated horsepower is the sum of both the friction horsepower and the brake horsepower.

$$\text{ihp} = \text{bhp} + \text{fhp}$$

Four-Stroke Cycle Engine

The four-stroke cycle engine requires four engine strokes or 720 degrees of crankshaft rotation to complete the basic functions of intake, compression, power, and exhaust. All flow into and away from the cylinders is controlled by valves directly operated by a camshaft that is driven at ½ engine speed. Figures 16-1 and 16-2 illustrate a cross section and an idealized P-V diagram for a four-cycle spark-ignited engine, respectively.

Figure 16-1. Cross section of 4-cycle, spark-ignited engine.

Intake Stroke (Point 1 to Point 2)

With the intake valve open, the piston movement to the right creates a low pressure region in the cylinder, which causes air and fuel to flow through the intake valve to fill the cylinder.

Compression Stroke (Point 2 to TDC)

The intake valve is now closed as the piston moves from the bottom dead center (BDC) to top dead center (TDC), compressing the fuel/air mixture. At Point 3, just prior to TDC, a spark ignites the fuel/air mixture and the resulting combustion causes the pressure and temperature to begin a very rapid rise within the cylinder.

Power Stroke (TDC to Point 4)

Burning continues as the piston reverses at TDC and pressure rises through the first portion of the "power" or "expansion" stroke. It is the increase in pressure due to burning the fuel that forces the piston to the right to produce useful mechanical power. The piston moves to the right until BDC is reached.

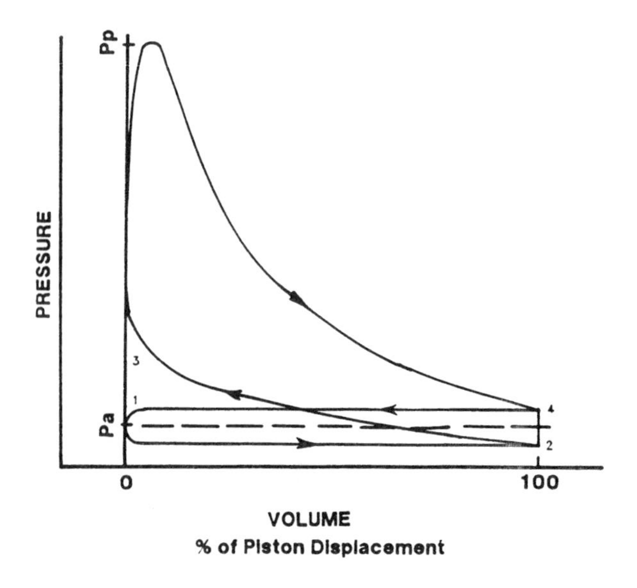

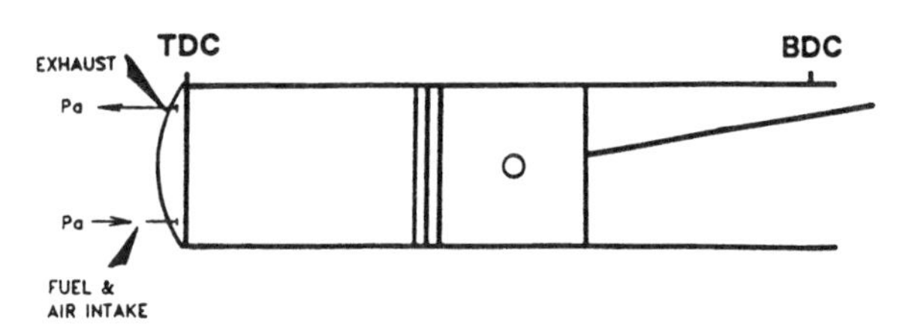

Figure 16-2. Idealized P-V diagram for a 4-cycle, spark-ignited engine.

Exhaust Stroke (Point 4 to Point 1)

With the exhaust valve open, the upward stroke from BDC to TDC creates a positive pressure within the cylinder, which forces combustion products from the cylinder on the "exhaust" stroke.

Two-Stroke Cycle Engine

Two-stroke cycle engines require two engine strokes or 360 degrees of crankshaft rotation to complete the basic functions of intake, compression, power, and exhaust. Figures 16-3 and 16-4 illustrate a cross section and a P-V diagram for a two-cycle engine, respectively. In this common type of engine, the piston in its traverse covers and uncovers passages or

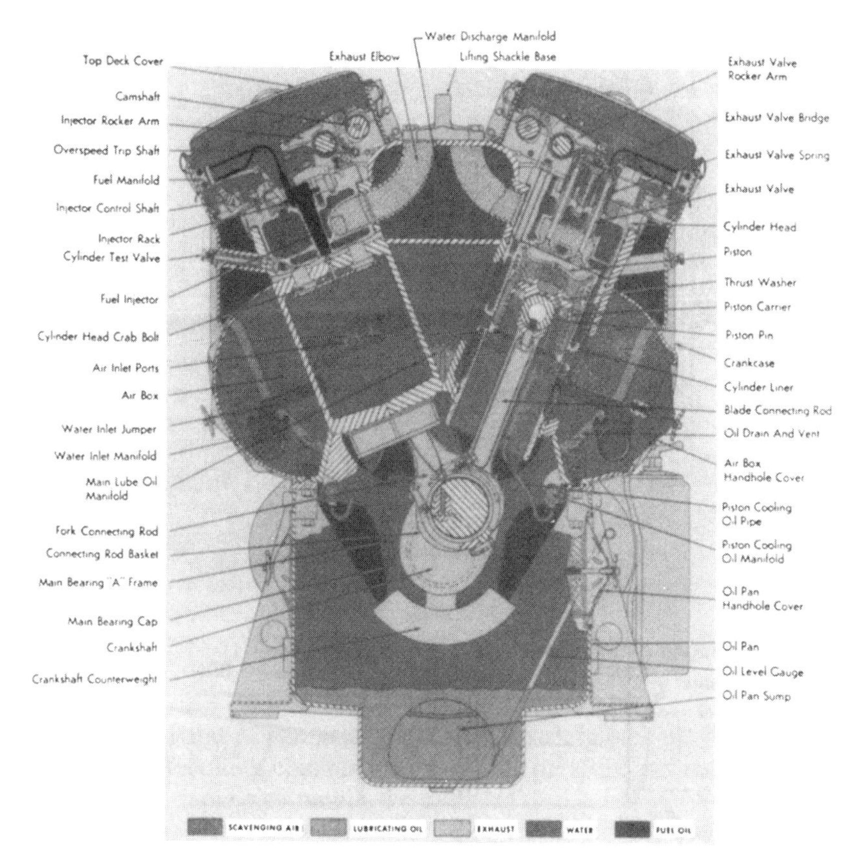

Figure 16-3. In this cross section of a two-cycle engine, only a trained eye can identify the engine porting that distinguishes it from a four-cycle engine. (*Courtesy of Electro-Motive Division, General Motors Corp., and Stewart and Stevenson Services, Inc.*)

ports in the lower cylinder wall that control the inflow of air and the outflow of exhaust gases. This type of engine is called a piston-ported engine. Inasmuch as both intake and exhaust ports are opened at every piston traverse, there are no periods of negative pressure to induce air nor of high positive pressure to completely expel exhaust gases. While a two-cycle engine has the ability to produce power with each down motion of the piston, it is at the expense of some external means of compressing enough air to fill the cylinder and to expel the combustion products from the previous cycle ("scavenge" the cylinder) as well.

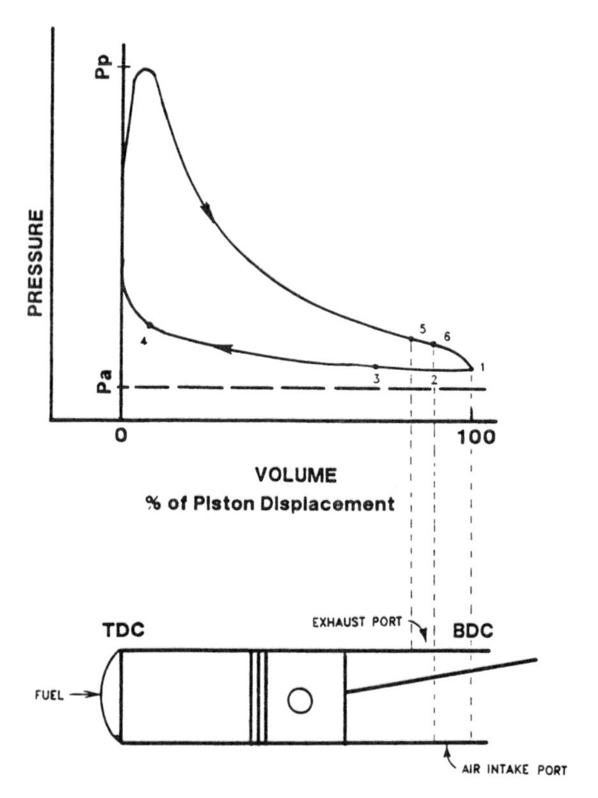

Figure 16-4. Idealized P-V diagram for a 2-cycle engine.

The Compression Stroke

As the piston begins its leftward stroke from bottom dead center (BDC), both inlet and exhaust ports are uncovered and air from some external source is flowing through the cylinder. Directional control is provided through port and/or piston design to ensure the most complete cylinder scavenging possible.

At Point 2, the air intake is closed, but compression does not begin until the exhaust port is covered also. Shortly after the exhaust port is closed and compression of the trapped air begins, fuel is injected at Point 3 into the cylinder through a high pressure fuel valve. At Point 4, just prior to completion of the compression stroke, a spark ignites the fuel/air mixture and the pressure rises rapidly through the remainder of the compression stroke and the beginning of the "power" stroke.

Power Stroke

During the power stroke, the piston is forced to the right, producing power. At Point 5, the exhaust port opens and terminates the power producing portion of the cycle as exhaust gases escape to the atmosphere. As the intake port is uncovered, Point 6, air begins to flow through the cylinder and out of the exhaust port "scavenging" the remaining exhaust gases and filling the cylinder with fresh air to begin the cycle again. On piston ported engines both ports are open during the part of the cycle when the intake port is open. Due to this fact, the large industrial two-strokc cycle engines use fuel injection valves in lieu of carburetion to maintain fuel economy. Otherwise, unburned fuel would be passed through the cylinder directly to the exhaust manifold.

The area contained within the P-V diagram represents the total mechanical work performed by the piston. Some of this work is required to sustain the cycle and must be subtracted from the work calculated from the P-V diagram to determine the total external work available from the piston.

Comparison of Two-Cycle and Four-Cycle Engines

Both two- and four-stroke cycle engines are used in commercial applications. For the purpose of this discussion, however, comments will be limited to applications normally encountered in the oil and gas industry. Some of the advantages and disadvantages associated with each engine type are as follows:

Two-Stroke Cycle Engines

Advantages

1. Relatively smaller size for comparable horsepower machines.
2. Reduced weight over similar horsepower four-stroke engines.
3. Fewer mechanical parts.
4. Reduced maintenance.
5. Generally simplified maintenance procedures.
6. Reduced overall installation cost due to size and weight.

Disadvantages

1. Scavenging system required to allow self-starting.
2. Prone to detonation at high ambient temperatures.

3. Lower exhaust temperature reduces available waste heat.
4. Power cylinders require frequent balancing.
5. Very sensitive to lube oil to prevent excessive port carboning.

Four-Stroke Cycle Engines

Advantages

1. Substantial exhaust heat available for waste heat recovery.
2. Reduced detonation tendency at high ambient temperatures.
3. Requires infrequent power cylinder balancing.

Disadvantages

1. Higher comparable package weight and space requirements.
2. More complex maintenance.
3. More expensive facility costs.

Both engine types are available in horsepower ranges to satisfy most any application. Two-stroke engines are generally of slow speed (300–600 rpm) design with horsepower exceeding 2000 bhp. The four-stroke engines are available over all speed ranges. Four-stroke engines tend to be used for lower horsepower applications although some are available in sizes exceeding 3000 bhp.

Engine Speed

The selection of a machine must also include the desired speed at which it will operate. The normally accepted classifications are:

Slow speed—300–600 rpm
Intermediate speed—600–900 rpm
High speed— > 900 rpm

In general, slower speed units will be larger, heavier, and more costly than higher speed units. Slower speed units have the advantages of higher reliability, greater fuel efficiency, and lower maintenance costs than higher speed machines. The overall economics of initial capital cost, reliability, fuel cost, maintenance costs, etc., must be considered to determine the most appropriate version for a particular installation.

Naturally Aspirated vs. Supercharged Engines

Air is supplied to the power cylinders by either natural air flow associated with the engine or by some external means. Engines that use no external means of air supply are termed to be naturally aspirated. Those with some external air supply are generally termed "supercharged." The horsepower developed by an engine is dependent on its supply of air. The more air mass contained in a cylinder at ignition the more fuel that can be burned and the more horsepower that will be developed by the cylinder.

Four-stroke cycle machines are available in either naturally aspirated or supercharged versions. Superchargers available include engine-driven blowers or "turbochargers." Except for some small engines, the turbocharger is the most common method of supercharging these engines.

Turbochargers use the expansion of exhaust gas to pump combustion air to an engine. Exhaust gas is directed through a set of nozzles to drive a turbine wheel. Directly connected to the exhaust turbine is an air compressor turbine that delivers combustion air to the power cylinders. Thus, back-pressure is put on the engine exhaust, reducing power slightly, but the net effect of the increase in air mass flow available for combustion is to increase horsepower.

Care must be exercised when turbocharging an engine. As more air is forced into the cylinders, the cylinder pressures associated with combustion will increase. Controls must be provided to limit these pressures to prevent engine damage. Excessive pressures will result in gas igniting prior to the spark activating, and/or overstress of the engine. Substantial engine damage can result from either.

Although turbocharging an engine increases the bhp for a nominal additional capital cost, it has the disadvantage of increasing maintenance costs and decreasing engine reliability.

Two-stroke cycle machines must be equipped with a scavenge air system that may include a separate scavenge air cylinder, gear-driven blower, or turbochargers. It should be noted, however, that a method must be provided to supply a turbocharged two-stroke cycle engine with air to start.

Carburetion and Fuel Injection

Reciprocating engines are classified as either spark ignited or compression ignited. Fuel is supplied to spark ignited engines by either a carburetor or fuel injection. Fuel is supplied to compression ignited (diesel) engines by fuel injection.

Carburetion

Fuel must be supplied to an engine in a manner determined by operating conditions. As engine load changes, changes in fuel flow are required to maintain correct fuel mixtures. The flow of the air is regulated by a throttling device as shown in Figure 16-5. The throttle device is controlled by an engine speed control device or by a "governor," which controls engine speed to a pre-set value. The pressure differential created by air flow through the venturi is used to regulate fuel flow to the range of 0.06 to 0.07 lb of fuel per lb of air.

Fuel Injection

This is another method of supplying an engine with the correct quantity of fuel. Three types of fuel injection are used:

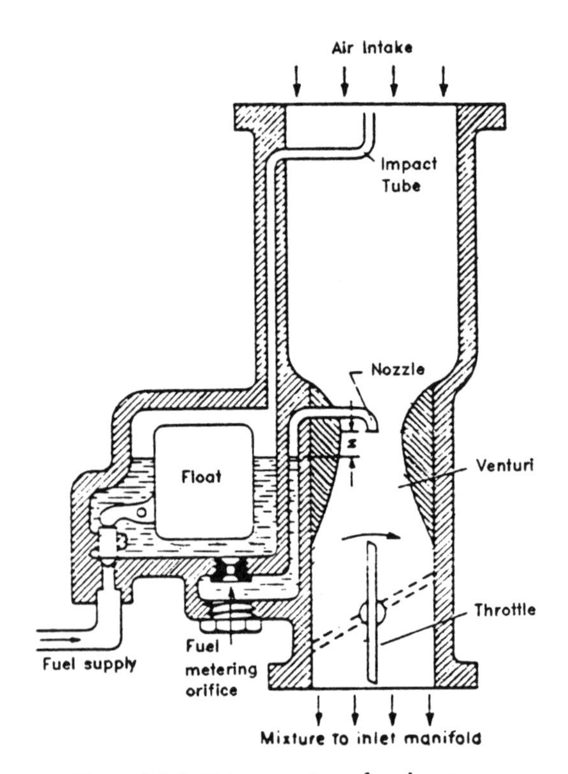

Figure 16-5. Cutaway view of carburetor.

1. Inlet port injection is used with liquid fuel, spark ignition engines only. Fuel is injected into the inlet port and mixed with the inlet air. The injection process may be either timed or continuous.
2. Early cylinder injection is used only with spark ignition engines. Fuel is admitted into the cylinder during the intake or compression stroke. This is the injection method used on the large two-stroke cycle engines to prevent loss of fuel during the scavenging process.
3. Late cylinder injection is typically used for diesel engines. Fuel is admitted to the cylinder as the piston is nearing top center. Very high injection pressures are required for proper fuel atomization and combustion control. Pressures can exceed 20,000 psi for this type of injection.

Engine Shutdown System

Most engines are operated in an unattended manner. Shutdown safety devices must be provided to prevent engine damage in case of a malfunction. Some of the more common engine shutdowns are:

1. Low lube oil pressure
2. Engine overspeed
3. Engine high temperature—jacket water
4. Engine high temperature—lube oil
5. Low jacket water pressure
6. High vibration
7. Low lube oil flow—lube oil to the stroke power cylinder
8. High bearing temperature
9. Low fuel gas pressure

This is not to be taken as a complete shutdown list. Installation and operating requirements may dictate other safety precautions. This list does not include any safety devices that would be associated with the driven equipment.

GAS TURBINE ENGINES

Gas turbines differ from conventional internal-combustion engines in the manner in which the expanded gases are employed. The principle of operation is to direct a stream of hot gases against the blading of a turbine rotor. As shown in Figures 16-6 and 16-7, the gas turbine consists of

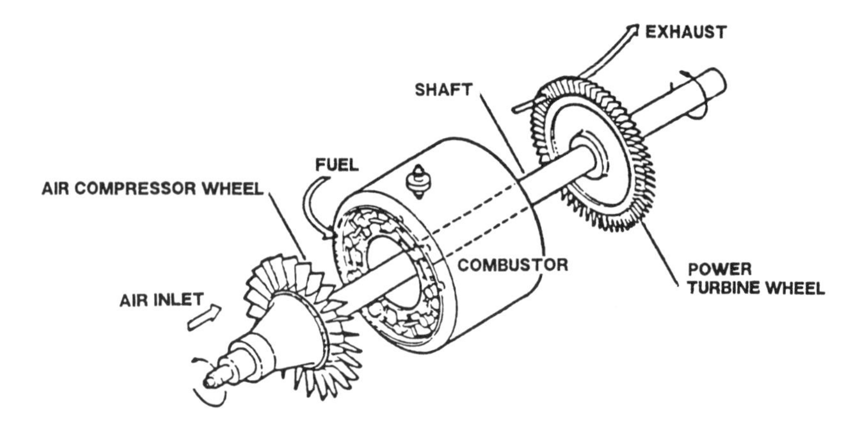

Figure 16-6. Schematic of single-shaft gas turbine.

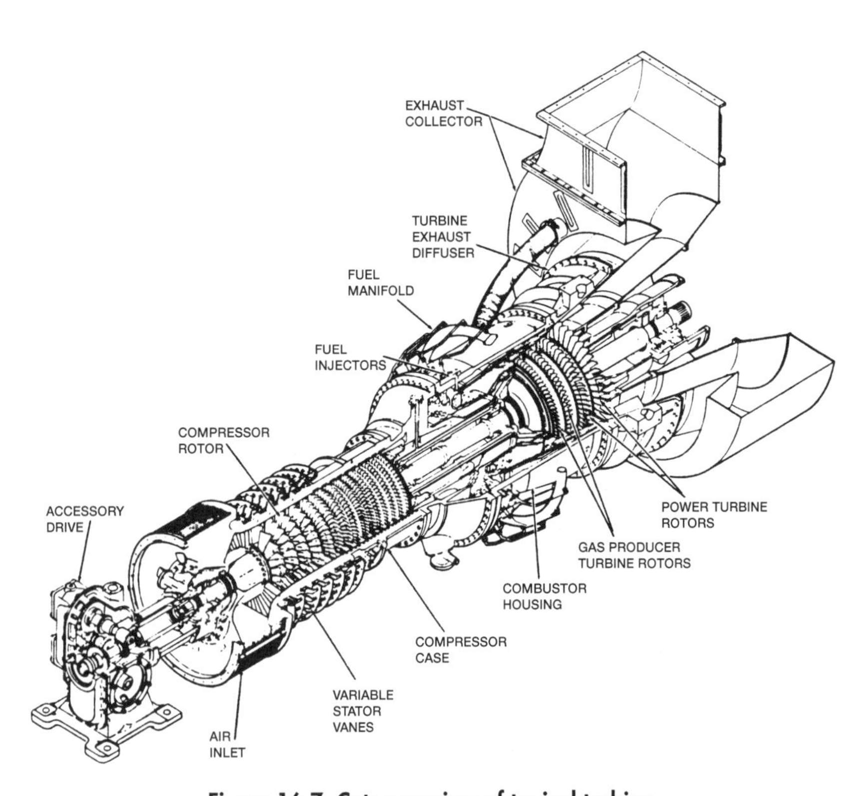

Figure 16-7. Cutaway view of typical turbine.

three basic sections: the air compressor section in which air is compressed in a compressor, the combustor section in which fuel is mixed with the compressed air and burned, and the turbine section where work is extracted from the hot gases. The expansion of the gases against the blades of the turbine provide power to drive an external load such as a pump or compressor, as well as power to drive the air compressor. The result is a smooth running, steady flow machine.

A significant difference between gas turbines and reciprocating engines is that gas turbines use much more air. For example, an 1100-hp gas turbine (Solar Saturn) handles approximately 12 lb_m/sec or almost 22 tons of air per hour. A comparable reciprocating engine will use only about ¼ that amount. Piston engines use almost all the air for combustion (a small amount may be used for cylinder scavenging), while turbines use only about 25% of the air flow for combustion. The remainder, or about 75%, is used for cooling and to obtain the mass flow required to operate the turbine.

Fundamentals

The basic gas turbine engine is described by the idealized Brayton air cycle as shown in Figure 16-8. In this cycle, air enters the air compressor (also called the "gas producer") at Point 1 under normal atmospheric pressure and temperature, P1 and T1. It is then isentropically compressed to Point 2 where the pressure and temperature are now P2 and T2. From Point 2 the air flows into the combustion chamber where fuel is injected and burned at constant pressure, raising the temperature to T3 and expanding the volume to V3. From the combustion chamber the heated gases enter the power turbine where they perform work by turning the output power shaft. These gases expand to near atmospheric pressure and are exhausted at greater than atmospheric temperature at Point 4. Ideally, it would be possible to have the same fluid going through this circuit all the time, and the step from Point 4 to Point 1 would be a cooling process. Actually, this step is accomplished by exhausting to atmosphere and taking in a new charge of air.

In this cycle, approximately 30% of the fuel consumed is available as power output. In addition, approximately 30% is used to drive the air compressors, 30% is contained in the hot exhaust gases, and 10% is lost to radiation and the lube oil system.

Simple cycle industrial gas turbines burn more fuel than comparable reciprocating machines. There are, however, several methods available to

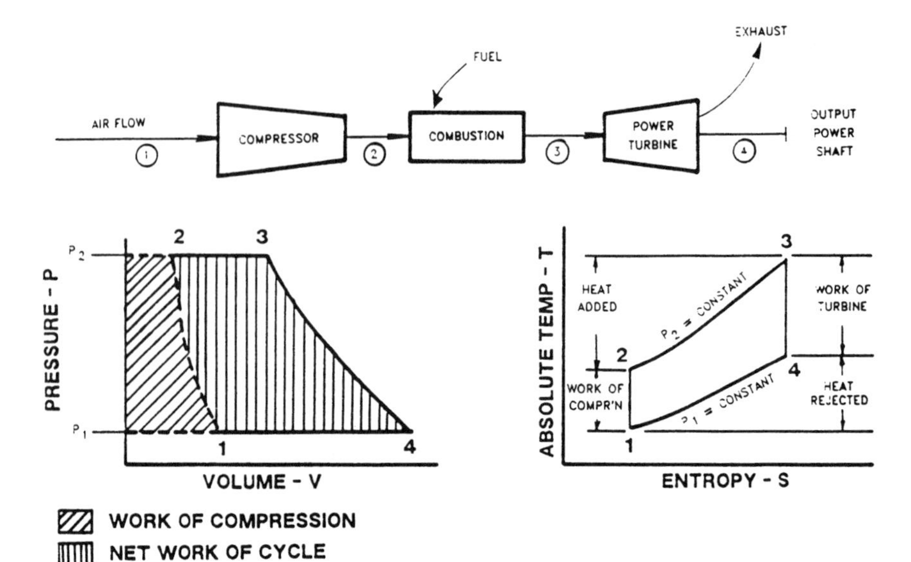

Figure 16-8. This Brayton cycle describes the basic operation of a gas turbine.

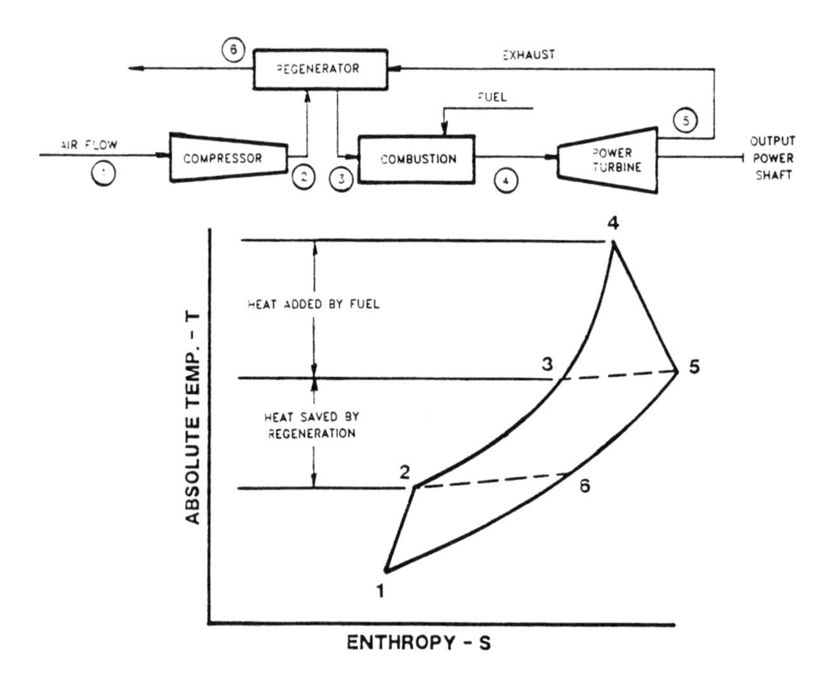

Figure 16-9. The regeneration cycle of a gas turbine uses recovered exhaust heat to preheat the compressed air prior to combustion.

either reduce direct fuel consumption or use the available exhaust heat to reduce overall facility fuel requirements. Three common methods available are regeneration, waste heat recovery, and combined cycle operation.

Regeneration or recouperation uses a heat exchanger in which exhaust heat is recovered to preheat the compressed inlet air prior to the combustion chamber as shown in Figure 16-9. The increased combustion air temperature reduces the fuel requirement to maintain power turbine inlet temperatures.

This heat recovery can increase the turbine's thermal efficiency to approximately 35 to 40%. The available horsepower output of the unit will drop slightly because of increased pressure losses due to the regenerator. Initial costs and additional maintenance expense of a regenerative cycle usually outweigh its advantages for most producing installations. However, where fuel prices are high and there are no waste heat users, the use of the regenerative cycle should be considered.

Another method of increasing the overall cycle efficiency is to use the waste heat energy in the exhaust air to heat process fluids as depicted in Figure 16-10. This is a direct savings in fuel gas that would otherwise be consumed in direct-fired heaters. Overall thermal efficiencies can be as high as 50 to 60% in this type of installation.

The energy in the turbine exhaust stream can also be used to generate steam as shown in Figure 16-11. This is called a combined cycle since mechanical energy is available both from the power turbine output and from the output of the steam turbine. The energy used to drive the steam turbine contributes to the overall thermal efficiency as the steam is generated without the expense of any additional fuel consumption. A combined cycle system can increase overall thermal efficiency to the 40% range.

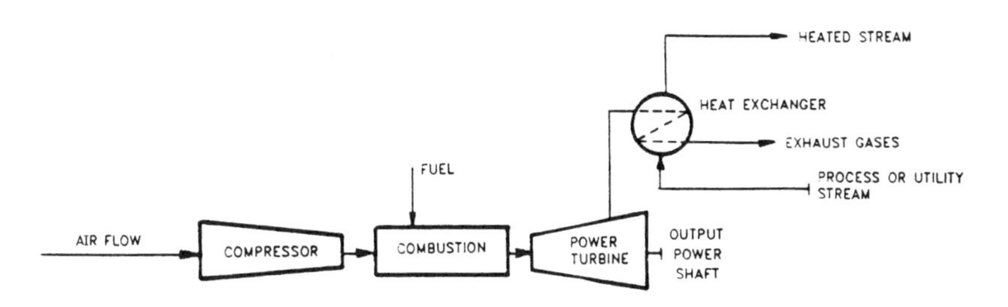

Figure 16-10. This schematic illustrates how waste heat can be recovered and used to heat process fluids.

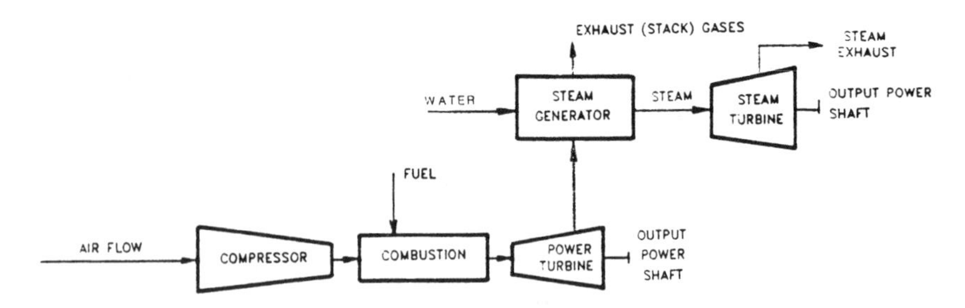

Figure 16-11. The combined cycle of waste heat recovery can be used to generate steam.

Effect of Ambient Conditions

Available horsepower from a gas turbine is a function of air compressor pressure ratio, combustor temperature, air compressor and turbine efficiencies, ambient temperature, and barometric pressure. High ambient temperatures and/or low barometric pressure will reduce available horsepower while low ambient temperatures and/or high barometric pressure will increase available horsepower. All industrial turbines will have high-temperature protection, but in areas subject to very low ambient temperatures horsepower limiting may be required.

Figure 16-12 shows the effect of ambient temperature on the horsepower output of a typical two-shaft gas turbine engine. At high temperatures the horsepower is limited by the maximum allowable power turbine inlet temperature. At low ambient temperatures, the available horsepower is limited by the maximum allowable air compressor speed.

Where hot ambient temperatures are expected, overall turbine efficiency and horsepower output can be increased by installing an evaporative cooler in the inlet. Inlet air flows through a spray of cold water. The temperature of the water and the cooling effect caused by the inlet air evaporating some of the water cools the inlet air. In desert areas where the inlet air is dry and thus able to evaporate more water before becoming saturated with water vapor, this process is particularly effective at increasing turbine efficiency.

Effect of Air Compressor Speed

The horsepower output of a gas turbine is a direct function of flow rate across the power turbine, which is a function of air compressor speed and combustion temperature. Figure 16-13 shows the relationship between

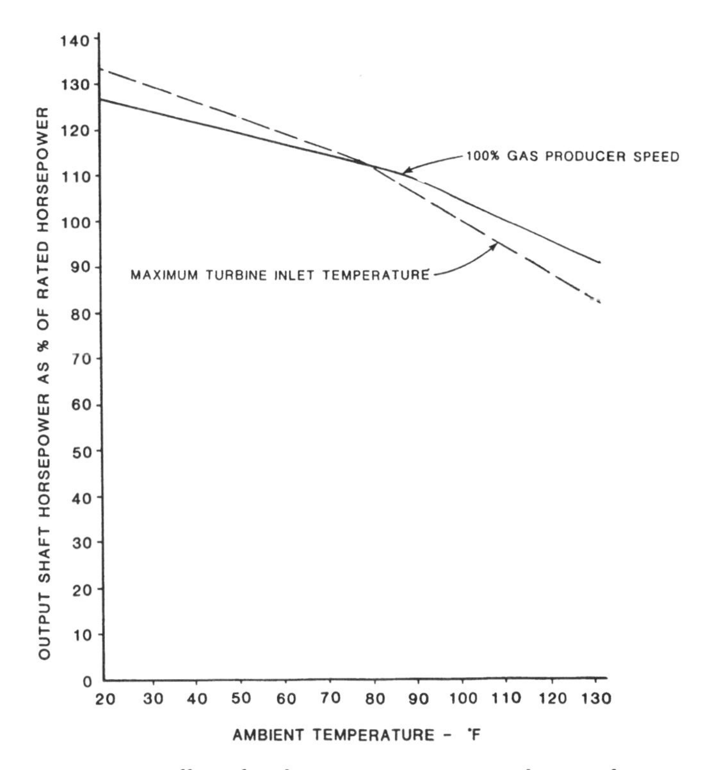

Figure 16-12. Effect of ambient temperature on turbine performance.

air compressor speed and available horsepower. For example, a 10% drop in air compressor speed will result in a drop of almost 50% in available horsepower. Thus, air compressor speed is critical to output horsepower and small speed changes can result in large changes in available horsepower.

Single- vs. Multi-Shaft Turbines

Industrial gas turbines are available as either single-shaft or multi-shaft engines. The turbine illustrated in Figure 16-6 has a single shaft. Both the air compressor and the power turbine section operate off the same shaft and thus rotate at the same speed. As illustrated in Figure 16-14, in a multi-shaft unit some of the power turbine wheels are on the same shaft as the air compressor, while the remainder of the power turbine wheels are on a separate shaft that provides power to the driven equipment. The speed of the wheels of the power turbine that provide the

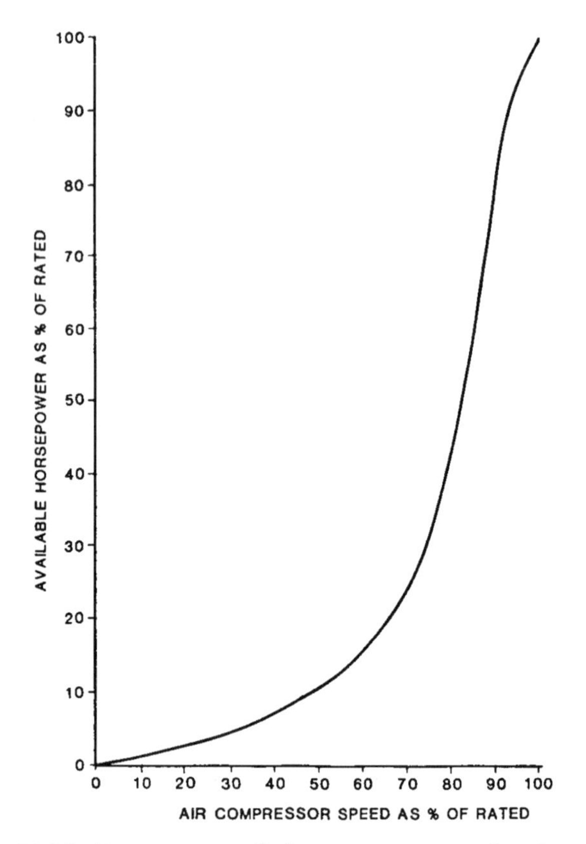

Figure 16-13. Air compressor discharge pressure as a function of its speed.

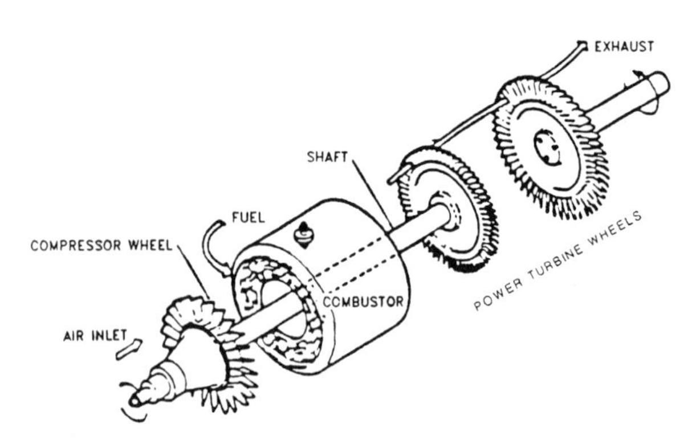

Figure 16-14. This two-shaft turbine illustrates how power turbine wheels may be placed on separate shafts.

output work and the speed of the wheels of the power turbine driving the air compressor are independent of each other. The output shaft speed can be faster or slower than the air compressor depending upon the requirements of the external load.

Figure 16-15 illustrates the performance characteristics of a single-shaft turbine. In this case the minimum speed is governed by the surge limit of the air compressor and the maximum power is governed by the maximum allowable temperature at the inlet of the power turbine section. Thus, the power turbine has a very narrow speed range over which it can operate. At speeds below 80% there is a significant loss in power output. This type of machine is more desirable in constant-speed, variable-load applications, such as powering a generator, since the speed of the generator must be held constant no matter what the load. In an application requiring variable output speeds, such as a pump or gas compressor, when the load declines below the capacity of the pump or compressor at the minimum power turbine operating speed, it becomes necessary to bypass gas from the discharge to the suction of the pump or gas compressor to maintain the minimum flow. This condition is not desirable as horsepower and fuel gas are wasted in recycling the process stream. Another disadvantage of single-shaft turbines is that the starting horsepower requirements may be large. To keep this starting requirement as

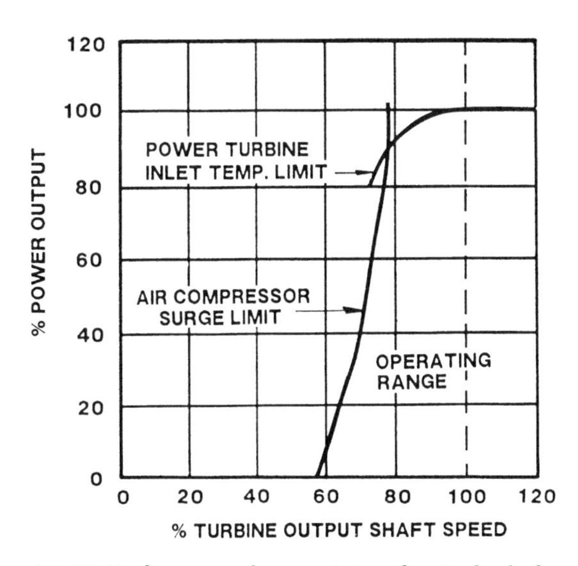

Figure 16-15. Performance characteristics of a single-shaft turbine.

low as possible, it may be necessary to unload the driven equipment during start-up.

Figure 16-16 shows the performance characteristic of a split-shaft turbine where the only power output limitation is the maximum allowable temperature at the inlet of the turbine section. In actual practice a torque limit, increased exhaust temperature, loss of turbine efficiency, and/or a lubrication problem on the driven equipment usually preclude operating at very low power turbine speeds. The useful characteristic of the split-shaft engine is its ability to supply a more or less constant horsepower output over a wide range of power turbine speeds. The air compressor essentially sets a power level and the output shaft attains a speed to provide the required torque balance. Compressors, pumps, and various mechanical drive systems make very good applications for split-shaft designs.

Effect of Air Contaminants

The best overall efficiency of a turbine can be ensured by maintaining the efficiency of the air compressor section. Conversely, allowing the air compressor efficiency to deteriorate will deteriorate the overall thermal efficiency of the turbine. Air compressor efficiency can be drastically reduced in a very short time when dirt, salt water mist, or similar air con-

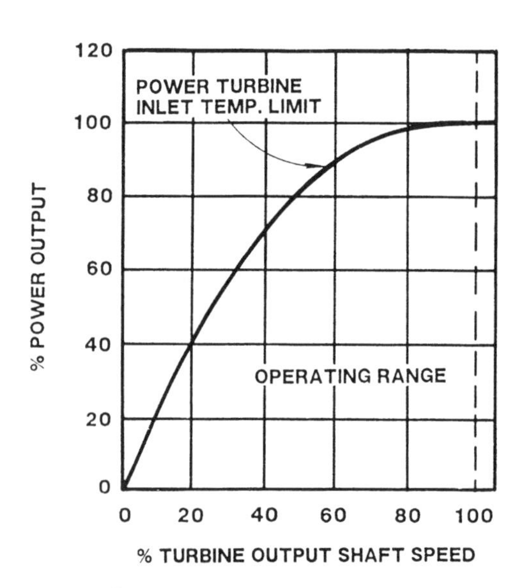

Figure 16-16. Performance characteristics of a multi-shaft turbine.

taminants enter the inlet air. Contaminants will accumulate in the air compressor and reduce its compression efficiency. The effect will be decreased mass flow, reduced compressor discharge pressure, reduced horsepower, and higher-than-normal engine temperatures.

Effective inlet air filtration is required to ensure satisfactory operation of the engine. The location of the unit determines the most appropriate filter system to use. Desert environments where a large amount of sand particles could be expected in the ambient air may use an automatic roll type of filter that allows new filter material to be rolled in front of the inlet without frequent shut-downs to change filters. Arctic or extremely cold locations may use pad type filters, snow hoods to prevent blockage, and exhaust recirculation to prevent icing. Filter assemblies for offshore marine environments may include weather louvers, demister pads, and barrier elements for salt and dirt removal. Screens may be used for insect removal prior to filtration in areas with bug problems.

Cleaning the air compressor can be accomplished by injecting water, steam, detergent, and/or abrasive material (such as walnut hulls) into the air inlet. Engine life and performance will be improved if cleaning is done on a periodic basis so as to keep any hard deposits of oil, dirt, etc. from forming. In general, frequent detergent washing will ensure compressor cleanliness. Steam cleaning with an appropriate detergent is also very effective. Abrasive cleaning should be avoided and only be necessary as the result of improper frequency or technique of detergent washing.

ENVIRONMENTAL CONSIDERATIONS

Air Pollution

Exhaust emissions will vary with the type and age of engine and the fuel used. Current environmental regulations must be consulted. It may be necessary to submit a permit to install the new equipment. Each country, state, or county has variations of the maximum emissions.

In general, liquid fueled engines tend to have increased emission levels of particulates and unburned hydrocarbons over those of gaseous fueled engines. Due to the large quantities of excess air, gas turbines tend to have lower emission levels of particulates and unburned hydrocarbons than reciprocating engines. Gas turbines do, however, tend to produce greater quantities of nitrogen oxides (NO_X). The formation of NO_X depends on combustion temperature and residence times at high temperatures, both of which are higher in gas turbines than in engines. Engines, on the other hand, tend to have greater concentrations of carbon monoxide, CO, in their exhausts.

Fuel quality will greatly affect emissions and can also have considerable effect on engine life. Manufacturers' specifications will generally specify fuel quality for proper operation.

In addition to carbon monoxide (CO) and unburned hydrocarbons (UHC), the most significant products of combustion are the oxides of nitrogen (NO_X). At high temperatures, free oxygen not consumed during combustion reacts with nitrogen to form NO and NO_2 (about 90% and 10% of total NO_X, respectively).

Improvements in engine and turbine design, along with the use of auxiliary equipment such as catalytic converters, selective catalytic reduction (SCR) units and the use of steam and water injection into turbines, combine to reduce overall emission levels.

When a hydrocarbon fuel such as natural gas is burned in an engine or turbine, the concentration of pollutants is dependent on the air to fuel (A/F) ratio as shown in Figure 16-17. If pollution was not a concern, in order to obtain maximum thermodynamic efficiency, the engine would be designed for a slightly greater than stoichiometric mixture. Because air and fuel are never perfectly mixed at the time of ignition, excess air must be present to burn all the fuel. The normal amount of excess air that achieves this efficiency is around 15–20%. Under these conditions, Figure 16-17 shows that a relatively large amount of NO_X will be formed.

NO_X emission controls in large engines and turbines are based on the same principles. However, special designs must be applied to accommodate differences in the combustion process. Methods to control NO_X include the following.

NO_x Reduction in Engines

1. **Lean Burn**

As shown in Figure 16-17, at very high A/F ratios—greater than 30:1—the production of NO_X can be very low. The problem with simply increasing A/F ratio is that, because the air/fuel mix is not uniform, increasing A/F ratio in the cylinder increases the probability that the mixture at the point of the spark plug may be too lean, thus leading to a misfire.

Installing a pre-combustion chamber (PCC) in the engine design solves this problem. In this design, a normal A/F ratio fuel is introduced into a PCC at the time of ignition. This creates ignition torch-

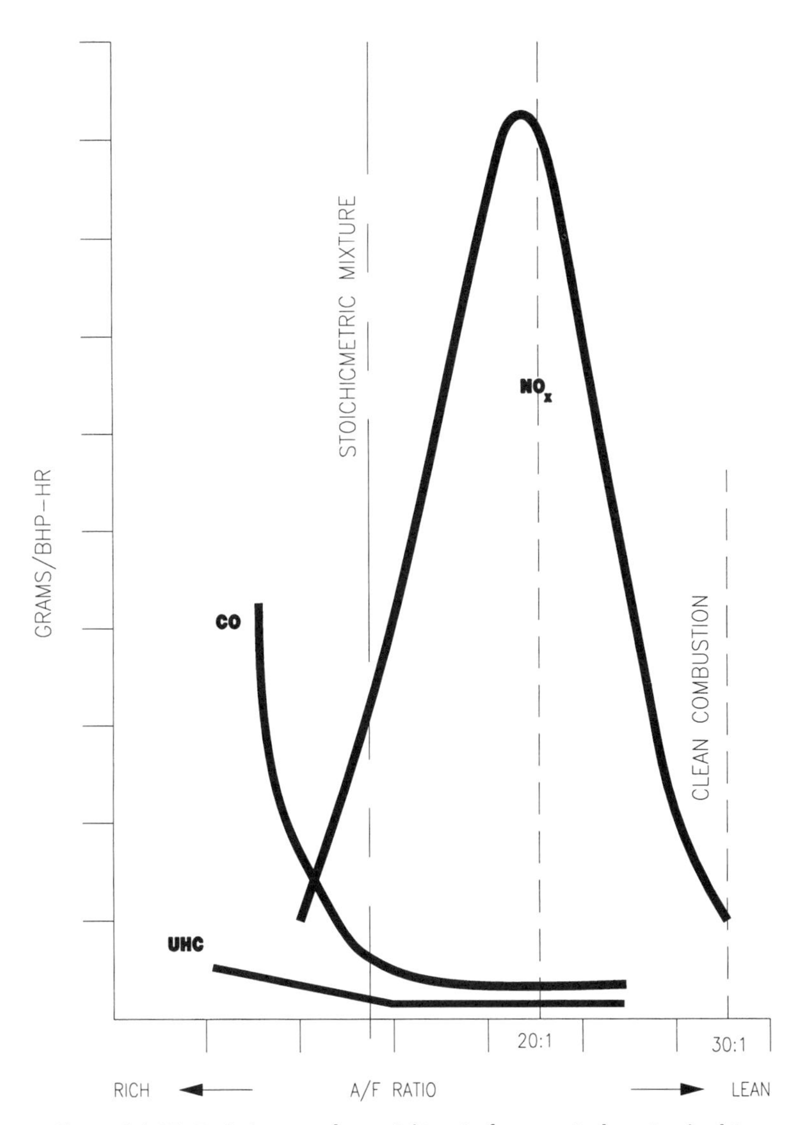

Figure 16-17. Emission trends vs. A/F ratio for a typical engine/turbine.

es that enter the main cylinder, which has the lean A/F ratio, and ignites the fuel.

2. Catalytic Converter

Catalytic converters are designed to oxidize the unburned UHCs and CO. The resulting combustion (oxidation) converts them into water and CO_2. A recent catalytic converter design called three-way converters also controls NO_x using a reduction process. Three-way converters contain two catalytic "bricks," one for reduction and the other one for oxidation. The oxidation process with CO takes place as:

$$2CO + O_2 \leftrightarrow 2CO_2$$

The oxidation of UHC is:

$$2O_2 + CH_4 \leftrightarrow CO_2 + 2H_2O$$

And finally, the reduction of NO_x with CO results in N_2 and CO_2:

$$NO_2 + CO \leftrightarrow NO + CO_2$$

$$2NO + 2CO \leftrightarrow N_2 + 2CO_2$$

Catalytic reactions occur when the temperature exceeds 500°–600°F (260°–316°C). Normal converter operating temperatures are 900°–1200°F (482°–649°C).

Excessively rich A/F ratio causes converter operating temperatures to rise dramatically, thus causing converter meltdown. On the other hand, if the A/F ratio is too lean, the excess O_2 will react with the CO, and the reduction of nitrogen with CO will not take place. Thus, catalytic converters cannot be used where there is excess air.

3. **Selective Catalytic Reduction (SCR)**

Selective catalytic reduction is based on selective reactions of a continuous gaseous flow of ammonia or similar reducing agents with the exhaust stream in the presence of a catalyst. The reaction that occurs is as follows:

$$4NH_3 + 6NO \leftrightarrow 5N_2 + 6H_2O$$

SCR units require handling, storage, and continuous injection of the reducing agent. The temperature level is critical because the SCR operates in a narrow temperature range between 550°–750°F (260°–399°C), and thus an exchanger is necessary to cool the exhaust stream. This leads to a complicated and costly process system that must be added to the engine exhaust.

NO_x Reduction in Turbines

1. **Inject Steam or Water**

   This system is called wet NO_x control. Water or steam is injected into the primary combustion zone. This method has been used effectively in the past. Current installations are using this system when the water or steam is readily available or if they are already part of the process. Maintenance costs are higher when compared with dry control, because this method requires high quality water. If high quality water is not used, the corrosion associated with dissolved minerals in the water may prematurely damage the turbine.

2. **Lean Premixed Combustion**

   When air and fuel are mixed and burned in standard turbine combustion systems, incomplete mixture occurs. Areas of rich A/F ratios exist, which cause local high temperatures called "hot spots." Normal turbine combustion temperatures can reach 2800°F (1538°C). Because NO_x formation rate is an exponential function of temperature, decreasing the combustor temperature can substantially reduce NO_x production. One method of reducing hot spots is to premix a lean A/F ratio prior to combustion. Lower temperature levels are achieved by using stages (multiple sets of air and fuel injectors) and by adding special instrumentation to control the appropriate proportion of air and fuel. Excess air is used to further reduce the overall flame temperature. Using this approach, most gas turbine manufacturers are able to guarantee about 25 ppmv.

3. **Selective Catalytic Reduction (SCR)**

   SCR is described above. It is important to note that SCRs require a lot of space, are relatively expensive, and use toxic metals. Therefore, they may not be practical and may be too costly to install and operate compared with other methods.

4. **Catalytic Combustor**

   A recent innovation includes use of a catalytic combustor. Several tests with large turbines indicate that this alternative can reduce NO_x emission to less than 5 ppmv. The catalyst inside the combustion chamber actually causes a portion of the fuel to burn without the presence of a flame. This significantly reduces combustion temperature.

Based on the results of the catalytic combustor, there is an effort under way to develop this concept into a practical, field-proven technology.

Noise Pollution

Increased public awareness of noise as an environmental pollutant and as a hazard to the hearing of personnel requires that attention be given to this problem during the design phase. When a prime mover installation is planned, enough silencing should be installed to ensure that the noise level will be acceptable to the community and meets all governmental requirements. The requirements will vary substantially depending upon such factors as location, population density, operating personnel in the area, etc.

CHAPTER

17

Electrical Systems*

This chapter introduces some concepts concerning electrical system design and installation that are particularly important from the standpoint of safety and/or operational considerations for production facilities. The reader is referred to texts in electrical engineering and to the various codes and standards listed at the end of this chapter for a more detailed description concerning the design of electrical circuits, sizing conductors, and circuit breakers, etc. This chapter is meant merely as an overview of this complex subject so that the project facilities engineer will be able to communicate more effectively with electrical design engineers and vendors who are responsible for the detail design of the electrical system.

SOURCES OF POWER

The required power for production facilities is either generated on site by engine- or turbine-driven generator units or purchased from a local utility company. For onshore facilities the power is generally purchased from a utility. However, if the facility is at a remote location where there

*Reviewed for the 1999 edition by Dinesh P. Patel of Paragon Engineering Services, Inc.

is no existing utility power distribution, an on-site generating unit may be considered. A standby generator may be required if utility power is not sufficiently reliable. The standby generator may be sized to handle either the total facility load or only essential loads during periods of utility power failure.

In the case of an offshore facility, electrical power is generally generated on site by engine- or turbine-driven generator sets using natural gas or diesel as fuel. Most installations are designed to handle the total electrical load even if one generator is out of service. To minimize the size of standby equipment, some facilities have a system to automatically shed non-essential loads if one generator is out of service. Some offshore facilities are furnished power from onshore via high-voltage cables. The cables are generally laid on the ocean floor and are buried in shallow water. In some cases a single cable (usually three-phase) is used for such applications to minimize initial project cost. However, if a fault develops in this single cable, the facility could be shut in for extended periods of time. To avoid extended shut-ins, either a spare or alternate cable can be installed or standby generators can be installed on the offshore platform.

The choice of whether to purchase or generate electricity and decisions on generator or cable configuration and sparing are often not obvious. An economic study evaluating capital and operating costs *and* system reliability of several alternatives may be required.

Utility Power

Utility companies have a power system network including large generating plants, overhead transmission lines, power substations which reduce transmission line voltages to distribution line voltages, and overhead/underground distribution lines which carry power to the end users (such as a production facility).

The power from the distribution line voltage is converted to facility distribution voltage using a "step-down" transformer, providing power to facility switchgear and motor control centers. The facility distribution voltage selection depends upon the length of the distribution system, the size, and location of the electrical loads to be served. Most oil field electrical distribution systems in the United States are 4,160 or 2,300 volts. Typically, 480 volts is used for motor and other three phase loads; 240/120 volts usually is used for lighting and other single phase loads, and 120 volts usually is used for control circuits. Step-down transformers deliver these voltages from the facility distribution system.

The electrical distribution system design and equipment selection must consider requirements of the utility company for protection and metering. Available short circuit currents from the utility distribution network to the primary of the facility's main transformer must be considered in selecting circuit protection devices for the facility distribution system.

Electrical Generating Stations

Where electricity is generated in the facility, generator sizing should consider not only connected electrical loads, but also starting loads and anticipated and non-anticipated expansions. In most installations this is done by developing an electrical load list itemizing the various loads as either continuous, standby or intermittent service. Examples of continuous loads are electric lighting, process pumps and compressors required to handle the design flow conditions, and either quarters heating or air conditioning, whichever is larger. Intermittent loads would include quarters kitchen equipment, washdown pumps, cranes, air compressors and similar devices which are not in use at the same time. The total demand is normally taken as 100% of the continuous loads, 40 to 60% of the intermittent loads and an allowance for future demand. Standby loads do not add to generator demand as they are activated only when another load is out of service.

Generators must be sized to handle the starting current associated with starting the largest motor. On large facilities with many small motors, starting current usually can be neglected unless all the motors are expected to start simultaneously. However, if the total load is dominated by several large motors, the starting load must be considered.

In calculating generator loads it must be remembered that each motor will only draw the load demanded by the process. It is this load and not the nameplate rating of the motor that should be used in the load list. For example, even though a pump is driven by a 100 hp motor, if the process conditions only demand 75 hp, the total load that will be demanded from the generator is 75 hp.

Generators are normally provided with static voltage regulators capable of maintaining 1% voltage regulation from no load to full load. While random ("mush") wound stators are acceptable for smaller units, formed coils are normally preferred for generators of approximately 150 kW or larger. Vacuum-pressure-impregnated (VPI) windings are recommended for all units operating in high-humidity environments.

Smaller generators, typical of those frequently used at production facilities, often cannot provide enough current to operate the instantaneous trip of magnetic circuit breakers used as main circuit breakers under certain conditions. Manufacturers' data should be obtained for units under consideration and, if necessary, a short-circuit boost option or a permanent magnet rotor (PMG) option should be considered. These options will assist the voltage regulator in delivering full exciter voltage and current during periods of severe generator overload and short circuit conditions. This helps assure that the generators are capable of delivering enough current to trip the main circuit breaker.

When generators are specified, it should be realized that both mechanical and electrical requirements differ between units which will be used for standby service and units which will be operated continuously. Typically, standby units have less copper in their windings than continuous duty units, causing standby units to reach higher temperatures if operated continuously, and thus reducing life. Standby units, as classified by most manufacturers, are not to be confused with units which are alternated weekly (or on some other regular basis), but which are operated continuously when they are "on line." This operating mode should be considered "continuous duty."

POWER SYSTEM DESIGN

Three-Phase Connections

Electrical power systems are normally three-phase systems connected using "wye" or "delta" connections. In "wye" connections the three phases are connected to form a letter "Y" with a neutral point at the intersection of the three phases. In "delta" connections the three phases are connected to form a Greek letter "delta" (Δ). Delta systems do not have a neutral; hence delta systems are 3-wire systems. A Y connection has a neutral and thus it is a 4-wire system.

Generators typically are Y-connected to provide three phases and a neutral for a 3-phase, 4-wire system. The neutral can be grounded or ungrounded, but a grounded neutral is usually preferred.

Transformers can be delta-connected on both the primary side and the secondary side for a 3-phase, 3-wire system or delta on the primary side and Y on secondary side for a 3-phase, 4-wire system. Transformers can also be Y connected on both primary and secondary sides, but such is not recommended unless special precautions are taken to accommodate third harmonic currents.

Figure 17-1 shows a typical system with a Y-connected generator, a delta-connected motor, and a transformer with a delta-connected primary and Y-connected secondary.

Power

Apparent power is the total power of a circuit and is measured in VA or kVA (1,000 VA). It is obtained by multiplying voltage and current.

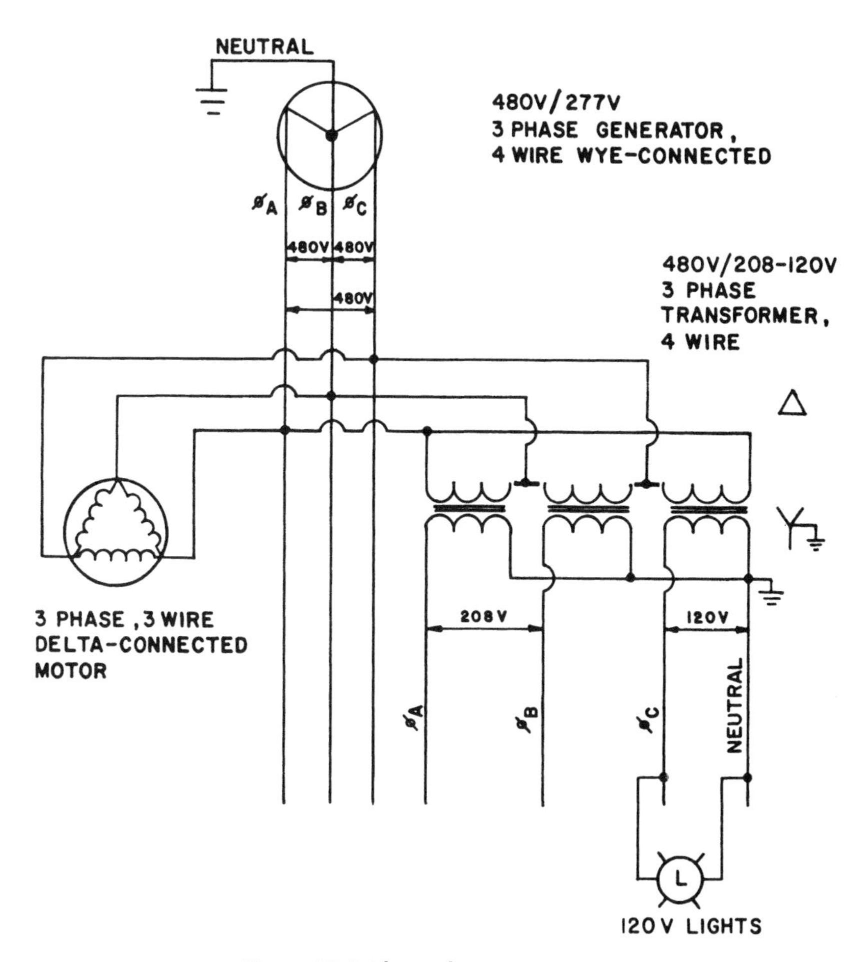

Figure 17-1. Three-phase connections.

$$\text{Single-phase apparent power} = EI \tag{17-1}$$

$$\text{Three-phase apparent power} = (3)^{1/2}\ EI \tag{17-2}$$

where E = line-to-line voltage, volts
I = line current, amperes (amps)

"Active power" is the portion of the apparent power that is consumed by the load to produce work. It is obtained by multiplying the current through a load by the voltage across the load and is expressed in watts or kilowatts (1,000 watts).

$$\text{Single-phase active power} = EI \cos\theta \tag{17-3}$$

$$\text{Three-phase active power} = (3)^{1/2} EI \cos\theta \tag{17-4}$$

where θ = power factor angle, radians or degrees

"Reactive power" is the portion of the apparent power necessary to produce magnetic flux in motors. This power is not transmitted to the load to produce real work. It is expressed in VAR or kVAR (1,000 VAR).

$$\text{Single-phase reactive power} = EI \sin\theta \tag{17-5}$$

$$\text{Three-phase reactive power} = (3)^{1/2} EI \sin\theta \tag{17-6}$$

Power Factor

Power factor is the ratio of the active power to the apparent power:

$$\text{Power Factor} = \frac{EI \cos\theta}{EI} = \cos\theta$$

The power factor indicates the portion of total power which is consumed by the load. The power factor is "leading" in capacitive loads, "lagging" in inductive loads, and "unity" in resistive loads. Figure 17-2 shows "power triangles" to illustrate power factor terms.

When the power factor is less than unity, there is reactive power present. In other words, more power is required to produce the work than is absolutely necessary. This translates into higher energy cost and larger

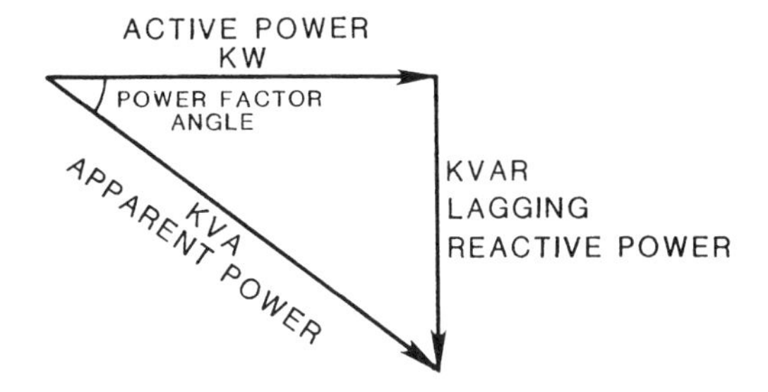

a) LAGGING POWER FACTOR

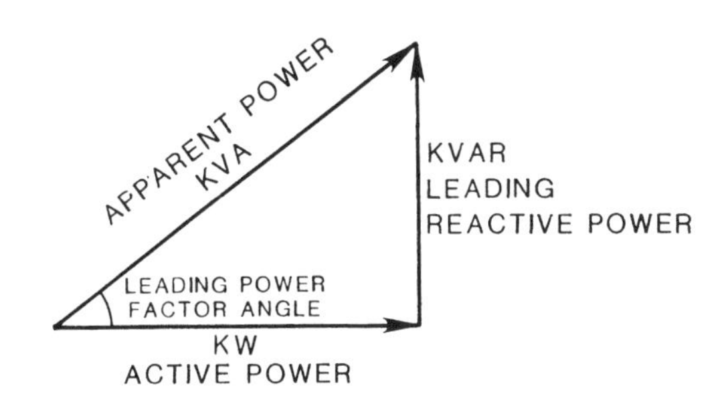

b) LEADING POWER FACTOR

c) UNITY POWER FACTOR

Figure 17-2. Power factor terms.

generating units and transformers. Also, low power factor may reduce voltage, causing motors to operate in a sluggish manner or lights to dim.

The power factor can be made to approach unity by removing reactive power from the system. Lightly loaded induction motors are particularly guilty of producing low power factors since the reactive power does not change with the load on the motor. This causes high phase angles at low load—creating low power factors. By keeping motors at or near full load, system power factors can be improved. Power factors can also be improved by using leading-power-factor synchronous motors or power capacitors.

Short Circuit Currents

To avoid damage to equipment and harm to personnel, electrical components of the facility power system must be selected to withstand available short circuit currents and to isolate facility circuits quickly.

When a short circuit occurs, load impedance no longer limits current. Only the power source capability and internal impedance limit the amount of short circuit current. In a facility powered by generators, the magnitude of these currents depends on generator capacity and internal impedance and the number of units operating at the time of the short circuit. In a facility powered by a utility company (with large available short circuit currents), the internal impedance of the facility's main transformer and the interconnecting wiring determines the magnitude of short circuit currents.

All electrical protective equipment (e.g., circuit breakers, fuses, bus bars, and motor starters) is rated for maximum short circuit currents by NEMA standards. Proper selection of equipment must be based on available short circuit currents.

HAZARDOUS AREA (LOCATION) CLASSIFICATION

In order to properly select and install electrical equipment in a location which contains, or may contain, flammable gases or vapors, combustible dust, or easily ignitible fibers, the location must first be "classified." The classification process is threefold: first, one must designate the type of hazard or "class" that may be present—gas, dust, or fiber; second, one must designate the specific "group" of the hazardous substance; third, one must determine the probability that the hazardous substance will be present. Figure 17-3 is a diagram showing the various classes and groups that are used in the U.S. and Canada. In oil and gas production facility design almost all classifications will be Class I, Group D.

In addition to being toxic, hydrogen sulfide-air mixtures are flammable in concentrations of approximately 4.3% to 45.5% hydrogen sulfide by volume. If air containing more than 4.3% hydrogen sulfide is possible, Class I, Group C, electrical equipment must be installed, although specific laboratory tests have not been performed to determine the percentage level of hydrogen sulfide in a hydrogen sulfide-natural gas mixture at which the mixture should be considered Group C. API RP 500

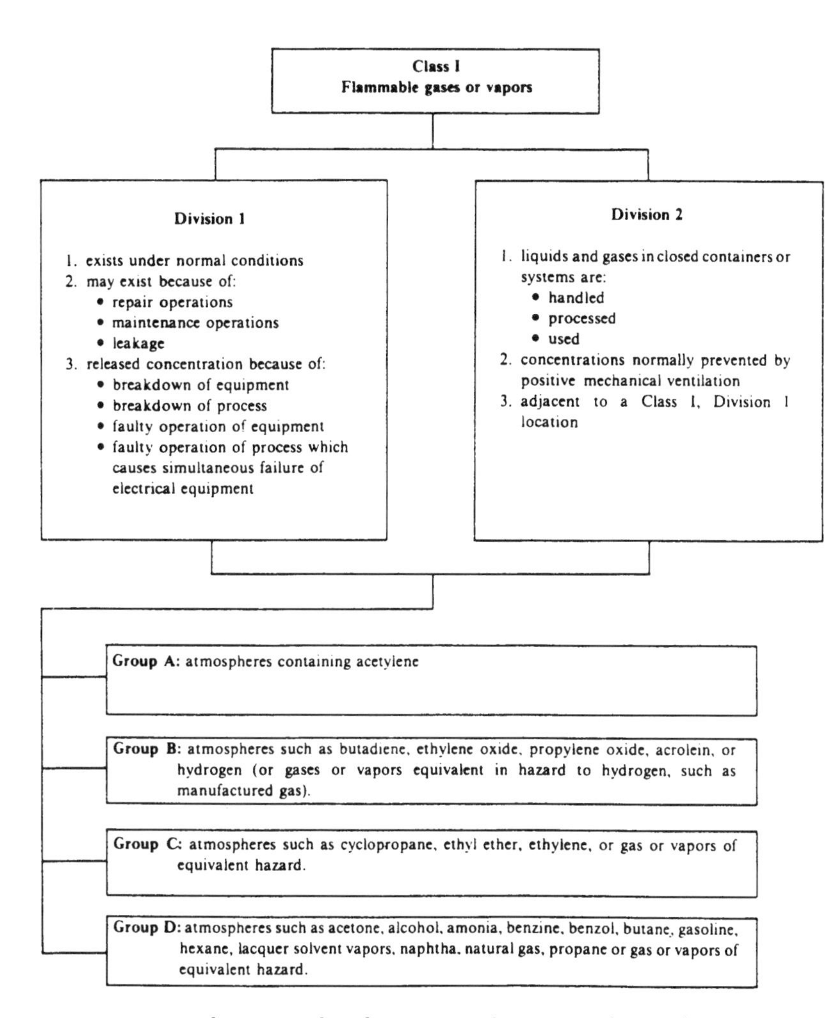

Figure 17-3. Hazardous area classifications used in U.S. and Canada, in accordance with Article 500, NEC Code—1984. (*Courtesy of R. Stahl, Inc.*)

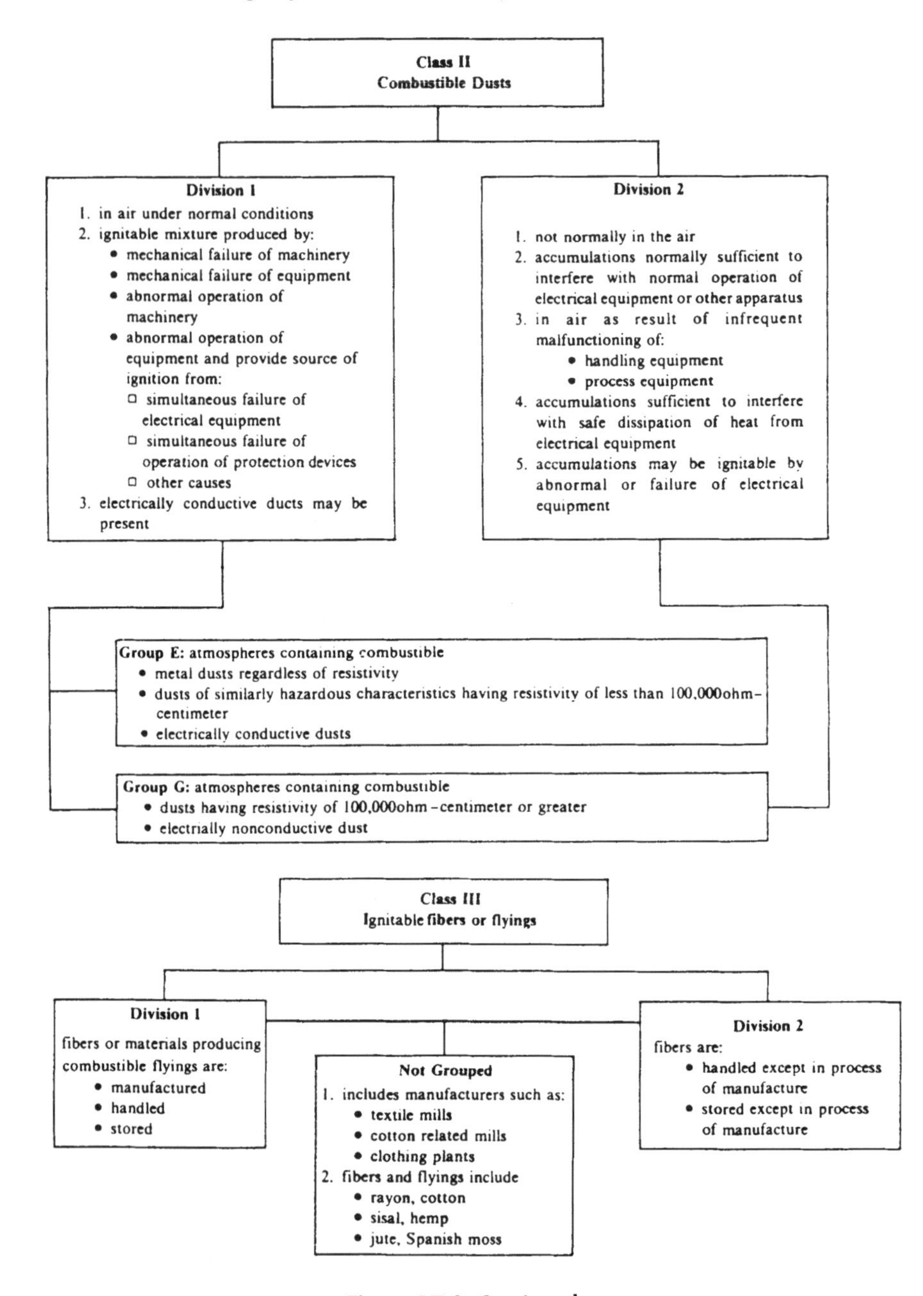

Figure 17-3. Continued.

recommends that mixtures containing less than 25% hydrogen sulfide be considered Group D; above this level they should be considered Group C. Designating the "group" of a specific material is easily accomplished by referencing either the National Electrical Code or the National Fire Protection Association's documents numbered 325M and 497M.

Deciding the probability that ignitible concentrations of the hazardous substance will be present is the most difficult of the steps in classifying an area. Class I and II areas are referred to as Division 1 areas when ignitible concentrations of hazardous materials are anticipated during normal operations, on a continuous or intermittent basis. Areas are referred to as Division 2 areas when ignitible concentrations are anticipated only during abnormal operations. Additionally, areas are considered Division 1 if a process failure is likely to cause ignitible levels of hazardous material and cause an electrical fault in a mode that could result in an electrical arc. Also, Class II areas are considered Division 1 if the hazardous dust involved is metallic. Class III areas are specified Division 1 if the easily ignitible materials are handled, manufactured, or used, and Division 2 if they are stored or handled in a non-manufacturing environment. Any area which is not Division 1 or 2 is termed "unclassified" or "non-hazardous."

The European philosophy on area classification varies from that of the United States and Canada. Specifically, in Europe and most other international areas, the "Zone" concept is utilized. An area in which an explosive gas-air mixture is continuously present, or present for long periods of time, is referred to as Zone 0. The vapor space of a closed, but vented, process vessel or storage tank is an example. An area in which an explosive gas-air mixture is likely to occur in normal operations is designated Zone 1. An area in which an explosive gas-air mixture is less likely to occur, and if it does occur will exist only for a short time, is designated Zone 2. Zone 0 and Zone 1 correspond to Division 1 in the U.S. and Canada System. Zone 2 is equivalent to Division 2.

Europeans characterize substances by Groups designated as I, IIA, IIB, and IIC. Group I is applicable to below-ground installations where methane may be present. Group IIA is applicable to above-ground installations where hazards due to gases or vapors with flammability properties similar to those of propane may exist. Group IIA most closely matches the U.S. and Canada Group D. Group IIB is applicable to above-ground installations where hazards may exist due to gases or vapors with flammability properties similar to those of ethylene, and most closely matches the U.S. and Canada Group C. Group IIC is applicable to above-ground installations where hazards may exist due to gases or vapors with flammability properties similar to those of hydrogen and acetylene. Group IIC includes both U.S. and Canada Groups A and B.

To promote uniformity in area classifications for oil and gas drilling and producing facilities, the American Petroleum Institute developed RP 500, "Recommended Practice for Classification of Locations for Electrical Installations at Petroleum Facilities Classified as Class I, Division 1, and Division 2." Figures 17-4 to 17-14 show some common recommended classifications surrounding common production facility equipment as given in RP 500. API RP 500 also provides valuable tutorial information on the philosophies of area classification and the reader is encouraged to become familiar with this publication.

Based on the information contained in these figures it is possible to draw an area classification diagram of the facility. Figure 17-15 shows an example for a typical offshore production platform.

Ventilation Requirements

In order to properly classify areas surrounding production equipment, not only must the specific items of equipment (separators, pumps, compressors, etc.) be identified, but also the degree of ventilation must be

(text continued on page 509)

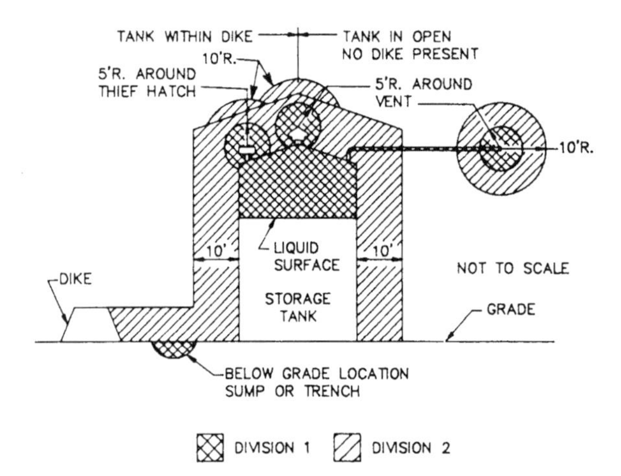

Figure 17-4. Flammable liquid storage tank in a nonenclosed, adequately ventilated area. ***(Reprinted with permission from API RP 500.)***

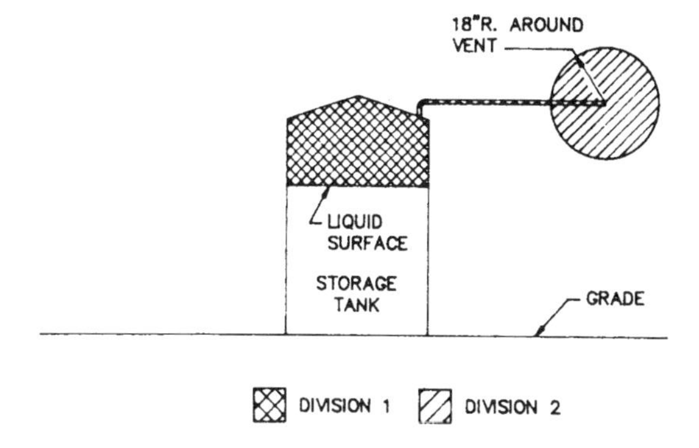

Figure 17-5. Combustible liquid storage tank in a nonenclosed, adequately ventilated area. ***(Reprinted with permission from API RP 500.)***

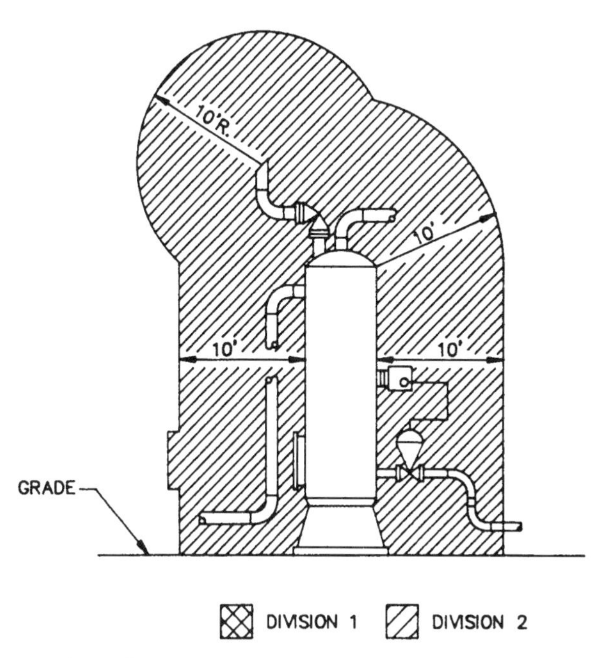

Figure 17-6. Hydrocarbon pressure vessel or protected fired vessel in a nonenclosed, adequately ventilated area. ***(Reprinted with permission from API RP 500.)***

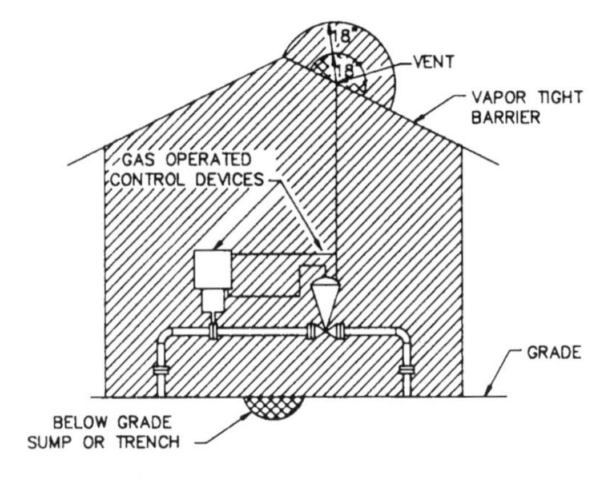

Figure 17-7. Flammable gas-operated instruments in an adequately ventilated or limited ventilated building or enclosure with all devices vented to the outside. ***(Reprinted with permission from API RP 500.)***

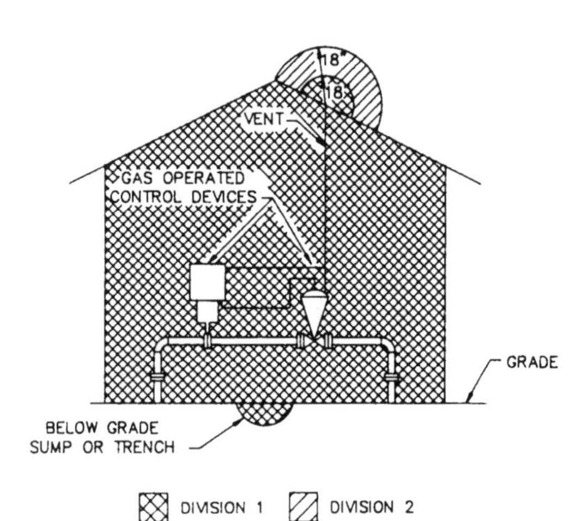

Figure 17-8. Flammable gas-operated instruments in an inadequately ventilated building or enclosure. ***(Reprinted with permission from API RP 500.)***

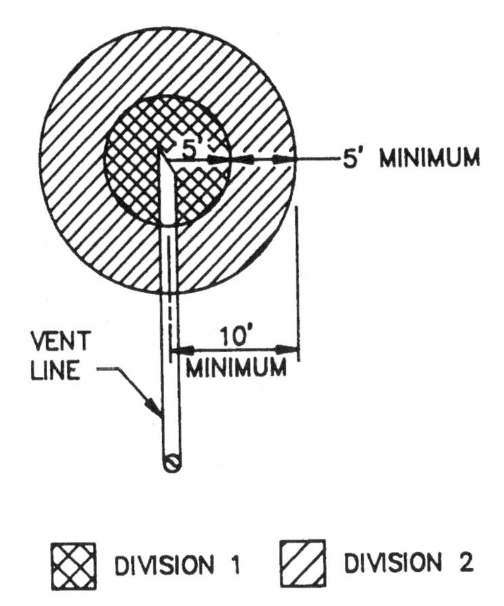

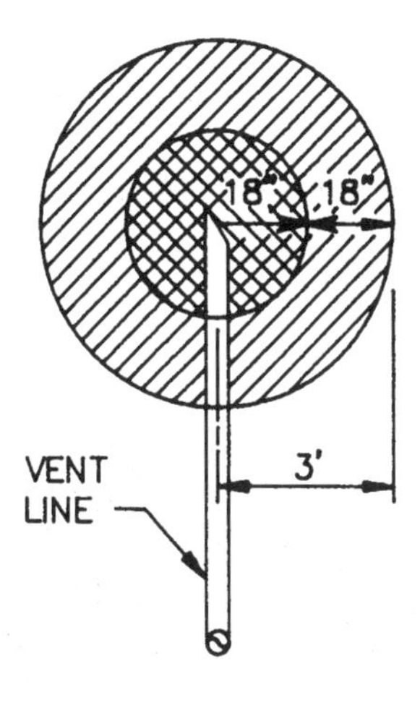

Figure 17-9. Process equipment vent in a nonenclosed, adequately ventilated area (top), and instrument or control device vent in a nonenclosed, adequately ventilated area (bottom). ***(Reprinted with permission from API RP 500.)***

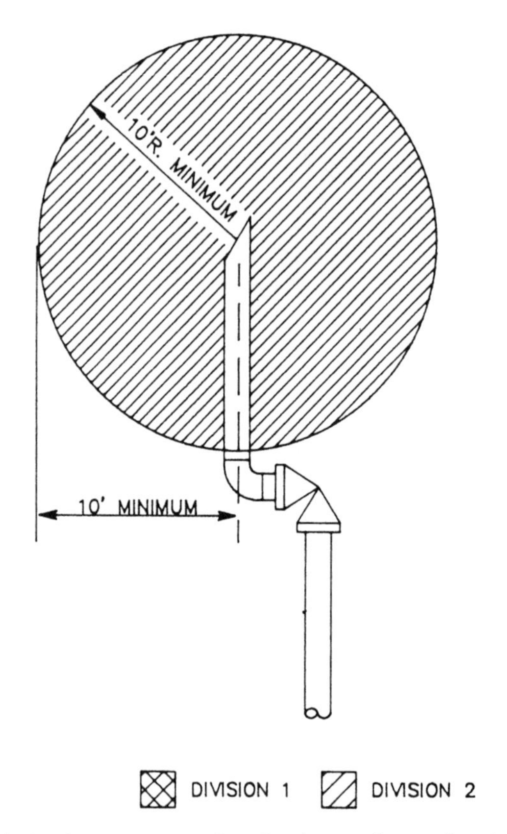

Figure 17-10. Relief valve in a nonenclosed, adequately ventilated area. ***(Reprinted with permission from API RP 500.)***

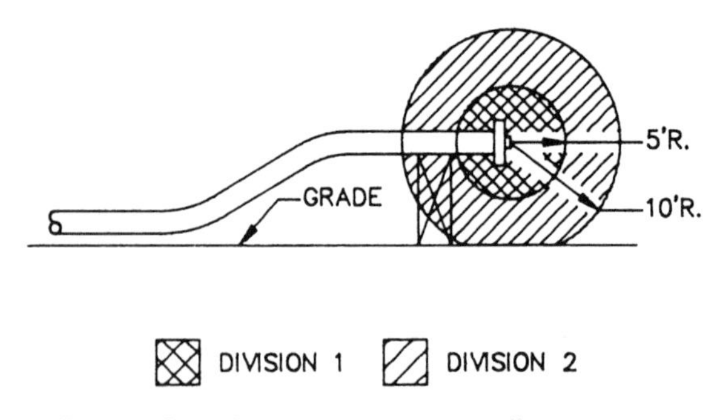

Figure 17-11. Ball or pig launching or receiving installation in a nonenclosed, adequately ventilated area. ***(Reprinted with permission from API RP 500.)***

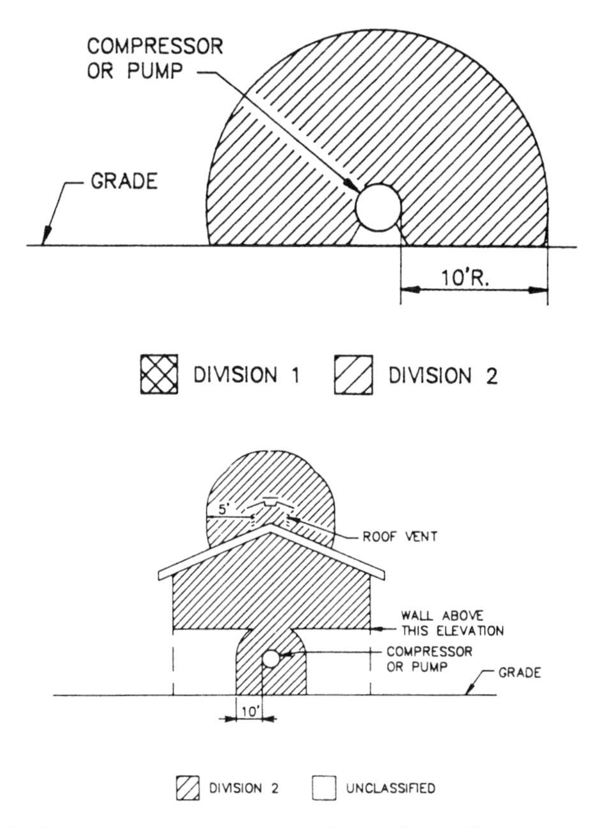

Figure 17-12. Compressor or pump in an adequately ventilated, nonenclosed area (top), and compressor or pump in an adequately ventilated, enclosed area (bottom). ***(Reprinted with permission from API RP 500.)***

(text continued from page 504)

determined. Specifically, most recommended practices and standards consider ventilation as either "adequate" or "inadequate." API RP 500, like most recommended practices for area classification in the United States, uses the basic principles of the definition of adequate ventilation in NFPA No. 30, "Flammable and Combustible Liquids Code." Adequate ventilation may be defined as ventilation (natural or artificial) that is sufficient to prevent the *accumulation* of significant quantities of vapor-air mixtures in concentrations above 25% of their lower flammable (explosive) limit (LFL).

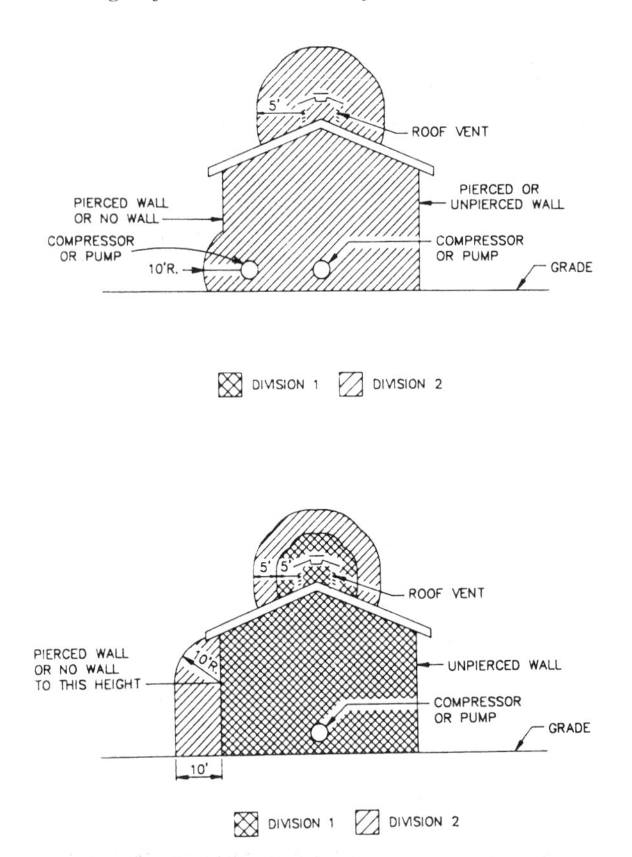

Figure 17-13. Compressor or pump in an adequately ventilated, enclosed area (top), and compressor or pump in an inadequately ventilated, enclosed area (bottom). ***(Reprinted with permission from API RP 500.)***

Although the above definition is quite definitive, it is difficult to apply to actual facilities. API RP 500 recommends several specific methods of achieving adequate ventilation. All methods require that there be no unventilated areas where flammable vapors or gases might accumulate. The most common method of providing adequate ventilation to larger buildings is to provide at least 1.0 cubic foot of air volume flow per minute per square foot of floor area, but at least six (6) air changes per hour, by either natural or mechanical ventilation. Some operators install ventilation equipment to obtain less than six air changes per hour under normal circumstances and increase ventilation to six or more air changes per hour if gas detectors sense the presence of gas above 20–25% of the lower flammable limit.

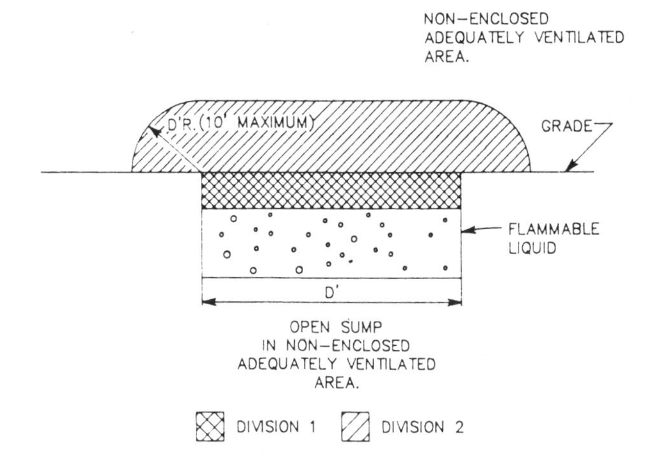

Figure 17-14. Open sump in a nonenclosed, adequately ventilated area. ***(Reprinted with permission from API RP 500.)***

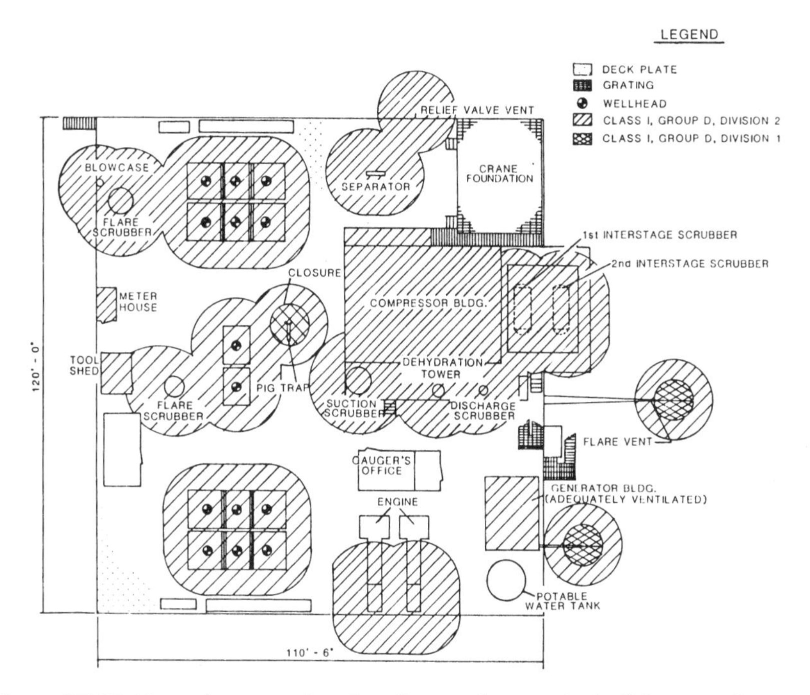

Figure 17-15. Hazardous area location diagram for a typical offshore production platform.

For buildings of 1,000 ft^3 or less (such as a typical meter house), API RP 500 defines the building as being adequately ventilated if it has sufficient openings to provide twelve air changes per hour due to natural thermal effects. Assuming the building has no significant internal resistance and that the inlet and outlet openings are the same size and are vertically separated and on opposite walls, the required free area of the inlet or outlet is:

$$A = \frac{Vol}{1,200\,[H'\,T_1 - T_2)/T_1]^{1/2}} \qquad (17\text{-}8)^*$$

where A = free area of inlet (or outlet) openings (includes a 50 percent effectiveness factor), ft^2
Vol = volume of building to be ventilated, ft^3
T_1 = temperature of indoor air, °R
T_2 = temperature of outdoor air, °R
H' = height from the center of the lower opening to the Neutral Pressure Level (NPL), ft

NPL is the point on the vertical surface of a building where the interior and exterior pressures are equal. It is given by:

$$H' = \frac{H}{1 + [(A_1/A_2)^2 (T_1/T_2)]} \qquad (17\text{-}9)$$

where H = vertical (center-to-center) distance between A_1 and A_2, ft
A_1 = free area of lower opening, ft^2
A_2 = free area of upper opening, ft^2

For example, assume a building with inside dimensions of 8 ft wide, 10 ft long, and 8 ft high, an outside temperature of 70°F, inside temperature of 80°F, $A_1 = A_2$ and the vertical (center-to-center) distance between A_1 and A_2 of 6 ft. The height from the center of the lower opening to the NPL is:

$$H' = \frac{6}{1 + 540/530} = 2.97 \text{ ft}$$

---

*Equation derived from *1985 ASHRAE Handbook of Fundamentals,* Chapter 22, assuming an air change every five (5) minutes. Refer to the ASHRAE Handbook, Chapter 22, for additional information on naturally ventilated buildings.

Therefore, the minimum area required is:

$$A = \frac{(8)(10)(8)}{1,200\,[2.97(10/540)]^{1/2}}$$

$$= 2.27 \text{ ft}^2 \text{ for both the inlet and the outlet}$$

GAS DETECTION SYSTEMS

Combustible gas detection systems are frequently used in areas of poor ventilation. By the early detection of combustible gas releases before ignitible concentration levels occur, corrective procedures such as shutting down equipment, deactivating electrical circuits and activating ventilation fans can be implemented prior to fire or explosion. Combustible gas detectors are also used to substantiate adequate ventilation. Most combustible gas detection systems, although responsive to a wide range of combustible gases and vapors, are normally calibrated specifically to indicate concentrations of methane since most natural gas is comprised primarily of methane.

Gas detectors are also used to sense the presence of toxic gases—primarily hydrogen sulfide (H_2S). These detectors often activate warning alarms and signals at low levels to ensure that personnel are aware of potential hazards before entering buildings or are alerted to don protective breathing apparatus if they are already inside the buildings. At higher levels, shut-downs are activated.

Consensus performance standards and guidance for installation are provided for combustible gas detectors by ISA S12.13 and RP 12.13 and for hydrogen sulfide gas detectors by ISA S12.15 and RP 12.15.

Required locations of gas detectors (sensors) are often specified by the authority having jurisdiction. For example, API RP 14C recommends certain locations for combustible detectors. These recommendations have been legislated into requirements in U.S. Federal waters by the Minerals Management Service. RP 14C should be referred to for specific details, but, basically, combustible gas detectors are required offshore in all inadequately ventilated, classified, enclosed areas. The installation of sensors in nonenclosed areas is seldom either required or necessary. Ignitible or high toxic levels of gas seldom accumulate and remain for significant periods of time in such locations.

When specifying locations for gas detector sensors, consideration should be given to whether the gases being detected are heavier than air or lighter than air. Hydrogen sulfide is heavier than air and therefore, hydrogen sulfide detectors are normally installed near the floor. Since sensors may be adversely affected (even rendered ineffective) if coated with water, they normally should be installed 18 to 36 inches above the floor if they may be subjected to flooding or washdown.

Most combustible gas detector sensors are installed in the upper portions of buildings for the detection of natural gas. However, in many cases the vapor which flashes off oil in storage tanks can be heavier than air. Below grade areas should be considered for sensor installations where heavier-than-air vapors might collect.

Sensing heads should be located in draft-free areas where possible, as air flowing past the sensors normally increases drift of calibration, shortens head life, and decreases sensitivity. Air deflectors are available from sensor manufacturers and should be utilized in any areas where significant air flow is anticipated (such as air conditioner plenum applications). Additionally, sensors should be located, whenever possible, in locations which are relatively free from vibration and easily accessed for calibration and maintenance. Obviously, this cannot always be accomplished. It usually is difficult, for example, to locate sensors in the tops of compressor buildings at locations which are accessible and which do not vibrate.

It generally is recommended, and often required, that gas detection systems be installed in a fail-safe manner. That is, if power is disconnected or otherwise interrupted, alarm and/or process equipment shutdown (or other corrective action) should occur. All specific systems should be carefully reviewed, however, to ensure that non-anticipated equipment shutdowns would not result in a more hazardous condition than the lack of shutdown of the equipment. If a more hazardous situation would occur with shutdown, only a warning should be provided. As an example, a more hazardous situation might occur if blowout preventers were automatically actuated during drilling operations upon detection of low levels of gas concentrations than if drilling personnel were only warned.

Concentration levels where alarm and corrective action should occur vary. If no levels are specified by the authority having jurisdiction, most recommend alarming (and/or actuating ventilation equipment) if combustible gas concentrations of 20 percent LEL (lower explosive limit) or more are detected. Equipment shutdowns, the disconnecting of electrical power, production shut-in, or other corrective actions usually are recommended if 60 percent LEL concentrations of combustible gas are detected. Hydrogen sulfide concentrations of 5 ppm usually require alarms and actuation of ventilation equipment and levels of 15 ppm usually dictate corrective action.

Special attention should be given to grounding the sheathes and shields of cables interconnecting sensing heads to associated electronic controllers. To avoid ground loops, care should be taken to ground shields only at one end, usually at the controller. If cables are not properly grounded, they may act as receiving antennas for radio equipment and other RF generators at the location, transmitting RF energy to the electronic controller. This RF energy can cause the units to react as if combustible or toxic gas were detected, causing false alarms or unwarranted corrective action. The use of RF-shielded enclosures is recommended where RF problems are experienced or anticipated.

GROUNDING

A ground, as defined by the National Electrical Code, is a conducting connection, whether intentional or accidental, between an electrical circuit or equipment and the earth. Proper grounding of electrical equipment and systems in production facilities is important for safety of operating personnel and prevention of equipment damage. The term "grounding" includes both electrical supply system grounding and equipment grounding.

The basic reasons for grounding an electrical supply system are to limit the electrical potential difference (voltage) between all uninsulated conductive equipment in the area; to provide isolation of faults in the system; and to limit overvoltage on the system under various conditions. In the case of a grounded system it is essential to ground at each separately derived voltage level.

Electrical Supply System Grounding

The electrical supply system neutral can be grounded or ungrounded, but there is an increasing trend in the industry toward grounded systems. Ungrounded power systems are vulnerable to insulation failures and increased shock hazards from transient and steady state overvoltage conditions. Grounding of an electrical supply system is accomplished by connecting one point of the system (usually the neutral) to a grounding electrode. The system can be solidly grounded, or the ground can be through a high or low resistance. A resistance ground is more suited for certain systems—particularly when process continuity is important.

Equipment Grounding

Equipment grounding is the grounding of non-current carrying conductive parts of electrical equipment or enclosures containing electrical components. This provides a means of carrying currents caused by insulation failure or loose connections safely to ground to minimize the danger of shock to personnel.

The following equipment (not all inclusive) requires adequate equipment grounding:

1. Housings for motors and generators
2. Enclosures for switchgear and motor control centers
3. Enclosures for switches, breakers, transformers, etc.
4. Metal frames of buildings
5. Cable and conduit systems
6. Conductive cable tray systems
7. Metal storage tanks

Grounding for Static Electricity

A discharge to ground of static electricity accumulated on an object can cause a fire or explosion. A static charge can have a potential of 10,000 volts, but because it has a very small current potential, it can be safely dissipated through proper bonding and grounding. Bonding two objects together (connecting them electrically) keeps them at the same potential (voltage), minimizing spark discharge between them. Generally, equipment bonded to nearby conducting objects is adequate for static grounding. The equipment grounding conductor carries static charges to ground as they are produced.

Grounding for Lightning

Elevated structures such as vent stacks, buildings, tanks, and overhead lines must be protected against direct lightning strikes and induced lightning voltages. Lightning arrestors or rods are installed on such objects and connected to ground to safely dissipate the lightning charges.

Grounding Methods

Onshore, grounding is generally provided by installing a ground loop, made of bare copper conductors, below the finished grade of the facility.

Individual equipment grounding conductors and system grounding conductors are then connected to this ground loop, usually by a thermoweld process. A number of grounding electrodes, generally ⅝-in. to ¾-in. diameter and 8–10 ft long copper or copper-clad steel rods, are driven into the earth and connected to the ground loop. The number of ground rods required and the depth to which they should be driven are calculated based on the resistivity of the soil and the minimum required resistance of the grounding system.

Most grounding systems are designed for less than 5 ohms resistance to ground. A continuous underground metallic water piping system can provide a satisfactory grounding electrode. The National Electrical Code, Article 250, covers requirements for sizing ground loops and equipment/system grounding conductors.

Offshore, the equipment and system ground conductors are connected to the facility's metal deck, usually by welding. The metal deck serves the function of the ground loop and is connected to ground by virtue of solid metal-to-metal contact with the platform jacket.

D.C. POWER SUPPLY

Generally, electrical control systems are designed "Fail-Safe." If power is temporarily lost, unnecessary shutdown of the process may occur. Thus, most safety systems such as fire and gas detectors, Nav-Aids, communications, and emergency lighting require standby D.C. power.

Most D.C. power systems include rechargeable batteries and a battery charger system which automatically keeps the batteries charged when A.C. power is available. In some systems, a D.C.-to-A.C. inverter is provided to power some A.C. emergency equipment such as lighting. Solar cells can also be used for charging batteries. Solar cells are frequently used at unmanned installations without on-site power generation. Sometimes non-rechargeable batteries are also used at such locations.

Batteries

Numerous types of batteries are available. A comparison of batteries by cell type is shown in Table 17-1. Rechargeable batteries emit hydrogen to the atmosphere, and hence must be installed such that hydrogen does not accumulate to create an explosion hazard. Ventilation should be provided for battery compartments.

Batteries should normally be installed in an unclassified area. However, if installed in Division 2 areas, a suitable disconnect switch must be installed to disconnect the load prior to removing the battery leads and thus avoid a spark if the battery leads are disconnected under load conditions. Batteries should not be installed in Division 1 areas.

Battery Chargers

Battery chargers are selected based on cell type and design ambient conditions. Chargers connected to self-generated power should be capable of tolerating a 5% frequency variation and a 10% voltage variation. Standard accessories of chargers include equalizing timers, A.C. and D.C. fuses or circuit breakers, current-limiting features, and A.C. and D.C. ammeters and voltmeters. Optional accessories such as low D.C. voltage alarms, ground fault indications, and A.C. power failure alarms are usually available.

Chargers are normally installed in unclassified areas. However, it is possible to purchase a charger suitable for installation in a classified area.

CATEGORIES OF DEVICES

Electrical switches, relays, and other devices are described for safety reasons by several general categories. Since these devices are potential sources of ignition during normal operation (for example, arcing contacts) or due to malfunction, the area classification limits the types of devices which can be used.

High-Temperature Devices

High-temperature devices are defined as those devices that operate at a temperature exceeding 80 percent of the ignition temperature (expressed in Celsius) of the gas or vapor involved. The ignition temperature of natural gas usually is considered to be 900°F (482°C). Therefore, a device is

Table 17-1
Comparison of Batteries by Cell Type

Type	Projected Useful Life (Years)	Projected Cycle Life* (Number of Cycles)	Wet Shelf Life** (Months)	Comments***
Primary	1–3	1	12	Least maintenance. Periodic replacement. Cannot be recharged.
SLI (Starting, Lighting & Ignition) (Automotive Type)	½–2	400–500	2–3	High hydrogen emission. High maintenance. Not recommended for float service or deep discharge. Low shock tolerance. Susceptible to damage from high temperature.
Lead Antimony	8–15	600–800	4	High hydrogen emission. Periodic equalizing is required for float service and full recharging. Low shock tolerance. Susceptible to damage from high temperature.
Lead Calcium	8–15	40–60	6	Low hydrogen emission if floated at 2.17 volts per cell. Periodic equalizing charge is not required for float service if floated at 2.25 volts per cell. However, equalizing is required for recharging to full capacity. When floated below 2.25 volts per cell, equalizing is required. Susceptible to damage from deep discharge and high temperature. Low shock tolerance.

(*table continued on next page*)

Table 17-1 (Continued)
Comparison of Batteries by Cell Type

Type	Projected Useful Life (Years)	Projected Cycle Life* (Number of Cycles)	Wet Shelf Life** (Months)	Comments***
Lead Selenium	20+	600–800	6	Low hydrogen emission if floated at 2.17 volts per cell. Periodic equalizing charge is not required for float service if floated at 2.25 volts per cell. However, equalizing is required for recharging to full capacity. When floated below 2.25 volts per cell, equalizing is required. Low shock tolerance. Susceptible to damage from high temperature.
Lead Planté (Pure Lead)	20+	600–700	4	Moderate hydrogen emission. Periodic equalizing charge is required for float service and full recharging. Low shock tolerance. Susceptible to damage from high temperature.
Nickel Cadmium (Ni-Cad)	25+	1000+	120+	Low hydrogen emission. Periodic equalizing charge is not required for float service, but is required for recharging to full capacity. High shock tolerance. Can be deep cycled. Least susceptible to temperature. Can remain discharged without damage.

Courtesy of API RP 14F

**Cycle life is the number of cycles at which time a recharged battery will retain only 80% of its original ampere-hour capacity. A cycle is defined as the removal of 15% of the rated battery ampere-hour capacity.*

***Wet shelf life is defined as the time that an initially fully charged battery can be stored at 77°F until permanent cell damage occurs.*

****Float voltages listed are for 77°F.*

considered a high-temperature device in a natural gas environment if the temperature of the device exceeds 726°F (385°C). The ignition temperature of hydrogen sulfide is usually considered to be 518°F (270°C). In classified areas, high-temperature devices must be installed in explosion-proof enclosures unless the devices are approved for the specific area by a nationally recognized testing laboratory (NRTL).

Weather-Tight Enclosures

Electrical equipment can be mounted in various types of enclosures. A weather-tight enclosure normally has a gasket and does not allow air (and the moisture contained in the air) to enter the enclosure. Offshore, such an enclosure, if properly closed, will help protect the enclosed electrical equipment from corrosion due to salt water spray. These types of enclosures can be used in Division 2 areas provided they do not enclose arcing, sparking or high temperature devices.

Explosion-Proof

Equipment described as "explosion-proof" is equipment installed in enclosures that will withstand internal explosions and also prevent the propagation of flame to the external atmosphere. As the gases generated by the explosion expand, they must be cooled before reaching the surrounding atmosphere.

Equipment may be rated explosion-proof for certain gases but not others. For example, an enclosure may be rated as suitable for Group D gases, but not for Group B gases. Therefore, it is not satisfactory to merely state that equipment must be "explosion-proof"; one must specify "Explosion-proof for Class I, Group D," as an example. Because explosion-proof enclosures must have a path to vent the expanding gases created by the explosion, explosion-proof enclosures "breathe" when the temperature inside the enclosure is different from that outside. That is, they cannot be weather tight. As a result, moisture frequently is introduced into explosion-proof enclosures. Unless suitable drains are provided in low spots, water can accumulate inside the enclosure and damage enclosed electrical equipment.

The surface temperature of explosion-proof enclosures cannot exceed that of high-temperature devices. Equipment can be tested by nationally recognized testing laboratories and given one of 14 "T" ratings, as indicated in Table 17-2. This equipment may exceed the "80 percent rule," but the "T" rating must be below the ignition temperature of the specific gas or vapor involved. As an example, equipment rated T1 has been verified not to exceed 842°F and, therefore, is suitable for most natural gas applications.

Hermetically Sealed Devices

Hermetically sealed devices are devices sealed to prevent flammable gases from reaching enclosed sources of ignition. These devices are suitable for use in Division 2 and unclassified areas.

Hermetically sealed electrical devices must be verified by a testing laboratory to meet mechanical abuse and to withstand aging and exposure to expected chemicals. Devices "potted" with common silicones and similar materials by an end user or even a manufacturer, without testing, and devices merely provided with O-rings seldom meet acceptable criteria. Normally, hermetically sealed devices must be sealed through metal-to-metal or glass-to-metal fusion. Many electrical relays, switches, and sensors are available as hermetically sealed devices for common oil and gas producing facility applications. Hermetically sealed devices are often desirable to protect electrical contacts from exposure to salt air and other contaminants.

Table 17-2
Temperature Ratings of Explosion-Proof Enclosures

Maximum Temperature °C	Maximum Temperature °F	Identification Number
450	842	T1
300	572	T2
280	536	T2A
260	500	T2B
230	446	T2C
215	419	T2D
200	392	T3
180	356	T3A
165	329	T3B
160	320	T3C
135	275	T4
120	248	T4A
100	212	T5
85	185	T6

Purged Enclosures

Purged enclosures are those enclosures provided with a purge (static or dynamic) of air or other inert gas to prevent enclosed electrical equipment from coming in contact with surrounding atmospheres which might be flammable. NFPA Publication No. 496 provides detailed requirements for the design of purged enclosures. Requirements are different for different size enclosures. Enclosures can be as small as a box for a single electrical switch or as large as a control room. Requirements vary for three recognized types of purging: Type X, the reduction from Division 1 to unclassified; Type Y, the reduction from Division 1 to Division 2; and Type Z, the reduction from Division 2 to unclassified. If purging is utilized in areas of high humidity or in areas where the atmosphere may contain flammable gases or contaminants, clean dehydrated air or an inert gas should be used as a purge to prevent explosions or damage to enclosed electrical equipment. Properly designed purged enclosures can eliminate the need for explosion-proof enclosures.

Nonincendive Devices

A nonincendive device is one which will not release sufficient energy under *normal* operating conditions to ignite a specific substance. Under abnormal conditions, such as a malfunction of the device, it may release enough energy to cause ignition. Because of this, such devices are suitable for use only in Division 2 and unclassified areas.

Intrinsically Safe Systems

Intrinsically safe systems are electrical systems which are incapable of releasing sufficient electrical or thermal energy under normal *or* abnormal equipment operating conditions to cause ignition of a specific flammable mixture in its most easily ignitible state. For example, intrinsically safe equipment suitable for a Class I, Group D application cannot ignite a mixture of methane and air in its most easily ignitible state (approximately 10 percent methane by volume), even, for example, if adjacent wiring terminals are accidentally shorted with a screwdriver.

The design of intrinsically safe equipment is governed by the rules of NFPA Publication No. 493, "Standard for Intrinsically Safe Apparatus and Associated Apparatus for Use in Class I, II, and III, Division 1, Hazardous Locations." It is cautioned, however, that the design of intrinsically safe equipment is a highly specialized skill and normally best left to those specifically trained in that art. The installation of equipment which has been rated by a testing organization as intrinsically safe should follow the guidelines of ISA RP 12.6, "Recommended Practice for Installation of Intrinsically Safe Systems for Hazardous (Classified) Locations." It must be realized that there is no such thing as an intrinsically safe temperature transmitter, pressure switch, or other such sensor; these devices must be properly installed in intrinsically safe systems to ensure safety from ignition of flammable gas or vapor.

The mere fact that voltage, current, or even both, are at low levels does not guarantee a circuit to be intrinsically safe, even though intrinsically safe circuits do utilize relatively low voltage and current levels. Intrinsically safe systems employ electrical barriers to assure that the system *remains* intrinsically safe. The barriers limit the voltage and current combinations so as not to present an ignition hazard should a malfunction develop. Typically, devices "upstream" of barriers are not intrinsically safe and are installed in control rooms or other unclassified locations. All devices and wiring on the "downstream" side of the barriers are intrinsically safe and can be installed in classified areas.

An additional benefit of intrinsically safe systems is the *reduction* of electrical shock hazards. It is cautioned, however, that intrinsically safe systems are not necessarily tested specifically for personnel shock hazards.

Circuit capacitance and inductance, including the values of these parameters for interconnecting wiring, are integral parts of the overall analysis. It is not always possible to assure that the system will be maintained as designed with only approved intrinsically safe components and with circuits of the capacitance and inductance as originally installed. For this reason, intrinsically safe systems are used primarily at locations where there are sufficiently trained personnel to assure that the intrinsic safety of the system is always maintained.

LIMITATIONS ON INSTALLATION OF ELECTRICAL DEVICES IN HAZARDOUS AREAS

Transformers

In Division 1 areas, transformers must be installed in approved vaults if they contain a flammable liquid. If they do not contain a flammable liquid, they must either be installed in vaults or be approved explosion-proof. In Division 2 areas, "standard" transformers are acceptable, but they must not contain circuit breakers or other arcing devices. Thus, common self-contained transformer/distribution panel packages which contain circuit breakers are not suitable for Division 2 areas. Standard practice is to provide separate units—installing the breakers in explosion-proof enclosures or in unclassified areas.

Meters, Instruments, and Relays

In Division 1 areas, meters, instruments, relays, and similar equipment containing high-temperature or arcing devices must be installed in approved explosion-proof or purged enclosures. Unless such devices are specifically labeled as suitable for Class I, Division 1 areas, it is best to assume they are not suitable.

Arcing contacts in Division 2 areas must be installed in explosion-proof enclosures, be immersed in oil, be hermetically sealed, or be non-incendive. High-temperature devices must be installed in explosion-proof enclosures. Fuses must be enclosed in explosion-proof enclosures unless the fuses are preceded by an explosion-proof, hermetically sealed, or oil-immersed switch and the fuses are used for overcurrent protection of instrument circuits not subject to overloading in normal use.

Figure 17-16 depicts typical devices containing arcing contacts enclosed in explosion-proof enclosures. Figure 17-17 shows typical explosion-proof alarm devices. A telephone instrument suitable for Class I, Divisions 1 and 2, Group D classified areas is shown by Figure 17-18.

Motors and Generators

In Division 1 areas, motors and generators must be either explosion-proof or approved for the classification by meeting specific requirements for a special ventilation system, inert gas-filled construction, or a special submerged unit. Although explosion-proof motors are expensive, they normally are available. Explosion-proof generators normally are not available.

Standard Open or Totally-Enclosed Fan-Cooled (TEFC) generators and motors are acceptable in Division 2 areas if they do not contain brushes or other arcing contacts or high-temperature devices. Three-phase TEFC motors are acceptable in Division 2 locations, but single-phase motors usually contain arcing devices and are not acceptable

(text continued on page 529)

THERMOSTAT

LINE STARTER ENCLOSURE
SIDE OPERATED

LINE STARTER ENCLOSURE
WITH LOAD LIMIT SWITCH

FACTORY
SEALED MANUAL
MOTOR STARTING
SWITCH WITH
ENCLOSURE

AIR BREAK CIRCUIT BREAKER
ENCLOSURE

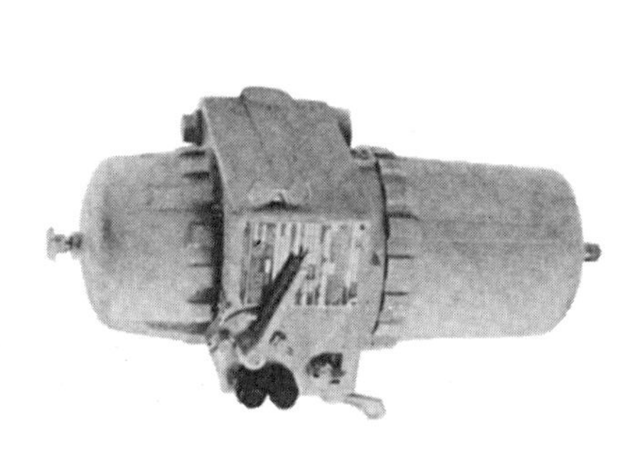

COMBINATION CIRCUIT BREAKER
AND MAGNETIC STARTER

SELECTOR SWITCH
SURFACE MOUNTING

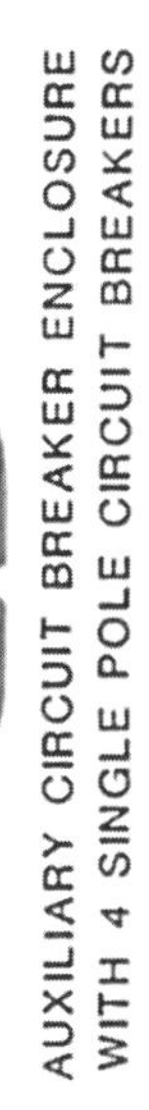

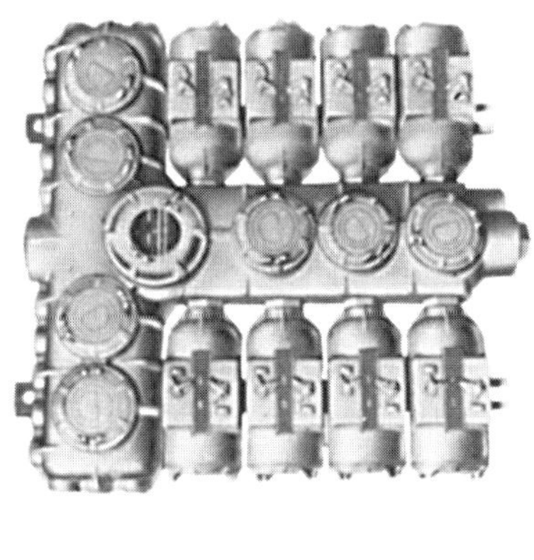

DIV. 1 CIRCUIT BREAKER PANELBOARD

FRONT - OPERATED PUSHBUTTON

Figure 17-16. Typical devices containing arcing contacts in explosion-proof enclosures. *(Courtesy of Crouse-Hinds Electrical Construction Materials, a division of Cooper Industries, Inc.)*

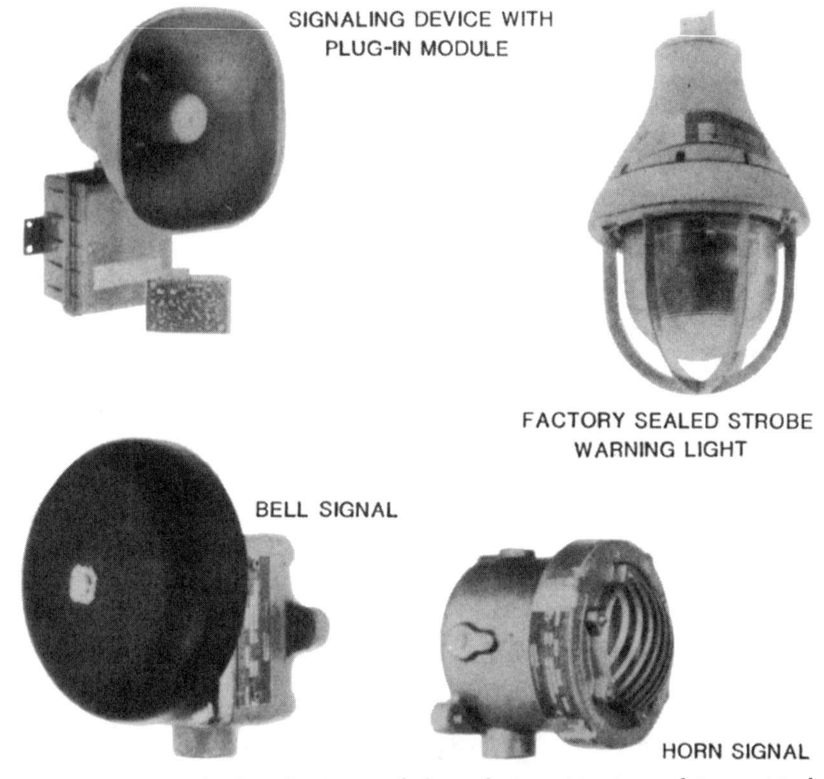

Figure 17-17. Standard explosion-proof alarm devices. (*Courtesy of Crouse-Hinds Electrical Construction Materials, a division of Cooper Industries, Inc.*)

Figure 17-18. Typical telephone instrument suitable for Class I, Divisions 1 and 2, Group D areas. (*Courtesy of Crouse-Hinds Electrical Construction Materials, a division of Cooper Industries, Inc.*)

(text continued from page 525)

unless the arcing devices are installed in explosion-proof enclosures. D.C. motors contain brushes and are not acceptable for classified areas unless they are provided with approved purged enclosures. If motor space heaters are provided, their surface temperature must not exceed 80 percent of the ignition temperature, expressed in degrees Celsius, of the potential gas or vapor which could be present.

Lighting Fixtures

Lighting fixtures installed in Division 1 areas must be explosion-proof and marked to indicate the maximum wattage of allowable lamps. Also, they must be protected against physical damage by a suitable guard or by location.

In both Division 1 and Division 2 areas pendant fixtures must be suspended by conduit stems and provided with set screws to prevent loosening. Stems over 12 inches in length must be laterally braced within 12 inches of fixtures.

All portable lamps in Division 1 areas must be explosion-proof. Figures 17-19, 17-20, and 17-21 show typical explosion-proof lighting fixtures.

Lighting fixtures for Division 2 locations must be either explosion-proof or labeled as suitable for Division 2 for the particular Class and Group involved. Figure 17-22 shows typical Division 2 lighting fixtures.

WIRING METHODS

Documents specified by local authorities having jurisdiction provide very explicit rules for the specific types of electrical equipment that are permitted in the various hazardous (classified) areas and the methods by which the equipment must be installed. Since it is rare to encounter Class II and Class III hazardous (classified) areas in oil and gas producing operations, only Class I requirements will be addressed further. In the United States, the National Electrical Code is referenced by most enforcing agencies. For fixed platforms in the Outer Continental Shelf (OCS), API RP 14F is referenced as the primary design and installation document by the Minerals Management Service (MMS), the enforcing agency. RP 14F deviates somewhat from the National Electrical Code in wiring methods required, relying heavily on U.S. Coast Guard philoso-

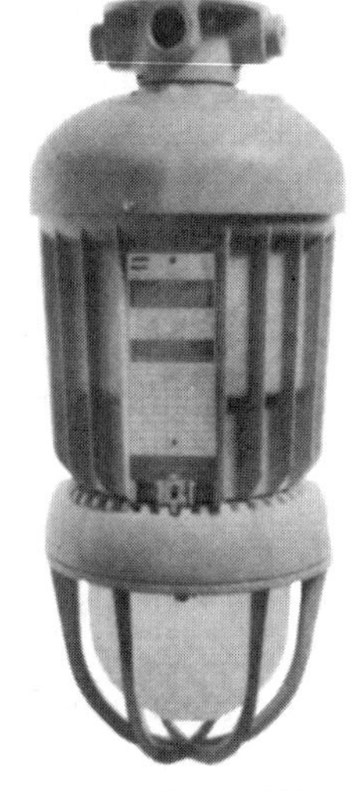

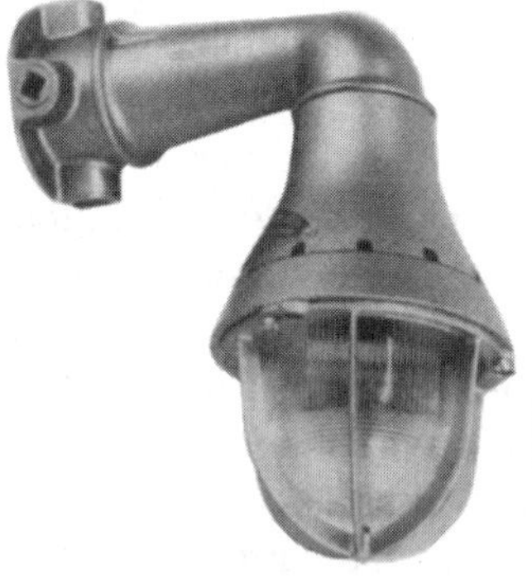

INTEGRALLY BALLASTED HID LIGHTING FIXTURE

EXPLOSION-PROOF LIGHTING FIXTURE INCANDESCENT AND MERCURY VAPOR (FOR EXTERNAL BALLAST)

Figure 17-19. Typical Class I, Division 1 lighting fixures. (*Courtesy of Crouse-Hinds Electrical Construction Materials, a division of Cooper Industries, Inc.*)

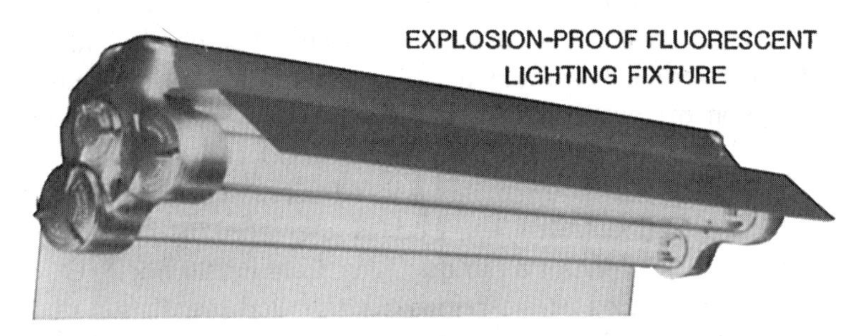

Figure 17-20. Typical explosion-proof fluorescent lighting fixtures. (*Courtesy of Crouse-Hinds Electrical Construction Materials, a division of Cooper Industries, Inc.*)

Figure 17-21. Standard explosion-proof portable lamp suitable for Class I, Divisions 1 and 2, Group D areas. (*Courtesy of Crouse-Hinds Electrical Construction Materials, a division of Cooper Industries, Inc.*)

phy. Except for these specific wiring deviations, however, RP 14F references the National Electrical Code.

Division 1 Areas

In Division 1 areas, the National Electrical Code (NEC) allows only the following wiring methods:

1. Threaded rigid metal conduit
2. IMC (Intermediate Metal Conduit)
3. MI cable (Mineral Insulted Cable)
4. Explosion-proof (XP) flexible connections

Threaded rigid metal conduit must be threaded with an NPT standard conduit cutting die that provides ¾-in. taper per foot, must be made up

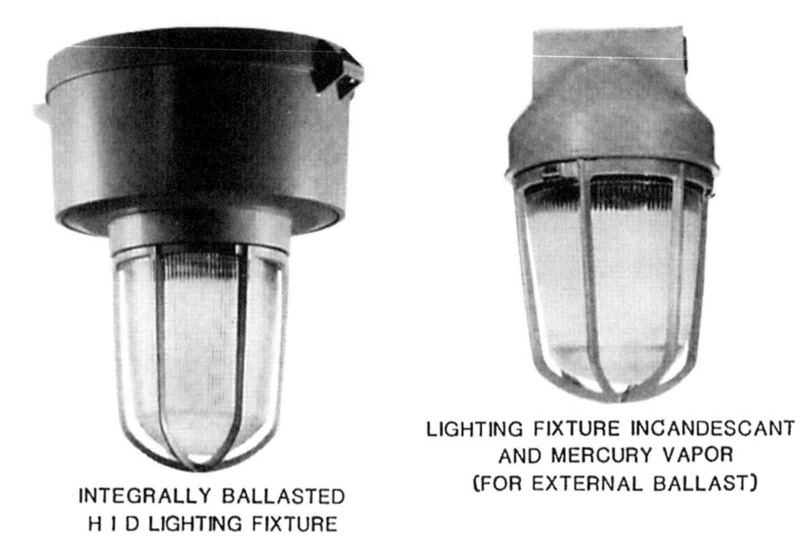

Figure 17-22. Standard lighting fixtures suitable for Class I, Division 2, Group D areas. (*Courtesy of Crouse-Hinds Electrical Construction Materials, a division of Cooper Industries, Inc.*)

wrench tight or provided with bonding jumpers at joints, and must have at least five full threads engaged. These precautions are necessary to minimize sparking across threads when a fault current flows through the conduit system and to provide proper distance for escaping gases to cool if an explosion occurs in the conduit. Pipe designed for fluids, and not approved as electrical equipment, must *not* be used—regardless of wall thickness.

For offshore locations where ignitible gas-air concentrations are neither continuously present nor present for long periods, API RP 14F also allows type MC cable with a continuous aluminum sheath and an outer impervious jacket (such as PVC) and armored cables satisfying ANSI/ Institute of Electrical and Electronic Engineers (IEEE) Standard No. 45. API RP 14F does not recommend IMC for offshore installations and cautions users that installations of MI cable require special precautions. The insulation of MI cable is hygroscopic (able to absorb moisture from the atmosphere).

The use of certain non-armored cables is acceptable to API RP 14F in Division 1 areas on drilling and workover rigs where ignitible concentrations of gases and vapors do not occur for appreciable lengths of time.

However, the non-armored cables must satisfy IEEE 383 or IEEE 45 flammability requirements and other requirements specified in API RP 14F.

NEC allows the use of portable cord connecting *portable* lighting equipment and other *portable* utilization equipment with the fixed portion of its supply circuit in Class I, Division 1 and 2 areas, provided:

1. The cord is rated for extra-hard service (frequently referred to as "Heavy-Duty SO Cord"), and
2. An approved grounding connector is provided *inside* the cord's outer jacket.

For all other applications (temporary or permanent) in *Division 1* areas, portable cord (whether rated for extra-hard service or not) is specifically disallowed.

Division 2 Areas

In Division 2 areas, the NEC allows the wiring methods that follow:

1. Threaded rigid metal conduit,
2. IMC,
3. Enclosed gasketed busway,
4. Enclosed gasketed wireway,
5. PLTC cable (Power Limited Tray Cable),
6. MI, MC, MV, TC, or SNM cable with approved termination fittings, and
7. Flexible cord approved for extra-hard service, flexible metal conduit, and liquidtight flexible conduit for limited flexibility. A suitable grounding conductor must be provided inside the flexible cord's outer jacket. Flexible conduit must be bonded with an external jumper or an approved internal system jumper; external bonding jumpers are disallowed for flexible conduit exceeding six feet. Typical liquidtight and flexible cord connectors and an explosion-proof flexible connection are shown in Figure 17-23.

Wiring System Selection

The designer must decide at inception whether to provide a cable system or a conduit system. Although both systems have specific advantages, the present trend is toward the installation of cable systems rather than conduit systems. Offshore, the vast majority of new systems are

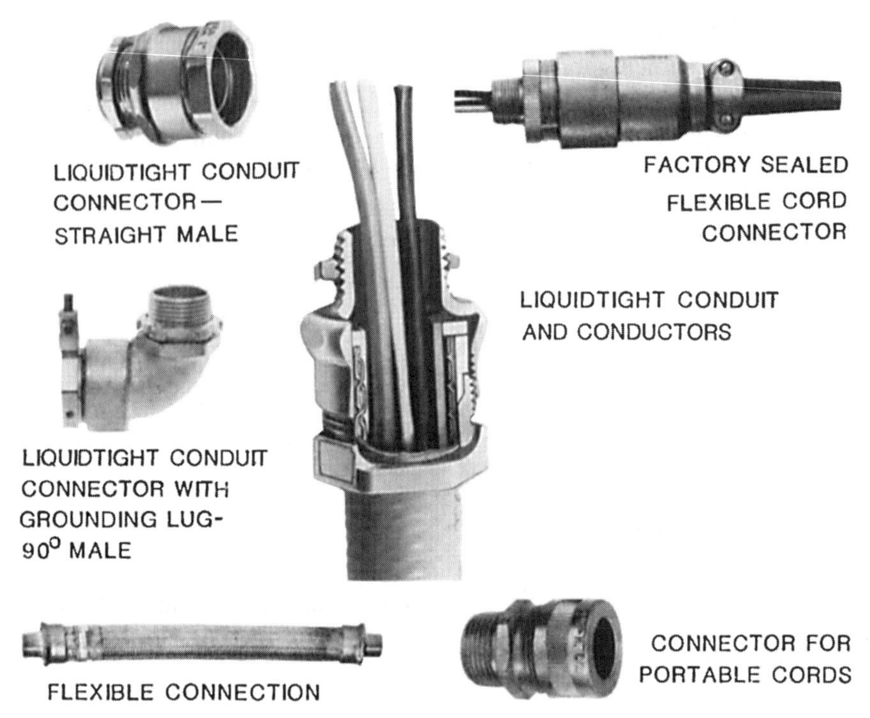

Figure 17-23. Liquidtight and flexible cord connectors. (*Courtesy of Crouse-Hinds Electrical Construction Materials, a division of Cooper Industries, Inc.*)

cable—particularly a cable with a gas/vapor-tight continuous corrugated aluminum sheath, rated Type MC, and with an overall jacket (usually PVC)—and normally installed in cable tray. This cable/cable tray system is usually less expensive to install than a conduit system (even though the materials may be more expensive) and offers the advantage of not requiring sealing fittings at area classification boundaries. Additionally, the cable tray system normally lends itself to easier and less expensive expansions than the conduit system. Cable trays are generally made of fiberglass, aluminum, stainless steel, or galvanized steel materials.

Conduit systems have the advantage of offering greater mechanical protection to enclosed conductors, but they can easily lose this advantage through corrosion if not properly maintained. A conduit system which corrodes on the inside can provide false security; although the outside appears completely sound, the system may not contain an internal explosion. Extremely rapid corrosion will occur in salt-air environments for most conduit systems of ferrous materials. In offshore environments, conduit of copper-free aluminum (normally defined as 0.4% or less copper) is normally preferred. If aluminum conduit is supported by ferrous supports, the aluminum must be carefully isolated from the ferrous material, or rapid corrosion will occur due to galvanic action. Ferrous conduits, normally satisfactory in heated areas (such as buildings containing operating engines) and, of course, in areas not subjected to corrosive elements, offer the advantage of magnetic sheilding—particularly desirable for communications and instrumentation circuits.

Conduits coated with PVC and other materials, preferably coated on the inside as well as on the outside, should be considered for corrosive environments. It should be noted, however, that coated conduit is significantly more expensive than non-coated conduit, and serves no real overall deterrent to corrosion unless coated fittings are also provided and extreme care is taken during installation to avoid damaging the coating. The threaded ends of the conduit, which cannot be coated without eliminating electrical continuity, are probably the weakest link of most coated conduit systems, even though some manufacturers provide couplings designed to prevent moisture intrusion.

Junction Boxes and Conduit Fittings

A box or fitting must be installed at each conductor splice connection point, receptacle, switch, junction point, or pull point for the connection of conduit system. In Division 1 areas only explosion-proof boxes or fittings are allowed. General purpose gasketed cover type fittings are allowed in Division 2 areas.

Boxes and fittings made of copper-free aluminum are generally used in offshore application as they provide better corrosion resistance. Galvanized steel or metal fittings with a PVC coating are also used in offshore applications. It should be noted that, although PVC-coated fittings provide resistance to exterior corrosion, they do not stop interior corrosion. Also, the cost of PVC-coated fittings is appreciably higher, and they require careful handling and installation to assure that the PVC coating is not damaged. Typical explosion-proof junction boxes and conduit fittings are shown in Figure 17-24.

Sealing Fittings

Conduit and cable sealing fittings as shown in Figures 17-25 and 17-26 are provided for the following purposes:

Figure 17-24. Standard explosion-proof junction boxes and conduit fittings. *(Courtesy of Crouse-Hinds Electrical Construction Materials, a division of Cooper Industries, Inc.)*

1. Confine internal explosions to explosion-proof enclosures and conduit systems.
2. Minimize the passage of gases, and prevent the passage of flame, through conduit or cable.
3. Prevent process gas or liquid in process piping from entering conduit or cable systems.
4. Prevent "pressure piling."

Pressure piling is a phenomenon caused by the fact that ignition in an enclosure can first pre-compress gases in a conduit or other enclosure to

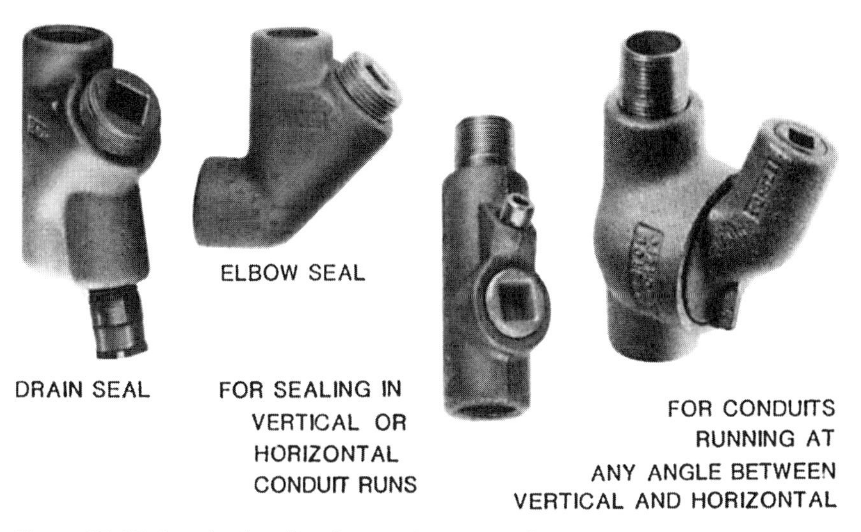

Figure 17-25. Standard sealing fittings. (*Courtesy of Crouse-Hinds Electrical Construction Materials, a division of Cooper Industries, Inc.*)

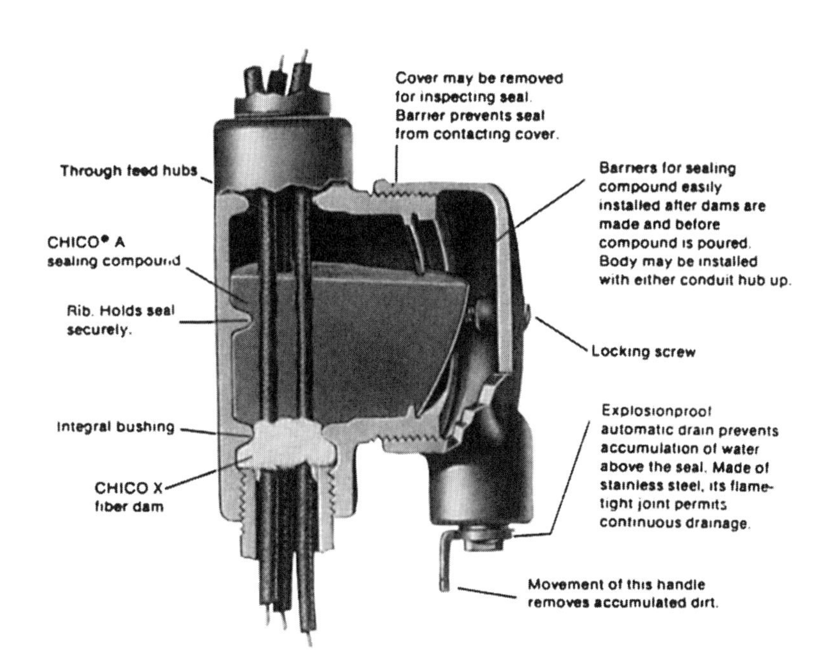

Figure 17-26. Cutaway drawing of a properly installed sealing fitting. (*Courtesy of Crouse-Hinds Electrical Construction Materials, a division of Cooper Industries, Inc.*)

which it is connected. When pre-compressed gases are then ignited in the second enclosure, pressures exceeding those for which it has been tested can be reached.

Receptacles and Attachment Plugs

Receptacles and attachment plugs for Class I, Division 1 and 2 areas must be approved for the area. They must provide a means of connection to the grounding conductor of a flexible cord. Typical Class I receptacles and attachment plugs are shown by Figure 17-27.

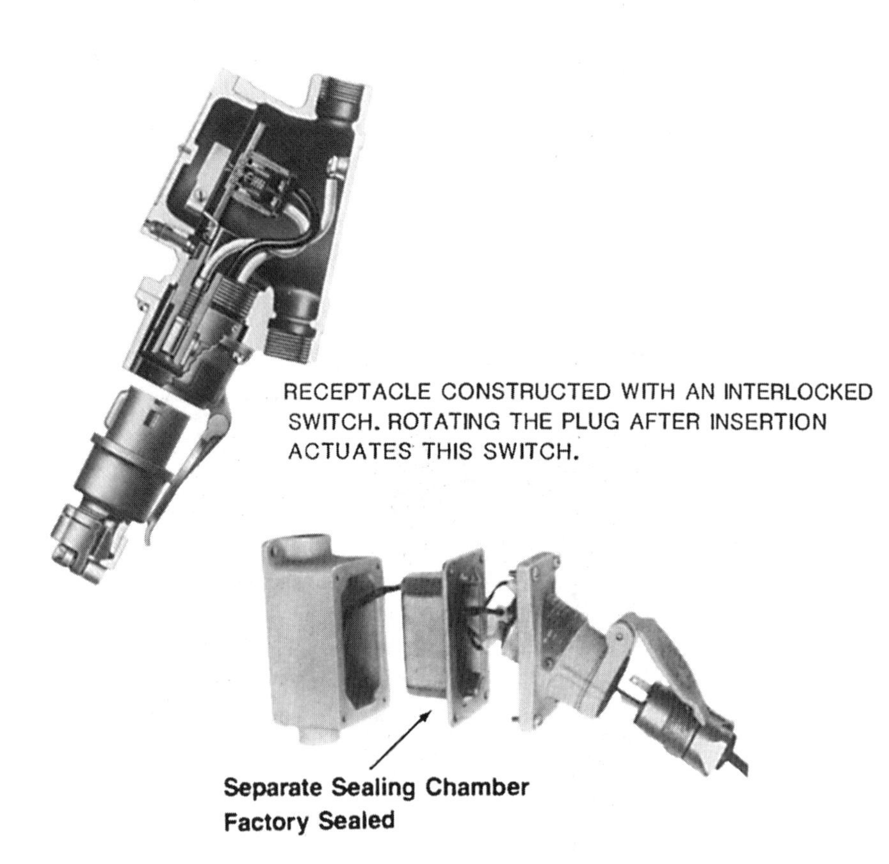

Figure 17-27. Typical Class I, Division 1 and 2 receptacles and attachment plugs. *(Top, courtesy of Crouse-Hinds Electrical Construction Materials, a division of Cooper Industries, Inc.; bottom, courtesy of Appleton Electric Co., a division of Emerson Electric Co.)*

Seal Locations

Seals must be installed in the following locations:

1. Seals are required at entries by conduit or cable to explosion-proof enclosures containing arcing or high-temperature devices in Division 1 and Division 2 locations. It is not required to seal 1½ in. or smaller conduits into explosion-proof enclosures in Division 1 areas housing switches, circuit breakers, fuses, relays, etc., if their current-interrupting contacts are hermetically sealed or under oil (having a 2-in. minimum immersion for power contacts and 1-in. for control contacts).
2. Seals are required where 2 in. or larger conduits enter explosion-proof enclosures containing taps, splices, or terminals in Division 1 areas (but not Division 2 areas).
3. Seals are required in conduits leaving Division 1 areas or traversing from Division 2 areas into unclassified areas, on either side of the boundary. No union, coupling, junction box, or fitting is allowed between the seal and the boundary. Metal conduits that pass completely through a Division 1 or a Division 2 location without a union, coupling, junction box, or fitting within 12 in. of the Division 1-Division 2 or Division 2-unclassified boundary do not require a sealing fitting at the boundary.
4. Except for conduit or cable entries into explosion-proof enclosures containing arcing or high-temperature devices (as described in Item 1 above), cables that will leak gas through the core at a rate of less than 0.007 ft^3/hr at 6 in. of water pressure need not be sealed if they are provided with a continuous gas/vapor-tight sheath. Cables with such a sheath that will transmit gas at or above this rate must be sealed if connected to process equipment that may cause a pressure of 6 in. of water at the cable end.
5. Cables without a continuous gas/vapor-tight sheath must be sealed at classified-unclassified area boundaries.
6. *All* cable terminations in Division 1 areas must be sealed. This requirement is imposed by API RP 14F, when specific cables are allowed in Division 1 areas.
7. Special sealing fittings (not yet commercially available) are required for cables and conduits connected to process connections that depend on a single seal, diaphragm, or tube to prevent process fluid

from entering the conduit or cable system. Single barrier devices probably are best avoided if multiple barrier devices are available.

Sealing fittings must be installed as close as practicable to explosion-proof enclosures, but in no case more than 18 inches from the enclosures. Although junction boxes and other devices which materially increase the cross-sectional area of the conduit system connecting the enclosure and the seal may not be installed between an enclosure and a seal, explosion-proof unions, couplings, elbows, capped elbows, and conduit bodies such as an "L," "T," or "cross" are allowed if the conduit bodies are not larger than the trade size of the connecting conduit. A single seal may suffice for two enclosures if it is installed no more than 18 in. from either enclosure.

Certain devices may be obtained which are "factory-sealed"—that is, interconnecting wiring is sealed by the manufacturer where it enters/exits enclosures. These devices do not require an additional (external) seal, and often can be utilized to advantage in lessening installation time and reducing space requirements (for an external seal).

These devices are tested only for internal explosions and *not* for external explosions pressurizing the devices from the outside. As an example, a factory-sealed push-button start/stop station connected to an explosion-proof motor starter *cannot* suffice as a seal for the motor starter conduit entry. A separate seal must be installed at the point of conduit entry.

Cable termination fittings are available which also are approved as sealing fittings, and often incorporate a union in their design. Particularly for space-limited installations, the difference in length requirements for such dual-purpose devices, as compared to standard sealing fitting/cable terminator/union combinations, can be consequential.

Seal Fittings Installation

When seal fittings are installed, certain mechanical practices must be followed. The following requirements are often overlooked by both installers and inspectors:

1. Accessibility. Sealing fittings should not be installed behind walls or in other inaccessible locations.
2. Orientation. Certain fittings are designed specifically for either horizontal or vertical mounting; others may be installed either horizontally or vertically, or even at oblique angles.

3. Approved compound. Only approved compound and damming fiber may be used. Rags or putty materials may not be used to construct dams. Some cable seals use a self-hardening putty-like material and do not require damming fiber.
4. Splices. Splices and taps may not be made in sealing fittings. Most sealing compounds are poor insulators, and electrical shorts could occur.
5. Drains. Drains or drain seals must be provided in locations necessary to prevent water accumulation.
6. Thickness. Completed conduit seals must have a seal which is at least as thick as the trade size of the conduit. In no case can the seal be less than ⅝-in. thick.

Specific Equipment Considerations

Transformers

Although transformers suitable for other industrial installations are generally suitable for producing applications, certain options may be desirable—primarily due to environmental considerations. At locations subject to harsh environmental conditions, and particularly at locations subject to washdown with high-pressure hoses, non-ventilated enclosures are desirable, if not necessary. Likewise, at locations subjected to salt water and salt-laden air, it often is desirable to specify copper windings and lead wires. Most manufacturers provide standard units with aluminum windings and lead wires. Even if aluminum coils are used, it is almost always desirable to require stranded copper lead wires. This will lessen corrosion and loose terminal problems when transformers are interconnected to the facility electrical system with copper conductors. If the transformers are to be installed outdoors in corrosive environments, cases should be of corrosion-resistant material (e.g., stainless steel) or be provided with an exterior coating suitable for the location.

Many producing facilities are located offshore or in other environmentally sensitive areas. In these areas, the use of dry (versus liquid-filled) transformers will eliminate the necessity of providing curbing and other containment systems to prevent pollution. Dry transformers are normally preferred for most production facility applications. Liquid-filled transformers should be considered, however, for high voltage and large units (particularly over several hundred kVA).

Electric Motors

Apart from considerations given to corrosion resistance and suitability for hazardous (classified) areas, the selection of electric motors for oil field applications is the same as the selection of electric motors for other industrial applications. One exception may be the selection of motors for areas where electric power is self-generated. Frequency and voltage variations may occasionally occur at such locations. For such locations, consideration should be given to specifying motors which are tolerant to at least 10% voltage variations and 5% frequency variations.

It is cautioned that NEMA Design B motors (normal starting torque) may not be suitable for applications requiring high starting torque such as positive displacement pumps. NEMA Design C motors should be used in this service.

Most standard motors are manufactured using non-hygroscopic NEMA Class B insulation. For added protection in an offshore environment, open drip-proof or weather protected motors should be specified with a sealed insulation system. NEMA Class F insulation is also available in most motor sizes and is advisable to provide an improved service factor.

A motor used in standby operation mode should be equipped with a space heater to keep the motor windings dry. In classified areas these space heaters must meet the surface temperature requirement of the specific hazardous area.

Lighting Systems

Lighting systems are installed both to provide safety to operating personnel and to allow efficient operations where natural light is insufficient. The lighting required for safety to personnel depends on the degree of the hazard requiring visual detection and the normal activity level. It typically varies from 0.5 to 5.0 footcandles. Lighting levels required for efficient operations vary from as low as 5 footcandles to as high as 100 footcandles, or more. Table 17-4 contains some general lighting guidelines.

The first step in the design of a lighting system is the determination of the various lighting levels required for the specific areas of the facility. Typically, the majority of the fixtures are high intensity discharge (HID) fixtures and fluorescent fixtures. Certain applications may require incandescent fixtures as well.

HID fixtures include those using mercury vapor and sodium vapor lamps. Mercury vapor fixtures are usually less expensive than sodium

Table 17-4A
Minimum Recommended Levels of Illumination for Efficient Visual Tasks

Area	Minimum Lighting Level (Footcandles)
Offices, General	50
Offices, Desk Area	70
Recreation Rooms	30
Bedrooms, General	20
Bedrooms, Individual Bunk Lights	70
Hallways, Stairways, Interior	10
Walkways, Stairways, Exterior	2
Baths, General	10
Baths, Mirror	50
Mess Halls	30
Galleys, General	50
Galleys, Sink & Counter Areas	100
Electrical Control Rooms	30
Storerooms, Utility Closets	5
Walk-in Freezers, Refrigerators	5
TV Rooms (lights equipped with dimmers)	Off to 30
Work Shops, General	70
Work Shops, Difficult Seeing Task Areas	100
Compressor, Pump and Generator Buildings, General	30
Entrance Door Stoops	5
Open Deck Areas	5
Panel Fronts	10
Wellhead Areas	5

fixtures initially, and are readily available in most styles. However, sodium vapor fixtures are more efficient in the use of electricity.

Because of quite poor color rendition and difficulty in safe disposal of expended lamps, low pressure sodium fixtures are less desirable than high pressure sodium fixtures and are seldom recommended for production facilities. High pressure sodium fixtures are particularly attractive for illuminating large open areas. At locations where power cost is low and where many fixtures are required due to equipment shadowing, mercury vapor fixtures often are preferred because of their lower initial cost, lower replacement lamp cost, and better color rendition.

The low profile of fluorescent fixtures often dictates their use in areas with low headroom, such as in wellbays on offshore platforms and in

Table 17-4B
Minimum Recommended Levels of Illumination for Safety

Area	Minimum Lighting Level (Footcandles)
Stairways	2.0
Offices	1.0
Exterior Entrance	1.0
Compressor and Generator Rooms	5.0
Electrical Control Rooms	5.0
Open Deck Areas	0.5
Lower Catwalks	2.0

buildings with conventional ceiling heights. The relatively short life, low efficiency, and susceptibility to vibration exclude incandescent lamps from serious consideration for many applications, particularly for general area lighting. In areas free from vibration and easily accessible for maintenance, however, incandescent fixtures may be quite acceptable.

When designing lighting systems, particular attention should be given to locating fixtures where relamping can be performed safely and efficiently. Poles which can be laid down, as opposed to climbed, are often preferred—particularly at offshore locations. This feature offers less advantage, of course, at land locations where bucket trucks or the like can be used for relamping. In locations subject to vibration, it normally is prudent to install lighting fixtures with flexible cushion hangers or flexible fixture supports (hanger couplings) to increase lamp life. Remotely mounted ballasts for HID fixtures are frequently desirable, particularly when the fixtures themselves must be installed in locations of high temperature and locations difficult to access for maintenance. The ceilings of large compressor and pump buildings are examples of locations where remote ballasts often are attractive.

Motor Control Center

The engineer providing the initial design of major facilities is faced with the decision of providing a motor control center building or individual (usually rack-mounted) motor starters and corresponding branch circuit protection devices. For installations using only several motors it frequently is more economical to provide individual (usually explosion-proof) motor starters and circuit protection devices.

For facilities that include large numbers of motors and other electrical equipment, it normally is both more economical and more convenient to furnish a building to enclose the required motor starters and distribution panels. This building is normally referred to as a motor control center (MCC). In addition to typically allowing less expensive non-explosion-proof equipment, these buildings are frequently environmentally controlled (air conditioned, and possibly heated in colder climates) to reduce equipment corrosion and enhance reliability. Maintenance is more easily performed indoors than if the equipment were installed outside and maintenance personnel were subject to extreme cold, rain, snow, or other adverse weather conditions.

If air conditioning systems are designed for buildings housing electrical equipment, the heat generated by the electrical equipment must be considered when sizing the air conditioning equipment. Artificial heat is seldom required in all but the coldest of climates.

Enclosures

The selection of equipment enclosures involves consideration of environmental conditions as well as the possibility of exposure to flammable gases and vapors. The National Electrical Manufacturers Association (NEMA) provides a list of designations for enclosures that is adequate to specify many enclosure requirements. As an example, enclosures designated as NEMA 7 are explosion-proof, suitable for Class I areas for the gas groups labeled. NEMA 7 enclosures may be labeled for only one group (such as Group D) or for several groups (such as Groups B, C, and D). NEMA 1 enclosures are designed to perform little other purpose than to prevent accidental personnel contact with enclosed energized components, but are suitable for most unclassified areas. NEMA 4X enclosures, watertight and constructed of corrosion-resistant material, are often preferred for outdoor non-explosion-proof applications in areas subjected to harsh environmental conditions or high pressure hose washdown.

CORROSION CONSIDERATIONS

Even though the electrical design details of a system may be well specified, the system will not endure or continue to provide safety to personnel unless proper materials are selected and certain installation procedures followed. Most land-based facilities are not subjected to the same harsh environmental conditions as offshore and marshland locations, but even they must be given careful consideration in material selection and installation procedures.

At locations where salt-laden air is present, aluminum should be specified as containing 0.4% or less copper. Such aluminum is often referred to as "copper-free" or "marine grade." Also, galvanic action will occur if aluminum and steel (or other dissimilar metals) are in direct contact. Galvanic action is accelerated in the presence of salt and moisture. Rapid corrosion of uncoated aluminum will occur if it is exposed to materials of high or low pH (less than 4.5 or greater than 8.5). Drilling fluids may fall into the class of high pH materials. Additionally, if aluminum is allowed to contact common fireproofing materials containing magnesium oxychloride, rapid corrosion will occur in the presence of moisture.

To prevent the accumulation of moisture in conduits and enclosures, drains should be installed at all low points. In classified areas, breathers and drains must be explosion-proof. Figure 17-28 shows typical explosion-proof breathers and drains.

Space heaters, particularly in electrical motors and generators which may be idle for significant periods of time, can also help prevent the accumulation of moisture. Space heaters installed in classified areas must operate at temperatures below "high temperature" devices.

To retard corrosion and to facilitate future maintenance (e.g., allow the non-destructive removal of threaded junction box covers), all threaded connections should be lubricated with an antiseize compound which will not dry out in the environment. If lubricant is applied to the threaded (or flanged) portion of covers of explosion-proof enclosures, the lubricant must have been tested and approved as suitable for flame path use. It is cautioned that some lubricants contain silicone, which will poison most catalytic gas detector sensors and should not be used near gas detectors.

Figure 17-28. Standard explosion-proof drains and breathers. (*Courtesy of Crouse-Hinds Electrical Construction Materials, a division of Cooper Industries, Inc.*)

Many materials are subject to deterioration by ultraviolet light (UV), particularly many of the "plastics" and fiberglass materials. Fiberglass materials for outside use should be specified as UV-stabilized, and most plastics installed outdoors should be carbon-impregnated (black in color). It is particularly recommended that plastic cable ties, which secure cables in cable trays, be carbon-impregnated if installed outdoors.

In areas where electrical equipment is exposed to contaminants, the selection of equipment whose contacts are oil-immersed or hermetically sealed can increase reliability and equipment life. Similarly, providing environmentally-controlled equipment rooms can greatly increase equipment life at locations where contaminants are prevalent. In offshore and other areas exposed to salt, type 316 stainless steel is often preferred over types 303 and 304, which will pit with time. Likewise, in similar locations, equipment fabricated from galvanized steel will corrode much more rapidly than equipment hot-dip galvanized after fabrication.

ELECTRICAL STANDARDS AND CODES

American National Standards Institute (ANSI)
1430 Broadway
New York, NY 10018
C84.1 Voltage Ratings for Electrical Power Systems and Equipment (60 Hz)
Y 14.15 Electrical and Electronics Diagrams

American Petroleum Institute (API)
2101 L Street, NW
Washington, DC 20037
RP 14F Recommended Practice for Design and Installation of Electrical Systems for Offshore Production Platforms
RP 500 Recommended Practice for Classification of Locations for Electrical Installations at Petroleum Facilities Classified As Class I, Division 1 and Division 2.

American Society of Heating, Refrigerating and Air Conditioning Engineers, Inc. (ASHRAE)
1791 Tullie Circle, NE
Atlanta, GA 30329
ASHRAE Fundamentals Handbook

British Standards Institute (BSI)
Newton House
101 Pentionville Road
London, N1 9ND
or,
c/o American National Standards Institute,
1430 Broadway
New York, NY 10018

BS5501	Part 1, General Requirements
	Part 2, Oil Immersion "o"
	Part 3, Pressurized Apparatus "p"
	Part 4, Power Filling "q"
	Part 5, Flameproof Enclosure "d"
	Part 6, Increased Safety "e"
	Part 7, Intrinsic Safety "i"
BS5345	Code of Practice in the Selection, Installation, and Maintenance of Electrical Apparatus for Use in Potentially Explosive Atmospheres (Other than Mining Applications or Explosive Processing and Manufacture)

Canadian Standards Association (CSA)
178 Rexdale Boulevard
Rexdale, Ontario M9W 1R3
Canada

C22.1, Part I	Canadian Electrical Code
C22.2, No. 30	Explosion-proof Enclosures for Use in Class I Hazardous Locations
C22.2, No. 14	Motors and Generators for Use in Hazardous Locations
C22.2, No. 157	Intrinsically Safe and Nonincendive Equipment for Use in Hazardous Locations
C22.2, No. 174	Cables and Cable Glands for Use in Hazardous Locations
C22.2, No. 213	Equipment for Use in Class I, Division 2 Hazardous Locations, A Guide for the Design, Construction and Installation of Electrical Equipment; John Bossert and Randolph Hurst

European Committee for Electrotechnical Standardization (CENELEC)

Rue Brederode
2 Boite No. 5
8-1000 Bruxelles
Beligique

EN50014	General Requirements
EN50015	Oil Immersion "o"
EN50016	Pressurized Apparatus "p"
EN50017	Power Filling "q"
EN50018	Flameproof Enclosure "d"
EN50019	Increased Safety "e"
EN50020	Intrinsic Safety "i"
EN50039	Intrinsically Safe Electrical Systems "i"

Factory Mutual Research Corporation (FM)

1151 Boston-Providence Turnpike
Norwood, MA 02062

Std. 3615	Explosion-proof Electrical Equipment
Std. 3610	Electrical Intrinsically Safe Apparatus and Associated Apparatus for Use in Class I, II, and III, Division 1, Hazardous Locations

Illuminating Engineering Society (IES)

345 East 47th Street
New York, NY 10017

RP-7	American National Standard Practice for Industrial Lighting
	IES Lighting Handbook

Institute of Electrical and Electronic Engineers (IEEE)

345 East 47th Street
New York, NY 10017

Std. 45	Recommended Practice for Electrical Installation on Shipboard
Std. 142	Recommended Practice for Grounding of Industrial and Commercial Power Systems
Std. 141	Recommended Practice for Electric Power Distribution for Industrial Plants

Std. 303 Recommended Practice for Auxiliary Devices for Motors in Class I, Groups A, B, C, and D, Division 2 Locations

RP 446 Recommended Practice for Emergency and Standby Power Systems for Industrial and Commercial Applications

Instrument Society of America (ISA)

P. O. Box 12277

Research Triangle Park, NC 27709

S5.1 Instrumentation Symbols and Identification

RP12.1 Recommended Practice for Electrical Instruments in Hazardous (Classified) Atmospheres

S12.1 Intrinsic Safety

S12.4 Instrument Purging for Reduction of Hazardous Area Classification

RP12.6 Installation of Intrinsically Safe Instrument Systems for Hazardous (Classified) Locations

S12.12 Electrical Equipment for Use in Class I, Division 2 Hazardous (Classified) Locations

S12.13 Part I, Performance Requirements, Combustible Gas Detectors

RP12.13 Part II, Installation, Operation, and Maintenance of Combustible Gas Detection Instruments

S12.15 Part I, Performance Requirements, Hydrogen Sulfide Gas Detectors (Draft Standard)

RP12.15 Part II, Installation, Operation, and Maintenance of Hydrogen Sulfide Gas Detection Instruments (Draft Standard)

S51.1 Process Instrumentation Terminology

S71.01 Environmental Condition for Process Measurement and Control Systems: Temperature and Humidity

Electrical Instruments in Hazardous Locations, Ernest C. Magison, 1978

Electrical Safety Abstracts. Edited by Alfred H. McKinney and Harry G. Conner

National Electrical Manufacturers Association (NEMA)

2101 L Street, NW

Washington, DC 20037

ICS 2 Standards for Industrial Control Devices, Controllers and Assemblies

ICS 6 Enclosures for Industrial Controls and Systems
MG 1 Motors and Generators
MG 2 Safety Standard for Construction and Guide for Selection, Installation, and Use of Electric Motors and Generators
MG 10 Energy Guide for Selection and Use of Polyphase Motors
VE 1 Cable Tray Systems

National Fire Protection Association (NFPA)
Batterymarch Park
Quincy, MA 02269
No. 30 Flammable and Combustible Liquids Code
No. 37 Standard for the Installation and Use of Stationary Combustion Engines and Turbines
No. 70 National Electrical Code
No. 77 Recommended Practice on Static Electricity
No. 78 Lightning Protection Code
No. 321 Basic Classification of Flammable and Combustible Liquids
No. 325M Fire Hazard Properties of Flammable Liquids, Gases, and Volatile Solids
No. 493 Standard for Intrinsically Safe Apparatus and Associated Apparatus for Use in Class I, II, and III, Division 1 Hazardous Locations
No. 496 Standard for Purged and Pressurized Enclosures for Electrical Equipment in Hazardous (Classified) Locations
No. 497 Recommended Practice for Classification of Class I Hazardous Locations for Electrical Installations in Chemical Plants
No. 497M Classification of Gases, Vapors and Dusts for Electrical Equipment in Hazardous (Classified) Locations

Underwriters Laboratories, Inc. (UL)
33 Pfingsten Road
Northbrook, IL 60062
UL 58 An Investigation of Fifteen Flammable Gases or Vapors with Respect to Explosion-proof Electrical Equipment
UL 58A An Investigation of Additional Flammable Gases or Vapors with Respect to Explosion-proof Electrical Equipment
UL 58B An Investigation of Additional Flammable Gases or Vapors with Respect to Explosion-proof Electrical Equipment

UL 595 Standard for Marine-type Electric Lighting Fixtures
UL 674 Electric Motors and Generators for Use in Hazardous Locations, Class I, Groups C and D, and Class II, Groups E, F, and G
UL 698 Safety Standard for Electric Industrial Controls Equipment for Use in Hazardous (Classified) Locations
UL 844 Standard for Electric Lighting Fixtures for Use in Hazardous Locations
UL 877 Circuit-Breakers and Circuit-Breaker Enclosures for Use in Hazardous Locations, Class I, Groups A, B, C, and D, and Class II, Groups E, F, and G
UL 886 Outlet Boxes and Fittings for Use in Hazardous Locations in Class I, II, and III, Division 1, Hazardous Locations
UL 1010 Receptacle—Plug Combinations for Use in Hazardous (Classified) Locations
UL 1203 Explosion-proof and Dust-ignition-proof Electrical Equipment for Use in Hazardous Locations
UL 1604 Electrical Equipment for Use in Hazardous Locations, Class I and II, Division 2, and Class III, Divisions 1 and 2

United States Code of Federal Regulations
c/o U.S. Government Printing Office
Washington, D.C. 20402
Title 29, Occupational Safety and Health Standards, Subpart S, Part 1910 Electrical
Title 30, Oil and Gas and Sulfur Operations in the Outer Part 250 Continental Shelf
Title 33, Subchapter C, Aids to Navigation Part 67
Title 46, Shipping Subchapter J, Electrical Engineering, Parts 110–113 (United States Coast Guard, CG259)

Index